Aspects of Anomalous Transport in Plasmas

Series in Plasma Physics

Series Editors:

Steve Cowley, Imperial College, UK
Peter Stott, CEA Cadarache, France
Hans Wilhelmsson, Chalmers University of Technology, Sweden

Other books in the series

Non-Equilibrium Air Plasmas at Atmospheric Pressure
K H Becker, U Kogelschatz, K H Schoenbach and R J Barker (eds)

**Magnetohydrodynamic Waves in Geospace: The Theory of ULF Waves
and their Interaction with Energetic Particles in the Solar–Terrestrial
Environment**
A D M Walker

Plasma Waves, Second Edition
D G Swanson

Microscopic Dynamics of Plasmas and Chaos
Y Elskens and D Escande

Plasma and Fluid Turbulence: Theory and Modelling
A Yoshizawa, S-I Itoh and K Itoh

The Interaction of High-Power Lasers with Plasmas
S Eliezer

Introduction to Dusty Plasma Physics
P K Shukla and A A Mamun

The Theory of Photon Acceleration
J T Mendonça

Laser Aided Diagnostics of Plasmas and Gases
K Muraoka and M Maeda

Reaction-Diffusion Problems in the Physics of Hot Plasmas
H Wilhelmsson and E Lazzaro

The Plasma Boundary of Magnetic Fusion Devices
P C Stangeby

Non-Linear Instabilities in Plasmas and Hydrodynamics
S S Moiseev, V N Oraevsky and V G Pungin

Collective Modes in Inhomogeneous Plasmas
J Weiland

Series in Plasma Physics

Aspects of Anomalous Transport in Plasmas

R Balescu

Université Libre de Bruxelles, Belgium

CRC Press
Taylor & Francis Group
Boca Raton London New York

CRC Press is an imprint of the
Taylor & Francis Group, an **informa** business

CRC Press
Taylor & Francis Group
6000 Broken Sound Parkway NW, Suite 300
Boca Raton, FL 33487-2742

First issued in paperback 2019

© 2005 by Taylor & Francis Group, LLC
CRC Press is an imprint of Taylor & Francis Group, an Informa business

ISBN-13: 978-0-7503-1030-7 (hbk)
ISBN-13: 978-0-367-39309-0 (pbk)

British Library Cataloguing-in-Publication Data

A catalogue record for this book is available from the British Library.

Library of Congress Cataloging-in-Publication Data are available

Commissioning Editor: John Navas
Editorial Assistant: Leah Fielding
Production Editor: Simon Laurenson
Production Control: Sarah Plenty
Cover Design: Victoria Le Billon
Marketing: Louise Higham, Kerry Hollins and Ben Thomas

Visit the Taylor & Francis Web site at
http://www.taylorandfrancis.com

and the CRC Press Web site at
http://www.crcpress.com

Contents

Preface

Plasma is often called "the fourth state of matter", assuming that solid, liquid and gas are, respectively, the first, the second and the third state. If, however, the criterion of classification is changed, plasma should be called the "first state of matter", given that more than 99% of matter in the universe is in the plasma state. This trivial remark underscores the importance of plasma physics. Its applications are in the most various fields. Some of these are not accessible to experiment, but only to observation: solar (or stellar) physics, geophysics (or planetary physics). Other types of plasmas are, indeed, open to human manipulation and experimentation.

The earliest studies, dating from the beginning of the 20th century, and still fully active today are related to electrical discharges in gases: these conventionally relate to "low temperature plasma physics". Their theoretical treatment is rather complex, because these plasmas are only partially ionized, and thus consist of a multicomponent system of electrons, ions of various charges and neutral particles in various states of excitation. Moreover, their composition is variable in time.

An even colder type of plasmas is found in metals and in semiconductors. Here one cannot avoid a quantum-mechanical description. This group of plasmas finds at present numerous important applications in the realm of condensed matter physics, nanotechnology, information and computing.

Last, but not least, the plasma state is the necessary environment of matter (essentially hydrogen isotopes) in the quest of controlled nuclear fusion as a source of energy of the future. The nuclear reactions require very high temperatures, at which all the atoms are stripped of their electrons. The most promising fusion devices (at present) consist of a plasma that is confined spatially by a magnetic field in a toroidal geometry (tokamaks, stellarators,...). This plasma is (at least in the core of the reactor) fully ionized. The magnetic confinement is, however, never perfect: the plasma is more or less slowly leaking through the magnetic field. The success of the fusion programme therefore strongly depends on understandinging and controlling the transport processes of matter and energy in these magnetically confined high-temperature plasmas.

The transport theory of neutral gases has been well understood since

Boltzmann's basic work in kinetic theory, and codified in the "bible" of the field, the book of Chapman and Cowling, dating from the early thirties. Its extension to magnetized plasmas has been, however, disappointing. Although it contributes to an interpretation of some important aspects of non-equilibrium processes, in particular of the important influence of the magnetic field geometry on transport, it greatly underestimates most of the relevant transport coefficients. Something important, and specific to plasmas, was missing: it was called, prudently, *anomalous transport.*

It appeared gradually that the very description of the plasma as a set of particles moving in undisturbed trajectories and undergoing rare quasi-instantaneous collisions was inadequate. Because of the long range of the electrostatic interactions, the charged particles tend to move *collectively*, producing waves. It is the collective interaction of these waves that predominantly control the transport properties of the high-temperature plasmas. These *nonlinear* processes affecting principally the unstable plasma modes lead to a profound restructuring of the system, leading it into a state of *turbulence* and, possibly, to the formation of non-equilibrium collective structures. Anomalous transport theory in plasmas is thus necessarily related to a theory of turbulence.

In the early eighties I started an attempt to collect in the form of books the known results of plasma transport theory, that were scattered in the literature. I thus wrote two volumes: one on Classical transport theory, the second on Neoclassical theory: they were published in 1988. Both of these theories constitute at present a complete, well-organized and elegant corpus. These books could therefore be considered as practically complete, at least as far as the principles are concerned. I then promised a third volume about anomalous transport.

It appears, however, that the latter field is by orders of magnitude more complex than classical or neoclassical theory. Turbulence is, indeed, an open problem even for neutral fluids. It took me sixteen more years (with some interruptions) for selecting and working out the matter presented here. Although there is a continuity (in the spirit) with my two preceding books on transport, the situation here is quite different. Anomalous transport theory is not, at present (or in the foreseeable future) a complete, self-contained corpus. One could therefore consider that today is not yet the right time for writing a book on anomalous transport theory. (Maybe there never will be a "right time" for a complete theory!) This kind of scruple partly explains the long delay in my presenting "the third volume" of the series. Nevertheless, a number of significant, though punctual, advances have been accomplished in recent years in the field. I thus finally decided to present a selection of these approaches. I would not use the misleading title (that was appropriate for the first two volumes) "Anomalous transport theory", but rather the correct title: *"Aspects of anomalous transport theory"*.

The problems treated here are inspired from fusion physics. The results are not, however, *directly* applicable to a tokamak (say). Given the enormous complexity of the underlying nonlinear problems, I decided to study only the simplest possible situation. I thus consider throughout a fully ionized plasma with a single species of singly-charged ions, in the presence of a homogeneous and stationary magnetic field (which could be subject to fluctuations): this defines the *shearless slab geometry*. This allowed me to concentrate on the essence of the molecular - or stochastic - mechanisms producing transport in a turbulent plasma. The methods used are those of nonequilibrium statistical mechanics and of its offspring, kinetic theory. In the second part of the book, they are combined with the methods of stochastic (Langevin) equations and random walk theory.

The application of these basic building stones to fusion theory require a generalization involving the action of an inhomogeneous and curved (toroidal) magnetic field. This is the same philosophy as the passage from classical to neoclassical theory. This aspect is not considered here.

The work presented in this book has benefited over the years from the collaboration of many colleagues and friends. I should like to mention in the first place Jacques Misguich (from Cadarache, France). We have been working together on many aspects of plasma turbulence since the seventies and his contribution to most subjects presented in this book has been invaluable. He also read critically many chapters of the manuscript. The following colleagues (cited in more or less chronological order) contributed greatly to the subjects discussed here: Hans Pecseli (from Risö, Denmark, presently in Oslo), Karl Spatschek (from Düsseldorf, Germany), Qiu Xiao-Ming, Zhang Weiyan and Haida Wang (from China), Eric Vanden Eijnden (from Brussels, presently in New York), Madalina Vlad and Florin Spineanu (from Bucharest, Romania), Marian Negrea and Iulian Petrisor (from Craiova, Romania), Boudewijn van Milligen and Raul Sanchez (from Madrid, Spain) and Ben Carreras (from Oak Ridge, Tennesee). The bibliographical list given at the end is far from complete; I apologize to all physicists who could not be cited explicitly.

All the work leading to the present book was done under the aegis of the Association Euratom-Belgian State for Controlled Thermonuclear Fusion, which provided for many years the framework of the research done in my group at the University of Brussels. Donato Palumbo, Charles Maisonnier, Hardo Bruhns were among our strongest supporters at the European Commission.

Chapter 1

Introduction

1.1 The Problem of Anomalous Transport

Understanding the mechanisms of transport of matter, electric charge, energy and momentum is one of the most important goals of research in the field of plasma physics. Practically all applications of plasmas are limited in some way or other by the transport phenomena taking place under specific circumstances. Thus, in the so-called "low-temperature" plasmas, the large variety of phenomena occurring in electric discharges through gases are determined by the electrical and thermal conductivity of weakly ionized plasmas under different conditions of electric field, chemical composition, degree of ionization and geometry of the vessel. Transport of energy from the centre of a star (such as the sun) to its edge is essential in understanding the stellar evolution. Its subsequent diffusion into space determines the influence of the star on its environment (for instance the influence of the sun on terrestrial phenomena). In the neighbourhood of the earth, the magnetic field determines the fate of the stream of solar wind and produces the very peculiar shape of the magnetosphere. The latter is an essentially dynamic structure; in particular, such dramatic events as magnetic field line reconnections produce important events on our planet (such as magnetic storms). These phenomena are again controlled by transport phenomena. Last but not least, the efforts undertaken with the goal of realizing controlled nuclear fusion, in order to produce an energy source that would (theoretically) satisfy mankind's needs for many centuries, are limited by an understanding of the transport of matter and heat across the magnetic field that confines the high-temperature plasma.

Transport processes in gases have been studied since the very beginning of kinetic theory by its founding fathers, Maxwell and Boltzmann. It would be even more correct to say that kinetic theory was designed precisely with the purpose of explaining from first principles (i.e., from a molecular view of matter) the non-equilibrium phenomena appearing in

1

the gases; among these, the transport processes are the most important and also the most readily measurable. After several decades of efforts, pioneered by the works of Hilbert, Chapman, Enskog and Grad, a systematic and complete theory of transport in dilute or moderately dense neutral gases has been constructed. This theory is very clearly reviewed in several textbooks: CHAPMAN & COWLING 1952, HIRSHFELDER et al., 1954 CERCIGNANI 1988, FERZIGER & KAPER 1972, GRAD 1949. Its success in predicting transport coefficients such as diffusion, heat conduction and viscosity coefficients, as well as cross-coefficients such as thermodiffusion, has been remarkable. The basic idea here is that the motor of irreversibility and dissipation is the existence of frequent collisions (mainly binary collisions) between individual molecules. The latter produce a "randomization" of the particle orbits ("molecular chaos"). The effect of the collisions is balanced by the free flow of the particles in a macroscopically inhomogeneous medium, as well as, possibly, by some external force field. As a result of this balance, the system reaches a quasi-stationary state in which steady macroscopic fluxes of matter, energy and momentum are present. The latter are, in first approximation, linearly related to the gradients of density, temperature and velocity. From these relations, the transport coefficients are readily determined. The values of the latter depend on the structure of the molecules, as well as on the nature of the intermolecular forces.

When the interest in high-temperature, fully ionized plasmas appeared in the middle of the 20th century as a result of the controlled fusion programme, the importance of transport was immediately recognized. The first theoretical attempts consisted, of course, in trying to adapt the successful formalism which existed for neutral gases. There were two main differences, both due to the fact that the constituents of the plasma, electrons and positive ions, are electrically charged. As a result, on one hand, the intermolecular forces are Coulomb interactions and, on the other hand, the particles are acted upon by external electric and magnetic fields. The Chapman-Enskog-Grad formalism can be readily applied to the magnetized plasmas: it results in the so-called *classical transport theory of plasmas* (BRAGINSKII 1965, BALESCU 1988 a).

When the classical predictions were compared to experiments, the result was quite disappointing: the classical theory grossly underestimated the transport coefficients in a fusion plasma, sometimes by several orders of magnitude. The efforts of the physicists were then directed towards understanding which assumptions of classical theory were invalid. It first appeared that the collisions in a fully ionized plasma cannot be considered as binary interactions. The Coulomb force has a very long range, which produces infinite results in the quantities (integrals) appearing in kinetic theory. This problem could, however, be solved in the case of weakly nonideal plasmas (such as the high-temperature fusion plasmas) by deriving a collision term which takes explicitly into account the many-particle nature of the collision process (BALESCU 1960, 1963, LENARD 1960). In this picture, the collisions take place between particles surrounded by a dynamical polarization cloud, whose nature is determined by the instantaneous state of the plasma. The new kinetic equation allowed us to understand the basic nature of the collision processes in the plasma, but in practice it did not lead to a substantial improvement of the prediction of transport coefficients.

The attention was then directed to the second part of the balance: the effect of the magnetic field on the free motion of the particles. The magnetic fields produced in fusion devices for the purpose of confining the plasma have a very peculiar structure: they are spatially inhomogeneous and have a globally toroidal topology. In astrophysical situations the interplanetary, interstellar or intergalactic magnetic fields have complicated, inhomogeneous and curved geometries. The particle orbits in such magnetic fields have very different shapes compared to the straight lines existing in absence of the fields (or even in a constant magnetic field). They are either helices encircling (but also drifting away from) the field lines, or else (depending on their velocity) they may be trapped in low-field regions because of the field inhomogeneity (magnetic mirror effect). It turns out that this effect of the global magnetic configuration has a surprisingly strong effect on the transport properties of the plasma. When the classical theory (based on collisions) is combined with these effects of the geometry one finds transport coefficients that are enhanced as compared to the classical ones. This theory has been worked out in great detail for the toroidal fields of the fusion machines (such as the *tokamak* or the *stellarator*) and has produced a complete and mathematically very elegant construction, called the *neoclassical transport theory of plasmas* (GALEEV & SAGDEEV 1968, HINTON & HAZELTINE 1976, BALESCU 1988 b).

Neoclassical transport theory is a significant step forward towards describing transport in magnetically confined plasmas. In some cases the agreement with experimental data is satisfactory. But in general, even the neoclassical enhancement is still far from explaining the large observed "leaks" in tokamaks. Attention had to be directed toward a completely different factor that would strongly enhance transport. The most plausible explanation comes from the well-documented fact that a real plasma is never in a "quiescent" state as assumed in neoclassical theory. Because of the long range of the Coulomb forces, the collective nature of the plasma dominates its behaviour. At high temperatures the individual particle collisions - even corrected by the dynamical polarization effect - play a relatively minor (though sometimes relevant!) role in the dynamics. The particles act synchronously by organizing themselves as "waves" or "modes" or "vortices" or "streamers" of various types, which may interact among themselves or with individual particles. These more or less coherent structures are much more efficient in transporting matter and energy than individual particles. The situation is similar to the dynamics of a crystal: the vibrating particles attached to the lattice sites act together, producing a spectrum of normal modes of vibration of the crystal as a whole. The transport properties of a dielectric crystal are thus (partly) described by the action of these collective degrees of freedom ("phonons"). The waves in a plasma, especially in the presence of a magnetic field, are of a variety of different kinds, each one characterized by a band of frequencies and by a specific relation between frequency and wave vector (a *dispersion relation*).

When the amplitudes of the modes (determined by external factors) are sufficiently small, the interactions among them result in rather simple processes; for instance, two waves are transformed into one or two waves of a different kind. Some of these may, however, become unstable as their amplitude starts growing (through an internal, intrinsic mechanism) with a characteristic time scale. Such a growth cannot proceed indefinitely. As the amplitudes become significantly large, the nonlinear interactions become more and more important. Their result is a transformation into a large number of new modes as well as a saturation of their amplitudes. The evolution of the plasma thus becomes more and more complex and unpredictable (because it is very sensitive to small differences in positions or velocities): the motion becomes *turbulent* (or *chaotic*). Just as the quasi-random individual particle collisions produce a redistribution of the energy leading to the thermal equilibrium in an ordinary ("quiescent") situation, the collective interactions among modes produce a dissipation with a tendency toward a *non-equilibrium* steady state. This process is discussed qualitatively in some more detail in Sec. 1.5, and will be the main subject of the present book.

Clearly, the turbulent processes will strongly affect the mechanism of transport. In addition to the collisional dissipation that produces the classical and neoclassical transport, the turbulent dissipation produces a supplementary contribution to the transport coefficients. In many cases the turbulent contribution largely dominates the classical and neoclassical ones: in this case we are in presence of **anomalous transport.** In some instances the macroscopic linear relations between fluxes and thermodynamical forces is maintained. The *anomalous transport coefficients* entering these relations are now functions of the characteristic parameters of the modes responsible for the dissipation. In some situations, the effect of turbulence can be even deeper. The usually defined transport coefficients are then either zero or infinite. In these cases the transport laws are qualitatively modified: these new phenomena are called **strange transport.** There is some confusion in the literature concerning the terminology. The term *anomalous transport* adopted here is in agreement with the usage in the plasma physics literature: it designates a situation in which the macroscopic laws are the same as in classical transport, but the transport coefficients are determined by the characteristic parameters of turbulence.

On the other hand, in the vast modern literature on *nonlinear dynamical systems*, such a situation would still be called "normal". The name "anomalous transport" is given there to the processes which we call here "strange transport". We believe that it is important to make a clear distinction between the concepts of *anomalous transport* and *strange transport*.[1] The problems related to strange transport will be discussed in Chaps. 11, 12, 17.

1.2 Microscopic and Macroscopic Levels of Description

The basic microscopic model of a fully ionized plasma adopted in this book is a set of N_e electrons and N_i positive ions. For simplicity we restrict ourselves to the simple case of a single species of singly charged ions.[2] The plasma is idealized as being of *infinite extension*. In other words, we do not consider here the effects of material boundaries: the latter are important in specific applications (such as tokamaks, in which the "scrape-off layer" has a rather different structure from the core region). Their study is, however, very complicated because it is no longer "pure" plasma physics: a number of impurities originating from the walls are now present in the form of heavy neutral and excited atoms that interact with the charged particles through non-Coulomb forces. For the study of high-temperature fusion plasmas or of low-density astrophysical or geophysical plasmas we may restrict ourselves to classical (non-quantum) mechanics.[3] We shall even restrict ourselves in this book to a non-relativistic description, although relativistic effects do appear in some specific problems in these fields.

The plasma is globally neutral, hence $N_e = N_i = N$. The species (electron or ion) is labelled by a greek index that can take two possible values, $\alpha = (e, i)$. A particle of species α has mass m_α and charge e_α [$e_e = -e$, $e_i = +e$, where e is the absolute value of the electron charge]. The plasma is generally embedded in an electric field $\mathbf{E}(\mathbf{x}; t)$ and a magnetic field $\mathbf{B}(\mathbf{x}; t)$. These derive from a scalar potential $\phi(\mathbf{x}; t)$ and a vector potential $\mathbf{A}(\mathbf{x}; t)$ by the well-known relations (We use Gaussian units for the electromagnetic quantities):

$$\mathbf{E} = -\boldsymbol{\nabla}\phi - c^{-1}\partial_t\mathbf{A}, \quad \mathbf{B} = \boldsymbol{\nabla} \times \mathbf{A}. \tag{1.1}$$

---

[1] The term "strange transport" was first introduced by SHLESINGER, ZASLAVSKY AND KLAFTER in 1993. It was used systematically in my own book on statistical dynamics BALESCU 1997.

[2] We do not discuss in this book the impurity transport, though this problem is a nonnegligible aspect of transport in tokamaks (most results are easily adapted to the case of multiply charged ions).

[3] In the scrape-off layer the study of excitation, ionization and recombination processes requires quantum mechanics.

The dynamics of the plasma at the microscopic level is basically determined by the *Hamiltonian*, a dynamical function of the *phase space* variables, *i.e.*, the positions $\mathbf{q}_i$ and the momenta $\mathbf{p}_i$, as well as the time t. (By convention, the indices $j = 1, .., N$ denote electrons and $j = N + 1, .., 2N$ denote ions.) It has the following well-known form (GOLDSTEIN 1980):

$$H(\mathbf{q}_1, \mathbf{p}_1, ..., \mathbf{q}_{2N}, \mathbf{p}_{2N}; t) = \sum_{\alpha=e,i} \sum_{j=1}^{2N} H^\alpha(\mathbf{q}_j, \mathbf{p}_j; t),$$

$$H^\alpha(\mathbf{q}, \mathbf{p}, t) = \frac{1}{2m_\alpha} \left| \mathbf{p} - \frac{e_\alpha}{c} \mathbf{A}(\mathbf{q}, t) \right|^2 + e_\alpha \phi(\mathbf{q}, t). \tag{1.2}$$

At first sight, this additive form of the Hamiltonian would imply that the particles move independently. This is, indeed, an illusion. The potentials, or equivalently the electric and magnetic fields are determined by Maxwell's equations:

$$\nabla \cdot \mathbf{E} = 4\pi\sigma, \tag{1.3}$$

$$\nabla \times \mathbf{E} = -\frac{\partial \mathbf{B}}{\partial t}, \tag{1.4}$$

$$\nabla \cdot \mathbf{B} = 0, \tag{1.5}$$

$$\nabla \times \mathbf{B} = \frac{\partial \mathbf{E}}{\partial t} + \frac{4\pi}{c} \mathbf{j}. \tag{1.6}$$

Here $\sigma(\mathbf{x}; t)$ and $\mathbf{j}(\mathbf{x}; t)$ are, respectively, the charge density and the electric current density. For a set of point particles, they are highly singular functions:

$$\sigma(\mathbf{x};t) = \sum_{\alpha=e,i} \sum_{j=1}^{2N} e_\alpha \delta[\mathbf{x} - \mathbf{q}_j(t)],$$

$$\mathbf{j}(\mathbf{x};t) = \sum_{\alpha=e,i} \sum_{j=1}^{2N} e_\alpha \mathbf{v}_j \delta[\mathbf{x} - \mathbf{q}_j(t)], \tag{1.7}$$

where, in the presence of a magnetic field, the velocity $\mathbf{v}$ depends on the momentum and on the position through the well-known relation:

$$\mathbf{v} = \frac{1}{m_\alpha} \left[\mathbf{p} - \frac{e_\alpha}{c} \mathbf{A}(\mathbf{q}, t) \right]. \tag{1.8}$$

Thus, the electric and magnetic fields at any point x in physical space depend on the instantaneous position of all particles; as a result, the Hamiltonian (1.2) is a sum of terms, each depending non-additively on all phase-space variables: it therefore describes the interactions among particles, as it should.

The equations of motion of the particles are expressed most generally in terms of *Poisson brackets* with the Hamiltonian. In the present case, where q and p are canonically conjugate variables, the latter reduce to the simple standard form of *Hamilton's equations*:

$$\dot{q}_j = [q_j, H]_P = \frac{\partial H}{\partial p_j}, \quad \dot{p}_j = [p_j, H]_P = -\frac{\partial H}{\partial q_j}, \tag{1.9}$$

where the Poisson bracket of any two dynamical functions A and B is defined as usual:

$$[A, B]_P = \sum_{n=1}^{2N} \left(\frac{\partial A}{\partial q_n} \cdot \frac{\partial B}{\partial p_n} - \frac{\partial A}{\partial p_n} \cdot \frac{\partial B}{\partial q_n} \right). \tag{1.10}$$

The solution of this set of $\sim 10^{23}$ coupled differential equations is, of course an impossible task. This type of problem must be treated by the methods of *nonequilibrium statistical mechanics* (BALESCU 1975, 1997). Instead of tracking each individual particle, we construct a statistical ensemble: each configuration of the system is possible, but each one is assigned a certain probability. The state of the system is described in statistical mechanics by a phase-space distribution function $F(q_1, p_1, ..., q_{2N}, p_{2N}; t)$ obeying the Liouville equation. Under certain (widely realized) conditions, the description can be reduced to a description in terms of one-particle reduced distribution functions for each species (briefly called *distribution functions*) $f^\alpha(q, p; t)$. The evolution of the distribution function of species α is determined by the kinetic equation:

$$\partial_t f^\alpha = [H^\alpha, f^\alpha] + \mathcal{K}^\alpha, \tag{1.11}$$

where the "one-particle" Hamiltonian H^α is defined in Eq. (1.2); and $\mathcal{K}^\alpha$ denotes the *collision term* of particles α. It should be noted that the time variable t is treated as a parameter (rather than as a dynamical variable).[4]

---

[4] Actually, the proper Hamiltonian treatment requires an extension of the phase space: the time t is treated as a canonical variable, conjugate to an additional variable h, the total energy; t and h obey a Poisson bracket relation: $[t, h] = 1$. The procedure is well known (GOLDSTEIN 1980, KRUSKAL 1962, LITTLEJOHN 1981, BALESCU 1988 a). The extended Hamiltonian is defined as $\tilde{H}^\alpha(q, p; t, h) = H^\alpha(q, p; t) + h$. It is shown, however, that for the derivation of the Liouville equation (or of the kinetic equation) it is not necessary to use an extended phase space: the time is treated as a parameter, not as a phase space variable, just as in the case of a time-independent Hamiltonian. This surprisingly simple feature is actually the result of a careful and highly non-trivial

We now discuss in more detail the nature of the electromagnetic fields. In a quiescent plasma, the total fields can be split into two parts:

$$\mathbf{B} = \mathbf{B}_{ex} + \mathbf{B}_s, \quad \mathbf{E} = \mathbf{E}_{ex} + \mathbf{E}_s. \tag{1.12}$$

Here $\mathbf{B}_{ex}, \mathbf{E}_{ex}$ are *external fields* produced by sources outside the plasma (such as the currents in the coils of a tokamak): they obey the homogeneous Maxwell equations with appropriate boundary conditions. $\mathbf{B}_s, \mathbf{E}_s$ are *self-consistent fields,* obeying the Maxwell equations in which the source terms (charge density and electric current density) are determined by the state of the plasma itself. In the present statistical description of the plasma, Eqs. (1.7) are replaced by:

$$\sigma(\mathbf{x};t) = \sum_{\alpha=e,i} e_\alpha \int d\mathbf{q}\, d\mathbf{p}\, \delta(\mathbf{x} - \mathbf{q}) f^\alpha(\mathbf{q}, \mathbf{p}; t),$$

$$\mathbf{j}(\mathbf{x};t) = \sum_{\alpha=e,i} e_\alpha \int d\mathbf{q}\, d\mathbf{p}\, \mathbf{v}\, \delta(\mathbf{x} - \mathbf{q}) f^\alpha(\mathbf{q}, \mathbf{p}; t). \tag{1.13}$$

As a result, $\mathbf{B}_s, \mathbf{E}_s$ are functionals of f^α and, when substituted back into Eq. (1.11), make the first term *nonlinear* in f^α. This is the essence of the ***Vlasov equation***.

Although the microscopic description based on the kinetic equation (1.11) is the most fundamental description of the state and the dynamics of the plasma, it is in most cases very difficult to handle. In order to obtain specific results, one often starts the theory from a *macroscopic picture*. Instead of describing the state of the system by the distribution functions $f^\alpha(\mathbf{q}, \mathbf{p}; t)$, one uses the *moments* of this function, i.e. the average values of the powers of $\mathbf{p}$ (or of some convenient linear combinations thereof, such as a basis of orthogonal polynomials). The advantage is that, instead of working with functions on the 6-dimensional phase space $(\mathbf{q}, \mathbf{p})$, we now work with *fields* in the 3-dimensional physical space $(\mathbf{x} = \mathbf{q})$. If the complete, infinite set of moments were used, the description would be, in principle, equivalent to the microscopic one. But, in practice, it is only feasible to retain a finite, small subset of moments. An unavoidable difficulty appears at this stage: the equations of evolution of the moments have a hierarchical structure: the determination of a moment of order n requires the knowledge of the one of order $n + 1$. We must thus introduce additional simplifying assumptions that allow us to *truncate* the hierarchy. Thus, the *macroscopic*, or *"fluid picture"* of the plasma is a *model*, which necessarily will miss some features of the exact description. The most complete

analysis which was developed in detail in BALESCU 1988 a, Sec. 2.3, and will not be repeated here. Eq. (1.11) is thus valid as it stands, with H^α defined as the "ordinary", unextended Hamiltonian.

fluid picture is based on the set of five *plasmadynamical moments* for each species, i.e., the *particle density* $n_\alpha(\mathbf{x}; t)$, the three components of the *local velocity field* $u_r^\alpha(\mathbf{x}; t)$ and the *thermal energy density*, related to the *pressure*: $\frac{3}{2}p_\alpha$. They are determined, respectively, by the zeroth, the first and the second centred moment (*i.e.*, the average of $|\mathbf{v} - \mathbf{u}^\alpha|^2$) of the distribution function (BALESCU 1988 a) and obey the well-known *plasmadynamical* (or *two-fluid*) balance equations:

$$\partial_t n_\alpha = -\nabla \cdot \mathbf{\Gamma}^\alpha, \qquad (1.14)$$

$$\partial_t m_\alpha n_\alpha u_r^\alpha = -\nabla_s \left(m_\alpha n_\alpha u_r^\alpha u_s^\alpha + p_\alpha \delta_{rs} + \pi_{rs}^\alpha \right)$$
$$+ e_\alpha n_\alpha \left(E_r + c^{-1} \epsilon_{rsn} u_s^\alpha B_n \right) + R_r^\alpha, \qquad (1.15)$$

$$\partial_t p_\alpha = -\nabla_r \left[\tfrac{5}{3} p_\alpha u_s^\alpha \delta_{rs} + \tfrac{2}{3} \left(\pi_{rs}^\alpha u_s^\alpha + q_r^\alpha \right) \right]$$
$$+ \tfrac{2}{3} u_r^\alpha \nabla_s \left(p_\alpha \delta_{rs} + \pi_{rs}^\alpha \right) + \tfrac{2}{3} h^\alpha, \qquad (1.16)$$

where $\mathbf{\Gamma}^\alpha = n_\alpha \mathbf{u}^\alpha$ is the particle flux, π_{rs}^α is the dissipative pressure tensor (related to viscosity), q_r^α is the heat flux, R_r^α is the collisional friction force and h^α is the collisional ion-electron heat exchange. These quantities represent the additional moments which have to be expressed in terms of the plasmadynamical ones in order to close the hierarchy. The various closure assumptions will be discussed below.

Before continuing this discussion, we note that in anomalous transport theory it is often useful to consider, instead of the thermal energy, the *total internal energy density* $\frac{3}{2}P_\alpha$, defined as the non-centred second moment, i.e., the average of $\frac{1}{2}m_\alpha v^2$. This quantity differs from p_α by including the average macroscopic kinetic energy $\frac{1}{2}m_\alpha u^{\alpha\,2}$. Its balance equation, derived from the exact kinetic equation, is:

$$\partial_t \tfrac{3}{2} P_\alpha = -\nabla \cdot \mathbf{Q}^\alpha + \mathbf{E} \cdot \mathbf{j}^\alpha, \qquad (1.17)$$

where $\mathbf{Q}^\alpha$ is the total internal energy flux, *i.e.*, the average of $(v^2\,\mathbf{v})$ and $\mathbf{j}^\alpha$ is the electric current density carried by particles of species α, the average of $e_\alpha \mathbf{v}$. Note that the electric field enters this equation, whereas in Eq.(1.16) this contribution is cancelled by a term originating from the momentum balance (1.15); the latter introduces a contribution of the dissipative pressure tensor $\overleftrightarrow{\pi}^\alpha$ in (1.16).

The plasmadynamical equations involve the electric and magnetic fields: they are thus coupled to Maxwell's equations. The complete macroscopic description thus involves $2 \times 5 + 8$ equations: still a rather formidable mathematical problem. Various reduced models have been proposed in the literature for studying specific problems; some of them will be discussed in forthcoming chapters.

1.3   Closure of the Plasmadynamical Equations

It is well known that the macroscopic plasmadynamical equations (1.14)
- (1.16) are the result of a truncation of an infinite hierarchy of moments
of the kinetic equations. As a result, there remain in these equations a
number of undetermined quantities: the equations are *not closed*. These
quantities are of four kinds: *thermodynamic quantities (T_α), electromag-
netic fields* ($\mathbf{E}$, $\mathbf{B}$), *momentum- and energy-exchanges* ($\mathbf{R}^\alpha$, h^α) *and fluxes*
($\mathbf{\Gamma}^\alpha$, $\mathbf{q}^\alpha$, $\overset{\leftrightarrow}{\pi}{}^\alpha$).

The *temperatures* T_α do not appear explicitly in the plasmadynamical
equations; they are nevertheless important, measurable quantities. For
dilute high-temperature plasmas they are related to the pressures and the
densities by the well-known ideal gas equation of state:

$$p_\alpha = n_\alpha T_\alpha. \tag{1.18}$$

In this book we will always use energy units (Joules or electron-volts)
for measuring temperatures: the relation between T_α and the temperature
t_α measured in Kelvins is: $T_\alpha = k_B t_\alpha$, where k_B is the Boltzmann constant.
It should be recalled that, as a result of the inefficiency of the electron-ion
collisions, the plasma is able to live for a long time in a non-equilibrium
state where the electron and the ion temperatures are different.

The electric and the magnetic fields are determined in terms of the
plasmadynamical quantities by the Maxwell equations (1.3) - (1.6). This
is at least the case in the ideal case of a *self-consistent* theory. In the first
part of this book such type of theories will be considered. It will appear,
however, that they are generally very complex and difficult to apply in
practice. Therefore, an approximate approach is very often used in plasma
transport analyses, in which the electromagnetic fields are *prescribed*, either
as given functions (as in neoclassical transport theory) or statistically (in
anomalous transport theory). The retroaction of the particle motion on the
fields is thus neglected. This type of non-selfconsistent theories requires a
degree of physical intuition; its validity must be evaluated *a posteriori*.

The exchange terms are essentially due to collisions, and belong mainly
to the realm of classical and neoclassical theories. They will not be systematically discussed here.

The most important closure relations are the so-called *transport equa-
tions*, relating the *dissipative fluxes* $\mathbf{\Gamma}^\alpha$, $\mathbf{q}^\alpha$, $\mathbf{j}$, $\overset{\leftrightarrow}{\pi}{}^\alpha$ to the *thermodynamic forces* that produce them. The latter are related to the spatial
inhomogeneity, and are expressed as gradients of thermodynamic quanti-
ties: ∇n_α, ∇T_α, $\nabla \Phi$, $\nabla \mathbf{u}_\alpha$. The study of these relations is the object of
Nonequilibrium Thermodynamics (DE GROOT & MAZUR 1984, BALESCU
1988 a). In full generality, these relations can be of an almost arbitrary
form. Grouping all the relevant fluxes into a "supervector" $\mathbf{J}$, and all ther-

modynamic forces into $\mathbf{X}$, the general transport equations can be written as:

$$\mathbf{J} = \mathbf{F}(\mathbf{X}) \qquad (1.19)$$

where $\mathbf{F}(\mathbf{X})$ is constrained by the condition $\mathbf{F}(\mathbf{0}) = \mathbf{0}$. This function could be *nonlinear*, it could also depend on higher order spatial derivatives of the thermodynamic functions, etc. Its form should be deducible from a microscopic kinetic theory.

In the vast majority of cases studied at present in plasma transport theory, be it classical, neoclassical or anomalous, it is assumed that the transport equations are *linear*. In that case any vector flux (such as $\mathbf{\Gamma}^\alpha$) can only be related to the vector forces (such as ∇n_α), whereas tensor fluxes ($\overset{\leftrightarrow}{\pi}{}^\alpha$) are related to tensor forces ($\nabla \mathbf{u}_\alpha$): this symmetry constraint is called *Curie's principle*. We will limit ourselves here to the discussion of vector fluxes.[5] In the linear regime, the flux vector components J_r are related to the force vector components X_r by a *transport matrix* $\overset{\leftrightarrow}{\mathbf{L}}$:

$$\mathbf{J} = \overset{\leftrightarrow}{\mathbf{L}} \cdot \mathbf{X} \quad \Longleftrightarrow \quad J_a = L_{ab}\, X_b. \qquad (1.20)$$

The constant components of the transport matrix are called *transport coefficients*. Some of these (in particular, the diagonal elements) are quite familiar: thus, the *diffusion coefficient* D, relating the particle flux to the density gradient, the *heat conductivity* (or *heat diffusivity*) $\kappa_\alpha = n_\alpha \chi_\alpha$, relating the heat flux to the temperature gradient, and the *electrical conductivity* σ relating the electric current density to the electric field. It should be stressed, however, that the transport matrix is, in general, non-diagonal: every flux is determined by *all* the gradients present in the system.

Let us finally mention that the transport equation may take even more general forms, while still remaining linear. The fluxes and the forces can (or must) be defined locally as fields depending on the space coordinates and on time. The most general linear transport relation then takes the form:[6]

$$J_a(\mathbf{x},t) = \int d\mathbf{x}' \int_0^t dt'\, L_{ab}(\mathbf{x}',t')\, X_b(\mathbf{x}-\mathbf{x}',t-t'). \qquad (1.21)$$

This type of *nonlocal and non-Markovian* equation expresses the fact that the flux at a given point $\mathbf{x}$ and at a given time t could be influenced by the values of the forces in its spatial environment (nonlocality) and by its past history (non-Markovianity). Equations of this form will be met

---

[5] The pressure tensor will be briefly discussed in Sec.16.4.

[6] Here and in all forthcoming equations of this book, an integral over the spatial coordinates $\int d\mathbf{x}$ without explicit specification of the limits, represents an integral over the whole volume enclosing the plasma (which may be infinite).

towards the end of our study. The space- and time-dependent coefficients $L_{ab}(\mathbf{x}', t')$ are called "nonlocal" transport coeficients: they should not be confused with the constant coeficients L_{ab} (they do not have the same dimension). Whenever the spatial and temporal ranges of influence (expressed by the width of the nonlocal coefficients) are sufficiently small, the delocalization and the retardation of the force can be neglected under the integral: $X_b(\mathbf{x} - \mathbf{x}', t - t') \approx X_b(\mathbf{x}, t)$; and the transport equation (1.21) reduces to:

$$J_a(\mathbf{x}, t) = L_{ab}(t)\, X_b(\mathbf{x}, t). \tag{1.22}$$

The time-dependent quantity $L_{ab}(t)$ called the *running transport coefficient* is defined as:

$$L_{ab}(t) = \int d\mathbf{x}' \int_0^t dt'\, L_{ab}(\mathbf{x}', t'). \tag{1.23}$$

Whenever $L_{ab}(\mathbf{x}', t')$ decays to zero sufficiently fast (*e.g.* exponentially) in order to ensure the convergence of the time integral, $L_{ab}(t)$ tends, for times longer than this decay time, to a finite, constant asymptotic value, which is none other than the "usual" transport coefficient L_{ab} appearing in Eq. (1.20):

$$L_{ab}(\infty) = L_{ab}. \tag{1.24}$$

Thus, Eq. (1.20) appears as an asymptotic approximation of the more general equation (1.21). The process described above, leading from the latter to the former, is called a **Markovianization process**: it will be met with quite often in the forthcoming work.

The determination of the transport coefficients from first (or "second"!) principles of microscopic dynamics is the object of transport theory. Whereas the classical and the neoclassical theories can be considered as closed fields, the anomalous transport theory is by orders of magnitude more complex (but also more important!) and its completely general solution is still an open problem (and will perhaps remain open forever!). An enormous amount of work has been devoted to its study in recent years and much progress has been achieved in particular aspects. It is impossible to consider all these works in a single book like the present one: this is why we modestly entitled it: "*Aspects* of Anomalous Transport in Plasmas".

1.4 Experimental Aspects

We now give a brief overview of the experimental strategies used in measuring quantities relevant for transport. This section is very schematic: the present book is devoted to transport *theory*. Its volume would double if experimental studies were considered in the same degree of detail![7] We consider here the experiments performed in magnetic fusion devices (tokamaks and stellarators). There exist many excellent review papers on the experimental aspects of plasma transport (in particular, anomalous transport), among which we quote only some of the more recent ones: BURRELL et al. 1990, WAGNER & STROTH 1993, LOPES CARDOZO 1995, STROTH 1998.

The first aspect to be kept in mind when considering the experimental data is that there are two classes of techniques for measuring "transport coefficients". These yield in general *different results*, because they really measure different quantities. More precisely, the differences arise from different *interpretations* of the measured quantities. A quite thorough discussion of this problem is found in the paper of LOPES CARDOZO 1995. Let us discuss for simplicity the simple case of a single-species particle flux ($J_1 = \Gamma$) and a heat flux ($J_2 = q$) and the forces $X_1 = -\nabla n$, $X_2 = -\nabla T$ (the generalization to several forces and fluxes is straightforward).

A first class comprises so-called *"power balance methods"*. These are "static" methods acting on plasmas in a quasi-steady state. In order to produce such a state, the presence of source terms is needed. These are generated by external heating and fuelling processes, which are always present in experiments. Eqs. (1.14) and (1.16) thus take the general form:

$$\partial_t g + \nabla \cdot \mathbf{F} = S, \qquad (1.25)$$

where g is the particle- or the thermal energy density, $\mathbf{F}$ the corresponding flux, and S the source described above, including also internal processes, such as the electron-ion energy exchange. In steady state ($\partial_t g = 0$, $\mathbf{u}^\alpha = 0$) an accurate measurement of the sources, integrated over the interior of a magnetic surface provides the flux across this surface. On the other hand, a detailed measurement of the density- and temperature profiles yields the values of the local gradients. One then deduces "power balance transport coefficients" by simply dividing the net local fluxes (*i.e.*, their value at a given radial distance from the magnetic axis) by the corresponding gradients, *i.e.*, by assuming the simplest possible Fick and Fourier laws:

$$\Gamma = D^{pb} (-\nabla n), \qquad q = n\chi^{pb} (-\nabla T). \qquad (1.26)$$

This interpretation of the results may be completely inadequate in certain cases, as will be shown below.

The second experimental strategy, called *"perturbative transport analysis"* consists of producing a small, well localized perturbation of the forces (*i.e.*, the gradients) in a steady-state plasma, and following the evolution in space and time of the effect of this perturbation. The perturbations can

---

[7] Moreover, the author does not feel competent in discussing critically the experimental procedures.

be internal ones, such as the "sawtooth crashes" which are always present near the core of a tokamak, and whose (diffusive) propagation on their way towards the edge are followed in time. Alternatively, local perturbations can be produced by the injection of solid pellets into the plasma. This method requires, of course, very precise diagnostics, with very good spatial and temporal resolution; it was therefore only realized rather recently, since the end of the '80s. This method allows a separate measurement of the influence of each thermodynamic force on a given flux: it thus produces (in principle) the complete transport *matrix* L_{ab} (1.20).

It is easily seen that, even in the case of strictly linear transport equations with constant transport coefficients, the two methods give different results. Consider, for instance, the particle flux. It is derived from Eq. (1.20) as:

$$\Gamma = D\,(-\nabla n) + a\,(-\nabla T) \tag{1.27}$$

where the true diffusion coeficient D and the thermodiffusion coefficient a - a *non-diagonal* element of the transport matrix - can be obtained separately by perturbative experimental techniques. If we were to interpret the same flux by a power balance method as in Eq. (1.26), it clearly leads to an inconsistency, because $\Gamma \neq 0$ when $\nabla n = 0$. The difference (or "off-set") $[a\,(-\nabla T)]$ is then interpreted as a falsely convective flux ("pinch flux"), usually directed towards the axis, *i.e.*, *against* the temperature gradient:

$$\Gamma = D\,(-\nabla n) - nV. \tag{1.28}$$

Thus, the steady state power balance analysis yields:

$$D^{pb} = -\frac{\Gamma}{\nabla n} = D + \frac{nV}{\nabla n} = D - n\frac{\nabla T}{\nabla n}\,a. \tag{1.29}$$

On the other hand, the perturbative (or *incremental*) transport analysis yields (see Fig. 1.1):

$$D^{inc} = -\frac{\partial \Gamma}{\partial(\nabla n)} = D. \tag{1.30}$$

Thus, the perturbative transport technique yields directly the true transport coefficients. The same ambiguity is resolved in the case of nonlinear transport equations. The reader is referred to the detailed discussions and presentation of the experimental techniques and results in the excellent review paper of LOPES CARDOZO 1995.

The perturbative techniques (mainly performed by the groups of Gentle on the TEXT tokamak of the University of Texas at Austin, of Lopes Cardozo and his group at JET, of Kissick and Callen on TFTR (Princeton), and many others) have led to precise determinations of the transport coefficients, but also to unexpected surprises. The latter will be discussed in Chap. 17.

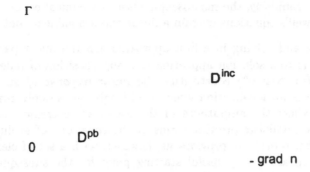

Figure 1.1. Power balance and Incremental diffusion coefficients.

1.5   The Turbulent State of the Plasma

In the quiescent plasma transport theory, the total fields are most often not determined selfconsistently by the solution of the Vlasov-Maxwell equations, because this process is too difficult in realistic situations. For instance, in classical transport theory (BRAGINSKII 1965, BALESCU 1988 a) the fields are simply replaced by given external fields: $E \approx E_0$, $B \approx B_0$, where E_0, B_0 are stationary and homogeneous. In the neoclassical transport theory (BALESCU 1988 b), the self-consistency is better approximated, by taking for B and E an equilibrium solution of the *macroscopic* magnetohydrodynamic equations (or a model field approaching the latter, such as the standard model of a large aspect ratio tokamak).

In all these cases, the fields B and E are determined independently of the distribution functions. They are then injected into the kinetic equation (1.11) as *given* functions. These theories are therefore typical *nonselfconsistent* theories. As a result, the first term in that equation reduces to the ordinary *linear* one-particle Liouville operator for a particle in an *external* field. The main emphasis in the classical and neoclassical theories is on the role of the individual particle collisions as the driving mechanism of the dynamics, in presence of *given* electric and magnetic fields.

The Vlasov equation (with or without the addition of collisions) can be studied more thoroughly by combining the equation for the distribution function with the Maxwell equations for the electromagnetic field. The resulting coupled system of equations thus acquires an explicitly *nonlinear* contribution. Similarly, the macroscopic plasmadynamical equations, coupled to Maxwell's equations contain a linear and a nonlinear contribution.

Isolating and solving in a first approximation the linear part of the equation leads to a solution appearing as a superposition of independent *eigenmodes* (or "waves") constituting the *linear response of the plasma*. Such a state is an idealization which could only be possibly realized in a situation where the amplitudes of the waves are extremely small (in that case the nonlinear product terms in the equations of evolution are small of higher order). It provides us, however, with a set of elementary "bricks" which are a very useful starting point for the subsequent construction of a more realistic theory. The modes are obtained by solving the eigenvalue problem of the linear evolution operator. In the simplest situations the equations of evolution can be transformed into algebraic equations by a Fourier transformation. The eigenvectors are then of the form $\varphi_n(\mathbf{x}, t) \sim \exp(i\mathbf{k} \cdot \mathbf{x} - i\Omega t)$, characterized by a wave vector $\mathbf{k}$ and a frequency Ω. In a linear analysis the frequency is related to the wave vector through a *dispersion relation* $D(\Omega, \mathbf{k}) = 0$. Each distinct root of this equation defines an eigenmode, whose frequency is linked to the wave vector as $\Omega_r = \Omega_r(\mathbf{k})$. In a linear approximation these modes evolve independently of each other (by definition of an eigenvector). The corresponding eigenvalues are generally complex: $\Omega_r(\mathbf{k}) = \omega_r(\mathbf{k}) + i\gamma_r(\mathbf{k})$. When $\gamma_r = 0$, the mode is purely oscillating; when $\gamma_r < 0$ it decays to zero for $t \gg \gamma_r^{-1}$. The most important case is $\gamma_r > 0$: the amplitude of the mode is then growing exponentially in time. The corresponding mode is briefly called an *instability* in the plasma literature. After a sufficiently long time, the behaviour of the plasma is dominated by these unstable modes.

The exponential growth of the instabilities can, clearly, not last forever. As the amplitudes grow, the nonlinear contributions become more and more important: they produce *interactions among the waves and between waves and individual particles*. The result is a reorganization of the entities describing the plasma. The modern theory of dynamical systems has shown that, as the nonlinearity grows, the system undergoes a sequence of *bifurcations* leading to new states, which may be ordered ("dissipative structures") or, more often, be completely disordered. In the latter case, a *turbulent state* sets in: it is characterized by the presence of a large number of collective "degrees of freedom". These *fluctuations* evolve very irregularly in space and time; in their midst some more or less transient *coherent structures* may be present. In principle, these complicated processes are still described by the laws of statistical mechanics and their evolution is contained in the Liouville equation. In a fully developed turbulent state, however, the complexity of the interactions reaches such a level that the assumptions made in statistical mechanics in order to obtain tractable equations become notoriously insufficient.

1.6   Statistical Description of a Turbulent Plasma

In order to deal with the turbulent plasma we appeal to the usual procedure for complex systems: a *statistical description*. The electric and magnetic fields produced by the collective plasma modes (or by external sources) fluctuate so wildly and irregularly that their precise definition at each point and at each time becomes illusory in practice. We therefore decide that the only property that can be stated about them is the *probability density* of their having a certain value at every given point in space-time. We thus imagine a statistical ensemble of plasmas, each of its members (or *realizations*) being decribed by a distribution function f^α obeying Eqs. (1.11) - (1.13) [or (1.14) - (1.16)] with given fields $\mathbf{E}(\mathbf{x}, t)$ and $\mathbf{B}(\mathbf{x}, t)$. But in different realizations, these fields have different functional forms. Each of the realizations is weighted by a "distribution functional" defining the probability density for the functions $\mathbf{E}(\mathbf{x}, t)$ and $\mathbf{B}(\mathbf{x}, t)$ having their specific functional form. We thus introduce *two levels of statistics*: besides the ordinary statistical mechanical description of the many-body plasma, there is also a statistical description of the electromagnetic fields acting on the particles. It should be stressed again that this is a kind of shortcut. As stated above, the dynamics of the turbulent plasma is entirely contained in the N-body phase space distribution function obeying the Liouville equation. But in a turbulent state, the assumption of an "ordinary" kinetic regime in which the system is entirely described by the one-particle distribution function is certainly invalid: many-body correlations (which are not functionals of f^α) play a role in the evolution of the system. By introducing the second level of statistics, we assume that the kinetic description (1.11) can be restored, provided that the fields $\mathbf{E}(\mathbf{x}, t)$ and $\mathbf{B}(\mathbf{x}, t)$ are described by an independent distribution functional. The particle distribution functions $f^\alpha(\mathbf{q}, \mathbf{p}, t)$, which are determined by the fluctuating fields through the kinetic equations, become themselves stochastic quantities. Let us stress this fundamental point. In ordinary statistical mechanics, or in kinetic theory, the distribution functions f^α are simple "deterministic functions", that associate a real, positive number with every value of the variables $\mathbf{q}$, $\mathbf{p}$, t. This number represents the probability density of finding the particle in this configuration at the specified time, given a well-determined electromagnetic field. When the latter becomes a random quantity, $f^\alpha(\mathbf{q}, \mathbf{p}, t)$ can no longer be ascribed a definite numerical value, but only a (functional) probability. We thus need to define a new "deterministic" quantity characterizing the statistical ensemble of turbulent realizations. This quantity is "*a functional probability density of probability density functions f^α*", ascribing a real, positive number to each functional form $f^\alpha(\mathbf{q}, \mathbf{p}, t)$.

The observable quantities are assumed to be defined by a *double averaging* over both ensembles. Thus, the statistical-mechanical average produces plasmadynamical quantities. These obey Eqs. (1.14) - (1.16), which are still stochastic equations. In order to obtain deterministic quantities (to which numerical values can be assigned) we need to perform a second average, over the turbulence ensemble. *We postulate that only such double-averaged quantities are physically observable.* Just like in ordinary statistical mechanics, this strong assumption of a second statistics can only be judged by its results. It is also assumed that the two averages can be performed in the inverse order. Thus, it is meaningful to define stochastic distribution functions $f^\alpha(\mathbf{q}, \mathbf{p}, t)$. For instance, their average over the turbulent ensemble, $\langle f^\alpha(\mathbf{q}, \mathbf{p}, t)\rangle$, integrated over the momenta $\mathbf{p}$, represents the observable spatial density profile of the plasma.

The direct manipulation of the turbulent distribution functional would be an awkward task, because of the complexity of such an object. But this functional can certainly not be observed in its globality. Here lies an important difference with the ordinary distribution function. For instance, the integral of f^e over the momenta yields the electron density at each point in a tokamak, a quantity that can be measured experimentally. Even better, space probes orbiting the earth in the magnetospheric region have provided us for a couple of decades with complete velocity distributions of electrons and ions at closely separated times and at various positions. By contrast, the much more abstract turbulent functional distribution is not susceptible to direct measurement: only average quantities are observable. Thus, a single random function is characterized by an infinite number of deterministic quantities: the *moments* and *autocorrelations* of this function, *e.g.*, $\langle f^\alpha(\mathbf{q}, \mathbf{p}, t)\rangle$, $\langle f^{\alpha_1}(\mathbf{q}_1, \mathbf{p}_1, t_1) f^{\alpha_2}(\mathbf{q}_2, \mathbf{p}_2, t_2)\rangle$, etc. *Cross-correlations* of several random functions can be similarly defined. Clearly, only a limited number of such quantities can be actually measured. The basic theory of turbulent plasmas should provide us with an algorithm for the determination of the moments of f^α, from which the macroscopic plasmadynamical quantities, and in particular the transport laws, could be derived.

This discussion indicates various approaches to the study of turbulent plasmas and of the associated anomalous transport problem. The most fundamental approach is the study of the stochastic kinetic equation coupled to the (stochastic) Maxwell equations. Such a self-consistent theory should not require any arbitrary assumption: it should produce equations of evolution for all the moments (averages of products of f^α). In practice, however, an exact solution of this problem is impossible. Indeed, the nonlinearity of the equations implies that the moment equations have a hierarchical structure: the determination of any moment of a given order requires the knowledge of the higher order moments. The chain of moment equations must therefore be truncated at a finite level. The criterion for the truncation is not easily determined *a priori*: it must be evaluated by its results. Roughly speaking, this approach is limited to situations of weak turbulence (*i.e.*, small nonlinear effects). A similar study starting from the macroscopic ("fluid") equations is confronted with the same difficulties originating from the nonlinearity.

Because of these difficulties, the "fundamental" studies - in spite of their basic importance - cannot easily produce explicit results that can be directly compared to experiment. In order to obtain such results, one is led to make compromises. In particular, one may give up the ambitious requirement of self-consistency. In this approach, *the electromagnetic fields* E *and* B *are defined independently of the particle distribution.* Their statistical properties are given in advance (*e.g.*, assuming a Gaussian process); they depend on a number of parameters (*e.g.*, correlation lengths, correlation time) that may be related to empirical data. These properties will, in turn, determine the statistics of the distribution functions, via the kinetic equations. The latter thus become stochastic equations decoupled from the Maxwell equations: they will be called *"hybrid kinetic equations"*. Their characteristic equations are stochastic differential equations of the same nature as the well-known *Langevin equations.* One may then take advantage of the thoroughly established methods of the theory of stochastic processes. It will be shown that in this case one may override the limitation of weak turbulence.

We must finally mention the enormous volume of the *numerical simulations* of fluids and plasmas, especially in turbulent states, realized in recent years. This is due to the continuing development of computing hardware and software at a very high speed. This branch of plasma physics is considered in many laboratories as "theory". I believe that this characterization is not correct. One type of simulations are closely associated with direct *experimental studies.* In fusion physics, the confinement devices have become very large, and the numerous diagnostic methods are getting very sophisticated. In order to exploit the richness of these results it becomes necessary to compare the (computerized) direct experimental results with numerical models of various scenarios of the discharges. A second type of numerical studies is motivated by the verification of theoretical models. Starting from theory, one sets up a numerical program designed for testing its predictions. In both cases, the numerical study is really a *numerical experiment* to be classified as experimental physics or, possibly, as a link between experiment and theory. The analysis of the recent experiments and numerical simulations would require a separate book, written by an expert of the field (which I am not). Therefore, these aspects will only marginally be indicated in the present book.

Another necessary limitation of the present book was already mentioned in the preface. The theory of anomalous transport has grown to such an enormous volume, mainly in fusion physics, but also in astrophysics and geophysics, not to mention condensed matter physics, that it is literally impossible to encompass all aspects of this activity in a single book. We therefore decided to discuss turbulent transport only in its "purest" form. This implies the limitation to the simplest plasma model (one single ion species) and to the simplest form of the external magnetic field (constant, or possibly sheared **B**). In this way we exhibit the effects that are specifically due to turbulence, and therefore have some degree of universality. Clearly, in specific applications (such as, *e.g.*, tokamaks or the magnetosphere) the magnetic field is inhomogeneous and curved. These *geometric effects* may introduce important new features (as, for instance, in neoclassical transport theory) that modify the "pure" turbulence analysis. The theory exposed in the present book may therefore serve as a basic methodology, to be applied to various physical systems, in combination with specific geometric constraints. Such applications should be the subject of specialized monographs to be devoted to specific classes of plasmas.

1.7   The Averaging Operation

We now set up a general framework of notations for fluctuating quantities. The *turbulent ensemble-averaging* operation will be denoted by angular brackets $\langle ... \rangle$. In some problems there may be several fluctuating quantities that must be averaged upon, for instance A and B. Whenever necessary, the partial average of an expression over A will be denoted by a subscript: $\langle ... \rangle_A$. Such an expression is still fluctuating; a further average over B is necessary to define the complete (deterministic) average expression. It will be denoted by $\langle \langle ... \rangle_A \rangle_B$.

Let Q be a fluctuating quantity (it could be an electromagnetic field, or a distribution function, or a plasmadynamical quantity). Any such quantity can be uniquely decomposed into a sum of two terms:

$$Q = Q_0 + \delta Q. \tag{1.31}$$

The first term Q_0 denotes the *ensemble average of* Q; it is a deterministic quantity.

$$Q_0 = \langle Q \rangle. \tag{1.32}$$

The deviation δQ from the average will be called the *fluctuation of* Q. It has the obvious property:

$$\langle \delta Q \rangle = 0. \tag{1.33}$$

The averaging operator has the formal properties of a *projection operator*. When applied to a fluctuating quantity, it projects the latter on the "space of deterministic functions", while its complement $(I - \langle ... \rangle)$ projects it on the "space of fluctuations". The averaging is clearly an idempotent operation:

$$\langle\langle...\rangle\rangle = \langle...\rangle.$$ (1.34)

An exception to this system of notations will be made for the distribution functions. In order not to unnecessarily burden the forthcoming equations, we adopt the simpler notation F^α, instead of f_0^α:

$$f^\alpha = F^\alpha + \delta f^\alpha.$$ (1.35)

Consider now a collection of fluctuating quantities $P, Q, ...$ We note the following obvious property of the average of a product:

$$\langle P_0\ Q_0...\delta R\ \delta S...\rangle = P_0\ Q_0...\langle\delta R\ \delta S...\rangle.$$ (1.36)

It follows from Eqs. (1.31) and (1.36) that any moment $\langle P\ Q...\rangle$ can be decomposed into a product of average quantities and a sum of averages of products of fluctuations. For instance, recalling (1.33):

$$\langle P\ Q\rangle = P_0\ Q_0 + \langle\delta P\ \delta Q\rangle.$$ (1.37)

These properties will be extensively used in forthcoming chapters for the analysis of turbulent dynamics.

Chapter 2

Macroscopic Plasmadynamical Equations

2.1 The Inhomogeneous Plasma

In the present chapter we introduce the tools necessary for a study of waves in a plasma at the macroscopic level. According to the general philosophy of the present book, we consider only the simplest possible, non-trivial situations. Even in this case the general problem is very complicated. We will isolate the more important aspects of the problem, which will also be treated later on a microscopic basis. The latter refinement introduces additional features, that are "missed" in the macroscopic description. *Let us stress that the description in the present and in the next chapter is deterministic.* The treatment of a turbulent plasma will start in Chap. 4.

We describe our approach as follows. We first define a *reference state* which is time-independent, but inhomogeneous; the plasmadynamical quantities in this reference state will be denoted by A_0. These *unperturbed quantities* must be determined by solving the appropriate plasmadynamical equations. Whenever this state is submitted to a space- and time-dependent perturbation (external or internal) the plasmadynamical quantities are changed into $A = A_0 + \delta A$. The *perturbation*, or *response* δA must, of course, also satisfy the equations of motion. The present chapter is devoted to the derivation, the discussion and the possible simplification of the response equations. The solution of these equations (for the case of small fluctuations) will be the subject of the next chapter.

In the macroscopic picture, the plasma is described by the *two-fluid plasmadynamical balance equations.* In order to make this set of equations complete, an additional relation is needed: an *equation of state*, relating the pressure p_α to the density n_α and the temperature T_α (see Sec. 1.3). The introduction of the latter requires a new (and rather strong) assumption on the microscopic level, viz., that each component of the plasma (electrons, ions) is in a state close to *local thermal equilibrium*, characterized by a Maxwellian velocity distribution, with density, average velocity and temperature slowly varying in space and time. The coupling between the components is weak (because of the large difference in mass); as a result, each component is characterized by a separate local temperature. The electron temperature and the ion temperature will eventually be equalized, but this process is very slow. From these assumptions we deduce the equation of state of a perfect gas for each component:

$$p_\alpha = n_\alpha T_\alpha. \qquad (2.1)$$

In general, $T_e \neq T_i$. With these additional assumptions, the macroscopic plasmadynamical equations (1.14) - (1.16), (neglecting the collisional heat exchange h^α) are [see Eqs. (1.14) - (1.16)]:

$$\partial_t n_\alpha = -\nabla \cdot (n_\alpha \mathbf{u}^\alpha), \qquad (2.2)$$

$$\begin{aligned} \partial_t\, m_\alpha n_\alpha \mathbf{u}^\alpha = &- \nabla \cdot (m_\alpha n_\alpha \mathbf{u}^\alpha \mathbf{u}^\alpha + \overleftrightarrow{\pi}^\alpha) - \nabla p_\alpha \\ &+ e_\alpha n_\alpha \left(\mathbf{E} + c^{-1}\mathbf{u}^\alpha \times \mathbf{B}\right) + \mathbf{R}^\alpha, \end{aligned} \qquad (2.3)$$

$$\partial_t p_\alpha = -\mathbf{u}^\alpha \cdot \nabla p_\alpha - \gamma p_\alpha \nabla \cdot \mathbf{u}^\alpha - \frac{2}{3} \overleftrightarrow{\pi}^\alpha : \nabla \mathbf{u}^\alpha - \frac{2}{3} \nabla \cdot \mathbf{q}^\alpha, \quad \alpha = e, i, \quad (2.4)$$

where $\overleftrightarrow{\pi}^\alpha$ is the dissipative pressure tensor, $\mathbf{q}^\alpha$ is the heat flux, $\mathbf{R}^\alpha$ is the electron-ion friction coefficient, proportional to the electric current density (BALESCU 1988 a, Chaps. 3, 4), and

$$\gamma = \frac{c_P}{c_V} = \frac{5}{3} \qquad (2.5)$$

is the well-known ratio of specific heats at constant pressure and at constant volume. The balance equations (2.2), (2.3), (2.4) are coupled to the Maxwell equations (1.3) - (1.6) which determine the electric and magnetic fields that are self-consistently produced by the plasma.

We now introduce two important simplifications. The first one is purely geometrical. The simplest possible *unperturbed* magnetic field and plasma configuration are defined as follows:

• We assume that *the unperturbed magnetic field is **stationary**, (i.e., independent of time) and **homogeneous** (i.e., it has everywhere the same value)*. Its direction is chosen as the z-axis of a cartesian coordinate system in physical space $\mathbf{x} = (x, y, z)$:

$$\mathbf{B} = B\,\widehat{\mathbf{b}} = B\,\mathbf{e}_z \qquad (2.6)$$

where $\widehat{\mathbf{b}}$ is a *constant* unit vector.

- We consider that the unperturbed scalar plasmadynamical quantities, such as *the density $n_{\alpha 0}$, the temperature $T_{\alpha 0}$ and the pressure $p_{\alpha 0}$, are stationary, but spatially inhomogeneous, all varying along the same direction x:*

$$n_{\alpha 0} = n_{\alpha 0}(x), \quad \mathbf{u}_0^\alpha = \mathbf{u}_0^\alpha(x), \quad T_{\alpha 0} = T_{\alpha 0}(x), \quad p_{\alpha 0} = p_{\alpha 0}(x). \qquad (2.7)$$

This geometrical assumption considerably simplifies the mathematical expressions. The resulting model is called the *shearless slab model* (see Fig. 2.1).

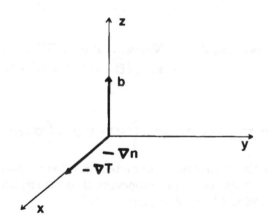

Figure 2.1. The shearless slab geometry.

In the context of fusion physics, the shearless slab geometry can be considered as a rough local approximation to the toroidal confining magnetic field of a tokamak or of a stellarator. It amounts to considering a parallelipipedic region tangent to a magnetic surface at the point P (Fig. 2.2). The fact that the density and the pressure are independent of y and z expresses the well-known fact that these are *surface quantities*, constant on each magnetic surface. The x-direction thus mimicks the radial direction in a toroidal reference frame centred on the magnetic axis of the torus (Fig. 2.2).

We now introduce an additional simplification, by considering only *electrostatic perturbations*. This implies the following assumptions:

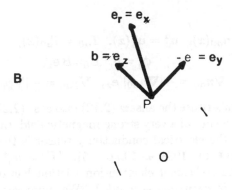

Figure 2.2. The shearless slab as a local approximation to a toroidal magnetic surface.

- The average scalar potential is zero:

$$\phi_0 = 0, \quad \delta\phi \neq 0; \quad \mathbf{E}_0 = \mathbf{0}, \quad \mathbf{E} = \delta\mathbf{E} = -\nabla \delta\phi. \tag{2.8}$$

- The vector potential perturbation is zero. In the shearless slab model, the average magnetic field is assumed constant:

$$\mathbf{A}_0 \neq \mathbf{0}, \quad \delta\mathbf{A} = \mathbf{0}; \quad \mathbf{B} = \mathbf{B}_0 = \nabla \times \mathbf{A}_0, \quad \delta\mathbf{B} = \mathbf{0}. \tag{2.9}$$

- The only relevant Maxwell equation is the Poisson equation (1.3). Thus, the electric field derives from a scalar potential ϕ $(= \delta\phi)$:

$$\mathbf{E} = -\nabla\phi, \tag{2.10}$$

and the Poisson equation takes the form:

$$\nabla^2\phi = -4\pi \sum_{\alpha=e,i} e_\alpha n_\alpha. \tag{2.11}$$

The validity of this approximation is discussed in Appendix A.

We now look for the possible existence of a *stationary reference state* in which all the plasmadynamical variables $n_{\alpha 0}$, $\mathbf{u}_0^\alpha$, $p_{\alpha 0}$, depend on a *single spatial coordinate* x, Eqs. (2.7). We assume that the densities and the temperatures decrease in the direction of increasing x (as in a tokamak). For convenience, all these conditions are rewritten here:

$$n_{\alpha 0} = n_{\alpha 0}(x), \quad \mathbf{u}_0^\alpha = \mathbf{u}_0^\alpha(x), \quad T_{\alpha 0} = T_{\alpha 0}(x), \quad p_{\alpha 0} = p_{\alpha 0}(x)$$

$$\phi_0 = 0, \quad \mathbf{B}_0 = B\,\mathbf{e}_z,$$

$$\boldsymbol{\nabla} n_{\alpha 0} = -|\boldsymbol{\nabla} n_{\alpha 0}|\,\mathbf{e}_x, \quad \boldsymbol{\nabla} p_{\alpha 0} = -|\boldsymbol{\nabla} p_{\alpha 0}|\,\mathbf{e}_x. \tag{2.12}$$

We now substitute the ansatz (2.12) into eqs. (2.2), (2.3), (2.4). We recall that, in presence of a very strong magnetic field, the only non-vanishing component of the electrical conductivity tensor is the scalar parallel conductivity (BALESCU 1988 a, Chap. 5): $\overleftrightarrow{\sigma} = \sigma_\parallel \mathsf{I}$, ($\mathsf{I}$ is the unit tensor), hence the collisional electron-ion friction has only a z-component: $\mathbf{R}^\alpha = R_\parallel^\alpha \, \mathbf{e}_z \propto \sigma_\parallel(n_{i0}u_{0z}^i - n_{e0}u_{0z}^e)$. We moreover neglect the dissipative pressure tensor $\overleftrightarrow{\pi}^\alpha$ which is small in a weakly collisional plasma. The unperturbed plasmadynamical equations then reduce to:

$$0 = -n_{\alpha 0}\frac{d}{dx}u_{0x}^\alpha - u_{0x}^\alpha\frac{d}{dx}n_{\alpha 0},$$

$$0 = -\frac{d}{dx}n_{\alpha 0}u_{0x}^\alpha u_{0x}^\alpha - \frac{1}{m_\alpha}\frac{d}{dx}p_{\alpha 0} + \Omega_\alpha n_{\alpha 0}u_{0y}^\alpha,$$

$$0 = -\frac{d}{dx}n_{\alpha 0}u_{0x}^\alpha u_{0y}^\alpha - \Omega_\alpha n_{\alpha 0}u_{0x}^\alpha,$$

$$0 = -m_\alpha\frac{d}{dx}n_{\alpha 0}u_{0x}^\alpha u_{0z}^\alpha + \sigma_\parallel(n_{i0}u_{0z}^i - n_{e0}u_{0z}^e),$$

$$0 = -\gamma\frac{d}{dx}p_{\alpha 0}u_{0x}^\alpha + \frac{2}{3}u_{0x}^\alpha\frac{d}{dx}p_{\alpha 0}. \tag{2.13}$$

We recall here the definition of the *Larmor frequency* of species α:

$$\Omega_\alpha = \frac{e_\alpha B}{m_\alpha c}, \tag{2.14}$$

and we underscore the fact that $\Omega_e < 0$, $\Omega_i > 0$.

The Poisson equation (2.11) reduces to the *global electroneutrality condition*:

$$0 = 4\pi\sum_\alpha e_\alpha n_{\alpha 0}. \tag{2.15}$$

The following solution of Eqs. (2.13) is readily checked:

$$n_{\alpha 0} = n_{\alpha 0}(x),$$

$$u_{0x}^\alpha = u_{0z}^\alpha = 0,$$

$$u_{0y}^\alpha \equiv u_{0y}^\alpha(x) = \frac{1}{\Omega_\alpha}\frac{p_{\alpha 0}(x)}{m_\alpha n_{\alpha 0}}\frac{d}{dx}\ln p_{\alpha 0},$$

$$p_{\alpha 0} = p_{\alpha 0}(x), \tag{2.16}$$

where $n_{\alpha 0}(x), p_{\alpha 0}(x)$ are functions of x; they must satisfy an additional constraint imposed by the *global electroneutrality condition (2.15)*. For simplicity, we consider in this book only ions that are singly charged ($Z = 1$). Then:

$$e_i = -e_e \equiv e, \quad n_{i0}(x) = n_{e0}(x) \equiv n_0(x). \tag{2.17}$$

It must be noted that the steady *inhomogeneous* state determined here is not motionless: both the ion fluid and the electron fluid have a velocity parallel to the y-direction. These velocities are in opposite directions for the two fluids (because the Larmor frequency Ω_α has the sign of the charge e_α): they produce an electric current in the y-direction (but no global centre-of-mass motion). The existence of this current, which is due to an inhomogeneity in the pressure $p_{\alpha 0}(x) = n_{\alpha 0}(x)T_{\alpha 0}(x)$, reflects the very peculiar phenomenon of the *diamagnetic drift*. Its origin is easily understood. In an inhomogeneous system at mechanical equilibrium, the pressure gradient must be balanced by another force, i.e. the Lorentz force. The equilibrium condition thus implies a non-vanishing electric current perpendicular both to the pressure gradient and to the magnetic field.

The inhomogeneity of the plasma is due to non-vanishing gradients of the density and/or of the temperatures, hence of the pressures. The strength of the inhomogeneity is then conveniently measured by five *gradient lengths* defined as follows:

$$\frac{1}{L_n} = -\frac{d \ln n_0}{dx}, \quad \frac{1}{L_{T_\alpha}} = -\frac{d \ln T_{\alpha 0}}{dx}, \quad \frac{1}{L_{p_\alpha}} = -\frac{d \ln p_{\alpha 0}}{dx}. \tag{2.18}$$

The minus sign makes these quantities positive under normal tokamak conditions (where the density and the temperatures decrease from core to edge). We also note that, because of (2.17), L_n is the same for both species. The unperturbed velocity of Eq. (2.16) can be transformed as follows:

$$\begin{aligned}
u^\alpha_{0y}(x) &= \frac{T_{\alpha 0}}{\Omega_\alpha m_\alpha} \frac{1}{n_0 T_{\alpha 0}} \left(T_{\alpha 0} \frac{dn_0}{dx} + n_0 \frac{dT_{\alpha 0}}{dx} \right) \\
&= \frac{T_{\alpha 0}}{\Omega_\alpha m_\alpha} \frac{1}{n_0} \frac{dn_0}{dx} \left(1 + \frac{d \ln T_{\alpha 0}/dx}{d \ln n_0/dx} \right).
\end{aligned}$$

We define the symbol η_α, very widely used in drift wave theory, as:

$$\eta_\alpha = \frac{d \ln T_{\alpha 0}/dx}{d \ln n_0/dx} = \frac{L_n}{L_{T_\alpha}}. \tag{2.19}$$

We also introduce the quantity $V_{d\alpha}$, called the **diamagnetic drift velocity** of species α: [1]

---

[1] More precisely, this should be called "the diamagnetic drift velocity at constant temperature"; but the last three words are omitted for simplicity, whenever no confusion is possible.

$$V_{d\alpha} = -\frac{T_{\alpha 0}}{\Omega_\alpha m_\alpha L_n}. \qquad (2.20)$$

Note that this quantity is independent of the mass m_α. The unperturbed velocity of Eq. (2.16) is now rewritten as:

$$u_{0y}^\alpha = V_{d\alpha} \, (1 + \eta_\alpha). \qquad (2.21)$$

2.2   Characteristic Quantities of a Fusion Plasma

It is well known that in a magnetized plasma two *characteristic frequencies* (for each species) play an important role: the *Larmor frequencies* Ω_α, Eq. (2.14) and the *plasma frequencies* (or *Langmuir frequencies*) $\omega_{p\alpha}$:

$$\omega_{p\alpha} = \left(\frac{4\pi e_\alpha^2 n_{\alpha 0}}{m_\alpha} \right)^{1/2}. \qquad (2.22)$$

In some cases we will also consider the *electron collision frequency* ν_e (or its inverse, the electron collisional relaxation time). Its value is (see e.g., BALESCU 1988 a, Chap. 4):

$$\nu_e = \frac{4\sqrt{2\pi}}{3} \frac{e^4 n_0}{m_e^{1/2}} \frac{\ln \Lambda}{T_{e0}^{3/2}}, \qquad (2.23)$$

where $\ln \Lambda$ is the Coulomb logarithm: $\Lambda = \frac{3}{2} e^{-2} (T_{e0} + T_{i0}) \lambda_D$, where λ_D is defined in Eq. (2.32).

Next, we consider various *characteristic velocities*. Besides the universal *speed of light in vacuo*, c, these are the *thermal velocities* $V_{T\alpha}$:

$$V_{T\alpha} = \left(\frac{2T_{\alpha 0}}{m_\alpha} \right)^{1/2}. \qquad (2.24)$$

Related to these is the *ion-acoustic velocity* c_s :

$$c_s = \left(\frac{T_{e0}}{m_i} \right)^{1/2}. \qquad (2.25)$$

The name of this quantity stems from its formal resemblance to the speed of sound in a neutral fluid; it exhibits, however, an important difference. It is evaluated with the *electron temperature* and with the *ion mass*: this hybrid character reflects the difference between the two fluids composing the plasma.

The magnetic field introduces the *Alfvén velocity* V_A :

$$V_A = \frac{\Omega_i}{\omega_{pi}} c. \qquad (2.26)$$

Finally, in an inhomogeneous plasma we have met the two *diamagnetic drift velocities*, $V_{d\alpha}$, defined in (2.20). These quantities are independent of the mass, they can therefore be conveniently rewritten in terms of a single (*ion-acoustic*) *diamagnetic drift velocity*:

$$V_{ds} = \frac{\rho_s}{L_n} c_s. \qquad (2.27)$$

[ρ_s is defined in Eq. (2.34)]. We then have:

$$V_{de} = V_{ds}, \qquad V_{di} = -\frac{T_{i0}}{T_{e0}} V_{ds}. \qquad (2.28)$$

From these velocities, we can construct another set of "composite frequencies" by multiplication with a wave vector. These frequencies will naturally appear in the normal mode dispersion equations. The most important of these are the (simple) *drift wave frequency*:

$$\omega^e_{*\mathbf{k}} = k_y V_{de} \qquad (2.29)$$

and the *ion-acoustic frequency*:

$$\omega_{sk} = k_z c_s. \qquad (2.30)$$

Next, we recall the *characteristic lengths*, starting with the *Debye lengths* $\lambda_{D\alpha}$, which are defined via the inverse *Debye wave vectors*, $k_{D\alpha}$:

$$k_{D\alpha} = \frac{2\pi}{\lambda_{D\alpha}} = \left(\frac{4\pi e_\alpha^2 n_{\alpha 0}}{T_{\alpha 0}}\right)^{1/2}. \qquad (2.31)$$

We also use the *global Debye wave vector* k_D :

$$k_D = \frac{2\pi}{\lambda_D} = \left(\sum_\alpha k_{D\alpha}^2\right)^{1/2}. \qquad (2.32)$$

The magnetic field introduces the *Larmor radii*:

$$\rho_{L\alpha} = \frac{V_{T\alpha}}{|\Omega_\alpha|}. \qquad (2.33)$$

We also consider the *ion-acoustic Larmor radius*:

$$\rho_s = \frac{c_s}{\Omega_i}. \qquad (2.34)$$

For the purpose of illustration we collected in Table 2.1 the numerical values of a variety of characteristic quantities corresponding to the following values, typical for the core of a fusion plasma confined in a tokamak:[2]

---

[2] The value $L_n = 1$ m adopted here corresponds to an average value equal to a typical minor radius of a (large) tokamak. It should be noted, however, that the density and the temperature profiles are not linear across a section of the torus. There are "flat" regions where L_n is very long, and narrow regions where the density and the temperature drop abruptly, with very small L_n.

$$n_0 = 10^{20}m^{-3}, \quad T_{e0} = T_{i0} = 10\,keV, \quad B = 5\,T, \quad L_n \approx L_{T\alpha} = 1\,m.$$

$$(2.35)$$

We included in this table the values of a number of important "composite" frequencies, corresponding to typical values of a wave vector **k** with components:

$$k_x = 0, \quad k_y = \rho_s^{-1}, \quad (k_\perp = k_y), \quad k_z = 10^{-2}k_y : \qquad (2.36)$$

this choice is representative of the drift wave mode structure.

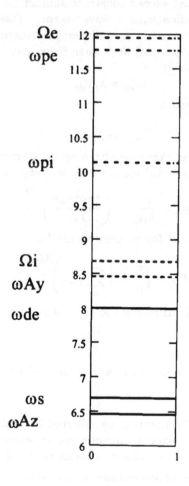

Figure 2.3. Characteristic frequencies of a fusion plasma (logarithmic scale). (Note: $\omega de \equiv \omega_{*e}$). HF: dash-dotted,   LF: dashed,   VLF: solid.

We represented the "spectrum" of characteristic frequencies of this typical fusion plasma on a logarithmic scale in Fig. 2.3. The frequencies may be classified in three groups. *High frequencies (HF)* include the electronic modes ω_{pe}, Ω_e, as well as the ion plasma frequency ω_{pi}. *Low frequencies (LF)* include the Alfvén frequency $\omega_{A\perp} = k_y V_A$ (for perpendicular propagation) and the ion Larmor frequency Ω_i. Finally, the drift wave frequency $\omega^e_{*k} = k_y V_{de}$, the (parallel) ion-acoustic frequency $\omega_{sk} = k_z c_s$ and the parallel Alfvén frequency $\omega_{A\parallel} = k_z V_A$ appear as *Very Low frequencies (VLF)*. In view of the importance of drift waves in transport theory, we shall concentrate on the study of the LF and VLF part of the spectrum.

A mathematical "trick" allows us to "eliminate" the HF electronic modes from the equations, right from the start: it consists of setting the electron mass equal to zero. This approximation is very often used in plasma physics. It has the following effects on the main physical quantities:

$$\omega_{pe} \approx \infty, \quad \Omega_e \approx \infty, \quad V_{Te} \approx \infty, \quad \rho_{Le} \approx 0. \quad [m_e \approx 0]. \tag{2.37}$$

Thus, the high frequencies are rejected to infinity. The effect of this trick is to allow us to "look through a microscope" at the low frequency part of the spectrum. There remains one high frequency after this simplification. In the calculations we are therefore justified in using, at the appropriate moment, the assumption:

$$\frac{\Omega_i}{\omega_{pi}} \ll 1. \tag{2.38}$$

Another frequent approximation is the so-called *cold ion approximation* (see Secs. 2.5, 2.6 below), in which a situation where the ion temperature is much smaller than the electron temperature is modelled by setting $T_i \approx 0$. In this case, the following quantities are affected:

$$V_{Ti} \approx 0, \quad \rho_{Li} \approx 0, \quad V_{di} \approx 0, \quad \lambda_{Di} \approx 0 \quad [T_i \approx 0]. \tag{2.39}$$

Note that the following quantities are *insensitive* to both these approximations:

$$c_s, \quad V_{ds}, \quad V_A, \quad \omega_{pi}, \quad \Omega_i, \quad \omega_{sk}, \quad \omega^e_{*k}, \quad \omega_A, \quad \lambda_{De}, \quad \rho_s. \tag{2.40}$$

Table 2.1: **Characteristic Parameters of a Typical Fusion Plasma**

Velocities		
Thermal velocity (e)	$V_{Te} = (2T_{e0}/m_e)^{1/2}$	$5.93\ 10^9\ m/s$
Thermal velocity (i)	$V_{Ti} = (2T_{i0}/m_i)^{1/2}$	$1.38\ 10^8\ m/s$
Ion-acoustic velocity	$c_s = (T_{e0}/m_i)^{1/2}$	$9.79\ 10^7\ m/s$
Ion-acoustic diamagnetic velocity	$V_{ds} = (\rho_s/L_n)\ c_s$	$2.0\ 10^7\ m/s$
Alfvén velocity	$V_A = (\Omega_i/\omega_{pi})^{1/2}c$	$5.72\ 10^7\ m/s$

Intrinsic frequencies		
Plasma frequency (e)	$\omega_{pe} = (4\pi e^2 n_0/m_e)^{1/2}$	$5.64\ 10^{11}\ s^{-1}$
Plasma frequency (i)	$\omega_{pi} = (4\pi e^2 n_0/m_i)^{1/2}$	$1.32\ 10^{10}\ s^{-1}$
Larmor frequency (e)	$|\Omega_e| = eB/m_e c$	$8.79\ 10^{11}\ s^{-1}$
Larmor frequency (i)	$\Omega_i = eB/m_i c$	$4.79\ 10^8\ s^{-1}$
Collision frequency (e)	ν_e, Eq. (2.23)	$1.1\ 10^{-2}\ s^{-1}$

Composite frequencies		
Ion-acoustic frequency (z)	$\omega_{sk} = k_z c_s$	$4.8\ 10^6\ s^{-1}$
Drift wave frequency (e)	$\omega^e_{*k} = k_y V_{de}$	$9.78\ 10^7\ s^{-1}$
Alfvén frequency (y)	$\omega_{A\perp} = k_y V_A$	$2.8\ 10^8\ s^{-1}$
Alfvén frequency (z)	$\omega_{A\|} = k_z V_A$	$2.8\ 10^6\ s^{-1}$

Wave vector	
Debye wave vector (e) = (i)	$k_{De} = (4\pi e^2 n_0/T_{e0})^{1/2}$	$1.34\ 10^2\ m^{-1}$

Lengths		
Debye length (e) = (i)	$\lambda_{De} = 2\pi/k_{De}$	$5\ 10^{-2}\ m$
Larmor radius (e)	$\rho_{Le} = V_{Te}/|\Omega_e|$	$6.74\ 10^{-3}\ m$
Larmor radius (i)	$\rho_{Li} = V_{Ti}/\Omega_i$	$2.9\ 10^{-1}\ m$
Larmor radius (ion-acoustic)	$\rho_s = c_s/\Omega_i$	$2.04\ 10^{-1}\ m$

2.3   Reduced Macroscopic Equations

The next step in our study is the determination of what happens when the state of the plasma is *slightly* disturbed away from the steady state (2.16). The standard method for this problem is the *linearization* of the plasmadynamical equations (2.2) - (2.4) and of the Poisson equation (2.11). After

Fourier transformation, the linearized equations reduce to a set of linear, homogeneous algebraic equations. The solubility condition of the latter generates the dispersion equation for the eigenmodes. This operation is very simple in principle, but it leads to practically intractable equations. Indeed, we end up with a set of 10 equations for the plasmadynamical variables, and thus to a dispersion equation of tenth degree. Clearly, further approximations will be needed for the analytical treatment of these equations.

An alternative strategy, which will be followed here, is to attempt a reduction of the number of equations at the very beginning. Instead of treating the fully general problem, we may concentrate on a specific problem, for instance, on drift waves. By introducing the specific features of that problem, we will be able to bring down the number of equations. The result is a set of *reduced macroscopic equations*. A list and a brief description of the known sets of reduced equations can be found in the books of ITOH et al. 1999, Chap. 7, and YOSHIZAWA et al. 2003, Chap. 11. The number of necessary macroscopic quantities (or fields) required in these models varies from 7 fields (YAGI & HORTON 1994) to one single field (HASEGAWA & MIMA 1978). These are still nonlinear equations, but much simpler than the complete set. Therefore, they can be (and have been) developed to a high degree of sophistication.

In the forthcoming sections, we consider three of these reduced sets of equations, which are widely used in the literature:

- The HAMAGUCHI-HORTON equations for four or three fields,
- The HASEGAWA-WAKATANI equations for two fields,
- The HASEGAWA-MIMA equation for one field,

The Hamaguchi-Horton equations are the more general ones, and contain the two others as special cases.

2.4   The Hamaguchi-Horton Equations

The starting point for the derivation of these reduced macroscopic equations is the following set of plasmadynamical equations, in which collisional terms are neglected:

- *The continuity equation for the ion number density* (2.2)

$$\partial_t n_i + \nabla \cdot n_i \, \mathbf{u}^i = 0. \tag{2.41}$$

- *The ion equation of motion* (2.3):

$$m_i n_i \left(\partial_t + \mathbf{u}^i \cdot \nabla \right) \mathbf{u}^i = -\nabla p_i - \nabla \cdot \overleftrightarrow{\pi}^i + e n_i \left(\mathbf{E} + c^{-1} \mathbf{u}^i \times \mathbf{B} \right) + \mathbf{R}^i. \tag{2.42}$$

- *The ion energy equation* (2.4):

$$\partial_t p_i = -\mathbf{u}^i \cdot \nabla p_i - \gamma p_i \, \nabla \cdot \mathbf{u}^i + \frac{2}{3}(\nabla \cdot \mathbf{q}^i - \overset{\leftrightarrow}{\pi}{}^i : \nabla \mathbf{u}). \qquad (2.43)$$

We now introduce the following simplifying assumptions.

A) *Local* (not only global!) *electroneutrality*:

$$n_e \approx n_i = n. \qquad (2.44)$$

B) *The electron mass is negligible*, $m_e \ll m_i$; we actually set $m_e \approx 0$ in the equations (see Sec. 2.2).

C) *The collisional dissipative tensor* $\overset{\leftrightarrow}{\pi}{}^i$ *and the collisional heat flux* $\mathbf{q}^i$ *are neglected.*

D) *The friction force has only a parallel component* $\mathbf{R}^i = R^i_\parallel \, \mathbf{e}_z$.

E) *The electric potential, the density and the temperatures* are split, as usual, into a reference part and a perturbation. The various quantities have, however, specific properties. The unperturbed electrostatic potential is zero (2.12), hence:

$$\phi(x, y, z, t) = \delta\phi(x, y, z, t). \qquad (2.45)$$

The number density and the ion temperature have both an unperturbed part and a perturbation:

$$n(x, y, z, t) = n_0(x) + \delta n(x, y, z, t),$$
$$T_i(x, y, z, t) = T_{i0}(x) + \delta T_i(x, y, z, t). \qquad (2.46)$$

The electron temperature is supposed to be unperturbed, and taken to be constant:

$$T_e(x, y, z, t) = T_{e0}. \qquad (2.47)$$

The latter assumption (which may seem strange) is motivated by the fact that anomalous transport in fusion plasmas appears to be mainly driven by the so-called *ion temperature gradient mode;* the present set of reduced equations is adapted to this problem, by introducing this simplification.

F) The following ordering is assumed:

$$\epsilon = \frac{\rho_s}{L_M} \ll 1, \qquad (2.48)$$

where L_M is any one of the macroscopic length scales L_n, $L_{T\alpha}$, $L_{P\alpha}$;

$$\frac{\delta n}{n_0} \sim \frac{|\delta \mathbf{u}^{\alpha}|}{c_s} \sim \frac{\delta p_i}{p_{i0}} \sim \frac{e\,\delta\phi}{T_{e0}} \sim \epsilon, \tag{2.49}$$

$$\frac{T_{i0}}{T_{e0}} \sim \epsilon^0, \tag{2.50}$$

$\rho_s \nabla_{\perp} \sim \epsilon^0, \quad \rho_s \nabla_{\parallel} \sim \epsilon, \quad \Omega_i^{-1}\partial_t \sim \epsilon$ when acting on perturbations

$\rho_s \nabla_{\perp} \sim \epsilon, \quad \rho_s \nabla_{\parallel} \sim \epsilon^2, \quad \Omega_i^{-1}\partial_t \sim \epsilon^3$ when acting on reference. (2.51)

We now implement these assumptions in the plasmadynamical equations. We follow here the treatment of HAMAGUCHI & HORTON 1990 a (Additional details of this calculation can be found in the report by HAMAGUCHI & HORTON 1990 b). We start with the ion equation of motion, Eq. (2.42), which we multiply vectorially with the unit vector $\hat{\mathbf{b}}$, and extract from the result an expression of the perpendicular ion velocity $\mathbf{u}_{\perp}^i \equiv \hat{\mathbf{b}} \times (\mathbf{u}^i \times \hat{\mathbf{b}})$:

$$\mathbf{u}_{\perp}^i = \delta \mathbf{u}^E + \mathbf{u}_d^i + \delta \mathbf{u}_p^i. \tag{2.52}$$

This very important formula exhibits the three types of macroscopic drift motions in a constant magnetic field (a similar formula holds for the electrons). The first term is the well-known *electrostatic drift velocity:*

$$\delta \mathbf{u}^E = \frac{c}{B} \left(\hat{\mathbf{b}} \times \nabla \delta\phi \right). \tag{2.53}$$

We stress the fact that this velocity is the same for the ions and for the electrons. The second term is the *diamagnetic drift velocity of species* α (here: the ions):

$$\mathbf{u}_d^{\alpha} = \frac{c}{e_{\alpha} n_0 B} \left(\hat{\mathbf{b}} \times \nabla p_{\alpha} \right). \tag{2.54}$$

Clearly, this is equivalent, in the reference state, to Eq. (2.16). The diamagnetic drift is a purely macroscopic effect, as explained after Eq. (2.17). It should be noted that this term has both an unperturbed component and a perturbation.

The last term in Eq. (2.52) is called the *polarization drift velocity.* It must be defined perturbatively, by approximating the perpendicular velocity by its first term in Eq. (2.52):

$$\delta \mathbf{u}_p^{\alpha} = \frac{1}{\Omega_{\alpha}} \hat{\mathbf{b}} \times (\partial_t + \mathbf{u}^{\alpha} \cdot \nabla) \mathbf{u}^{\alpha} \approx \frac{1}{\Omega_{\alpha}} \hat{\mathbf{b}} \times (\partial_t + \delta \mathbf{u}^E \cdot \nabla)\,\delta \mathbf{u}^E. \tag{2.55}$$

Under the scaling (2.51) this term is of higher order, and will be neglected here.

Consider next the ionic continuity equation. Hamaguchi and Horton make the additional assumption ("for simplicity") that the convection is produced only by the electrostatic drift velocity. Noting that $\nabla \cdot \delta \mathbf{u}^E = 0$, Eq. (2.41) reduces to:

$$\partial_t \delta n_i = -\delta \mathbf{u}^E \cdot \nabla(n_0 + \delta n_i) - n_0 \nabla_\parallel \delta u_\parallel^i. \qquad (2.56)$$

We note:

$$-\delta \mathbf{u}^E \cdot \nabla n_0 = -\frac{c}{B}(\hat{\mathbf{b}} \times \nabla \delta \phi)_x \frac{dn_0}{dx} = -\frac{c}{B} \frac{\partial \, \delta \phi}{\partial y} \, n_0 L_n^{-1},$$

and:

$$-\delta \mathbf{u}^E \cdot \nabla \delta n_i = -\frac{c}{B} \left(\frac{\partial \, \delta \phi}{\partial x} \frac{\partial \, \delta n_i}{\partial y} - \frac{\partial \, \delta \phi}{\partial y} \frac{\partial \, \delta n_i}{\partial x} \right). \qquad (2.57)$$

We introduce the very useful notation in terms of a Poisson bracket:

$$[A, B] = \frac{\partial A}{\partial x} \frac{\partial B}{\partial y} - \frac{\partial A}{\partial y} \frac{\partial B}{\partial x}. \qquad (2.58)$$

The continuity equation is conveniently expressed in terms of the dimensionless quantities

$$\delta \tilde{n}_i = \delta n_i / n_0, \qquad \delta \tilde{\phi} = e \, \delta \phi / T_{e0}; \qquad (2.59)$$

using Eq. (2.27) we obtain:[3]

$$\partial_t \delta \tilde{n}_i = -V_{ds} \frac{\partial \, \delta \tilde{\phi}}{\partial y} - \nabla_\parallel \delta u_\parallel^i - \frac{c_s^2}{\Omega_i} [\delta \tilde{\phi}, \delta \tilde{n}_i]. \qquad (2.60)$$

We now consider the perpendicular momentum balance for the ions:

$$m_i n_i \left(\partial_t + \mathbf{u}^i \cdot \nabla \right) \mathbf{u}_\perp^i = -\nabla_\perp p_i - e n_i \nabla_\perp \delta \phi + e c^{-1} n_i \left(\mathbf{u}^i \times \mathbf{B} \right).$$

This equation is multiplied vectorially by $\hat{\mathbf{b}}$, thus transforming it into an equation for the *vorticity* $\delta \Psi = \hat{\mathbf{b}} \times \delta \mathbf{u}^E$. The divergence of this quantity is directly related to the perpendicular Laplacian of the potential:

$$\nabla \cdot \delta \Psi = \frac{c}{B} \nabla \cdot [\hat{\mathbf{b}} \times (\hat{\mathbf{b}} \times \nabla \delta \phi)] = -\frac{c}{B} \nabla_\perp^2 \, \delta \phi, \qquad (2.61)$$

where $\nabla_\perp^2 = \nabla_x^2 + \nabla_y^2$. The equation for the vorticity is, to dominant order:

---

[3] In $n_0^{-1} \nabla_\perp (n_0 + \delta n)$, the term $(\delta n / n_0^2) \nabla_\perp n_0$ is considered of higher order, and thus neglected.

$$m_i n_i \nabla \cdot \left\{ \hat{\mathbf{b}} \times \left[\partial_t + \left(\delta \mathbf{u}^E + \mathbf{u}_{d0}^i \right) \cdot \nabla \right] \delta \mathbf{u}^E \right\}$$
$$= -\nabla \cdot (\hat{\mathbf{b}} \times \nabla p_i) - en_i \nabla \cdot (\hat{\mathbf{b}} \times \nabla \delta \phi). \tag{2.62}$$

When the calculations are done explicitly, using Eqs. (2.53) and (2.54), we find:

$$\partial_t \, \nabla_\perp^2 \delta \phi = -\frac{1}{\Omega_i m_i n_0} \frac{dp_{io}}{dx} \frac{\partial}{\partial y} \, \nabla_\perp^2 \delta \phi - \frac{c}{B} \left[\delta \phi, \nabla_\perp^2 \delta \phi \right]. \tag{2.63}$$

We note the following important relation:

$$\frac{1}{n_0 T_{e0}} \frac{dp_{i0}}{dx} = \frac{1}{n_0 T_{e0}} \left(T_{i0} \frac{dn_0}{dx} + n_0 \frac{dT_{i0}}{dx} \right)$$
$$= \frac{T_{i0}}{T_{e0}} \frac{1}{n_0} \frac{dn_0}{dx} \left[1 + \frac{d \left(\ln T_{i0} \right)/dx}{d \left(\ln n_0 \right)/dx} \right] = - L_n^{-1} \theta_i \left(1 + \eta_i \right), \tag{2.64}$$

where η_i was defined in Eq. (2.19). We also introduce the abbreviations:

$$\theta_i = \frac{T_{i0}}{T_{e0}}, \qquad \kappa_i = \theta_i \left(1 + \eta_i \right); \tag{2.65}$$

Eq. (2.63) thus becomes:

$$\partial_t \, \nabla_\perp^2 \delta \tilde{\phi} = -V_{ds} \, \kappa_i \, \frac{\partial}{\partial y} \, \nabla_\perp^2 \delta \tilde{\phi} - \frac{c_s^2}{\Omega_i} \left[\delta \tilde{\phi}, \nabla_\perp^2 \delta \tilde{\phi} \right]. \tag{2.66}$$

It turns out to be convenient to combine Eqs. (2.60) and (2.66). To this purpose, $\nabla_\perp^2 \delta \tilde{\phi}$ is made dimensionless by multiplication with ρ_s^2; Eq. (2.66) is then subtracted from (2.60):

$$\partial_t \left(\delta \tilde{n}_i - \rho_s^2 \nabla_\perp^2 \delta \tilde{\phi} \right) = -V_{ds} \frac{\partial}{\partial y} \left(1 - \kappa_i \rho_s^2 \nabla_\perp^2 \right) \delta \tilde{\phi}$$
$$- \nabla_\parallel u_\parallel^i - \frac{c_s^2}{\Omega_i} \left[\delta \tilde{\phi}, \left(\delta \tilde{n}_i - \rho_s^2 \nabla_\perp^2 \delta \tilde{\phi} \right) \right]. \tag{2.67}$$

The parallel ion momentum equation is easily obtained. It is shown that the contribution of the ion diamagnetic velocity $(\mathbf{u}^i \cdot \nabla) \, u_\parallel^i$ is negligible, and [using Eq. (2.57)] we are left with:

$$\partial_t \delta u_\parallel^i = -\frac{1}{m_i n_0} \nabla_\parallel \delta p_i - c_s^2 \delta \tilde{\phi} - \frac{c_s^2}{\Omega_i} \left[\delta \tilde{\phi}, \delta u_\parallel^i \right]. \tag{2.68}$$

Finally, the equation for the ion pressure is obtained to the same order of approximation:

$$\partial_t \delta p_i = -V_{ds}\,\kappa_i\,\frac{\partial\,\widetilde{\delta\phi}}{\partial y} - \gamma p_{i0}\,\nabla_\parallel \delta u_\parallel^i - \frac{c_s^2}{\Omega_i}\,\left[\widetilde{\delta\phi}, \delta p_i\right].\qquad(2.69)$$

Eqs. (2.60), (2.67), (2.68), (2.69) constitute a closed set of four equations for the four plasmadynamical fluctuations $\delta\widetilde{n}$, $\widetilde{\delta\phi}$, $\delta u_\parallel^i$ and δp_i. These are the *four-fields reduced **Hamaguchi-Horton equations***. They first appeared in HAMAGUCHI & HORTON 1990 a. A similar set of equations was also derived in HORTON et al. 1980. It is remarkable that within the range of validity of the present assumptions the set of ten plasmadynamical equations is reduced to four. These equations are widely used in the study of the so-called *ion-temperature gradient (ITG) modes*, also called *eta-i modes* [from the name of the parameter η_i defined in Eq. (2.19)]. They will be discussed in Chap. 3.

2.5   The Hasegawa-Wakatani Equations

We now consider an alternative set of reduced macroscopic equations, adapted to the description of a different (simpler) type of drift waves. This will be made explicit in the simplifying assumptions made now. In the derivation, we follow the same line as in Sec. 2.4.

The starting point is the set of the following three plasmadynamical equations (including collisional effects):

- *The continuity equation for the electron number density* (2.2):

$$\partial_t n_e + \nabla \cdot n_e\,\mathbf{u}^e = 0.\qquad(2.70)$$

- *The electron equation of motion* (2.3):

$$\partial_t m_e n_e \mathbf{u}^e + \nabla \cdot [m_e n_e \mathbf{u}^e \mathbf{u}^e + n_e T_i\,\mathsf{I} + \overleftrightarrow{\pi}^e] = -e n_e\left(\mathbf{E} + c^{-1}\mathbf{u}^e \times \delta\mathbf{B}\right) + \mathbf{R}^e.\qquad(2.71)$$

- *The centre-of-mass equation of motion* [resulting from a combination of the ion and electron equations of motion; see e.g. BALESCU 1988 a, Chap. 3]:

$$\partial_t \rho \mathbf{u} + \nabla \cdot \left[\rho\,\mathbf{u}\,\mathbf{u} + \frac{\rho}{m_i}\left(T_e + T_i\right)\mathsf{I} + \overleftrightarrow{\pi}^e + \overleftrightarrow{\pi}^i\right] = \frac{1}{c}\mathbf{j} \times \mathbf{B}.\qquad(2.72)$$

We recall the well-known relations defining the mass density ρ and the centre-of-mass velocity **u**:

$$\rho = \sum_{\alpha=e,i} m_\alpha n_\alpha, \quad \rho\mathbf{u} = \sum_{\alpha=e,i} m_\alpha n_\alpha \mathbf{u}^\alpha. \tag{2.73}$$

We now introduce the following simplifying assumptions:
a) *The temperatures are constant, the ions are "cold"*:

$$T_e = T_{e0} = const., \quad T_i \approx 0. \tag{2.74}$$

Here lies the main difference with the situation considered in Sec. 2.4: assumption (2.50) is no longer valid. As a result, we do not need to consider the temperature (or pressure) balance equation; indeed, the pressure perturbations are simply related to the density perturbations: $\delta p_e = T_{e0}\delta n_e$, $\delta p_i = 0$.

b) *The electron mass is negligible*, $m_e \ll m_i$; as a result, the mass density and the centre-of-mass momentum are identified, respectively, with the ion mass density and the ion momentum:

$$\rho \approx m_i n_i, \quad \rho\mathbf{u} \approx m_i n_i \mathbf{u}^i. \tag{2.75}$$

c) The parallel ion velocity is much smaller than the electron parallel velocity; it is therefore neglected:

$$u_z^i \approx 0, \quad \mathbf{u}^i \approx \mathbf{u}_\perp^i. \tag{2.76}$$

As a result, we do not need a separate equation for the ion parallel velocity. It also implies that the *parallel component of the electric current is mainly carried by the electrons:*

$$j_z = \sum_{\alpha=e,i} e_\alpha n_\alpha u_z^\alpha \approx -e n_e u_z^e. \tag{2.77}$$

d) *Transport equations.*

* dα) The parallel electric current density is related to the electric field and to the electron pressure gradient by *Ohm's law* [BALESCU 1988 a, Chap. 5, Eq. (3.8)]:

$$j_z = \sigma_\| \cdot [E_z + (e n_e)^{-1} \nabla_z (n_e T_e)]. \tag{2.78}$$

where $\sigma_\|$ is the parallel classical (collisional) electrical conductivity. The perpendicular electrical conductivity, proportional to B_0^{-2}, will be neglected.

- dβ) The electron-ion friction $\mathbf{R}^e$ is proportional to the electric current density $\mathbf{j}$ [BALESCU 1988 a, Chaps. 3 and 4]; it therefore is directed along the magnetic field (see Sec. 2.1):

$$\mathbf{R}^e \approx R_z^e\, \mathbf{e}_z. \tag{2.79}$$

- dγ) The dissipative pressure tensor $\overset{\leftrightarrow}{\pi}{}^{\alpha}$ (in presence of a magnetic field) is related to the traceless velocity gradient tensor $\mathbf{u}\,\hat{}\,\mathbf{u}$ by a fourth-order viscosity tensor [BALESCU 1988 a, Chap. 5, Sec. 5.5]. In many cases, the latter can be reduced to a single scalar "parallel" viscosity coefficient $\eta_\parallel^\alpha$.[4] In the forthcoming treatment we shall only be interested in the 2-dimensional perpendicular velocity $\mathbf{u}_\perp^\alpha$, hence:[5]

$$\pi_{rs}^\alpha = -\eta_\parallel^\alpha \left(u_r^\alpha u_s^\alpha + u_s^\alpha u_r^\alpha - \delta_{rs}\, u_n^\alpha u_n^\alpha \right), \quad r, s = x, y. \tag{2.80}$$

We also note that the electron viscosity is smaller than the ion viscosity by a factor $(m_e/m_i)^{1/2}$, and will therefore be neglected. We also define the *ion kinematic viscosity*, which will be useful below:

$$\mu = \frac{1}{m_i n_i}\, \eta_\parallel^i. \tag{2.81}$$

e) The plasmadynamical quantities are considered to be *perturbed quantities*. They can thus be represented as in Sec. 2.4, assumption E). The scalings assumed in Eqs. (2.49), (2.51) remain valid here too.

We now implement these approximations, starting with the electron continuity equation, which is rewritten as:

$$\partial_t n_e = -\nabla_\perp \cdot (n_e \mathbf{u}_\perp^e) - \nabla_z (n_e u_z^e). \tag{2.82}$$

In order to evaluate the first term in the right hand side, we need the perpendicular electron velocity, which is extracted from Eq. (2.71) [see Eqs. (2.52) - (2.55)]. The second term in the polarization drift will be neglected. Substituting Eq. (2.52) (for electrons) into the first term on the right hand side of Eq. (2.82) we find, to dominant order:

$$-\nabla_\perp \cdot (n_e \mathbf{u}_\perp^e) = -n_e \nabla_\perp \cdot \mathbf{u}_\perp^e - \mathbf{u}_\perp^e \cdot \nabla n_e$$

$$= -\frac{c}{B} \left\{ \left[\left(\hat{\mathbf{b}} \times \nabla_\perp \delta\phi \right) - \frac{1}{\Omega_e}\, \partial_t\, \nabla_\perp \delta\phi \right] \cdot \nabla_\perp n_e - \frac{n_e}{\Omega_e}\, \partial_t\, \nabla_\perp^2 \delta\phi \right\}, \tag{2.83}$$

---

[4] The notation $\eta_\parallel^i$ should, of course, not be confused with the parameter η_i defined in Eq. (2.19).

[5] Note that the "usual" coefficient $(2/3)$ in front of the subtracted trace is replaced here by $(2/2) = 1$ because the perpendicular velocity gradient matrix is 2×2 instead of 3×3.

where we used the properties: $\nabla_\perp \cdot \delta \mathbf{u}^E = 0$, $\nabla_\perp \cdot (n_e \mathbf{u}_d^e) = 0$. We now consider the second term in (2.82) and use Eqs. (2.76) and (2.77):

$$-\nabla_z (n_e u_z^e) \approx e^{-1} \nabla_z j_z = -\frac{\sigma_\parallel}{e} \nabla_z^2 \left(\delta \phi - \frac{T_{e0}}{e} \ln n_e \right). \qquad (2.84)$$

Substituting these results into Eq. (2.82) we find:

$$\partial_t n_e + \frac{c}{B} \left\{ \left[\left(\hat{\mathbf{b}} \times \nabla_\perp \delta \phi \right) - \frac{1}{\Omega_e} \partial_t \nabla_\perp \delta \phi \right] \cdot \nabla_\perp n_e - \frac{n_e}{\Omega_e} \partial_t \nabla_\perp^2 \delta \phi \right\}$$
$$= -\frac{\sigma_\parallel}{e} \nabla_z^2 \left(\delta \phi - \frac{T_{e0}}{e} \ln n_e \right). \qquad (2.85)$$

Using the decomposition (2.46), this is conveniently expressed in terms of $n_0[1 + (\delta n_e/n_0)]$ and of the electrostatic drift velocity (2.53):

$$\partial_t \, \delta \tilde{n}_e + \delta \mathbf{u}^E \cdot \nabla_\perp (\ln n_0 + \delta \tilde{n}_e) = -\bar{\sigma} \nabla_z^2 \left(\delta \tilde{\phi} - \delta \tilde{n}_e \right), \qquad (2.86)$$

with:

$$\bar{\sigma} = \frac{T_{e0}}{e^2 n_0} \sigma_\parallel. \qquad (2.87)$$

This is a first reduced equation, involving only the perturbation fields $\delta \tilde{n}_e$ and $\delta \tilde{\phi}$ [see Eq. (2.59)].

A second equation is obtained from the centre-of-mass equation of motion (2.72). The perpendicular ion velocity is obtained as in Eq. (2.52): $\rho \mathbf{u}_\perp \approx m_i n_i \mathbf{u}_\perp^i = m_i n_i (\delta \mathbf{u}^E + \delta \mathbf{u}_p^i)$ [the ion diamagnetic drift is zero because of the assumption (2.74)]. The following calculations follow closely those described in ITOH et al. 1999, Chap. 7. Eq. (2.72) becomes:

$$\partial_t \frac{c}{B} \left[(\mathbf{e}_z \times \nabla \delta \phi) - \frac{1}{\Omega_i} \partial_t \nabla_\perp \delta \phi \right]$$
$$= -\left(\delta \mathbf{u}^E + \delta \mathbf{u}_p^i \right) \cdot \nabla_\perp \left(\delta \mathbf{u}^E + \delta \mathbf{u}_p^i \right) - \frac{1}{m_i n} \nabla_\perp \cdot \overleftrightarrow{\pi}^i + \frac{1}{m_i n c} \mathbf{j} \times \mathbf{B}. \qquad (2.88)$$

A lengthy, but elementary calculation leads to the following result, expressed as a vector equation:

$$\mathbf{A} \equiv A_x \, \mathbf{e}_x + A_y \, \mathbf{e}_y = 0,$$

with:

$$A_x = -\frac{c}{B}\left[\partial_t \nabla_y \delta\phi + \frac{1}{\Omega_i}\partial_t^2 \nabla_x \delta\phi\right] + \left(\frac{c}{B}\right)^2 [\nabla_y \delta\phi \, (\nabla_x \nabla_y \delta\phi)$$

$$-\nabla_x \delta\phi \, (\nabla_y^2 \delta\phi)] + \frac{T_{i0}}{m_i}\nabla_x \ln n + \mu\,\frac{c}{B}\nabla_\perp^2 \delta\phi\left(\nabla_y \delta\phi + \frac{1}{\Omega_i}\nabla_x \delta\phi\right) - \frac{B}{m_i n e}\,j_y,$$

and a similar equation for A_y. We perform on this vector the operation: $-\mathbf{e}_z \cdot (\nabla \times \mathbf{A}) \equiv \nabla_y A_x - \nabla_x A_y$, with the result:

$$\frac{c}{B}\left[\partial_t + \delta\mathbf{u}^E \cdot \nabla_\perp\right] \nabla_\perp^2 \delta\phi - \mu\,\frac{c}{B}\nabla_\perp^4 \delta\phi + \frac{B}{c m_i n}\,\nabla_\perp \cdot \mathbf{j}_\perp = 0. \quad (2.89)$$

We now note that the electrical continuity equation for a quasi-neutral plasma implies:

$$\nabla_\perp \cdot \mathbf{j}_\perp = -\nabla_z j_z = \sigma_\parallel \nabla_z^2 \left\{\delta\phi - \frac{T_{e0}}{e}\ln\left[n_0\left(1 + \frac{\delta n}{n_0}\right)\right]\right\}.$$

After linearization of the logarithm, multiplication by ρ_s^2 and use of (2.59), Eq. (2.89) becomes:

$$\left(\partial_t + \delta\mathbf{u}^E \cdot \nabla_\perp\right) \rho_s^2 \nabla_\perp^2 \widetilde{\delta\phi} - \mu\,\rho_s^2 \nabla_\perp^4 \widetilde{\delta\phi} - \bar{\sigma}\,\nabla_z^2\left(\widetilde{\delta\phi} - \widetilde{\delta n}_e\right) = 0. \quad (2.90)$$

Finally, using the Poisson bracket notation (2.58), the two equations (2.86) and (2.90) are rewritten as:

$$\partial_t \widetilde{\delta n} = -V_{ds}\,\nabla_y \widetilde{\delta\phi} - \bar{\sigma}\,\nabla_z^2\,(\widetilde{\delta\phi} - \widetilde{\delta n}) - \frac{c_s^2}{\Omega_i}\,[\widetilde{\delta\phi}, \widetilde{\delta n}], \quad (2.91)$$

$$\partial_t\,\rho_s^2 \nabla_\perp^2 \widetilde{\delta\phi} = -\bar{\sigma}\,\nabla_z^2\,(\widetilde{\delta\phi} - \widetilde{\delta n}_e) + \mu\,\rho_s^2 \nabla_\perp^4 \widetilde{\delta\phi} - \frac{c_s^2}{\Omega_i}\left[\widetilde{\delta\phi}, \rho_s^2 \nabla_\perp^2 \widetilde{\delta\phi}\right]. \quad (2.92)$$

By inserting the values of the classical (collisional) values of the transport coefficients (BALESCU 1988 a, Sec. 5.4) one obtains the following alternative forms of the coefficients:

$$\bar{\sigma} = \frac{T_{e0}}{m_e \nu_e}, \quad \mu = \left(\frac{m_e}{m_i}\right)^{1/2}\left(\frac{T_i}{T_e}\right)^{5/2}\bar{\sigma}, \quad (2.93)$$

where ν_e is the electron collision frequency (inverse collisional electron relaxation time). [Note that both $\bar{\sigma}$ and μ have dimension (length)^2/(time).]

Eqs. (2.91), (2.92) are the **Hasegawa-Wakatani (HW) equations**, first derived by HASEGAWA & WAKATANI 1983. They form a closed set of equations for the two fields $\delta\phi$ and $\delta\tilde{n}$. As mentioned in the beginning of this section, they describe a different, simpler type of drift waves, present when the temperatures are constant, and $T_{i0} \approx 0$. They may be called *density-gradient drift waves*, as they are driven only by an inhomogeneous density profile. This is not a very realistic situation, but the equations have the advantage of providing a simple analytical model for the study of drift waves. They have been used by many authors, especially in numerical simulations, for which these equations provide a sufficiently simple non-linear model. Moreover, they include the action of the collisions through the transport coefficients $\bar{\sigma}$ and μ.[6]

It thus clearly appears that *the Hasegawa-Wakatani equations are a special case of the Hamaguchi-Hinton equations. They are obtained from the latter by assuming that the ion temperature is much smaller than the electron temperature ($\theta_i \ll 1$) and neglecting its inhomogeneity. This situation is modelled by setting $T_{i0} = 0$, hence $\kappa_i = 0$. It is, moreover, assumed that $u_z^e \approx (en_0)^{-1} j_z$, and the equations are closed by Ohm's law.*

2.6   The Hasegawa-Mima Equation

With the assumptions introduced at the beginning of Sec. 2.5, the complete set of macroscopic equations has been reduced to a set of two equations for the fluctuating fields $\delta\tilde{n}$ and $\delta\tilde{\phi}$. This set can be even further reduced by making some additional reasonable assumptions.

We start from the Hasegawa-Wakatani equations which we rewrite by replacing Eq. (2.92) by the difference between the two equations [see Eq. (2.67)]:

$$\partial_t \delta\tilde{n} = -V_{ds}\nabla_y \delta\tilde{\phi} - \bar{\sigma}\,\nabla_z^2\,(\delta\tilde{\phi} - \delta\tilde{n}) - \frac{c_s^2}{\Omega_i}[\delta\tilde{\phi}, \delta\tilde{n}], \qquad (2.94)$$

$$\partial_t(\delta\tilde{n} - \rho_s^2\nabla_\perp^2\delta\tilde{\phi}) = -V_{ds}\nabla_y\delta\tilde{\phi} - \mu\,\rho_s^2\nabla_\perp^4\delta\tilde{\phi} - \frac{c_s^2}{\Omega_i}[\delta\tilde{\phi}, (\delta\tilde{n} - \rho_s^2\nabla_\perp^2\delta\tilde{\phi})]. \quad (2.95)$$

It is well known that at very high electron temperature and $\theta_i \ll 1$ the effect of the collisions on transport becomes very small. In particular, the ion viscosity becomes very small, whereas the parallel electrical conductivity is very large (the plasma is an almost perfect conductor). In this situation we may set $\mu \approx 0$ in Eq. (2.95) and reduce Eq. (2.94) to its dominant term, $\bar{\sigma}\,\nabla_z^2\,(\delta\tilde{\phi} - \delta\tilde{n}) = 0$. Hence, provided $(\delta\tilde{\phi} - \delta\tilde{n}) = \nabla_z\,(\delta\tilde{\phi} - \delta\tilde{n}) = 0$ in $z = 0$, we obtain the solution:

---

[6] In the original paper by HAMAGUCHI & HORTON 1990 a the collisional terms are also included.

$$\delta \tilde{n} = \delta \tilde{\phi}, \qquad \delta n = n_0 \frac{e}{T_{e0}} \delta \phi.$$

The electrons are thus *adiabatic*, i.e., they are distributed according to a local equilibrium Maxwell-Boltzmann distribution:

$$n_e = n_0 \exp\left(-\frac{e_e \, \delta\phi}{T_{e0}}\right) \approx n_0 \left(1 + \frac{e \, \delta\phi}{T_{e0}}\right), \qquad (2.96)$$

which is easily understood, given their high mobility in the "collisionless" situation present when $\bar{\sigma} \gg 1$. Given that $[\delta\tilde{\phi}, \delta\tilde{n}] = [\delta\tilde{\phi}, \delta\tilde{\phi}] = 0$, Eq. (2.95) reduces to:

$$\partial_t \left(\delta\tilde{\phi} - \rho_s^2 \nabla_\perp^2 \delta\tilde{\phi}\right) = -V_{ds}\nabla_y \delta\tilde{\phi} + \frac{c_s^2}{\Omega_i}[\delta\tilde{\phi}, \rho_s^2 \nabla_\perp^2 \delta\tilde{\phi}]. \qquad (2.97)$$

This is the famous *Hasegawa-Mima Equation*. It was derived, chronologically, before the Hasegawa-Wakatani equation by HASEGAWA & MIMA 1978. We also write it in dimensional form:

$$\partial_t \left(1 - \rho_s^2 \nabla_\perp^2\right) \frac{e\delta\phi}{T_{e0}} + V_{ds}\nabla_y \frac{e\delta\phi}{T_{e0}} - \delta\mathbf{u}^E \cdot \nabla_\perp \rho_s^2 \nabla_\perp^2 \frac{e\delta\phi}{T_{e0}} = 0.$$

$$(2.98)$$

The **Hasegawa-Mima equation** is certainly the simplest non-trivial macroscopic equation describing density-gradient drift waves in terms of a single fluctuating field, the potential $\delta\phi$. It should be stressed that, in spite of its relatively simple appearance, it is a *nonlinear equation*: the nonlinearity resides in the last term of the left hand side, which is a quadratic functional of the potential. This equation will be further studied in Chaps. 8 and 16.

Chapter 3

Drift Waves

3.1 Introduction

It is well known that a variety of waves can develop and propagate in a plasma as a response to a perturbation of a quiet, stationary state. The number of different kinds of waves, or *"modes"*, is much larger than in the vacuum, where only *electromagnetic waves* can propagate, or in a neutral fluid, where there exist in addition *acoustic waves*. Each of the modes is characterized (in a Fourier description) by a specific relation between its frequency, Ω, and its wave vector, $\mathbf{k}$: $\Omega = \Omega(\mathbf{k})$. Such a relation is called a **dispersion equation** (see the discussion of Sec. 1.5). Note that the function $\Omega(\mathbf{k})$ may be complex (for real $\mathbf{k}$); in that case it will be represented in the standard form: $\Omega(\mathbf{k}) = \omega(\mathbf{k}) + i\gamma(\mathbf{k})$. The real part $\omega(\mathbf{k})$ is the (ordinary) *frequency*, whereas the imaginary part is either the *damping rate*, if $\gamma(\mathbf{k}) < 0$, or the *growth rate*, if $\gamma(k) > 0$. The latter case characterizes an unstable mode.

In a plasma the constituent particles are electrically charged; as a result, their motion is strongly affected by any external or internal electromagnetic field and, conversely, the fields are modified by the presence of the particles. The result is an alteration in the propagation properties of the electromagnetic and acoustic waves in a plasma (compared to a neutral fluid), as well as the emergence of new types of waves, specific of a plasma. The number of modes is even larger in a homogeneous, but magnetized, thus anisotropic plasma. Finally, if the plasma is spatially inhomogeneous because of the presence of density and temperature gradients, there are additional modifications of the "homogeneous" modes, as well as the appearance of new modes related to the gradients. The latter, called **drift waves**, play a crucial role in the mechanism of anomalous transport. The drift waves will be the object of a large fraction of our work. Their properties can be studied at various levels of sophistication and for different ranges of conditions in the plasma. In the present chapter we present a discussion at the macroscopic level.

Any plasmadynamical quantity Q (density, velocity, pressure, electric potential,...) can be represented in the form $Q = Q_0 + \delta Q$ [see Eq. (1.31)], where Q_0 is the reference value defined in Sec. 2.1. As long as the perturbations remain very small, their fate is determined by the plasmadynamical equations, *linearized around the reference state* determined in Sec. 2.1. The solutions of these linear equations allow us to identify the various possible "waves" or "modes": the latter define the *linear response* of the plasma. At this stage, there is no question of turbulence. If, however, some of these modes are linearly unstable, their amplitudes δQ will grow in time. The nonlinear terms that were previously neglected become more and more important and lead to interactions among the linear modes and, generally, towards a very complex rearrangement of the plasma structure: the state becomes truly turbulent. In the fully developed turbulent state, the fluctuations must be considered to be stochastic functions. The methodology described above has become quite standard. It is exposed in all textbooks of plasma physics [e.g.: CAIRNS 1985; CHEN 1974; DENDY 1990; DOLAN 1982; GOLDSTON & RUTHERFORD 1995; HAZELTINE & MEISS 1992; ICHIMARU 1992; NICHOLSON 1983; NISHIKAWA & WAKATANI 1990] as well as in more specialized monographs [e.g., AKHIEZER et al., 1975; BATEMAN 1978; BOOKER 1984; BRAMBILLA 1998; CAP 1976; KADOMTSEV 1965; MANHEIMER & LASHMORE-DAVIES 1989; MIKHAILOVSKII 1974, 1998; PRIEST 1982; STIX 1992; SWANSON 1989; WEILAND 1999; WENTZEL & TIDMAN 1968].

The study of the linearized plasmadynamical equations for the perturbations δQ (or of the corresponding reduced equations) is greatly facilitated by using a Fourier representation. Whenever the linearized plasmadynamical partial differential equations have constant coefficients, they are transformed into algebraic equations. There appears a problem, however, when the reference state is spatially inhomogeneous, as described in Chap. 2: the Fourier-transformed equations are then no longer algebraic, but rather integral (involving convolutions of Fourier transforms). This difficulty can be circumvented whenever there is a large separation of length scales between the variation of the perturbations and the variation of the reference quantities. The latter being described by a wave vector $\mathbf{k}$, the condition is $k \gg L_M^{-1}$ (where L_M is the largest of the gradient scales L_n, L_{P_a}). In this case we may use (in a shearless slab) the *local approximation*: the Fourier transformation only affects the fast space dependence. As a result, the unperturbed quantities Q_0 are treated approximately as constants in the linearized plasmadynamical equations and the Fourier transforms of the perturbations keep a slow dependence on x.[1]

---

[1] A more rigorous treatment (leading to the same result) would involve a "multiple-scale perturbation" analysis (BALESCU 1988 b).

We seek solutions of the linearized plasmadynamical equations in the form:

$$\delta Q(x, y, z, t) = \delta Q_{\mathbf{k}, \Omega}[x] \, \exp(ik_x x + ik_y y + ik_z z - i\Omega t), \qquad (3.1)$$

where the amplitude $\delta Q_{\mathbf{k}, \Omega}[x]$ is a *slowly varying* function of the coordinate x (in the direction of the unperturbed spatial gradients). This slow x-dependence will always be understood, but usually not be written down explicitly. When (3.1) is substituted into the linearized plasmadynamical equations, the latter are transformed into linear algebraic equations, whose coefficients (related to the unperturbed state Q_0) are considered as constants in the local approximation. Thus, as usual, $\partial_t \rightarrow -i\Omega$, $\nabla \rightarrow i\mathbf{k}$. We introduce the notation:

$$\mathbf{k}_\perp = (\mathbf{I} - \mathbf{e}_z \, \mathbf{e}_z) \cdot \mathbf{k}, \qquad k_\perp = (k_x^2 + k_y^2)^{1/2}. \qquad (3.2)$$

We now consider successive particular situations, exhibiting various types of waves.

3.2 Homogeneous plasmas

The simplest situation is obtained when the unperturbed state of the plasma is spatially homogeneous, and the ion temperature is constant and much smaller than the electron temperature. This case can be modelled by taking the limits: $L_n \rightarrow \infty$, $V_{ds} \rightarrow 0$, $T_i \rightarrow 0$, $\eta_i \rightarrow 0$

Assuming also that the electrons are adiabatic ($\delta \tilde{n} = \delta \tilde{\phi}$), the plasma linear response in this case is determined by the Fourier-transformed Hamaguchi-Horton equations (2.67) - (2.69), in which the nonlinear terms (*i.e.*, the Poisson brackets) are discarded:

$$-i\Omega \, \delta \tilde{\phi}_\kappa + ik_z \, \delta u_\kappa = 0,$$
$$ik_z c_s^2 \, \delta \tilde{\phi}_\kappa - i\Omega \, \delta u_\kappa + ik_z (m_i n_0)^{-1} \, \delta p_\kappa = 0,$$
$$-i\Omega \, \delta p_\kappa = 0, \qquad (3.3)$$

[where the following abbreviation is used: $\kappa = (\mathbf{k}, \Omega)$]. The solubility condition of this set of homogeneous linear algebraic equations is obtained by annulling the characteristic determinant, yielding the *dispersion equation*:

$$\Omega \, (\Omega^2 - k_z^2 c_s^2) = 0. \qquad (3.4)$$

This trivially simple equation has the three real solutions: $\Omega = 0$, and

$$\Omega_{\mathbf{k}} = \pm \, c_s \, k_z. \equiv \pm \, \omega_{sk}. \qquad (3.5)$$

The corresponding solutions represent two waves propagating in the directions parallel, resp. antiparallel, to the magnetic field with the speed c_s defined in Eq. (2.25): they are called *ion-acoustic waves*. They are analogous, but not identical to the sound waves propagating in a neutral fluid. There are two main differences between the two cases. On one hand, in the plasma there are no solutions corresponding to propagation perpendicular to the magnetic field, contrary to a homogeneous neutral fluid, where the sound waves propagate isotropically in all directions. This is due to the presence of a strong magnetic field, which inhibits the motion of the particles in the perpendicular direction. On the other hand, the ion-acoustic velocity c_s is a hybrid, constructed with T_{e0} and m_i: this is due to the smallness of the ratio (m_e/m_i) and to the cold ion assumption $T_{i0}/T_{e0} \approx 0$.

The two ion-acoustic waves are the only possible "very low frequency (VLF)" modes that can propagate in a homogeneous cold-ion magnetized plasma. One should not forget, however, that there also exist "high frequency (HF)" and "low frequency (LF)" modes in such a plasma. These are the Langmuir plasma waves with frequencies ω_{pe}, ω_{pi}, the cyclotron waves with frequencies Ω_e, Ω_i, and the Alfvén waves with frequencies Ω_A; they are illustrated schematically in Fig. 2.1. These modes were deliberately left out of consideration by making the assumptions (2.37) - (2.39), as discussed in Sec. 2.2.

3.3   Density Gradient (DG) Drift Waves

We now consider another very simple situation: a plasma in which the *density is spatially inhomogeneous, but both temperatures* T_{e0}, T_{i0} *are constant, with* $T_i \ll T_e$. Moreover, is is assumed that there are no temperature fluctuations, $\delta T_{\alpha 0} = 0$. This is precisely the situation modelled by the *Hasegawa-Mima equation* (2.97). The Fourier transform of its linearized approximation is:

$$-i\Omega \left(1 + \rho_s^2 k_\perp^2\right) \tilde{\phi}_\kappa = -i k_y V_{ds} \tilde{\phi}_\kappa,$$

which leads to the very simple dispersion equation:

$$\Omega_\mathbf{k} = \frac{k_y V_{ds}}{1 + \rho_s^2 k_\perp^2} \cdot \equiv \frac{\omega_{*\mathbf{k}}^e}{1 + \rho_s^2 k_\perp^2}, \qquad (3.6)$$

where V_{ds} is the *diamagnetic drift velocity at constant temperature* [see Eq. (2.28)]:

$$V_{ds} = \frac{\rho_s}{L_n} c_s. \qquad (3.7)$$

The corresponding solution is called the *density-gradient drift wave*. It is a mode propagating in the y-direction (poloidal direction in a torus).

It should be noted that it is a dispersive mode: its phase velocity $\Omega/k_y = V_{ds}(1 + \rho_s^2 k_\perp^2)^{-1}$ depends on the wave vector. For small $k_\perp$ the phase velocity is almost constant, but for larger $k_\perp$ it decreases, as shown in Fig. 3.1. For simplicity, in this and all other figures of this chapter, we set $k_x = 0$.

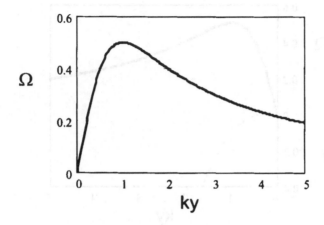

Figure 3.1. Frequency vs. perpendicular wave vector for collisionless density-gradient (DG) drift waves (dimensionless), Eq. (3.6). $k_x = 0$.

3.4   Collisional DG Drift Waves

We now consider the effect of collisional dissipation on the density-gradient (DG) drift waves. The starting point for this study is the set of linearized Hasegawa-Wakatani equations (2.91) - (2.92) which, in dimensionless Fourier representation are:

$$-i\Omega\,\delta\tilde{n}_\kappa = -ik_y V_{ds}\delta\tilde{\phi}_\kappa + \bar{\sigma}\,k_z^2\,(\delta\tilde{\phi}_\kappa - \delta\tilde{n}_\kappa)$$
$$-i\Omega(\delta\tilde{n}_\kappa + \rho_s^2 k_\perp^2\,\delta\tilde{\phi}_\kappa) = -ik_y V_{ds}\delta\tilde{\phi}_\kappa - \mu\,\rho_s^2 k_\perp^4\,\delta\tilde{\phi}_\kappa. \qquad (3.8)$$

The dispersion equation, obtained by annulling the characteristic determinant, yields a second-degree equation for Ω. We shall consider separately the effects of the viscosity and of the electrical conductivity.

A) Assuming $\bar{\sigma} \to \infty$, we obtain the adiabatic relation $\delta\tilde{n}_\kappa = \delta\tilde{\phi}_\kappa$, and after elimination of $\delta\tilde{n}_\kappa$, the second equation (3.8) yields the dispersion relation:

$$\Omega_k = \frac{1}{1 + \rho_s^2 k_\perp^2}\,\left(k_y V_{ds} - i\,\mu\,k_\perp^2\right). \qquad (3.9)$$

This very simple result is easily understood: it represents a DG drift wave whose (real) frequency (3.6) is unaffected by the dissipation, but which is now damped as an effect of the viscosity. The damping rate depends very strongly on $\tilde{k}_\perp$ (Fig. 3.2).

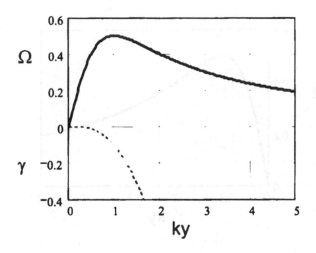

Figure 3.2. Frequency (solid) and damping rate (dotted) vs. perpendicular wave vector for density-gradient (DG) drift waves with finite viscosity $\bar{\mu} = 0.2$ (dimensionless), Eq. (3.9). $k_z = 0$.

B) In the case of finite electrical conductivity $\bar{\sigma}$ and zero viscosity $\bar{\mu}$ the dispersion equation is:

$$\rho_s^2 k_\perp^2 \, \Omega^2 + i \, \bar{\sigma} k_z^2 (1 + \rho_s^2 k_\perp^2) \, \Omega - i k_y V_{ds} \, \bar{\sigma} k_z^2 = 0. \qquad (3.10)$$

In order to avoid divergences, we consider Ω as a function of k_y and of $k_p = k_z/k_\perp$. The exact solution involves a square root of a complex quantity, which yields a rather cumbersome expression. A more transparent form is obtained in the limit $\bar{\sigma} k_p^2 \gg 1$. In this case the two roots are approximated as:

$$\Omega_+ = \Omega_{\mathbf{k}} = \frac{k_y V_{ds}}{1 + \rho_s^2 k_\perp^2},$$
$$\Omega_- = i \, \gamma_{\bar{\mathbf{k}}} = -i \, (1 + \rho_s^2 k_\perp^2) \, \rho_s^{-2} \, \bar{\sigma} k_p^2. \qquad (3.11)$$

The situation here is different from the previous one. Instead of a single damped drift wave, we find here two distinct modes. One of them is the same, *undamped* drift wave with frequency (3.6); the second one is a purely damped mode, whose amplitude very quickly decays to zero.

3.5   DG Drift-Ion Acoustic Waves

We now study again collisionless DG-drift waves, but consider a more realistic situation than in Sec. 3.3. We solve the full linearized Hamaguchi-Horton equations for inhomogeneous systems ($V_{ds} \neq 0$), but still assuming $T_{e0} = const$, $T_{i0} = 0$. Eqs. (2.60), (2.67), (2.68), (2.69) then yield, by the usual procedure, the following dispersion equation:

$$\Omega \left[(1 + \rho_s^2 k_\perp^2) \ \Omega^2 - k_y V_{ds} \Omega - k_z^2 c_s^2 \right] = 0.$$

This yields a zero-frequency solution and a couple of non-trivial real solutions obeying the following dispersion equation:

$$\Omega^2 - \omega_d \ \Omega - \omega_s^2 = 0. \tag{3.12}$$

The coefficients are:

$$\omega_d = \frac{k_y V_{ds}}{1 + \rho_s^2 k_\perp^2} \equiv \Omega_k, \tag{3.13}$$

$$\omega_s = \frac{k_z c_s}{(1 + \rho_s^2 k_\perp^2)^{1/2}}. \tag{3.14}$$

Eq. (3.12) was first obtained by KADOMTSEV 1965 [without the factor $(1 + \rho_s^2 k_\perp^2)$, see Sec. 7.8]. The coefficients have clear meanings. ω_d is simply the drift wave frequency (3.6), and ω_s is the ion-acoustic frequency (3.5), modified by a non-zero perpendicular wave number. It follows that Eq. (3.12) yields, for a wave propagating in the perpendicular plane ($k_z = 0$) a zero-frequency solution and the pure DG-drift wave; and for a wave propagating along the magnetic field ($k_y = 0$) the pair of ion-acoustic waves with frequency (3.5). These two modes appear as the building blocks which are involved in all types of waves in inhomogeneous plasmas. In the present case, upon considering waves propagating obliquely (*i.e.*, $k_\perp \neq 0$, $k_z \neq 0$), their frequency results from a (nonlinear) mixture of the two fundamental ones:

$$\Omega_\pm = \frac{k_y V_{ds}}{2(1 + \rho_s^2 k_\perp^2)} \left[1 \pm \sqrt{1 + 4(1 + \rho_s^2 k_\perp^2) \frac{k_z^2 c_s^2}{k_y^2 V_{ds}^2}} \right]. \tag{3.15}$$

In the case when $k_z c_s \ll k_y V_{de}$, the two roots can be approximated as:

$$\Omega_+ = \omega_d + \frac{k_z c_s}{k_y V_{ds}}\,\omega_s, \qquad \Omega_- = -\frac{k_z c_s}{k_y V_{ds}}\,\omega_s. \tag{3.16}$$

The two branches of the dimensionless frequency are shown in Fig. 3.3 as functions of $(k_z c_s / k_y V_{ds})$ for fixed $k_y V_{ds} = 1$, and in Fig. 3.4 as functions of $k_y V_{ds}$, for fixed $k_z c_s / k_y V_{ds} = 0.2$.

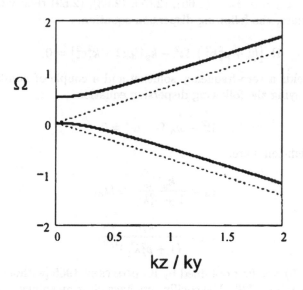

Figure 3.3. Dimensionless frequency of obliquely propagating drift-ion acoustic waves, vs. $(c_s k_z / V_{ds} k_y)$ for fixed $\rho_s k_y = 1$ and $k_x = 0$. The positive solid curve represents $\Omega_+ / k_y V_{ds}$, Eq. (3.15), the negative solid branch is $\Omega_- / k_y V_{ds}$. The dotted curves represent the corresponding ion-acoustic frequencies, Eq. (3.5).

3.6   Ion Temperature Gradient (ITG) Modes

We now consider the general case described by the Hamaguchi-Horton equations (2.67) - (2.69). The main difference with the previously discussed cases is the presence of a non-vanishing, x-dependent unperturbed ion temperature: $T_{i0} \neq 0$, $dT_{i0}/dx \neq 0$, $\delta T_i \neq 0$. Thus, in contrast with the previous cases, $\theta_i = T_{i0}/T_{e0} \neq 0$, $\kappa_i = \theta_i(1 + \eta_i) \neq 0$. (We still consider $T_{e0} = const \neq 0$, $\delta T_e = 0$ and $\delta \tilde{n} = \delta \tilde{\phi}$.) The linearized equations (2.67), (2.68) and (2.69) are:

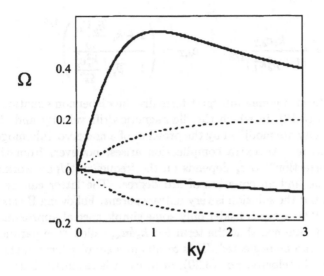

Figure 3.4. Dimensionless frequency of obliquely propagating drift-ion acoustic waves, vs. $k_y V_{ds}$, for fixed $(k_z c_s / k_y V_{ds}) = 0.2$ and $k_x = 0$. The positive solid curve represents $\Omega_+ / k_y V_{ds}$, Eq. (3.15), the negative solid branch is $\Omega_- / k_y V_{ds}$. The dotted curves represent the corresponding ion-acoustic frequencies Ω_s, Eq. (3.14).

$$\left[-\Omega \left(1 + \rho_s^2 k_\perp^2 \right) + k_y V_{ds} \left(1 - \kappa_i \rho_s^2 k_\perp^2 \right) \right] \delta \widetilde{\phi}_\kappa + k_z \delta u_\kappa = 0,$$

$$-\Omega \, \delta u_\kappa + k_z \left[(m_i n_0)^{-1} \delta p_\kappa + c_s^2 \delta \widetilde{\phi}_\kappa \right] = 0,$$

$$-\Omega \, \delta p_\kappa + \kappa_i \, k_y V_{ds} \, \delta \widetilde{\phi}_\kappa + p_{i0} \gamma \, k_z \delta u_\kappa = 0. \quad (3.17)$$

The dispersion equation can be obtained as above, by annulling the characteristic determinant. A more convenient form, however, is obtained by first eliminating successively δp_κ and δu_κ in terms of $\delta \widetilde{\phi}_\kappa$ and annulling the coefficient of the resulting equation. We then obtain an equation of the same form as (3.12), with modified coefficients:

$$\Omega^2 - \omega_{d,T} \, \Omega - \omega_{s,T}^2 = 0, \quad (3.18)$$

with:

$$\omega_{d,T} = \frac{k_y V_{ds,T}}{1 + \rho_s^2 k_\perp^2}, \qquad V_{ds,T} = \left(1 - \kappa_i \, \rho_s^2 k_\perp^2 \right) V_{ds}, \quad (3.19)$$

$$\omega_{s,T} = \frac{k_z c_{s,T}}{\left(1 + \rho_s^2 k_\perp^2\right)^{1/2}}, \qquad c_{s,T} = \left(\frac{1 + \kappa_i \frac{k_y V_{ds}}{\omega}}{1 - \theta_i \gamma \frac{k_z^2 c_s^2}{\Omega^2}}\right)^{1/2}. \tag{3.20}$$

In this form, we may interpret formally the dispersion equation in the same way as (3.12), in which the diamagnetic drift velocity and the ion-acoustic velocity are modified by the presence of a non-zero, inhomogeneous ion temperature. An extra complication arises, however, from the fact that the "correction" to c_s depends on the frequency. Upon working out Eq. (3.18) we find an equation of 3rd degree. The latter can be solved analytically, but the solution is very untransparent. Following HAMAGUCHI & HORTON 1990 a, we consider only some simple special approximations.

It is first assumed that the term $\theta_i \gamma \, k_z \delta u_\kappa$, called "the parallel compressibility", can be neglected. This results in a simpler form of the modified ion-acoustic velocity, Eq. (3.20), in which it is assumed that:

$$\theta_i \gamma \frac{k_z^2 c_s^2}{|\Omega^2|} \ll 1. \tag{3.21}$$

As this condition depends on the frequency, its validity must be determined *a posteriori*.[2] The dispersion equation is then written explicitly as:

$$\left(1 + \rho_s^2 k_\perp^2\right) \Omega^2 - \left(1 - \kappa_i \, \rho_s^2 k_\perp^2\right) k_y V_{ds} \, \Omega - k_z^2 c_s^2 \left(1 + \kappa_i \frac{k_y V_{ds}}{\Omega}\right) = 0. \tag{3.22}$$

1) A first regime is defined by considering a small perpendicular wave vector and a small frequency, such that:

$$\rho_s k_\perp \ll \kappa_i^{-1/2}, \qquad |\Omega| \ll |k_y| \, V_{ds}. \tag{3.23}$$

If a solution exists satisfying all these conditions, the first term in Eq. (3.22) can be neglected and the latter reduces to:

$$k_y V_{ds} \, \Omega + k_z^2 c_s^2 \, \kappa_i \frac{k_y V_{ds}}{\Omega} = 0,$$

with the roots:

$$\Omega_\pm = \pm \, i \, \gamma_k^{(1)} = \pm \, i \, \sqrt{\kappa_i} \, c_s k_z. \tag{3.24}$$

Substituting this solution into the conditions (3.23), we find:

---

[2] This condition is not discussed in HAMAGUCHI & HORTON 1990 a.

$$\kappa_i c_s^2 k_z^2 \ll k_y^2 V_{ds}^2 < k_\perp^2 V_{ds}^2 \ll \rho_s^{-2} \kappa_i^{-1} V_{ds}^2 = \kappa_i^{-1} c_s^2 L_n^{-2}.$$

On the other hand, the condition (3.21) requires $\theta_i \gamma \ c_s^2 k_z^2 / \theta_i (1 + \eta_i) \ c_s^2 k_z^2 \ll 1$. With $\gamma = 5/3$, these conditions combine to:

$$\rho_s k_\perp \ll \kappa_i^{-1/2}, \qquad L_n |k_z| \ll \kappa_i^{-1}, \qquad \eta_i > \frac{2}{3}. \qquad (3.25)$$

Eq. (3.24) exhibits an important new feature: the frequency of the ITG modes in the domain (3.25) is purely imaginary. Thus, besides a purely decaying mode, there appears an *instability*. This property turns out to be very important in anomalous transport theory.

2) A different regime is obtained for larger $\tilde{k}_\perp$: the first term of Eq. (3.22) becomes more important as compared to the second one. We consider in particular the following case, which replaces (3.23):

$$\rho_s k_\perp = \kappa^{-1/2}, \qquad |\Omega| \ll |k_y| V_{de}. \qquad (3.26)$$

The second term of Eq. (3.22) is then zero, and we obtain the equation:

$$(1 + \kappa_i^{-1}) \ \Omega^2 - k_z^2 c_s^2 \ \kappa_i \ \frac{k_y V_{de}}{\Omega} = 0. \qquad (3.27)$$

This 3-rd degree equation has one real root and two complex conjugate ones (resulting from the complex cubic roots of unity):

$$\Omega_0^{(2)} = \Omega_{0k} = \left(\frac{\kappa_i^2}{1 + \kappa_i} \right)^{1/3} \left(\frac{\rho_s}{L_n} \right)^{1/3} c_s \ k_y^{1/3} \ k_z^{2/3},$$

$$\Omega_\pm^{(2)} = \Omega_k^{(2)} \pm i \gamma_k^{(2)} = \frac{1}{2} \left(-1 \pm i \ \sqrt{3} \right) \omega_{0k} \qquad (3.28)$$

We must again check the validity conditions for this solution. For simplicity we consider the case $k_x = 0$, $k_y = k_\perp$. Substituting $\Omega = \Omega_0$ into the conditions (3.21) and (3.26) we find:

$$L_n |k_z| \ll (\theta_i \gamma)^{-3/2} \ \kappa_i^{3/2} \ (1 + \kappa_i)^{-1}, \qquad (3.29)$$

$$L_n |k_z| \ll \kappa_i^{-3/2} \ (1 + \kappa_i)^{1/2}. \qquad (3.30)$$

It is seen from Fig. 3.5 that these conditions, as well as (3.25), are easily satisfied for all reasonable values of $\eta_i > 2/3$ (and $\theta_i = 1$) by taking a small enough value of k_z.

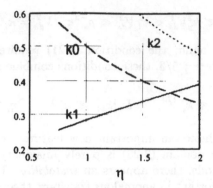

Figure 3.5. Domain of validity of Eqs. (3.24) and (3.28) as a function of η_i. For $\theta_i = 1$, the curves $k0$, $k1$ and $k2$ represent, respectively, the right hand sides of Eqs. (3.25), (3.29) and (3.30). The approximate forms (3.24) and (3.28) are valid for values of $\widetilde{k}_z = L_n k_z$ lying well below all three curves and for $\eta_i > 2/3$.

When these conditions are met, we find in this domain a real frequency mode and a pair of modes with complex conjugate frequencies. Thus, here too there exists an unstable mode with growth rate:

$$\gamma_{\mathbf{k}}^{(2)} = \frac{\sqrt{3}}{2}\left(\frac{\kappa_i^2}{1+\kappa_i}\right)^{1/3}\left(\frac{\rho_s}{L_n}\right)^{1/3}c_s\,k_y^{1/3}\,k_z^{2/3}. \qquad (3.31)$$

The two (dimensionless) growth rates $\widetilde{\gamma}_{\mathbf{k}}^{(1)}$ and $\widetilde{\gamma}_{\mathbf{k}}^{(2)}$ are shown as functions of $\widetilde{k}_z$ in Fig. 3.6. (Remember that the two curves correspond to different values of $\widetilde{k}_y$.)

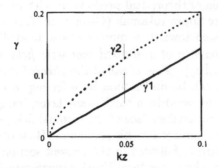

Figure 3.6. Dimensionless growth rate $\tilde{\gamma} = (L_n/c_s)\gamma$ of the ITG mode in the two regimes $\tilde{\gamma}^{(1)}$ (lower curve) and $\tilde{\gamma}^{(2)}$ (upper curve) *vs.* $\tilde{k}_z = L_n k_z$. Here $\theta_i = 1$, $\eta_i = 1$.

The main conclusion of this section is that *an ion temperature gradient can produce an instability*. The existence and the growth rate of the latter depend on the range of wave vectors k_y, k_z and of the parameters θ_i and η_i (or κ_i).

3.7 The ITG Mode in Different Geometries

A general conclusion of the present chapter is the strong dependence of the stability and of the values of frequencies and growth rates of the drift wave modes on the intrinsic properties of the plasma: unperturbed density and temperatures, unperturbed gradients of these quantities, collisionality, as well as on the external magnetic field. The study of the influence of these factors on the drift mode properties has been the subject of an enormous literature. Its exposition would require a complete book. According to the philosophy of the present work, we have studied in detail the simplest case, *i.e.*, a plasma (with constant electron temperature) in presence of a constant strong magnetic field (shearless slab geometry). We now give a few qualitative indications about the results obtained in more realistic situations, in particular in relation with magnetic fusion physics, addressing the interested reader to some key references. (It is impossible in the present framework to cite all relevant works, and we apologize for all omitted authors.)

A basic review paper on drift waves is HORTON 1999, which contains a very exhaustive bibliography. It treats many aspects related to drift waves in relation with anomalous transport: eigenmodes in toroidal geometry, ITG modes, nonlinear theory and turbulence,... A more recent work completing the previous one is HORTON et al. 2003: this paper also contains a detailed comparison of theoretical predictions with experimental results, mainly from the Tore Supra tokamak (Cadarache, France).

An additional step towards a more realistic modelling of a tokamak is to assume the presence of a *sheared magnetic field*, of the form $B = B\{e_z + [(x - x_0)/L_s] e_y\}$. This is called the *sheared slab model.* Its relation to the tokamak can be understood by referring to Fig. 2.2, in which the tangent plane is replaced by a thick plane layer, lying on both sides of a plane tangent to a rational surface of the toroidal field, located at $r = r_0$, *i.e.*, $x = x_0$. A basic reference for this problem is HAMAGUCHI & HORTON 1990 a (which was largely followed in the present section for the shearless case). Technically speaking, the local Fourier representation can no longer be used in this case, because of the x-dependence of B. The structure of the eigenmodes is thus obtained by solving a differential equation (of Weber type). The eigenfrequencies depend, of course, non-trivially on the shear parameter $s = (L_n/L_s)$. The ITG modes are not unstable in the whole range of parameters. There exists a critical value $\eta_i = \eta_{ic}(s)$, above which the instability appears. Just as in the shearless case, the collisionality (*i.e.*, finite values of the collisional viscosity and heat conductivity) is a stabilizing factor, which competes with the growth rate.

The fully toroidal geometry was studied analytically and numerically by many authors, among whom we cite HORTON et al. 1981, NORDMAN & WEILAND 1989, as well as ROMANELLI 1989 (for a microscopic theory). The toroidal geometry introduces a new dimensionless parameter $\epsilon_n = L_n/L_B$, with $L_B^{-1} = |d \ln B/dr|$. For $\eta_i > \eta_{ic}(\epsilon_n, \theta_i)$, the growth rate of the ITG instability is a growing function ($\sim \epsilon_n^{1/2}$) of the toroidicity parameter.

We limit here these very brief remarks, and refer the reader to Horton's review papers for additional bibliography.

Chapter 4

Microscopic Equations for Plasma Fluctuations

4.1   Guiding Centre Variables

We now enter the real problem of a turbulent plasma. The philosophy of the description of such a system was discussed in Chap. 1, especially in Secs. 1.5 and 1.7. Let us summarize the conclusions by saying that individual realizations of the macroscopic or microscopic state of a turbulent plasma are not reproducible, hence a *statistical description* becomes necessary. This implies the definition of an *ensemble of realizations* of the turbulent system. It is postulated that only quantities (both microscopic and macroscopic) averaged over this turbulence ensemble are physically meaningful. This implies an important difference in the interpretation of the quantities introduced in Eqs. (2.45) - (2.47) on one hand, and Eq. (1.31) on the other hand. In the former case (Chaps. 2 and 3), the description was fully deterministic: Q_0 represented a specific solution of the equations of evolution, and δQ was the (deterministic) departure from that state. In the turbulent case, the meaning of the representation $Q_0 + \delta Q$ is quite different, as explained in Sec. 1.7. Q_0 now represents the *ensemble-average* of the *stochastic* quantity Q, and δQ is the deviation of Q from the average, in a given realization [see Eqs. (1.31) - (1.35)]. As will be seen later, this interpretation influences the form of the equations (see, *e.g.* Chap. 8).

In the present chapter we derive the basic equations describing the microscopic state of a turbulent plasma and its evolution. As will be seen, this task is too wide for obtaining a workable set of mathematical tools. We will therefore introduce additional restrictions and/or approximations as we go on. These limitations relate to the external environment (essentially, the geometry of the external magnetic field) and to the type of turbulence present. We begin our study with the simplest possible situation, but keep track of the approximations and indicate at what points modifications are needed for the consideration of more general cases.

The plasma considered in this book (electrons and a single species of singly charged positive ions) was defined in Sec. 1.2. Its microscopic description involves the determination of the *distribution functions* $f^\alpha(\mathbf{q}, \mathbf{p}; t)$ ($\alpha = e, i$) obeying the kinetic equation (1.11). It is well-known that in very high temperature plasmas the collision frequency is very small. On the other hand the phenomena related to turbulence are mainly controlled by the collective interactions, that are adequately described by the first term of the kinetic equation (1.11) (although the collisions may play a relevant role, which will be considered in forthcoming developments, whenever necessary). We thus start our analysis with a *"collisionless plasma"*, for which the last term in the kinetic equation is neglected:

$$\partial_t f^\alpha = [H^\alpha, f^\alpha]. \qquad (4.1)$$

This is the well-known *Vlasov equation*: it is coupled to Maxwell's equations, because the Hamiltonian (1.2) contains the electromagnetic potentials. In a turbulent plasma, as explained in Sec. 1.5, these fields are *fluctuating quantities*, which can only be described statistically. Following the notation of Eq. (1.31), we decompose the fluctuating fields and the potentials as follows:[1]

$$\mathbf{E}(\mathbf{x}; t) = \mathbf{E}_0(\mathbf{x}) + \delta\mathbf{E}(\mathbf{x}; t), \quad \mathbf{B}(\mathbf{x}; t) = \mathbf{B}_0(\mathbf{x}) + \delta\mathbf{B}(\mathbf{x}; t), \qquad (4.2)$$

$$\phi(\mathbf{x}; t) = \phi_0(\mathbf{x}) + \delta\phi(\mathbf{x}; t), \quad \mathbf{A}(\mathbf{x}; t) = \mathbf{A}_0(\mathbf{x}) + \delta\mathbf{A}(\mathbf{x}; t). \qquad (4.3)$$

As a result, the Hamiltonian (1.2) becomes a fluctuating quantity. For simplicity, we assume that the *average fields are independent of time*.

Our next step is the choice of a convenient set of phase space variables for the description of the dynamics of charged particles. This is most elegantly done by performing a succession of transformations which leave unchanged the algebraic structure of the Hamiltonian theory [in particular, Eq. (4.1)]: these are called *pseudo-canonical transformations* (KRUSKAL 1962, LITTLEJOHN 1981, BALESCU 1988 a). They are invertible transformations connecting the initial canonical pair $(\mathbf{q}, \mathbf{p})$ to a new set of variables $(\mathbf{X}, \mathbf{Y})$ (in general, non-canonical) associated with a new table of Lie brackets $[X_i, Y_j]$ that are calculated from the old Poisson brackets by the usual algebraic rules.

The first of these transformations consists of using the *particle velocity* $\hat{\mathbf{v}}$ instead of the momentum $\mathbf{p}$ as a phase space variable. We treat this first

---

[1] The decomposition (4.2) is distinct from Eq. (1.12); indeed, the fluctuations may originate both from the external sources and from the internal instabilities of the plasma.

step in some detail, because it exhibits some important features. In absence of fluctuations, the transformation is defined as follows (BALESCU 1988 a):

$$\mathbf{q} = \mathbf{q}, \quad \hat{\mathbf{v}} = \frac{1}{m_\alpha}\left[\mathbf{p} - \frac{e_\alpha}{c}\mathbf{A}_0(\mathbf{q})\right], \tag{4.4}$$

where $\mathbf{A}_0(\mathbf{q})$ is the (time-independent) *average, deterministic* vector potential, evaluated at the particle's position $\mathbf{x} = \mathbf{q}$.

In presence of fluctuations, the true *physical velocity* $\hat{\mathbf{v}}^{PH}$ is:

$$\hat{\mathbf{v}}^{PH} = \frac{1}{m_\alpha}\left\{\mathbf{p} - \frac{e_\alpha}{c}\left[\mathbf{A}_0(\mathbf{q}) + \delta\mathbf{A}(\mathbf{q}, t)\right]\right\}. \tag{4.5}$$

As a result, $\hat{\mathbf{v}}^{PH}$ is a fluctuating quantity. Although nothing prevents us from considering $\hat{\mathbf{v}}^{PH}$ as a dynamical variable, its use in the present theory is not advisable, for the following reason. $\mathbf{q}$ and $\hat{\mathbf{v}}^{PH}$ are the co-ordinates labelling the phase space, *i.e.*, the framework of the "dynamical drama". In a statistical treatment of the turbulent fluctuations, it would be necessary to define a new phase space in each individual realization of the ensemble: this would be an awkward procedure. It is much preferable to define a *fixed phase space*, spanned by *deterministic coordinates*, the same in all realizations. The only thing which differs from one realization to another is then the functional form of the fluctuating quantities $\delta\phi(\mathbf{q}, t)$, $\delta\mathbf{A}(\mathbf{q}, t)$, as explained in Sec. 1.5.

Thus, in the presence of fluctuations, we choose to use the pseudo-canonical transformation (4.4), with $\mathbf{A}_0(\mathbf{q})$ taken to be the *average vector potential*: $\hat{\mathbf{v}}$ is thus a deterministic quantity. The price to be paid is that $\hat{\mathbf{v}}$ can no longer, in general, be interpreted as the true physical velocity of the particle; indeed:

$$\hat{\mathbf{v}}^{PH} = \hat{\mathbf{v}} - \frac{e_\alpha}{m_\alpha c}\delta\mathbf{A}(\mathbf{q}, t). \tag{4.6}$$

It should be noted, however, that in the case where $\delta\phi \neq 0$, $\delta\mathbf{A} = 0$, $\hat{\mathbf{v}}$ is again the true physical velocity. This situation, called *"electrostatic fluctuations"*, will often occur in the present book. More generally, it is clear that the ensemble-average of the fluctuating physical velocity coincides with $\hat{\mathbf{v}}$:

$$\langle\hat{\mathbf{v}}^{PH}\rangle = \hat{\mathbf{v}}. \tag{4.7}$$

For simplicity of language, we shall most often call $\hat{\mathbf{v}}$ "the velocity", although it must be kept in mind that it is really the "average velocity". The Lie brackets of the new variables are (BALESCU 1988 a):

$$[q_i, q_j] = 0, \quad [q_i, \hat{v}_j] = \frac{1}{m_\alpha}\delta_{ij}, \quad [\hat{v}_i, \hat{v}_j] = \frac{e_\alpha}{m_\alpha c}\epsilon_{ijk}B_k. \tag{4.8}$$

The Hamiltonian, expressed in terms of the new variables, becomes:

$$H^\alpha(\mathbf{q},\widehat{\mathbf{v}};t) = H_0^\alpha(\mathbf{q},\widehat{\mathbf{v}}) + \delta H^\alpha(\mathbf{q},\widehat{\mathbf{v}};t) \qquad (4.9)$$

with:

$$H_0^\alpha(\mathbf{q},\widehat{\mathbf{v}}) = \frac{1}{2}m_\alpha \widehat{v}^2 + e_\alpha \phi_0(\mathbf{q}), \qquad (4.10)$$

$$\delta H^\alpha(\mathbf{q},\widehat{\mathbf{v}};t) = -\frac{e_\alpha}{c}\widehat{\mathbf{v}}\cdot\delta\mathbf{A}(\mathbf{q},t) + e_\alpha \delta\phi(\mathbf{q},t) + \frac{e_\alpha^2}{2m_\alpha c^2}|\delta\mathbf{A}(\mathbf{q},t)|^2. \quad (4.11)$$

Clearly, as the Hamiltonian is no longer the same as in the deterministic case, the equations of motion have a different form. They are obtained, as usual, by calculating the Lie bracket of the basic variables with the Hamiltonian, using the Lie algebra relations (4.8):

$$\dot{q}_i = \left[q_i, H^{\alpha F}\right] = \left[q_i, q_j\right]\frac{\partial H^\alpha}{\partial q_j} + \left[q_i, \widehat{v}_j\right]\frac{\partial H^\alpha}{\partial \widehat{v}_j} \qquad (4.12)$$

and a similar equation for $\widehat{v}_i$. These equations will not be written explicitly here.

The next step consists of introducing components of the velocity, parallel and perpendicular to the local unperturbed magnetic field. More precisely, we introduce a *local fixed reference frame* $e_1, e_2, e_\parallel$, intrinsically determined by the geometry of the unperturbed magnetic field [Fig. 4.1]: the unit vector $e_\parallel(\mathbf{x}) \equiv \widehat{\mathbf{b}}(\mathbf{x})$ is parallel to the average magnetic field at point $\mathbf{x}$, $e_1(\mathbf{x})$ and $e_2(\mathbf{x})$ being directed, respectively, along the principal normal and the binormal of the average magnetic field line (a more detailed discussion is given in BALESCU 1988 a, Sec. 1.4)

Next, we introduce a *local moving reference frame*, attached to the particle; its orientation is determined by the dynamics of the latter: if $\widehat{\mathbf{v}} = \widehat{v}_\parallel \widehat{\mathbf{b}}(\mathbf{q}) + \widehat{\mathbf{v}}_\perp$, we choose a unit vector n_1 parallel to $\widehat{\mathbf{v}}_\perp$, and a second one $n_2 = \widehat{\mathbf{b}} \times n_1$; the third unit vector is unchanged: $n_\parallel \equiv e_\parallel$. The *gyrophase* $\widehat{\varphi}$ of the particle is, by definition, the angle between the fixed unit vector e_1 and the moving unit vector n_2. We then have:

$$n_1(\mathbf{q},\widehat{\varphi}) = -\sin\widehat{\varphi}\; e_1(\mathbf{q}) - \cos\widehat{\varphi}\; e_2(\mathbf{q}),$$
$$n_2(\mathbf{q},\widehat{\varphi}) = \cos\widehat{\varphi}\; e_1(\mathbf{q}) - \sin\widehat{\varphi}\; e_2(\mathbf{q}). \qquad (4.13)$$

We now define a pseudocanonical transformation to the new set of variables $(\mathbf{q}, \widehat{v}_\parallel, \widehat{v}_\perp, \widehat{\varphi})$ as follows:

$$\mathbf{q} = \mathbf{q}, \quad \widehat{\mathbf{v}} = \widehat{v}_\parallel \widehat{\mathbf{b}}(\mathbf{q}) + \widehat{v}_\perp n_1(\mathbf{q},\widehat{\varphi}). \qquad (4.14)$$

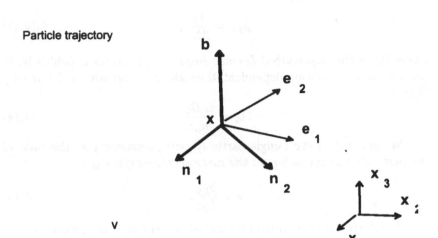

Figure 4.1. Absolute reference frame (x_1, x_2, x_3), Fixed local reference frame $(e_1, e_2, e_3 = \hat{b})$, Moving local reference frame $(n_1, n_2, n_3 = \hat{b})$.

This is the same as in BALESCU 1988 a, but involves again only the unperturbed magnetic field. The fundamental Lie brackets among these new dynamical variables are (BALESCU 1988 a, Table 1.4.3):

$$[q_i, q_j] = 0, \quad [\mathbf{q}, \hat{v}_{\parallel}] = m^{-1}\hat{\mathbf{b}}, \quad [\mathbf{q}, \hat{v}_{\perp}] = m^{-1}\mathbf{n}_1,$$

$$[\mathbf{q}, \hat{\varphi}] = -\frac{1}{m_{\alpha}\hat{v}_{\perp}}\mathbf{n}_2, \quad [\hat{v}_{\parallel}, \hat{v}_{\perp}] = m^{-1}\mathbf{n}_2 \cdot \mathbf{D},$$

$$[\hat{v}_{\parallel}, \hat{\varphi}] = \frac{1}{m_{\alpha}\hat{v}_{\perp}}\mathbf{n}_1 \cdot \mathbf{D}, \quad [\hat{v}_{\perp}, \hat{\varphi}] = -\frac{1}{m_{\alpha}\hat{v}_{\perp}}\left(\Omega_{\alpha} + \hat{\mathbf{b}} \cdot \mathbf{D}\right), \qquad (4.15)$$

where:

$$\mathbf{D} = \hat{v}_{\parallel}(\nabla \times \hat{\mathbf{b}}) + \hat{v}_{\perp}(\nabla \times \mathbf{n}_1). \qquad (4.16)$$

One of the main difficulties here is the occurrence of the gyrophase $\hat{\varphi}$ in the coefficients of the equations of motion of a charged particle. This difficulty can be (partly) overcome in the case of an intense, but weakly inhomogeneous average magnetic field. More precisely, let L_M be the smallest characteristic macroscopic length of the system. Here it is associated with the magnetic field geometry, such as $L_B = |\nabla \ln B|^{-1}$ or the radius of curvature of the magnetic field line. Let also $\hat{\rho}_{L\alpha}$ denote the *Larmor radius of a single particle of species* α, whose perpendicular velocity is $\hat{v}_\perp$:

$$\hat{\rho}_{L\alpha} = \frac{\hat{v}_\perp}{|\Omega_\alpha|}, \tag{4.17}$$

where Ω_α is the *unperturbed Larmor frequency of species* α [which is, in general, weakly position-dependent through the magnetic field intensity $B_0(\mathbf{x})$]:

$$\Omega_\alpha = \frac{e_\alpha B_0}{m_\alpha c}. \tag{4.18}$$

We now define the (single-particle) *drift parameter* $\hat{\epsilon}$ as the ratio of the particle's Larmor radius to the macroscopic length L_M :

$$\hat{\epsilon} = \frac{\hat{\rho}_{L\alpha}}{L_M}. \tag{4.19}$$

The **drift approximation** is valid whenever $\hat{\epsilon} \ll 1$. It is, moreover, assumed that the strength of the average magnetic field is of order $\hat{\epsilon}^{-1}$. In this case, the "odious"[2] dependence on the gyrophase in the equations of motion can be eliminated in a very elegant way by performing a pseudo-canonical transformation based upon the *Kruskal-Littlejohn Theorem*, which we recall here (BALESCU 1988 a, Sec. 1.7):

Let $\hat{z}^m$ ($m = 1,..,6$) be the set of *particle* phase space coordinates: $\hat{z}^m = \{\mathbf{q}, \hat{v}_\parallel, \hat{v}_\perp, \hat{\varphi}\}$. There exists a pseudo-canonical transformation to a new set of phase space coordinates $z^m = \{\mathbf{Y}, v_\parallel, v_\perp, \varphi\}$:

$$z^m = z^m \left(\hat{z}^1, .., \hat{z}^6\right), \quad m = 1, .., 6 \tag{4.20}$$

such that:

- The Lie brackets of the new phase space coordinates are all independent of the new gyrophase $\varphi = z^6$:

$$\left[z^k, z^m\right] = \Sigma^{km}\left(z^1, .., z^5\right), \quad k, m = 1, .., 6 \tag{4.21}$$

- The transformed **unperturbed** Hamiltonian is independent of the new gyrophase $\varphi = z^6$:

$$H^\alpha = H^\alpha\left(z^1, .., z^5\right). \tag{4.22}$$

The vector $\mathbf{Y} \equiv \{Y_1, Y_2, Y_3\} \equiv \{X, Y, Z\}$ represents the position of the *guiding centre* associated with the particle.[3] The phase space coordinates

---

[2] This very suggestive term is due to KRUSKAL 1962.

[3] The notations used here should be clearly understood. The vector $\mathbf{Y}$ (boldface) has the cartesian components $\{X, Y, Z\}$ (italic). The notation $\mathbf{Y}$ is chosen to denote globally the position of the guiding centre, a *dynamical*, phase space variable. It is to be distinguished from the notation $\mathbf{x} \equiv \{x, y, z\}$ [appearing, for instance, in Eqs. (4.2), (4.3), and in Chap. 2] which denotes the purely *geometrical* coordinates of a point in physical space.

z^k will thus be called **Guiding Centre variables** (in contrast to the original *Particle variables* $\hat{z}_k$). The relation between guiding centre position and particle position is given, to first order in $\hat{\epsilon}$, by:

$$q = Y + \frac{v_\perp}{\Omega_\alpha(Y)} n_2(Y, \varphi) + O\left(\hat{\epsilon}^2\right), \qquad (4.23)$$

where both the Larmor frequency Ω_α (4.18) and the basis vector n_2 (4.13) are evaluated at the guiding centre position Y and at the new gyrophase φ. The "velocity" $\hat{v}$ (4.14) is transformed into v:

$$\hat{v} \rightarrow v = v_\| \hat{b}(Y) + v_\perp n_1(Y, \varphi) \qquad (4.24)$$

where, again, $e_\|$ and n_1 are evaluated at Y, φ. As for the variables $v_\|, v_\perp, \varphi$, they are defined as functions of $\left(q, \hat{v}_\|, \hat{v}_\perp, \hat{\varphi}\right)$ in terms of series in powers of $\hat{\epsilon}$ (BALESCU 1988 a); in the forthcoming problems, we only need the dominant term of these series:

$$v_\| = \hat{v}_\| + O(\hat{\epsilon}), \quad v_\perp = \hat{v}_\perp + O(\hat{\epsilon}), \quad \varphi = \hat{\varphi} + 0(\hat{\epsilon}). \qquad (4.25)$$

The equations of motion and the Liouville equation can be readily transformed to the new variables. Their form is, however, very complicated for a general magnetic field configuration. In accordance with the general philosophy exposed in Sec. 1.5, we will use in this book a very simple model of the magnetic geometry.

4.2 Electrostatic Fluctuations in Shearless Slab Geometry

The simplest possible *unperturbed* magnetic field and plasma configuration was defined in Sec. 2.1. For convenience, we recall this definition here.

- We assume that *the average (unperturbed) magnetic field is stationary, (i.e., independent of time) and homogeneous (i.e., it has everywhere the same value)*. Its direction is chosen as the z-axis of a cartesian coordinate system.

$$B = B_0 \hat{b} = B_0 e_z \qquad (4.26)$$

where $\hat{b}$ is a *constant* unit vector.

- We also consider that the average scalar plasmadynamical quantities, such as *the density $n_{\alpha 0}$, the temperature $T_{\alpha 0}$ and the pressure $P_{\alpha 0}$, are spatially inhomogeneous, varying along the same direction x:*

$$n_{\alpha 0} = n_{\alpha 0}(X), \quad T_{\alpha 0} = T_{\alpha 0}(X), \quad P_{\alpha 0} = P_{\alpha 0}(X) \qquad (4.27)$$

This geometrical assumption considerably simplifies the mathematical expressions. The resulting model is called the *shearless slab model* (see Fig. 2.1).

The associated fundamental Lie brackets are then easily obtained (to dominant order in $\hat{\epsilon}$) from Eqs. (4.15), (4.23), (4.24) (BALESCU 1988 a, Table 1.8.1):

$$[Y_i, Y_j] = -\frac{1}{m_\alpha \Omega_\alpha} \epsilon_{ijk}\delta_{kz}, \quad [\mathbf{Y}, v_\parallel] = \frac{1}{m_\alpha}\,\hat{\mathbf{b}}, \quad [\mathbf{Y}, v_\perp] = 0,$$

$$[\mathbf{Y}, \varphi] = 0, \quad [v_\parallel, v_\perp] = 0, \quad [v_\parallel, \varphi] = 0, \quad [v_\perp, \varphi] = -\frac{\Omega_\alpha}{m_\alpha v_\perp}. \quad (4.28)$$

The Hamiltonian, expressed in terms of guiding centre variables, has the same form as in Eqs. (4.9) - (4.11), but now $\mathbf{q}$ is no longer an independent variable: it is to be considered as a function: $\mathbf{q}(\mathbf{Y}, v_\perp, \varphi)$, defined by Eq. (4.23).

Let us stress again that $(\mathbf{Y}, v_\parallel, v_\perp, \varphi)$ are *guiding centre variables*, not to be confused with the *particle variables* $(\mathbf{q}, \hat{v}_\parallel, \hat{v}_\perp, \hat{\varphi})$, even though they are *numerically* very close to the former [as in Eq. (4.25)]. Indeed, their dynamics is determined by the Lie brackets (4.28), not by (4.15).

In the context of fusion physics, the shearless slab geometry can be considered as a rough local approximation to the toroidal confining magnetic field of a tokamak or of a stellarator (Fig. 2.2). We recall that the x-direction mimicks the radial direction in a toroidal reference frame centred on the magnetic axis of the torus.

Next, we introduce an additional simplification, as in Chap. 2, by considering only *electrostatic fluctuations*. This implies the following assumptions:

• The average scalar potential is zero; hence the electric field is a purely fluctuating quantity:

$$\phi_0 = 0, \quad \delta\phi \neq 0; \quad \mathbf{E} = \delta\mathbf{E} = -\nabla\,\delta\phi. \quad (4.29)$$

• The vector potential fluctuation is zero, hence the magnetic field is deterministic. In the shearless slab model, it is even assumed constant:

$$\mathbf{A}_0 \neq 0, \quad \delta\mathbf{A} = 0; \quad \mathbf{B} = \mathbf{B}_0 = \nabla \times \mathbf{A}_0. \quad (4.30)$$

This implies, in particular, that there is no ∇B-drift in this simple model.

The Hamiltonian now reduces to a very simple form:

$$H^\alpha(\mathbf{Y}, v_\parallel, v_\perp, \varphi) = H_0^\alpha(v_\parallel, v_\perp) + \delta H^\alpha(\mathbf{Y}, v_\perp, \varphi), \quad (4.31)$$

$$H_0^\alpha(v_\parallel, v_\perp) = \frac{m_\alpha}{2}\left(v_\parallel^2 + v_\perp^2\right), \tag{4.32}$$

$$\delta H^\alpha(\mathbf{Y}, v_\perp, \varphi) = e_\alpha\, \delta\phi[\mathbf{Y} + \Omega_\alpha^{-1}v_\perp\, \mathbf{n}_2(\varphi)]. \tag{4.33}$$

The equations of motion $\dot{z} = [H, z]$ then become:

$$\dot{\mathbf{Y}} = v_\parallel\widehat{\mathbf{b}} + \frac{c}{B}\,\widehat{\mathbf{b}}\times\nabla\delta\phi, \tag{4.34}$$

$$\dot{v}_\parallel = -\frac{e_\alpha}{m_\alpha}\,\widehat{\mathbf{b}}\cdot\nabla\delta\phi, \tag{4.35}$$

$$\dot{v}_\perp = -\frac{e_\alpha}{m_\alpha}\,\mathbf{n}_1\cdot\nabla\delta\phi, \tag{4.36}$$

$$\dot{\varphi} = \Omega_\alpha + \frac{e_\alpha}{m_\alpha v_\perp}\,\mathbf{n}_2\cdot\nabla\delta\phi. \tag{4.37}$$

An important conclusion of the present treatment is that *the transformed Hamiltonian depends on the gyrophase φ*. As a result, *in presence of fluctuations, the coefficients of the equations of motion depend on the gyrophase, even after the averaging pseudo-canonical transformation*. Indeed,

$$\dot{z}^k = \left[z^k, H^{\alpha F}\right] = \sum_{j=1}^{6}\left[z^k, z^j\right]\frac{\partial H^\alpha}{\partial z^j}. \tag{4.38}$$

The brackets $\left[z^k, z^j\right]$ are independent of φ, but the derivatives $\partial H^\alpha(\varphi)/\partial z^j$ do depend on φ. This feature is a typical *finite Larmor radius (FLR) effect*. In some cases it produces small corrections which can be neglected in the final results. Eqs. (4.34) - (4.37) will be used in the next section for the derivation of the kinetic equations.

4.3 The Gyrokinetic Equation

We now go back to Eq. (4.4), *i.e.* the basic Liouville-Vlasov equation describing a collisionless plasma, and transform it into an equation well adapted to the study of turbulent fluctuations. We first note that the form (4.1), expressed as a Lie bracket with a Hamiltonian, is invariant under pseudo-canonical transformations. It is therefore valid as it stands when both the Hamiltonian and the distribution functions are expressed in terms of guiding centre variables, *i.e.*, when $f^\alpha = f^\alpha(\mathbf{Y}, v_\parallel, v_\perp, \varphi; t)$ and the Hamiltonian has the form (4.31) - (4.33).

Next, we note that, with the Hamiltonian depending on $\delta\phi$, which is a random fluctuation, the Liouville-Vlasov equation becomes a *stochastic equation*. In particular, its solution f^α must be interpreted as a random, fluctuating function. This is the main conceptual step taken in the description of a turbulent plasma (DAVIDSON 1972): it was discussed in detail in Sec. 1.5. As a result, the distribution function can be represented in the form (1.35), like any other fluctuating function:

$$f^\alpha = F^\alpha + \delta f^\alpha, \tag{4.39}$$

where F^α is the ensemble-average of f^α:

$$F^\alpha = \langle f^\alpha \rangle, \quad \langle \delta f^\alpha \rangle = 0. \tag{4.40}$$

Using Eqs. (4.38) and the algebra of Lie brackets, the Liouville-Vlasov Eq. (4.1) can be rewritten in the (self-evident) form:

$$\left(\partial_t + \dot{\mathbf{Y}} \cdot \frac{\partial}{\partial \mathbf{Y}} + \dot{v}_\parallel \frac{\partial}{\partial v_\parallel} + \dot{v}_\perp \frac{\partial}{\partial v_\perp} + \dot{\varphi} \frac{\partial}{\partial \varphi} \right) f^\alpha = 0, \tag{4.41}$$

where the coefficients are given by Eqs. (4.34) - (4.37). This equation can be written more compactly in terms of the variables z^m introduced in Eqs. (4.20) and (4.38):[4]

$$\left(\partial_t + \dot{z}^m \frac{\partial}{\partial z^m} \right) (F^\alpha + \delta f^\alpha) = 0. \tag{4.42}$$

Note that each coefficient $\dot{z}^m$ has an average part and a fluctuating part, the latter being easily recognized in Eqs. (4.34) - (4.37) as the term containing $\delta\phi$. By means of an ensemble-average, Eq. (4.42) is split into two coupled equations. The average distribution function obeys:

$$\partial_t F^\alpha + \langle \dot{z}^m \rangle \frac{\partial}{\partial z^m} F^\alpha + \left\langle \delta \dot{z}^m \frac{\partial}{\partial z^m} \delta f^\alpha \right\rangle = 0. \tag{4.43}$$

The first two terms yield the well-known *Vlasov equation for a quiescent plasma* (for the shearless slab situation). The effect of the fluctuations on the average distribution function is contained in the last term. Eq. (4.43) was first derived by VEDENOV et al. 1961, 1962 and by DRUMMOND & PINES 1964. A very thorough study of its properties is found in the books of SAGDEEV & GALEEV 1969 and of DAVIDSON 1972.

The equation for the fluctuation δf^α is obtained by subtracting Eq. (4.43) from (4.42). Using now explicitly Eqs. (4.34) - (4.37), we find (with $\nabla = \partial/\partial \mathbf{Y}$):

---

[4] A summation over the repeated indices m is understood.

$$\left(\partial_t + v_\parallel \, \hat{\mathbf{b}} \cdot \nabla + \Omega_\alpha \frac{\partial}{\partial \varphi}\right) \delta f^\alpha$$

$$+\mathcal{Q} \, \frac{c}{B}(\hat{\mathbf{b}} \times \nabla \delta \phi) \cdot \nabla \, \delta f^\alpha$$

$$-\mathcal{Q} \, \frac{e_\alpha}{m_\alpha} \left[(\hat{\mathbf{b}} \cdot \nabla \delta \phi) \frac{\partial}{\partial v_\parallel} + (\mathbf{n}_1 \cdot \nabla \delta \phi)\frac{\partial}{\partial v_\perp} - (\mathbf{n}_2 \cdot \nabla \delta \phi)\frac{\partial}{v_\perp \partial \varphi}\right] \delta f^\alpha$$

$$= -\frac{c}{B}(\hat{\mathbf{b}} \times \nabla \delta \phi) \cdot \nabla F^\alpha$$

$$+\frac{e_\alpha}{m_\alpha} \left[(\hat{\mathbf{b}} \cdot \nabla \delta \phi) \frac{\partial}{\partial v_\parallel} + (\mathbf{n}_1 \cdot \nabla \delta \phi)\frac{\partial}{\partial v_\perp} - (\mathbf{n}_2 \cdot \nabla \delta \phi)\frac{\partial}{v_\perp \partial \varphi}\right] F^\alpha. \quad (4.44)$$

We introduced here the operator $\mathcal{Q}$ acting on products of fluctuating quantities; this allows us to use a compact notation:

$$\mathcal{Q} \, (\delta A_1 \, \delta A_2...) = \delta A_1 \, \delta A_2... - \langle \delta A_1 \, \delta A_2...\rangle. \quad (4.45)$$

At this point, we note that the coefficients of the kinetic equations still depend on the gyrophase φ. As mentioned above, the averaging pseudo-canonical transformation does not eliminate completely this dependence in the case of a turbulent plasma. But in the drift approximation (very strong average magnetic field, that varies slowly in space and time) we wish to smooth out this fast gyrophase dependence. This requires an additional averaging of the distribution functions and of the kinetic equations over the gyrophase. This is done mathematically as follows. We split each phase space function $A(\varphi)$ (which is 2π-periodic in φ) into two complementary parts, like in Eq. (1.31):[5]

$$A(\varphi) = \overline{A} + \tilde{A}(\varphi), \quad (4.46)$$

The *gyro-average* part $\overline{A}$ is defined as:

$$\overline{A} = (2\pi)^{-1} \int_0^{2\pi} d\varphi \, A(\varphi). \quad (4.47)$$

This gyro-average is, of course, distinct and independent of the turbulent ensemble-average. Clearly, the gyro-average of a φ-independent quantity is this quantity itself, whereas the gyro-average of $\tilde{A}$ is zero. On the other hand:

$$(2\pi)^{-1} \int_0^{2\pi} d\varphi \frac{\partial}{\partial \varphi} A(\varphi) = 0. \quad (4.48)$$

All the terms in Eq. (4.44) and in the corresponding equation for the average distribution are now submitted to gyro-averaging. We note,

[5] We only write here explicitly the gyrophase dependence.

however, that terms involving products $\delta\phi\,\delta f^\alpha$ cannot be simply expressed in terms of $\overline{\delta\phi}\,\overline{\delta f^\alpha}$, hence the gyro-average of Eq. (4.44) is not, in general, a closed equation for the gyro-averaged function $\overline{\delta f^\alpha}$. The difficulty is, however, avoided by exploiting the characteristic features of the drift wave turbulence. The treatment is based on a multiple time scale perturbation theory, based on the assumption of smallness of three parameters: the *thermal drift parameter* ϵ_α:

$$\epsilon_\alpha = \frac{\rho_{L\alpha}}{L_M} \ll 1, \tag{4.49}$$

where $\rho_{L\alpha}$ is the *thermal* Larmor radius of species α [see (4.19)]:

$$\rho_{L\alpha} = \frac{V_{T\alpha}}{|\Omega_\alpha|}. \tag{4.50}$$

$V_{T\alpha}$ is the *thermal velocity of species* α:

$$V_{T\alpha} = \left(\frac{2T_{\alpha0}}{m_\alpha}\right)^{1/2}, \tag{4.51}$$

where $T_{\alpha0}$ is the unperturbed temperature of species α, measured in energy units. The macroscopic length L_M appearing in Eq. (4.49) is associated with the gradient lengths of the macroscopic quantities, such as the density, $L_n = |\nabla \ln n_0|^{-1}$, the temperature, $L_{T\alpha} = |\nabla \ln T_{\alpha0}|^{-1}$, etc. L_M is defined as the shortest of these lengths.

Next, we assume that we are in a regime of *weak turbulence*:

$$\frac{\delta f^\alpha}{F^\alpha} \approx \frac{e_\alpha\delta\phi}{T_{\alpha0}} = \lambda \ll 1. \tag{4.52}$$

Finally, we adopt the so-called *gyrokinetic ordering* involving the different wave-vectors of the perturbations in directions parallel $(k_\parallel)$ and perpendicular $(k\perp)$ to the magnetic field, as well as their frequencies:

$$k_\parallel\rho_{L\alpha} = \mu \ll 1, \quad k_\perp\rho_{L\alpha} = \left\{ \begin{array}{l} \mu \text{ for } \alpha = e \\ \mu^0 \text{ for } \alpha = i \end{array} \right\}, \quad \frac{\omega}{\Omega_\alpha} = \mu. \tag{4.53}$$

Usually, one assumes that all three parameters are of the same order: $\epsilon \approx \lambda \approx \mu$.

The general treatment of the kinetic equations involves a (multiple-scale) perturbation expansion limited to the lowest order terms. This approach is used in most works appearing in the literature. We shall not give here the details of this somewhat lengthy calculation, which is quite standard (BALESCU, 1988 b), and only quote the final results, valid to dominant order.

Up to this order, both the average distribution function F^α and the fluctuation δf^α are independent of the gyrophase φ; hence: $\overline{F^\alpha} = F^\alpha$, $\overline{\delta f^\alpha} = \delta f^\alpha$. As a result, the gyro-averaging of Eq. (4.44) only involves its coefficients. The gyro-averaged equation (4.44) is then, to dominant order:

$$\left[\partial_t + v_\parallel \, \widehat{\mathbf{b}} \cdot \boldsymbol{\nabla}\right] \delta f^\alpha$$

$$+ \mathcal{Q}\left[\frac{c}{B} \, (\widehat{\mathbf{b}} \times \boldsymbol{\nabla}\overline{\delta\phi}) \cdot \boldsymbol{\nabla}\delta f^\alpha - \frac{e_\alpha}{m_\alpha} \left(\widehat{\mathbf{b}} \cdot \boldsymbol{\nabla}\overline{\delta\phi}\right) \frac{\partial \delta f^\alpha}{\partial v_\parallel}\right]$$

$$= -\frac{c}{B} \, (\widehat{\mathbf{b}} \times \boldsymbol{\nabla}\overline{\delta\phi}) \cdot \boldsymbol{\nabla}F^\alpha + \frac{e_\alpha}{m_\alpha} \left(\widehat{\mathbf{b}} \cdot \boldsymbol{\nabla}\overline{\delta\phi}\right) \frac{\partial F^\alpha}{\partial v_\parallel}. \qquad (4.54)$$

In deriving this equation, we anticipated the result: $\overline{\mathbf{n}_1(\varphi) \cdot \boldsymbol{\nabla}\delta\phi(\varphi)} = 0$, derived in Sec. 4.4, Eq. (4.79). It appears that the "parallel nonlinearity" term $(e_\alpha/m_\alpha) \left(\widehat{\mathbf{b}} \cdot \boldsymbol{\nabla}\overline{\delta\phi}\right) (\partial \delta f^\alpha/\partial v_\parallel)$ yields but a small contribution; it will henceforth be neglected. The kinetic equation for the average distribution becomes:

$$\partial_t F^\alpha + v_\parallel \, \widehat{\mathbf{b}} \cdot \boldsymbol{\nabla} \, F^\alpha$$

$$= -\frac{c}{B} \left\langle (\widehat{\mathbf{b}} \times \boldsymbol{\nabla}\overline{\delta\phi}) \cdot \boldsymbol{\nabla}\delta f^\alpha \right\rangle + \frac{e_\alpha}{m_\alpha} \left\langle \left(\widehat{\mathbf{b}} \cdot \boldsymbol{\nabla}\overline{\delta\phi}\right) \frac{\partial \delta f^\alpha}{\partial v_\parallel} \right\rangle. \qquad (4.55)$$

This equation can also be written in a more compact form by introducing two quantities that have a clear physical meaning, through Eqs. (4.34) and (4.35). The first one is the (gyro-averaged) *electrostatic drift velocity* of the guiding centre (which is independent of the particle species):

$$\delta\mathbf{V}^E = \frac{c}{B} \, (\widehat{\mathbf{b}} \times \boldsymbol{\nabla}\overline{\delta\phi}); \qquad (4.56)$$

the other quantity is the *acceleration parallel to the (average) magnetic field:*

$$\delta a^\alpha = -\frac{e_\alpha}{m_\alpha} \, (\widehat{\mathbf{b}} \cdot \boldsymbol{\nabla}\overline{\delta\phi}). \qquad (4.57)$$

The average kinetic equation can then be written as follows:

$$\partial_t F^\alpha + v_\parallel \, \widehat{\mathbf{b}} \cdot \boldsymbol{\nabla} \, F^\alpha = - \left\langle \delta\mathbf{V}^E \cdot \boldsymbol{\nabla}\delta f^\alpha \right\rangle - \left\langle \delta a^\alpha \frac{\partial}{\partial v_\parallel} \delta f^\alpha \right\rangle. \qquad (4.58)$$

An alternative, sometimes very useful form, is obtained by noting that $\boldsymbol{\nabla} \cdot \delta\mathbf{V}^E = 0$ and $\partial a^\alpha/\partial v_\parallel = 0$; hence:

$$\partial_t F^\alpha = -\nabla \cdot (\hat{\mathbf{b}}\, v_\| F^\alpha) - \nabla \cdot \left\langle \delta \mathbf{V}^E\, \delta f^\alpha \right\rangle - \frac{\partial}{\partial v_\|} \left\langle \delta a^\alpha\, \delta f^\alpha \right\rangle . \qquad (4.59)$$

The equation for the fluctuation is written as:

$$\partial_t \delta f^\alpha + v_\| \,\hat{\mathbf{b}} \cdot \nabla \delta f^\alpha + \mathcal{Q}\, \delta \mathbf{V}^E \cdot \nabla \delta f^\alpha$$
$$= -\,\delta \mathbf{V}^E \cdot \nabla F^\alpha - \delta a^\alpha\, \frac{\partial F^\alpha}{\partial v_\|} . \qquad (4.60)$$

Eq. (4.54) [or (4.60)] is the **gyrokinetic equation** in shearless slab geometry. It is the basis of most works on the microscopic theory of anomalous transport in magnetized plasmas. Let us stress here again its meaning. In a *quiet state*, it describes the (deterministic) *response* of the plasma to a perturbation $\delta\phi$ of the electric potential. In a *turbulent state*, it yields the plasma response *in a given realization*. Indeed, we know that $\delta\phi$, hence also δf^α are stochastic functions; the solution of Eq. (4.60) allows us to calculate ensemble averages of the relevant physical quantities.

The gyrokinetic equation was first derived by RUTHERFORD AND FRIE-MAN 1968 and, independently by TAYLOR & HASTIE 1968. It was later rederived and generalized by numerous authors, among whom we cite: ANTONSEN & LANE 1980, CATTO et al. 1981, WONG 1982, LEE 1983, DUBIN et al. 1983, LITTLEJOHN 1984, BERNSTEIN & CATTO 1985, HAGAN & FRIEMAN 1985, YANG & CHOI 1985, ZHANG 1988, WEYSSOW 1990, ROSENBLUTH & HINTON 1998. In these works various effects neglected here were taken into account, such as: electromagnetic perturbations ($\delta \mathbf{A} \neq 0$), shear and inhomogeneity of the magnetic field, toroidal effects.

Finally, to make the equations self-consistent, we have to add the Poisson equation, determining the fluctuating electric potential in terms of the distribution function:

$$\nabla^2 \delta\phi = -4\pi \sum_{\alpha=e,i} e_\alpha \int d\mathbf{v}\, \delta f^\alpha . \qquad (4.61)$$

Here and in forthcoming formulae, we use the following abbreviation for the integration over the velocity space (in cylindrical coordinates):

$$\int d\mathbf{v} \equiv \int_{-\infty}^{\infty} dv_\| \int_0^\infty dv_\perp\, v_\perp \int_0^{2\pi} d\varphi . \qquad (4.62)$$

Eqs. (4.58), (4.60) and (4.61) constitute a *closed set of equations for the determination of the microscopic state of the turbulent plasma.*

4.4   The Fourier Representation

In the shearless slab geometry, the gyrokinetic equation (4.60) takes its simplest explicit form in the *Fourier representation*. It is well-known from elementary plasma physics that the Vlasov equation, linearized around a spatially homogeneous and stationary reference state, is readily solved by using the *Fourier transformation* with respect to both space and time, which transforms it into a purely algebraic equation. The application of this method leads to some difficulty in the present case, because the coefficients of the linear equation contain the average function $F^\alpha(X, v_\parallel, v_\perp, t)$, which is assumed here to depend on the single spatial coordinate X and on the time t. Hence the Fourier transformation would not lead to an algebraic equation. This question was discussed in Sec. 3.1 for the macroscopic description; it is readily adapted to the present problem. It will be assumed that the scale of the macroscopic gradients is much longer than the wavelength of the "microinhomogeneities"; similarly, the rate of change of average quantities, *i.e.*, F^α and also $\langle \delta \mathbf{V} \, \delta f^\alpha \rangle$, is supposed to be much slower than the frequencies of the fluctuations. As a result, the coefficients in the gyrokinetic equation for the fluctuations may be approximately treated as constants in space and time. In other words, the Fourier transformation only affects the "fast" dependence on $\mathbf{Y}$ and on t.

In most of the forthcoming developments we will use a representation that is Fourier-transformed with respect to the position variable $\mathbf{Y}$, but *not* with respect to time. At some point, however, the final results of some calculations will be Fourier- or Laplace-transformed with respect to time as well. We thus define:

$$\delta f^\alpha(\mathbf{Y}, \mathbf{v}, t) = \int d\mathbf{k} \, e^{i\mathbf{k} \cdot \mathbf{Y}} \, \delta f^\alpha_\mathbf{k}(\mathbf{v}; t). \tag{4.63}$$

It is understood that the Fourier transform $\delta f^\alpha_\mathbf{k}(\mathbf{v}, t)$ still contains a *slow* dependence on X, on the macroscopic scale. The present approximation is called the *local approximation* [see ICHIMARU 1973, KRALL & TRIVELPIECE 1986]. The slow X-dependence of $\delta f^\alpha_\mathbf{k}$ will usually not be written down explicitly. The fluctuating potential is similarly represented as follows:

$$\delta\phi(\mathbf{Y}, t) = \int d\mathbf{k} \, e^{i\mathbf{k} \cdot \mathbf{Y}} \, \delta\phi_\mathbf{k}(t). \tag{4.64}$$

The average distribution function, which only depends *slowly* on the X-coordinate, will not be Fourier-transformed.

Let us insist on the well-known features of the Fourier transformation. The integrations are performed on all values (from $-\infty$ to $+\infty$) of the three components of the wave vector $\mathbf{k}$. The choice of factors (2π) is to some degree a matter of convenience. This choice determines, however, uniquely the factors (2π) in the inverse transformation formula:

$$\delta\phi_{\mathbf{k}}(t) = \frac{1}{(2\pi)^3} \int d\mathbf{Y} \, e^{-i\mathbf{k}\cdot\mathbf{Y}} \, \delta\phi(\mathbf{Y}, t), \qquad (4.65)$$

and a similar formula for $\delta f_{\mathbf{k}}^{\alpha}$. The integration runs over the whole $\mathbf{Y}$-space. The specific feature of the gyrokinetic equation (4.54), (4.55) is the appearance of the *gyro-averaged potential* $\overline{\delta\phi}$. The latter is readily related to the unaveraged one in the Fourier representation.

The necessity of the gyro-averaging operation comes from the fact that the potential appearing in Eq. (4.33) is evaluated at the particle's position $\mathbf{q}$ (not the guiding centre's, $\mathbf{Y}$!), which depends on the gyrophase φ, as shown in Eq. (4.23). The gyro-averaged scalar potential is now evaluated as follows:

$$
\begin{aligned}
\overline{\delta\phi}(\mathbf{Y}, t) &= \frac{1}{2\pi} \int_0^{2\pi} d\varphi \, \delta\phi(\mathbf{q}, t) \\
&= \frac{1}{2\pi} \int_0^{2\pi} d\varphi \int d\mathbf{k} \, \exp\left\{ i\mathbf{k}\cdot\left[\mathbf{Y} + \frac{v_\perp}{\Omega_\alpha}\mathbf{n}_2(\varphi)\right]\right\} \delta\phi_{\mathbf{k}}(t).
\end{aligned}
$$

$$(4.66)$$

This can be rewritten in the form:

$$\overline{\delta\phi}(\mathbf{Y}, t) = \int d\mathbf{k} \, e^{i\mathbf{k}\cdot\mathbf{Y}} \, \overline{\delta\phi}_{\mathbf{k}}(t), \qquad (4.67)$$

with:

$$\overline{\delta\phi}_{\mathbf{k}}(t) = \frac{1}{2\pi} \int_0^{2\pi} d\varphi \, \exp\left[i\mathbf{k}\cdot\mathbf{n}_2(\varphi)\frac{v_\perp}{\Omega_\alpha}\right] \delta\phi_{\mathbf{k}}(t). \qquad (4.68)$$

From (4.28) we have:

$$
\begin{aligned}
\mathbf{n}_1(\varphi) &= -\sin\varphi\,\mathbf{e}_1 - \cos\varphi\,\mathbf{e}_2, \\
\mathbf{n}_2(\varphi) &= \cos\varphi\,\mathbf{e}_1 - \sin\varphi\,\mathbf{e}_2,
\end{aligned}
\qquad (4.69)
$$

where $\mathbf{e}_1, \mathbf{e}_2$ are the fixed (local) basis vectors, evaluated at the guiding centre position $\mathbf{Y}$ (not written explicitly). We represent the (local) wave vector in the form: $\mathbf{k} = k_\parallel \widehat{\mathbf{b}} + \mathbf{k}_\perp$ and write its perpendicular component in the same form as (4.69):

$$\mathbf{k}_\perp = k_\perp \left(\cos\alpha\,\mathbf{e}_1 - \sin\alpha\,\mathbf{e}_2\right), \qquad (4.70)$$

where α is the angle between the vectors $\mathbf{k}_\perp$ and $\mathbf{e}_1$ [see Fig. 4.1]. We then obtain:

$$\mathbf{k}\cdot\mathbf{n}_2(\varphi) = k_\perp \cos(\varphi - \alpha). \qquad (4.71)$$

The exponential factor in the integrand of Eq. (4.68) can now be Fourier-expanded with respect to the gyrophase in terms of Bessel functions of order m: $J_m(\xi)$, using the well-known formula [see, *e.g.* ABRAMOWITZ & STEGUN 1965]:

$$\exp(i\xi \sin\theta) = \sum_{m=-\infty}^{\infty} J_m(\xi)\, e^{-im\theta}. \tag{4.72}$$

We then find:

$$\exp\left[i\mathbf{k} \cdot \mathbf{n}_2(\varphi) \frac{v_\perp}{\Omega_\alpha}\right] = \sum_{m=-\infty}^{\infty} \exp[-im(\varphi - \alpha + \tfrac{1}{2}\pi)]\, J_m\left(k_\perp \frac{v_\perp}{\Omega_\alpha}\right). \tag{4.73}$$

The gyro-average of this expression is:

$$\frac{1}{2\pi}\int_0^{2\pi} d\varphi \, \exp\left[i\mathbf{k} \cdot \mathbf{n}_2(\varphi)\frac{v_\perp}{\Omega_\alpha}\right]$$

$$= \frac{1}{2\pi}\int_0^{2\pi} d\varphi \sum_{m=-\infty}^{\infty} \exp[-im(\varphi - \alpha + \tfrac{1}{2}\pi)]\, J_m\left(k_\perp \frac{v_\perp}{\Omega_\alpha}\right)$$

$$= \sum_{m=-\infty}^{\infty} \delta_{m,0}\, J_m\left(k_\perp \frac{v_\perp}{\Omega_\alpha}\right). \tag{4.74}$$

We thus obtain the useful result:[6]

$$(2\pi)^{-1}\int_0^{2\pi} d\varphi \, \exp\left[i\mathbf{k} \cdot \mathbf{n}_2(\varphi)\right] = J_0\left(k_\perp \frac{v_\perp}{\Omega_\alpha}\right) = J_0\left(k_\perp \hat{\rho}_{L\alpha}\right). \tag{4.75}$$

Hence, combining (4.68) and (4.75), we find the desired relation:

$$\overline{\delta\phi_\mathbf{k}}(t) = J_0\left(k_\perp \hat{\rho}_{L\alpha}\right) \delta\phi_\mathbf{k}(t). \tag{4.76}$$

We now prove the result announced after Eq. (4.54). In Fourier space we calculate:

$$\overline{\mathbf{n}_1(\varphi) \cdot \mathbf{k}\,\delta\phi_\mathbf{k}(\varphi;t)} = \frac{1}{2\pi}\int d\varphi \, \exp\left[i\mathbf{k} \cdot \mathbf{n}_2(\varphi)\frac{v_\perp}{\Omega_\alpha}\right] \mathbf{k} \cdot \mathbf{n}_1(\varphi)\, \delta\phi_\mathbf{k}(t). \tag{4.77}$$

Using Eqs. (4.69) and (4.70) we find:

---

[6] Note that at this final stage we may replace the variable $k_\perp v_\perp/\Omega_\alpha$, which has different signs for the electrons and for the ions, by its absolute value $k_\perp v_\perp/|\Omega_\alpha| = \hat{\rho}_{L\alpha}$ [Eq. (4.17)], *i.e.*, the single particle Larmor radius: indeed, the Bessel function $J_0(x)$ is an even function of x.

$$\mathbf{k} \cdot \mathbf{n}_1(\varphi) = -k_\perp \sin(\varphi - \alpha); \qquad (4.78)$$

hence:

$$\frac{1}{2\pi} \int d\varphi \, \exp\left[ik \cdot \mathbf{n}_2(\varphi)\frac{v_\perp}{\Omega_\alpha}\right] \, \mathbf{k} \cdot \mathbf{n}_1(\varphi)$$

$$= -\frac{1}{2\pi} \int d\varphi \, \exp\left[ik_\perp \frac{v_\perp}{\Omega_\alpha} \cos(\varphi - \alpha)\right] \, k_\perp \sin(\varphi - \alpha)$$

$$= \frac{1}{\hat{\rho}_{L\alpha}} \frac{1}{2\pi} \int d\varphi \, \frac{\partial}{\partial\varphi} \exp[ik_\perp \frac{v_\perp}{\Omega_\alpha} \cos(\varphi - \alpha)] = 0.$$

We thus find the announced result:

$$\overline{\mathbf{n}_1(\varphi) \cdot \mathbf{k} \, \delta\phi_{\mathbf{k}}(\varphi, t)} = 0. \qquad (4.79)$$

It is now easily shown that the coefficients of the gyrokinetic equations are related as follows to the gyro-averaged potential:

$$\delta\mathbf{V}_{\mathbf{k}}^E(t) = i\frac{c}{B} \, (\hat{\mathbf{b}} \times \mathbf{k}) \, \overline{\delta\phi}_{\mathbf{k}}(t) = i\frac{c}{B} \, (-\mathbf{e}_x \, k_y + \mathbf{e}_y \, k_x) \, \overline{\delta\phi}_{\mathbf{k}}(t) \qquad (4.80)$$

$$\delta a_{\mathbf{k}}^\alpha(t) = -i\frac{e_\alpha}{m_\alpha} \, \hat{\mathbf{b}} \cdot \mathbf{k} \, \overline{\delta\phi}_{\mathbf{k}}(t) = -i\frac{e_\alpha}{m_\alpha} k_\parallel \, \overline{\delta\phi}_{\mathbf{k}}(t). \qquad (4.81)$$

We will henceforth use the convenient abbreviation: $\partial_\parallel = \partial/\partial v_\parallel$. The kinetic equation for the average distribution function is now written as:

$$(\partial_t + v_\parallel \hat{\mathbf{b}} \cdot \boldsymbol{\nabla})F^\alpha(\mathbf{Y}, \mathbf{v}, t)$$

$$= -\, \langle \delta\mathbf{V}_{-\mathbf{k}}^E(t) \, \delta f_{\mathbf{k}}^\alpha(\mathbf{v}, t)\rangle - \partial_\parallel \, \langle \delta a_{-\mathbf{k}}^\alpha(t) \, \delta f_{\mathbf{k}}^\alpha(\mathbf{v}, t)\rangle. \qquad (4.82)$$

The gyrokinetic equation determining the fluctuating distribution function $\delta f_{\mathbf{k}}^\alpha(\mathbf{v}; t)$ is easily obtained by a Fourier transformation of Eq. (4.60):[7]

$$(\partial_t + ik_\parallel v_\parallel) \, \delta f_{\mathbf{k}}^\alpha(\mathbf{v}, t)$$

$$+ \int d\mathbf{k}' \frac{c}{B} i \left(\mathbf{k}' \times \hat{\mathbf{b}}\right) \cdot \, i\mathbf{k} \, \mathcal{Q} \, \overline{\delta\phi}_{\mathbf{k}'}(t) \, \delta f_{\mathbf{k}-\mathbf{k}'}^\alpha(\mathbf{v}, t)$$

$$= i\left[\frac{c}{B} \left(\mathbf{k} \times \hat{\mathbf{b}}\right) \cdot \boldsymbol{\nabla} F^\alpha(x, \mathbf{v}, t) + \frac{e_\alpha}{m_\alpha} k_\parallel \, \partial_\parallel F^\alpha(x, \mathbf{v}, t)\right] \, \overline{\delta\phi}_{\mathbf{k}}(t). \qquad (4.83)$$

Finally, these equations are completed with the *Poisson equation*, whose derivation requires some care. The Poisson equation tells us that

[7] We used here the obvious property: $(\mathbf{k}' \times \mathbf{b}) \cdot (\mathbf{k} - \mathbf{k}') = (\mathbf{k}' \times \mathbf{b}) \cdot \mathbf{k}$.

the Laplacian of the potential $\delta\phi(\mathbf{x}, t)$, *evaluated at an arbitrary point of coordinate* $\mathbf{x}$ *in physical space* (*not* in phase space) is proportional to the charge density at the same point. This density is produced by the charged particles located at the positions $\mathbf{q}$, not to be confused with the coordinates $\mathbf{Y}$ of the guiding centres. The latter are purely geometrical fictions: there is no particle, hence no source of electric field at the point $\mathbf{Y}$! Thus, we write the Poisson equation in the form of a statistical average in the guiding centre phase space:

$$\nabla^2 \delta\phi(\mathbf{x}, t) = 4\pi \sum_\alpha e_\alpha \int d\mathbf{v} \int d\mathbf{Y}\, \delta[\mathbf{x} - \mathbf{q}(\mathbf{Y})]\, \delta f^\alpha(\mathbf{Y}, \mathbf{v}, t)$$

$$= 4\pi \sum_\alpha e_\alpha \int d\mathbf{v} \int d\mathbf{Y}\, \delta\left[\mathbf{x} - \mathbf{Y} - \frac{v_\perp}{\Omega_\alpha} \mathbf{n}_2(\varphi)\right]\, \delta f^\alpha[\mathbf{Y}, \mathbf{v}, t]. \qquad (4.84)$$

The integral in the right hand side is the particle density fluctuation, which can be expressed in Fourier representation, to dominant order in ϵ, as:

$$\delta n_\alpha(\mathbf{x}, t) = \int d\mathbf{v} \int d\mathbf{Y}\, \delta\left[\mathbf{x} - \mathbf{Y} - \frac{v_\perp}{\Omega_\alpha} \mathbf{n}_2(\varphi)\right]\, \delta f^\alpha[\mathbf{Y}, \mathbf{v}, t]$$

$$= \int dv_\parallel \int dv_\perp v_\perp \int d\varphi \int d\mathbf{k}\, \exp\left\{i\mathbf{k} \cdot \left[\mathbf{x} + \frac{v_\perp}{\Omega_\alpha} \mathbf{n}_2(\varphi)\right]\right\}\, \delta f_\mathbf{k}^\alpha(\mathbf{v}, t). \qquad (4.85)$$

The gyrophase integral is performed by using the relation (4.75) (and recalling that δf^α is independent explicitly of φ to dominant order in ϵ):

$$\delta n_\alpha(\mathbf{x}, t) = \int d\mathbf{k}\, e^{i\mathbf{k} \cdot \mathbf{x}}\, \delta n_\mathbf{k}^\alpha(t), \qquad (4.86)$$

with:

$$\delta n_\mathbf{k}^\alpha(t) = \int d\mathbf{v}\, J_0(k_\perp \hat{\rho}_{L\alpha})\, \delta f_\mathbf{k}^\alpha(\mathbf{v}, t).$$

$$= \int d\mathbf{v}\, J_0(|\Omega_\alpha|^{-1} k_\perp v_\perp)\, \delta f_\mathbf{k}^\alpha(\mathbf{v}, t). \qquad (4.87)$$

We wrote down explicitly the expression of the Larmor radius $\hat{\rho}_{l\alpha}$ in order to underscore its dependence on the perpendicular velocity, which must be taken into account in the integration.

Thus, the Poisson equation in Fourier representation, taking into account the finite Larmor radius effect, is:

$$k^2 \, \delta\phi_{\mathbf{k}}(t) = 4\pi \sum_\alpha e_\alpha \, \delta n_{\mathbf{k}}^\alpha(t). \qquad (4.88)$$

The set of four coupled equations (4.82), (4.83), (4.87), (4.88) constitute the basis of the **gyrokinetic theory** for *electrostatic turbulence* in the *shearless slab geometry*.

We finish this section with some general remarks on the structure of the GKE. Eq. (4.83) can be written schematically in the form:

$$\mathcal{O}^\alpha \, \delta f^\alpha = \mathcal{S}^\alpha. \qquad (4.89)$$

The operator $\mathcal{O}^\alpha$ is defined by the left hand side of (4.83). Thus, strictly speaking, the GKE has the standard form of a *linear equation* for the unknown function δf^α, with a source term $\mathcal{S}^\alpha$ in the right hand side. One could therefore wonder why turbulence theory is such a difficult subject when its dynamics is controlled by such a "simple" equation!

The answer is twofold and can be understood by writing the GKE in a slightly more detailed form:

$$\mathcal{O}_L^\alpha \, \delta f^\alpha + \mathcal{O}_{NL}^\alpha \, \delta f^\alpha = \mathcal{S}^\alpha, \qquad (4.90)$$

where $\mathcal{O}_L^\alpha$ represents the first term and $\mathcal{O}_{NL}^\alpha$ the second term in the left hand side of (4.83). Both the operator $\mathcal{O}_{NL}^\alpha$ and the source term $\mathcal{S}^\alpha$ involve (linearly) the fluctuating potential $\delta\phi$. But the latter is determined self-consistently by the Poisson equation (4.88) as a linear functional of δf^α. Upon substitution in (4.90) it appears that $\mathcal{O}_{NL}^\alpha \, \delta f^\alpha$ is really a *nonlinear* expression in δf^α, and that $\mathcal{S}^\alpha$ is *linear* in δf^α. Thus, the gyrokinetic equation is a *nonlinear equation*, for the same reasons as the Vlasov equation. The nonlinearity introduces, as usual, an enormous complication into the problem, but also a fascinating richness which is still far from being fully exploited.

But this is not the end of the story. It must not be forgotten that the GKE is a *stochastic equation*. This implies that, even if the equation has been formally solved, the solution δf^α describes the response of the plasma in a *given single realization*. The only measurable quantities associated with a random function are its *moments* of various orders, *i.e.*, ensemble-averages of products of such functions $\langle \delta f^\alpha \delta f^\beta ... \rangle$. From Eq. (4.83) one can easily obtain the equation of evolution of any moment of order n by multiplying both sides by $(n-1)$ appropriate factors δf^α and ensemble-averaging both sides. In other words, (4.83) should really be regarded as a *generating equation* for the moment equations. It is clear that, because of the nonlinearity of $\mathcal{O}_{NL}^\alpha \, \delta f^\alpha$, the rate of change of any moment of order n depends on a moment of order $(n+1)$. We are thus confronted with the familiar difficulty appearing so often in statistical physics: an *infinite hierarchy of moment equations*.

Chapter 5

Definition of Anomalous Fluxes

5.1 Introduction

The final aim of transport theory is the determination of the *fluxes* of particles, energy and electric charge[1] in a non-equilibrium, inhomogeneous plasma. These fluxes are driven by *thermodynamic forces* related to the inhomogeneity of density, pressure, temperature, electric potential and are influenced by the magnetic field and by the geometry of the system. The relations between fluxes and forces are embodied in the so-called *transport equations* [see Sec. 1.3], describing the well-known processes of diffusion, energy (or heat) conductivity, electrical conductivity, as well as "cross-effects" such as thermodiffusion, Hall effect, bootstrap current, etc.

The various fluxes **J** described above can be split, theoretically, into three components, *viz.*, a *classical*, a *neoclassical* and an *anomalous* component

$$\mathbf{J} = \mathbf{J}^{CL} + \mathbf{J}^{NCL} + \mathbf{J}^{AN}. \tag{5.1}$$

The *classical fluxes* $\mathbf{J}^{CL}$ are determined by individual particle collision processes, just as in neutral gases. They have been extensively studied and their theory is perfectly understood (CHAPMAN & COWLING 1952, BRAGINSKII 1965, BALESCU 1988 a). The *neoclassical fluxes* $\mathbf{J}^{NCL}$ are also determined by the individual particle collision processes, but the latter are strongly influenced by the presence of inhomogeneous and curved magnetic fields (such as those confining the plasma in thermonuclear fusion devices like the tokamak, the stellarator, etc.). As a result, a far from negligible contribution must be added to the classical fluxes (which pertain to a constant magnetic field). Neoclassical transport theory is also at present an elegant and complete mathematical theory (PFIRSCH & SCHLUTER 1962, GALEEV

---

[1] We only consider in the present chapter the vector fluxes, and do not discuss here the tensor fluxes, such as the momentum flux, related to viscosity. See, however, Secs. 2.5 and 16.4.

& SAGDEEV 1968, HINTON & HAZELTINE 1976, HIRSHMAN & SIGMAR 1981, BALESCU 1988 b). The classical and neoclassical fluxes are, however, not sufficient for explaining most of the experimental results, both in fusion physics and in astrophysics. The consideration of *collective effects*, due to the fluctuations, appears to be absolutely necessary for the understanding of real plasmas (which are practically always turbulent to various degrees). This leads to the third term of Eq. (5.1), which represents the ensemble-average of the part of the flux due to the turbulent fluctuations. Anomalous transport theory is at present by far not as fully developed and unified as the previous two theories. This is due to various factors: the large variety of collective modes existing in a non-uniformly magnetized plasma, the poor understanding of the various possible couplings between these modes or between the modes and the individual particles, and the complexity of the nonlinear mathematics needed for their study. Even in the simpler neutral fluids, the turbulence problem is still largely open. There has been, however, an enormous amount of work done in the past forty years which has led to significant progress in specific situations. Large scale numerical simulations have played an important role in this research. It is impossible to collect all this work in a single book. We therefore restrict ourselves to a limited goal. We will treat here primarily the analytical methods which have been developed for the description of the simplest possible non-trivial idealization of the plasma. As described in Chap. 4, we consider as a starting point a turbulent plasma in presence of a strong, constant, deterministic magnetic field.

The first question to be answered is the unambiguous mathematical definition of the anomalous fluxes, *i.e.*, the part of the global fluxes related to the fluctuations. Two methods can be used to this purpose. The first one starts from the macroscopic (two-fluid) description of the plasma, more precisely from the particle balance (1.14) and the energy balance (1.16). One assumes that each plasmadynamical quantity is fluctuating[2] and writes it in the form (1.31), after which the equations are ensemble-averaged. The anomalous fluxes are then identified (and separated from the classical and neoclassical contributions) from the form of these average equations.

A more direct method starts from the microscopic kinetic equation for the average distribution function, Eq. (4.59) from which the balance equations for the number density and for the energy density are derived by the usual methods of kinetic theory. The definition of the fluxes is closely connected with the existence of dynamical quantities conserved by the kinetic equations.

---

[2] We stress again that the interpretation of the plasmadynamical quantities $A = A_0 + \delta A$ of Chap. 2 has changed: instead of describing a deterministic deviation from a reference state, δA represents the fluctuating deviation from the ensemble-average $A_0 = \langle A \rangle$.

5.2 The Particle Number Balance

The fluctuating particle number density is defined quite straightforwardly as follows:

$$n_\alpha(\mathbf{x}, t) = n_{\alpha 0}(\mathbf{x}, t) + \delta n_\alpha(\mathbf{x}, t). \tag{5.2}$$

It should be stressed that the *macrocopic densities*, such as the particle number density, arc *fields*, *i.e.*, continuous functions of time and of the physical space coordinates $\mathbf{x}$. The latter do *not* quite coincide with the guiding centre coordinates $\mathbf{Y}$. This delicate point was discussed and solved in Sec. 4.4, in connection with the Poisson equation. The particle number density is then defined as:

$$n_\alpha(\mathbf{x}, t) = \int d\mathbf{v} \int d\mathbf{Y}\, \delta[\mathbf{x} - \mathbf{q}(\mathbf{Y})]\, [F^\alpha(\mathbf{Y}, \mathbf{v}, t) + \delta f^\alpha(\mathbf{Y}, \mathbf{v}, t)]. \tag{5.3}$$

The first term involves the average distribution function, which is very slowly varying over a distance comparable to the Larmor radius; hence we may replace the particle position $\mathbf{q}$ by the guiding centre position $\mathbf{Y}$: $F^\alpha(\mathbf{Y}, \mathbf{v}, t) = F^\alpha(\mathbf{q}, \mathbf{v}, t)$ with an error of order $\rho_L/L_M = \epsilon$. This argument is not valid for the fluctuation, for which the error would be of order $k_\perp \rho_L = O(\epsilon^0)$ (at least for the ions). We thus define, to dominant order:

$$n_{\alpha 0}(\mathbf{x}, t) = \int d\mathbf{v} \int d\mathbf{Y}\, \delta(\mathbf{x} - \mathbf{Y})\, F^\alpha(\mathbf{Y}, \mathbf{v}, t). \tag{5.4}$$

The average particle number density is also briefly called the *density profile*.

For the fluctuation we must retain the finite Larmor radius effects to dominant order (at least for the ions), thus using Eq. (4.85):

$$\delta n_\alpha(\mathbf{x}, t) = \int d\mathbf{v} \int d\mathbf{Y}\, \delta\left[\mathbf{x} - \mathbf{Y} - \frac{v_\perp}{\Omega_\alpha} \mathbf{n}_2(\varphi)\right] \delta f^\alpha\, [\mathbf{Y}, \mathbf{v}, t]. \tag{5.5}$$

The Fourier transform of this quantity, averaged over the gyrophase φ, was calculated in Eqs. (4.85) - (4.87):

$$\delta n_{\mathbf{k}}^\alpha(t) = \int d\mathbf{v}\, J_0(k_\perp \hat{\rho}_{L\alpha})\, \delta f_{\mathbf{k}}^\alpha(\mathbf{v}, t). \tag{5.6}$$

The number of particles of each species is clearly a conserved quantity of the kinetic equation. This conservation law should be valid, in particular in a turbulent fluid, for the ensemble-averaged particle number density. It is expressed by a *balance equation* of the form [see Eq. (2.2)]:

$$\partial_t n_{\alpha 0}(\mathbf{x}, t) = -\boldsymbol{\nabla} \cdot \boldsymbol{\Gamma}^\alpha(\mathbf{x}, t). \tag{5.7}$$

This equation defines the *particle flux* $\Gamma^\alpha(\mathbf{x}, t)$. Using now Eq. (4.59), we find:

$$\partial_t n_{\alpha 0}(\mathbf{x}, t) = \int d\mathbf{v} \int d\mathbf{Y} \, \delta[\mathbf{x} - \mathbf{q}(\mathbf{Y})] \, \partial_t F^\alpha(\mathbf{Y}, \mathbf{v}, t)$$

$$= -\int d\mathbf{v} \int d\mathbf{Y} \, \delta[\mathbf{x} - \mathbf{q}(\mathbf{Y})] \, v_\parallel \nabla_Z \, F^\alpha(\mathbf{Y}, \mathbf{v}, t)$$

$$- \int d\mathbf{v} \int d\mathbf{Y} \, \delta[\mathbf{x} - \mathbf{q}(\mathbf{Y})]$$

$$\times \left\langle \left[\delta \mathbf{V}^E(\mathbf{Y}, v_\perp; t) \cdot \nabla_\mathbf{Y} + \delta a^\alpha(\mathbf{Y}; t) \frac{\partial}{\partial v_\parallel} \right] \delta f^\alpha(\mathbf{Y}, \mathbf{v}, t) \right\rangle. \qquad (5.8)$$

Note that the quantities $\delta \mathbf{V}^E$ and δa^α are already gyro-averaged [see Eqs. (4.56), (4.57)]; hence we may replace directly $\mathbf{Y}$ by $\mathbf{x}$ in these functions. From their definitions we note that $\nabla_\mathbf{Y} \cdot \delta \mathbf{V}^E = 0$, $\partial a_\parallel^\alpha / \partial v_\parallel = 0$, hence Eq. (5.8) is transformed into:

$$\partial_t n_{\alpha 0}(\mathbf{x}, t) = -\nabla \cdot \int d\mathbf{v} \left[v_\parallel \, \hat{\mathbf{b}} \, F^\alpha(\mathbf{x}, \mathbf{v}, t) - \langle \delta \mathbf{V}^E(\mathbf{x}, v_\perp, t) \, \delta f^\alpha(\mathbf{x}, \mathbf{v}, t) \rangle \right],$$

$$(5.9)$$

where $\nabla \equiv \partial / \partial \mathbf{x}$. Comparing this equation with the standard form (5.7), we find the expressions of the parallel and of the perpendicular particle fluxes as:

$$\Gamma_\parallel^\alpha(\mathbf{x}, t) = \int d\mathbf{v} \, v_\parallel \, F^\alpha(\mathbf{x}, \mathbf{v}, t), \qquad (5.10)$$

$$\Gamma_\perp^\alpha(\mathbf{x}, t) = \frac{c}{B} \int d\mathbf{v} \, \langle \delta \mathbf{V}^E(\mathbf{x}, t) \, \delta f^\alpha(\mathbf{x}, \mathbf{v}, t) \rangle$$

$$= \frac{c}{B} \int d\mathbf{v} \, \left\langle \left[\hat{\mathbf{b}} \times \nabla \overline{\delta \phi}(\mathbf{x}, v_\perp, t) \right] \delta f^\alpha(\mathbf{x}, \mathbf{v}, t) \right\rangle. \qquad (5.11)$$

If the Larmor radius is negligibly small (as for the electrons), Eq. (5.11) reduces to:

$$\Gamma_\perp^\alpha(\mathbf{x}, t) = \frac{c}{B} \left\langle \left[\hat{\mathbf{b}} \times \nabla \delta \phi(\mathbf{x}, t) \right] \delta n_\alpha(\mathbf{x}, t) \right\rangle, \quad (\hat{\rho}_{L\alpha} = 0). \qquad (5.12)$$

The first conclusion is that the parallel particle flux is purely deterministic: it has no contribution from the fluctuations. *The anomalous particle flux is thus entirely perpendicular to the magnetic field.* The asymmetry between parallel and perpendicular fluxes can be traced back to Eq. (4.34).

The velocity of the *guiding centre* has a deterministic parallel component $v_{\parallel}$ and a fluctuating perpendicular component $\delta \mathbf{V}^E$. This asymmetry is then transferred to the gyrokinetic equations (4.58) and (4.60).

Eq. (5.12) has been very well known and was derived by many authors, e.g., TANGE et al. 1979, HORTON 1985, LIEWER 1985. In the special geometry of the shearless slab, $\hat{\mathbf{b}} = \mathbf{e}_z$, (Fig. 2.1), the expression of the "radial" particle flux Γ_x^{α} is:

$$\Gamma_x^{\alpha}(\mathbf{x}, t) = \frac{c}{B} \left\langle \delta E_y(\mathbf{x}, t) \, \delta n_{\alpha}(\mathbf{x}, t) \right\rangle, \qquad (\hat{\rho}_{L\alpha} = 0). \qquad (5.13)$$

5.3 The Thermal Energy Balance

The starting point here is the balance equation for the *total thermal energy density,* $\mathcal{E}_{\alpha} \equiv \frac{3}{2} P_{\alpha}$, defined as the average kinetic energy of the particles (HINTON & HAZELTINE, 1976):

$$P_{\alpha} = P_{\alpha 0} + \delta P_{\alpha}, \qquad (5.14)$$

with:

$$\frac{3}{2} P_{\alpha 0}(\mathbf{x}, t) = \int d\mathbf{v} \, \frac{m_{\alpha}}{2} \, (v_{\parallel}^2 + v_{\perp}^2) \, F^{\alpha}(\mathbf{x}, \mathbf{v}, t), \qquad (5.15)$$

$$\frac{3}{2} \delta P_{\alpha}(\mathbf{x}, t) = \int d\mathbf{v} \int d\mathbf{Y} \, \delta[\mathbf{x} - \mathbf{q}(\mathbf{Y})] \, \frac{m_{\alpha}}{2} \, (v_{\parallel}^2 + v_{\perp}^2) \, \delta f^{\alpha}(\mathbf{Y}, \mathbf{v}, t). \qquad (5.16)$$

We recall that P_{α} is *not* the thermodynamic (or plasmadynamic) pressure p_{α}:

$$\frac{3}{2} p_{\alpha}(\mathbf{x}, t) = \frac{m_{\alpha}}{2} \int d\mathbf{v} \int d\mathbf{Y} \, \delta[\mathbf{x} - \mathbf{q}(\mathbf{Y})] \, |\mathbf{v} - \mathbf{u}^{\alpha}(\mathbf{x}, t)|^2 \, f^{\alpha}(\mathbf{Y}, \mathbf{v}, t)$$

$$= P_{\alpha} - \frac{1}{2} m_{\alpha} n_{\alpha} u^{\alpha \, 2}, \qquad (5.17)$$

where $\mathbf{u}^{\alpha}$ is the average velocity of species α ($\mathbf{u}^{\alpha} = n_{\alpha}^{-1} \Gamma^{\alpha}$). This question was discussed in Sec. 1.2. The last term in Eq. (5.17) is, however, usually very small and can be neglected. We shall therefore call P_{α} the "pressure", for simplicity. The internal energy balance equation is very easily obtained, as in the previous section, by multiplying Eq. (4.59) with $\frac{1}{2} m_{\alpha}(v_{\parallel}^2 + v_{\perp}^2)$ and integrating over the velocity:

$$\partial_t \tfrac{3}{2} P_{\alpha 0} = -\nabla \cdot \mathbf{e}_z \int d\mathbf{v} \int d\mathbf{Y} \, \delta[\mathbf{x} - \mathbf{q}(\mathbf{Y})] \, \frac{m_\alpha}{2} \left(v_\parallel^2 + v_\perp^2 \right) v_\parallel \, F^\alpha(\mathbf{Y}, \mathbf{v}, t)$$

$$-\nabla \cdot \int d\mathbf{v} \int d\mathbf{Y} \, \delta[\mathbf{x} - \mathbf{q}(\mathbf{Y})] \, \frac{m_\alpha}{2} \left(v_\parallel^2 + v_\perp^2 \right) \langle \delta \mathbf{V}^E \, \delta f^\alpha \rangle$$

$$-\int d\mathbf{v} \int d\mathbf{Y} \, \delta[\mathbf{x} - \mathbf{q}(\mathbf{Y})] \, \frac{m_\alpha}{2} \left(v_\parallel^2 + v_\perp^2 \right) \left\langle \delta a^\alpha \frac{\partial}{\partial v_\parallel} \delta f^\alpha \right\rangle. \qquad (5.18)$$

This equation has the standard form of the balance equation (1.17):

$$\partial_t \tfrac{3}{2} P_{\alpha 0} = -\nabla \cdot \mathbf{Q}^\alpha + S^\alpha. \qquad (5.19)$$

The presence of a source term shows that the thermal energy is not a conserved quantity: only the total energy, including the electromagnetic energy, is conserved (see Sec. 5.4). We now identify the components of the *energy flux vector* as follows, in a manner similar to the previous section:

$$Q_\parallel^\alpha(\mathbf{x}, t) = \int d\mathbf{v} \, \frac{m_\alpha}{2} \left(v_\parallel^2 + v_\perp^2 \right) v_\parallel \, F^\alpha(\mathbf{x}, \mathbf{v}, t), \qquad (5.20)$$

$$\mathbf{Q}_\perp^\alpha(\mathbf{x}, t) = \int d\mathbf{v} \, \frac{m_\alpha}{2} \left(v_\parallel^2 + v_\perp^2 \right) \langle \delta \mathbf{V}^E(\mathbf{x}, v_\perp, t) \, \delta f^\alpha(\mathbf{x}, \mathbf{v}, t) \rangle. \qquad (5.21)$$

If the Larmor radius can be neglected, this reduces to:

$$\mathbf{Q}_\perp^\alpha(\mathbf{x}, t) = \frac{3}{2} \frac{c}{B} \left\langle [\mathbf{e}_z \times \nabla \delta \phi] \, \delta P_\alpha(\mathbf{x}, t) \right\rangle, \qquad (\widehat{\rho}_{L\alpha} = 0). \qquad (5.22)$$

The first (classical) term represents the transport of kinetic energy with a velocity $v_\parallel$ along the magnetic field. The perpendicular component represents the transport of fluctuating internal energy $\tfrac{3}{2} \delta P_\alpha$ with the drift velocity $\delta \mathbf{V}^E$ across the magnetic field, averaged over the ensemble.

The *source term* is obtained after an integration by parts:

$$S^\alpha(\mathbf{x}, t) = \int d\mathbf{v} \int d\mathbf{Y} \, \delta[\mathbf{x} - \mathbf{q}(\mathbf{Y})] \, e_\alpha v_\parallel \left\langle \overline{\delta E_\parallel}(\mathbf{x}, v_\perp, t) \, \delta f^\alpha(\mathbf{Y}, \mathbf{v}, t) \right\rangle. \qquad (5.23)$$

Introducing the parallel component of the *partial electric current* carried by species α:

$$\delta j_\parallel^\alpha(\mathbf{x}, t) = \int d\mathbf{v} \int d\mathbf{Y} \, \delta[\mathbf{x} - \mathbf{q}(\mathbf{Y})] \, e_\alpha v_\parallel \, \delta f^\alpha(\mathbf{Y}, \mathbf{v}, t)$$

$$= \int d\mathbf{k} \, e^{i\mathbf{k} \cdot \mathbf{x}} \int d\mathbf{v} \, e_\alpha v_\parallel J_0(k_\perp \widehat{\rho}_{L\alpha}) \, \delta f_\mathbf{k}^\alpha(\mathbf{v}, t) \qquad (5.24)$$

the source term is rewritten as:

$$S^\alpha(\mathbf{x}, t) = \left\langle \overline{\delta E}_\parallel(\mathbf{x}, t)\, \delta j_\parallel^\alpha(\mathbf{x}, t) \right\rangle. \tag{5.25}$$

This source term also has a clear physical meaning: it represents the contribution of species α to the (positive or negative) ensemble-averaged dissipation of electrical energy into heat through the Joule effect [see Eq. (1.17)]. However, only the parallel electric field appears here after ensemble-gyro-averaging.

We note again, as in the case of the particle flux, that the parallel component of the energy flux is deterministic: *the anomalous energy flux is entirely perpendicular to the magnetic field.* In the standard shearless slab geometry, the "radial" energy flux is:

$$Q_x^\alpha(\mathbf{x}, t) = \frac{3}{2} \frac{c}{B} \left\langle \delta E_y(\mathbf{x}, t)\, \delta P_\alpha(\mathbf{x}, t) \right\rangle \qquad (\widehat{\rho}_{L\alpha} = 0). \tag{5.26}$$

Next, it should be stressed again that the energy flux should not be confused with the *heat flux* $\mathbf{q}^\alpha$, appearing in the macroscopic thermal energy balance, Eq. (1.16). The latter is defined as [BALESCU 1988 a: Eq. (3.2.20); HINTON & HAZELTINE 1976, Eq. (2.6); ROSS 1989, Eq. (6)]:

$$\mathbf{q}^\alpha(\mathbf{x}, t)$$
$$= \frac{m_\alpha}{2} \int d\mathbf{v} \int d\mathbf{Y}\, \delta[\mathbf{x} - \mathbf{q}(\mathbf{Y})]\, [\mathbf{v} - \mathbf{u}^\alpha(\mathbf{x}, t)]\, |\mathbf{v} - \mathbf{u}^\alpha(\mathbf{x}, t)|^2\, F^\alpha(\mathbf{Y}, \mathbf{v}, t) \tag{5.27}$$

Finally, it should be noted that the definition (5.22) of the energy flux is not unique. The derivation given here from the gyrokinetic equation yields a "natural" and well defined result. On the macroscopic level, however, Eq. (5.19) can be manipulated, as is well known in non-equilibrium thermodynamics (DE GROOT & MAZUR 1982). Whenever one is dealing with a non-conserved quantity, it is often possible to modify the source term by extracting a term in the form of a divergence, and add it to the flux term. In the particular problem under discussion, this has led to a certain amount of confusion in the past literature. In particular, a group of authors (among them the classical work of LIEWER 1985, and our own work BALESCU 1990) use a definition of the form (5.22) in which the factor $\frac{3}{2}$ is replaced by $\frac{5}{2}$. This puzzle was finally resolved by the work of ROSS 1989,1992. He showed (among other interesting relations) that any one of these definitions can be used consistently, provided the source term in the balance equation is correspondingly adapted. The "$\frac{3}{2} - \frac{5}{2}$ puzzle" is thus clarified; we do not further discuss it here, but refer the interested reader to the literature: DUCHS 1990, BALESCU 1990, STRINGER 1991, SMOLYAKOV & HIROSE 1993.

5.4   The Electrostatic Energy Balance

The total energy of the turbulent plasma includes, besides the average kinetic energy of the particles, the energy of the fluctuating electric field.[3] The total energy must be conserved under the gyrokinetic dynamics. We now check this important property by deriving a balance equation for the energy of the fluctuating electric field.

Our starting point is the gyrokinetic equation for the fluctuations, Eq. (4.83). We introduce the following compact notation for the Bessel functions:

$$J_{\mathbf{k}}^{\alpha}(\mathbf{v}) = J_0(|\Omega_\alpha|^{-1}k_\perp v_\perp). \tag{5.28}$$

We now multiply all terms of Eq. (4.83) by $(2\pi)^{-3}e_\alpha J_{\mathbf{k}''}^{\alpha}(\mathbf{v})\delta\phi_{\mathbf{k}''}(t)$, average over the ensemble, sum over α and integrate over $\mathbf{v}$, $\mathbf{k}$ and $\mathbf{k}_1$:

$$(2\pi)^{-3} \sum_\alpha e_\alpha \int d\mathbf{v} \int d\mathbf{k} \int d\mathbf{k}_1$$

$$\times \Big\{ J_{\mathbf{k}}^{\alpha}(\mathbf{v}) \big\langle [\partial_t \delta f_{\mathbf{k}}^{\alpha}(\mathbf{v},t)] \delta\phi_{\mathbf{k}_1}(t) \big\rangle + ik_\parallel v_\parallel J_{\mathbf{k}_1}^{\alpha}(\mathbf{v}) \big\langle \delta f_{\mathbf{k}}^{\alpha}(\mathbf{v},t)\, \delta\phi_{\mathbf{k}_1}(t) \big\rangle$$

$$-i\Big[\frac{c}{B}\left(\mathbf{k}\times\hat{\mathbf{b}}\right)\cdot\nabla F^\alpha(\mathbf{v},t) + \frac{e_\alpha}{m_\alpha}k_\parallel\,\partial_\parallel F^\alpha(\mathbf{v},t)\Big]J_{\mathbf{k}}^{\alpha}(\mathbf{v})J_{\mathbf{k}_1}^{\alpha}(\mathbf{v})\big\langle \delta\phi_{\mathbf{k}}(t)\delta\phi_{\mathbf{k}_1}(t)\big\rangle$$

$$+\int d\mathbf{k}'\,\frac{c}{B}\,i\left(\mathbf{k}'\times\hat{\mathbf{b}}\right)\cdot\, ik\, J_{\mathbf{k}'}^{\alpha}(\mathbf{v})J_{\mathbf{k}_1}^{\alpha}(\mathbf{v})\big\langle \delta\phi_{\mathbf{k}'}(t)\,\delta f_{\mathbf{k}-\mathbf{k}'}^{\alpha}(\mathbf{v},t)\delta\phi_{\mathbf{k}_1}(t)\big\rangle\Big\} = 0.$$

$$\tag{5.29}$$

We now recall a well-known property of Fourier transforms. Whenever the turbulent state is homogeneous, the average of products of fluctuations is proportional to a delta-function of the sum of the wave vectors.[4] In particular:

$$\left\{ \begin{array}{c} \langle \delta f_{\mathbf{k}}^{\alpha}\, \delta\phi_{\mathbf{k}_1}\rangle \\ \langle \delta\phi_{\mathbf{k}}\, \delta\phi_{\mathbf{k}_1}\rangle \end{array} \right\} = \delta(\mathbf{k}+\mathbf{k}_1)\left\{ \begin{array}{c} \langle \delta f_{\mathbf{k}}^{\alpha}\, \delta\phi_{-\mathbf{k}}\rangle \\ \langle \delta\phi_{\mathbf{k}}\, \delta\phi_{-\mathbf{k}}\rangle \end{array} \right\},$$

$$\langle \delta\phi_{\mathbf{k}'}\, \delta f_{\mathbf{k}-\mathbf{k}'}^{\alpha}\delta\phi_{\mathbf{k}_1}\rangle = \delta(\mathbf{k}+\mathbf{k}_1)\langle \delta\phi_{\mathbf{k}'}\, \delta f_{\mathbf{k}-\mathbf{k}'}^{\alpha}\delta\phi_{-\mathbf{k}}\rangle. \tag{5.30}$$

Performing the integration over $\mathbf{k}_1$ in Eq. (5.29), we are left with:

---

[3] There is no contribution of the magnetic field to the energy in the electrostatic turbulence considered here, where the magnetic field is considered to be deterministic: $\delta\mathbf{B} = 0$.

[4] A more detailed discussion of these properties will be given in the forthcoming Sec. 6.3.

$$\sum_{\alpha} e_{\alpha} \int dv \int dk \left\{ J_{\mathbf{k}}^{\alpha}(\mathbf{v}) \left\langle [\partial_t \delta f_{\mathbf{k}}^{\alpha}(\mathbf{v},t)] \, \delta\phi_{-\mathbf{k}}(t) \right\rangle \right.$$

$$+ ik_{\parallel} v_{\parallel} J_{\mathbf{k}}^{\alpha}(\mathbf{v}) \left\langle \delta f_{\mathbf{k}}^{\alpha}(\mathbf{v},t) \, \delta\phi_{-\mathbf{k}}(t) \right\rangle$$

$$-i \left[\frac{c}{B} \left(\mathbf{k} \times \hat{\mathbf{b}} \right) \cdot \boldsymbol{\nabla} F^{\alpha}(\mathbf{v},t) + \frac{e_{\alpha}}{m_{\alpha}} k_{\parallel} \, \partial_{\parallel} F^{\alpha}(\mathbf{v},t) \right] [J_{\mathbf{k}}^{\alpha}(\mathbf{v})]^2 \left\langle \delta\phi_{\mathbf{k}}(t) \delta\phi_{-\mathbf{k}}(t) \right\rangle$$

$$+ \int d\mathbf{k}' \, \frac{c}{B} i \left(\mathbf{k}' \times \hat{\mathbf{b}} \right) \cdot i\mathbf{k} \, J_{\mathbf{k}}^{\alpha}(\mathbf{v}) J_{\mathbf{k}'}^{\alpha}(\mathbf{v})$$

$$\left. \times \left\langle \delta\phi_{\mathbf{k}'}(t) \, \delta f_{\mathbf{k}-\mathbf{k}'}^{\alpha}(\mathbf{v},t) \delta\phi_{-\mathbf{k}}(t) \right\rangle \right\} = 0. \tag{5.31}$$

The third term in this equation is clearly zero, because its integrand is an odd function of $\mathbf{k}$:

$$\sum_{\alpha} e_{\alpha} \int dk \left[\frac{c}{B} \left(\mathbf{k} \times \hat{\mathbf{b}} \right) \cdot \boldsymbol{\nabla} F^{\alpha}(\mathbf{v}) + \frac{e_{\alpha}}{m_{\alpha}} k_{\parallel} \, \partial_{\parallel} F^{\alpha}(\mathbf{v}) \right]$$

$$\times [J_{\mathbf{k}}^{\alpha}(\mathbf{v})]^2 \left\langle \delta\phi_{\mathbf{k}}(t) \delta\phi_{-\mathbf{k}}(t) \right\rangle = 0. \tag{5.32}$$

We then consider the nonlinear term which contains the following double integral:

$$K = \int d\mathbf{k} \int d\mathbf{k}' \left(\mathbf{k}' \times \hat{\mathbf{b}} \right) \cdot \mathbf{k} \, J_{\mathbf{k}'}^{\alpha}(\mathbf{v}) J_{\mathbf{k}}^{\alpha}(\mathbf{v}) \left\langle \delta\phi_{\mathbf{k}'}(t) \, \delta f_{\mathbf{k}-\mathbf{k}'}^{\alpha}(\mathbf{v},t) \delta\phi_{-\mathbf{k}}(t) \right\rangle$$

$$= \int d\mathbf{k} \int d\mathbf{k}' \left(\mathbf{k} \times \hat{\mathbf{b}} \right) \cdot \mathbf{k}' \, J_{\mathbf{k}'}^{\alpha}(\mathbf{v}) J_{\mathbf{k}}^{\alpha}(\mathbf{v}) \left\langle \delta\phi_{\mathbf{k}}(t) \, \delta f_{\mathbf{k}'-\mathbf{k}}^{\alpha}(\mathbf{v},t) \delta\phi_{-\mathbf{k}'}(t) \right\rangle$$

$$= \int d\mathbf{k} \int d\mathbf{k}' \left(\mathbf{k} \times \hat{\mathbf{b}} \right) \cdot \mathbf{k}' \, J_{\mathbf{k}'}^{\alpha}(\mathbf{v}) J_{\mathbf{k}}^{\alpha}(\mathbf{v}) \left\langle \delta\phi_{-\mathbf{k}}(t) \, \delta f_{-\mathbf{k}'+\mathbf{k}}^{\alpha}(\mathbf{v},t)) \delta\phi_{\mathbf{k}'}(t) \right\rangle$$

$$= -K. \tag{5.33}$$

In the first step we interchanged the dummy variables $\mathbf{k}$ and $\mathbf{k}'$, and in the second step we made the substitution $\mathbf{k} \to -\mathbf{k}$, $\mathbf{k}' \to -\mathbf{k}'$. From the scalar product of the wave vectors it clearly appears that $K = -K = 0$. We are thus left with the following property, following from the *nonlinear* GKE (4.83):

$$0 = \sum_{\alpha} e_{\alpha} (2\pi)^{-3} \int dv \int dk \, J_{\mathbf{k}}^{\alpha}(\mathbf{v}) \left\{ \left\langle [\partial_t \delta f_{\mathbf{k}}^{\alpha}(\mathbf{v},t)] \, \delta\phi_{-\mathbf{k}}(t) \right\rangle \right.$$

$$\left. + ik_{\parallel} v_{\parallel} \left\langle \delta f_{\mathbf{k}}^{\alpha}(\mathbf{v},t) \, \delta\phi_{-\mathbf{k}}(t) \right\rangle \right\}. \tag{5.34}$$

A few simple transformations will bring out a very transparent physical meaning. We recall Eqs. (5.6) and (5.24), from which we derive the expressions of the electric charge density and of the parallel electric current density fluctuations (in Fourier representation):

$$\delta\sigma_{\mathbf{k}}(t) = \sum_\alpha e_\alpha \delta n_{\mathbf{k}}^\alpha(t) = \sum_\alpha e_\alpha \int d\mathbf{v}\, J_{\mathbf{k}}^\alpha(\mathbf{v})\, \delta f_{\mathbf{k}}^\alpha(\mathbf{v}, t), \qquad (5.35)$$

$$\delta j_{\|\mathbf{k}}(t) = \sum_\alpha e_\alpha\, \delta\Gamma_{\|\mathbf{k}}^\alpha(t) = \sum_\alpha e_\alpha \int d\mathbf{v}\, v_\|\, J_{\mathbf{k}}^\alpha(\mathbf{v})\, \delta f_{\mathbf{k}}^\alpha(\mathbf{v}, t). \qquad (5.36)$$

Noting also that the electric field fluctuation is given by:

$$\delta\mathbf{E}_{\mathbf{k}}(t) = -i\mathbf{k}\,\delta\phi_{\mathbf{k}}(t),$$

we rewrite Eq. (5.34) in the form:

$$\sum_\alpha e_\alpha (2\pi)^{-3} \int d\mathbf{k}\, \left\{ \langle [\partial_t \delta\sigma_{\mathbf{k}}(t)]\, \delta\phi_{-\mathbf{k}}(t)\rangle + \langle \delta j_{\|\mathbf{k}}(t)\, \delta E_{\|-\mathbf{k}}(t)\rangle \right\} = 0.$$

Using also the Poisson equation $\delta\sigma_{\mathbf{k}}(t) = \left(k^2/4\pi\right)\delta\phi_{\mathbf{k}}(t)$, we obtain:

$$(2\pi)^{-3} \int d\mathbf{k}\, \left[(4\pi)^{-1} \tfrac{1}{2}\partial_t\, \langle \delta\mathbf{E}_{\mathbf{k}}(t)\cdot \delta\mathbf{E}_{-\mathbf{k}}(t)\rangle + \langle \delta j_{\|\mathbf{k}}(t)\, \delta E_{\|-\mathbf{k}}(t)\rangle \right] = 0.$$
$$(5.37)$$

This equation is now expressed in space-time coordinates by inverse Fourier transformation. We therefore recall some useful results about Fourier transforms.

By the convolution theorem we have, for arbitrary functions $\delta A(\mathbf{x})$, $\delta B(\mathbf{x})$:

$$\langle \delta A(\mathbf{x})\, \delta B(\mathbf{x})\rangle = \int d\mathbf{k}\, e^{i\mathbf{k}\cdot\mathbf{x}} \int d\mathbf{k}_1\, \langle \delta A_{\mathbf{k}_1}\, \delta B_{\mathbf{k}-\mathbf{k}_1}\rangle. \qquad (5.38)$$

We now derive an expression for the volume integral of such expressions:

$$\int d\mathbf{x} \int d\mathbf{k}\, e^{i\mathbf{k}\cdot\mathbf{x}} \int d\mathbf{k}_1\, \langle \delta A_{\mathbf{k}_1}\, \delta B_{\mathbf{k}-\mathbf{k}_1}\rangle$$
$$= \int d\mathbf{k}\, (2\pi)^3 \delta(\mathbf{k}) \int d\mathbf{k}_1\, \langle \delta A_{\mathbf{k}_1}\, \delta B_{\mathbf{k}-\mathbf{k}_1}\rangle.$$

Thus, Eq. (5.38) reduces to:

$$\int d\mathbf{x}\, \langle \delta A(\mathbf{x})\, \delta B(\mathbf{x})\rangle = (2\pi)^3 \int d\mathbf{k}\, \langle \delta A_{\mathbf{k}}\, \delta B_{-\mathbf{k}}\rangle. \qquad (5.39)$$

Eq. (5.37) can therefore be rewritten in the form:

$$\frac{d}{dt} \int d\mathbf{x} \, \frac{\langle \delta E^2(\mathbf{x}, t) \rangle}{8\pi} = - \int d\mathbf{x} \, \langle \delta j_\parallel(\mathbf{x}, t) \, \delta E_\parallel(\mathbf{x}, t) \rangle. \tag{5.40}$$

This global balance equation can be written in local form as:

$$\partial_t \, (8\pi)^{-1} \, \langle \delta E^2(\mathbf{x}, t) \rangle = - \, \langle \delta j_\parallel(\mathbf{x}, t) \, \delta E_\parallel(\mathbf{x}, t) \rangle. \tag{5.41}$$

This is nothing other than the *electrostatic fluctuation energy balance* for the turbulent plasma described by the nonlinear stochastic GKE (4.83). We may, indeed, identify $(8\pi)^{-1} \langle \delta E^2 \rangle$ with the average electrostatic energy density contained in the fluctuations, according to the well known Maxwell definition. Comparing this equation with Eqs. (5.19) and (5.25) we note that the source term in the internal energy balance equation precisely equals minus the source term of the electrostatic energy balance equation. This finalizes the proof of the *total energy conservation*:

$$\partial_t \left[\frac{3}{2} \sum_\alpha P_{\alpha 0}(\mathbf{x}; t) + \frac{1}{8\pi} \, \langle \delta E^2(\mathbf{x}, t) \rangle \right] = -\boldsymbol{\nabla} \cdot \sum_\alpha \mathbf{Q}^\alpha(\mathbf{x}; t). \tag{5.42}$$

Let us stress an important feature of this energy conservation equation. *The nonlinear term in the GKE (4.83) contributes zero to the energy balance.*

Chapter 6

Anomalous Fluxes and Correlations

6.1 Kinetic Equation and Langevin Equations

In the present chapter we obtain an exact, though formal solution of the kinetic equation for the fluctuations. This will lead to a closed average kinetic equation and to general formulae for the anomalous fluxes.

The starting point will be the couple of basic equations (4.59) and (4.60). We recall our simplifying assumptions about the geometry of the plasmas considered here. The magnetic field B is homogeneous and stationary. A Cartesian reference frame is used, whose z-axis is oriented along the magnetic field: $\mathbf{e}_z = \hat{\mathbf{b}}$. According to the degree of explicitness desired, we adopt one of the folowing notations for the coordinates of the guiding centre:

$$\mathbf{Y} = (\mathbf{Y}_\perp, Z) = (X, Y, Z). \tag{6.1}$$

The distinction between particle coordinates and guiding centre coordinates must always be kept in mind (especially for the ions). The average distribution function depends on the coordinates of the guiding centre, on the parallel velocity $v_\parallel$, on the perpendicular speed $v_\perp$, and on time: $F^\alpha = F^\alpha(\mathbf{Y}, v_\perp, v_\parallel, t) \equiv F^\alpha(\mathbf{Y}, \mathbf{v}, t)$. The fluctuating potential depends on the guiding centre coordinates and on time, but not on the velocity. For conciseness, we sometimes use the following abbreviation for the "fluctuating evolution operator":

$$\delta\mathcal{K}^\alpha(\mathbf{Y};t) = \delta\mathbf{V}^E(\mathbf{Y}, t) \cdot \nabla_\perp + \delta a^\alpha(\mathbf{Y}, t)\, \partial_\parallel, \tag{6.2}$$

where $\partial_\parallel = \partial/\partial v_\parallel$, and the quantities $\delta\mathbf{V}^E$ and δa^α were defined in Eqs. (4.56) and (4.57) in terms of the gyro-averaged potential $\overline{\delta\phi}$.

With these assumptions and notations, the kinetic equation for the average distribution function (4.58) becomes:

$$\left(\partial_t + v_\parallel \nabla_Z\right) F^\alpha(\mathbf{Y}, \mathbf{v}, t) = -\left\langle \delta \mathcal{K}^\alpha(\mathbf{Y}, t)\, \delta f^\alpha(\mathbf{Y}, \mathbf{v}, t)\right\rangle. \qquad (6.3)$$

The equation for the fluctuation is written as:

$$\partial_t \delta f^\alpha(\mathbf{Y}, \mathbf{v}, t) = -v_\parallel \nabla_Z\, \delta f^\alpha(\mathbf{Y}, \mathbf{v}; t)$$
$$-\left\{\delta \mathbf{V}^E(\mathbf{Y}, t) \cdot \nabla_\perp\, \delta f^\alpha(\mathbf{Y}, \mathbf{v}, t) - \left\langle \delta \mathbf{V}^E(\mathbf{Y}, t) \cdot \nabla_\perp \delta f^\alpha(\mathbf{Y}, \mathbf{v}, t)\right\rangle\right\}$$
$$-\delta \mathcal{K}^\alpha(\mathbf{Y}, t)\, F^\alpha(\mathbf{Y}, \mathbf{v}, t). \qquad (6.4)$$

We now adopt the following strategy. In a first step *we treat the potential fluctuation $\delta\phi$ as a given function.* Eq. (6.4) then appears as a *linear, inhomogeneous, first order partial differential equation for δf^α*. This equation is solved by using standard methods. Its solution is then substituted into Eq. (6.3), which becomes a closed kinetic equation for F^α. The latter, in turn, is used for the calculation of the anomalous fluxes. These fluxes are determined by the average distribution function F^α, but also by the potential fluctuation $\delta\phi$. The latter, in turn, depends on the particle distribution. Hence, in a second step, $\delta\phi$ must be eliminated. In a fully self-consistent theory, one uses the Poisson equation to determine $\delta\phi$ as a function of δf^α, hence of F^α. This usually leads to difficult calculations. The latter may sometimes be short-circuited by making reasonable intuitive assumptions about the statistical nature of the potential fluctuation. In the present chapter we only deal with the first step described above.

We follow here essentially a method used (for a simpler problem) in BALESCU 2000. Eq. (6.4) is rewritten in the form:

$$\left[\partial_t + v_\parallel \nabla_Z + \delta \mathbf{V}^E(\mathbf{Y}, t) \cdot \nabla_\perp\right] \delta f^\alpha(\mathbf{Y}, \mathbf{v}, t)$$
$$= S^\alpha(\mathbf{Y}, \mathbf{v}, t) + C^\alpha(\mathbf{Y}, \mathbf{v}, t). \qquad (6.5)$$

The source terms on the right hand side are:

$$S^\alpha(\mathbf{Y}, \mathbf{v}, t) = -\delta \mathcal{K}^\alpha(\mathbf{Y}, t)\, F^\alpha(\mathbf{Y}, \mathbf{v}, t), \qquad (6.6)$$

$$C^\alpha(\mathbf{Y}, \mathbf{v}, t) = \left\langle \delta \mathcal{K}^\alpha(\mathbf{Y}, t)\, \delta f^\alpha(\mathbf{Y}, \mathbf{v}, t)\right\rangle. \qquad (6.7)$$

(We wrote down explicitly the arguments of all functions, for clarity in the forthcoming calculations). It is well-known that the solution of a first order partial differential equation is expressed in terms of the solution of the associated set of ordinary differential equations defining the characteristic curves (see, *e.g.*, JONES 1982; we use here the notations of VANDEN EIJNDEN 1997). In the case of Eq. (6.5), the characteristic equations are:

$$\frac{d\mathbf{Y}_\perp(t|\mathbf{Y}',t')}{dt} = \delta\mathbf{V}^E[\mathbf{Y}(t|\mathbf{Y}',t'),t], \tag{6.8}$$

$$\frac{dZ(t|Z',t')}{dt} = v_\parallel, \tag{6.9}$$

$$\frac{d\mathbf{v}_\perp(t)}{dt} = \mathbf{0}, \tag{6.10}$$

$$\frac{dv_\parallel(t)}{dt} = 0. \tag{6.11}$$

These equations are to be integrated with the initial condition, at $t = t'$:

$$\mathbf{Y}_\perp(t'|\mathbf{Y}',t') = \mathbf{Y}'_\perp, \ \ Z(t'|Z',t') = Z'. \tag{6.12}$$

Thus, the notation $\mathbf{Y}(t|\mathbf{Y}',t')$ denotes the position of a guiding centre at time t, given that it was at point $\mathbf{Y}'$ at time t', hence it exhibits explicitly the information about the initial condition. This rather heavy notation is useful for a clear understanding of the forthcoming rather subtle relations; it will be simplified later on.

The following comments are useful for a correct understanding of these equations. The drift velocity $\delta\mathbf{V}^E = (c/B)\,[\hat{\mathbf{b}} \times \overline{\delta\phi}]$ depends both on $\mathbf{Y}$ and on $v_\perp$. The latter dependence is introduced by the **gyroaveraging** operation, see Eq. (4.76). By Eq. (6.10), the perpendicular velocity is constant in time, hence $v_\perp$ enters Eq. (6.8) just as a parameter of $\delta\mathbf{V}^E$, and will therefore not be written down explicitly.

The parallel velocity $v_\parallel$ is constant in the present approximation and is therefore also a mere parameter. This important simplification is due to the neglect of the parallel nonlinearity in Eq. (4.60). If the latter contribution were retained, Eq. (6.11) would be replaced by $dv_\parallel(t)/dt = \delta a^\alpha(\mathbf{Y}_\perp, Z, t)$. Instead of a constant, $v_\parallel$ would be a fluctuating quantity, and we would have a tightly coupled system of three equations for $\mathbf{Y}_\perp, Z, v_\parallel$, a much more complicated problem. Eq. (6.9) is trivially solved:

$$Z(t|Z',t') = Z' + v_\parallel(t - t'). \tag{6.13}$$

The difficulty of the problem is concentrated in Eq. (6.8): it determines the motion of a guiding centre *in a given realization of the turbulent field*. The right hand sides of these equations are, indeed, fluctuating quantities, thus Eq. (6.8) must be interpreted as a *stochastic equation*. Its solutions are therefore *random functions*, and have to be treated as such in the forthcoming developments. By analogy with the well-known theory of Brownian motion, we call Eqs. (6.8) - (6.11) the *Langevin equations* associated with Eq. (6.5).

6.2　The Non-Markovian Kinetic Equation

The solution of the initial value problem for the partial differential equation (6.5) can be expressed by means of a *propagator* in terms of the solution of the characteristic Langevin equations (even though this solution is not known explicitly). The propagator $g(\mathbf{Y}, \mathbf{v}, t|\mathbf{Y}', \mathbf{v}', t')$ is defined as follows:[1]

$$g(\mathbf{Y}, \mathbf{v}, t|\mathbf{Y}', \mathbf{v}', t') = \mathcal{G}(\mathbf{Y}, t|\mathbf{Y}', t')\,\delta(\mathbf{v} - \mathbf{v}'). \qquad (6.14)$$

We will be mainly interested in the non-trivial spatial part of this function:

$$\mathcal{G}(\mathbf{Y}, t|\mathbf{Y}', t') = \delta\left[\mathbf{Y} - \mathbf{Y}(t|\mathbf{Y}', t')\right]. \qquad (6.15)$$

We recall that $\mathcal{G}(\mathbf{Y}, t|\mathbf{Y}', t')$ also depends parametrically on $\mathbf{v}$; this variable is not written down explicitly, but must be kept in mind.

It is easily shown that, as a result of Eqs. (6.8), - (6.12), the propagator is a solution of the following equation:

$$\left[\partial_t + v_\parallel \nabla_Z + \delta \mathbf{V}^E(\mathbf{Y}, t) \cdot \nabla_\perp\right] \mathcal{G}(\mathbf{Y}, t|\mathbf{Y}', t') = 0, \qquad (6.16)$$

with the initial condition:

$$\mathcal{G}(\mathbf{Y}, t'|\mathbf{Y}', t') = \delta(\mathbf{Y} - \mathbf{Y}'). \qquad (6.17)$$

The solution of the initial value problem for Eq. (6.5) is then:

$$
\begin{aligned}
\delta f^\alpha(\mathbf{Y}, \mathbf{v}, t) \\
= \int_0^t dt' \int d\mathbf{Y}' \mathcal{G}(\mathbf{Y}, t|\mathbf{Y}', t') S^\alpha(\mathbf{Y}', \mathbf{v}, t') \\
+ \int_0^t dt' \int d\mathbf{Y}' \mathcal{G}(\mathbf{Y}, t|\mathbf{Y}', t') C^\alpha(\mathbf{Y}', \mathbf{v}, t') \\
+ \int d\mathbf{Y}' \mathcal{G}(\mathbf{Y}, t|\mathbf{Y}', 0)\, \delta f^\alpha(\mathbf{Y}', 0).
\end{aligned} \qquad (6.18)
$$

(It is easily checked, by direct calculation, that this quantity is indeed the solution of Eq. (6.5), *whatever the sign of t.*)

The integral over $\mathbf{Y}'$ is not easily evaluated explicitly in the general (nonlinear) case. Consider the first term in the right hand side of Eq. (6.18): it involves an integral of the following form, where $\Sigma(\mathbf{Y}, t)$ is an arbitrary function:

---

[1] It should be clear that the notation $\delta(\mathbf{y})$ denotes a three-dimensional Dirac delta function, *i.e.*, the product: $\delta(x)\delta(y)\delta(z)$. In the cylindrical coordinates used for the velocity we have: $\delta(\mathbf{v}) = (2\pi)^{-1} v_\perp^{-1} \delta(v_\perp)\,\delta(v_\parallel)$.

$$J(\mathbf{Y}, t, t') \equiv \int d\mathbf{Y}' \delta \left[\mathbf{Y} - \mathbf{Y}(t|\mathbf{Y}', t')\right] \Sigma(\mathbf{Y}', t'), \qquad (6.19)$$

In order to use the delta-function, we must solve the following equation for $\mathbf{Y}'$:

$$\mathbf{Y} - \mathbf{Y}(t|\mathbf{Y}', t') = 0. \qquad (6.20)$$

This step requires the knowledge of the solution of the associated Langevin equation (6.8), a difficult problem. Let this solution be written as:

$$\mathbf{Y}' = \mathbf{f}(\mathbf{Y}, t, t'). \qquad (6.21)$$

Although the solution may not be known explicitly, its physical meaning is clear. $\mathbf{f}(\mathbf{Y}, t, t')$ represents *the position at time t' of a point (guiding centre), given that it reaches at time t the position $\mathbf{Y}$, by moving along its orbit.*[2] Using our system of notations, we identify:

$$\mathbf{f}(\mathbf{Y}, t, t') = \mathbf{Y}(t'|\mathbf{Y}, t). \qquad (6.22)$$

(We stress again the fact that $\mathbf{Y}(t'|\mathbf{Y}, t)$ depends parametrically on $\mathbf{v}$.) Hence, we obtain the following important relation:

$$\int d\mathbf{Y}' \, \mathcal{G}(\mathbf{Y}, t|\mathbf{Y}', t')\Sigma(\mathbf{Y}', t') = \Sigma[\mathbf{Y}(t'|\mathbf{Y}, t), t']. \qquad (6.23)$$

Although this form does not help us in its explicit evaluation, it illuminates the physical meaning of this integral, as well as of the forthcoming results.

In the next step, the solution for the fluctuations (6.18) is substituted into Eq. (6.3). Before writing down the result, we note that the second term in the right hand side of (6.18) contributes zero to Eq. (6.3). We see, indeed, from Eq. (6.7) that it involves only averages of products of fluctuations, of the type $\langle \delta \mathcal{K}^\alpha \, \delta f^\alpha \rangle$; when such an expression is substituted into (6.3) it produces a contribution of the form $\langle \delta \mathcal{K}^\alpha \, \langle \delta \mathcal{K}^\alpha \, \delta f^\alpha \rangle \rangle = \langle \delta \mathcal{K}^\alpha \rangle \langle \delta \mathcal{K}^\alpha \, \delta f^\alpha \rangle = 0$.

Substituting now Eq. (6.18) into (6.3) and using the definition (6.6) we find:

$$(\partial_t + v_\parallel \, \nabla_z) \, F^\alpha(\mathbf{Y}, \mathbf{v}; t)$$

$$= \int d\mathbf{Y}' \left\{ \int_0^t dt' \, \langle \delta \mathcal{K}^\alpha(\mathbf{Y}, t) \, \mathcal{G}(\mathbf{Y}, t|\mathbf{Y}', t') \, \delta \mathcal{K}^\alpha(\mathbf{Y}', t') F^\alpha(\mathbf{Y}', \mathbf{v}, t') \rangle \right.$$

$$\left. - \langle \delta \mathcal{K}^\alpha(\mathbf{Y}, t) \, \mathcal{G}(\mathbf{Y}, t|\mathbf{Y}', 0) \, \delta f^\alpha(\mathbf{Y}', \mathbf{v}, 0) \rangle \right\}. \qquad (6.24)$$

---

[2] Note that t' may be either smaller or larger than t $(t' \lessgtr t)$.

At this point appears a serious difficulty. Using Eqs. (6.19), (6.23), the first term in the right hand side is:

$$\int d\mathbf{Y}' \int_0^t dt' \ \langle \delta \mathcal{K}^\alpha(\mathbf{Y},t)\, \mathcal{G}(\mathbf{Y},t|\mathbf{Y}',t')\, \delta\mathcal{K}^\alpha(\mathbf{Y}',t') F^\alpha(\mathbf{Y}',\mathbf{v},t')\rangle$$

$$= \int_0^t dt' \ \langle \delta\mathcal{K}^\alpha(\mathbf{Y},t)\, \delta\mathcal{K}^\alpha[\mathbf{Y}(t'|\mathbf{Y},t),t']\, F^\alpha[\mathbf{Y}(t'|\mathbf{Y},t),\mathbf{v},t']\rangle. \qquad (6.25)$$

Thus, the function F^α is evaluated at the point $\mathbf{Y}(t'|\mathbf{Y},t)$; but this is a *random function* obeying the stochastic Langevin equation (6.8). It involves the fluctuating velocity $\delta\mathbf{V}^E$, (hence $\overline{\delta\phi}$) in a highly nonlinear manner. As a result, the function F^α cannot be taken out of the average, and Eq. (6.24) represents a very complicated operator acting on F^α. This fact exhibits the intrinsic complexity of the turbulence process.

The situation can be more clearly understood by formally expanding the distribution function around its value at the given point $\mathbf{Y} = \mathbf{Y}(t|\mathbf{Y},t)$:

$$F^\alpha[\mathbf{Y}(t'|\mathbf{Y},t),\mathbf{v},t'] = F^\alpha(\mathbf{Y},\mathbf{v},t') + [\mathbf{Y}(t'|\mathbf{Y},t) - \mathbf{Y}] \cdot \nabla F^\alpha(\mathbf{Y},\mathbf{v},t') + \dots$$
$$(6.26)$$

The exact expression (6.25) thus appears formally as a differential operator of infinite order acting on $F^\alpha(\mathbf{Y},\mathbf{v},t')$.

An important simplification appears in the (realistic) case when the characteristic gradient length of the macroscopic quantities, $L_M \approx |d\ln F^\alpha /d\mathbf{Y}|^{-1}$ is much longer than the correlation length λ_c of the fluctuations:[3] $L_M \gg \lambda_c$. In that case (*localization*) we may retain, to dominant order, only the first term in the expansion (6.26):

$$F^\alpha[\mathbf{Y}(t'|\mathbf{Y},t),\mathbf{v},t'] = F^\alpha(\mathbf{Y},\mathbf{v},t') + O(\lambda_c/L_M). \qquad (6.27)$$

The function F^α in the right hand side of Eq. (6.24) is then aproximated by a *deterministic* function of a deterministic variable, which can be taken out of the average.

$$(\partial_t + v_\| \, \nabla_Z)\, F^\alpha(\mathbf{Y},\mathbf{v};t)$$

$$= \int d\mathbf{Y}' \left\{ \int_0^t dt' \ \langle \delta\mathcal{K}^\alpha(\mathbf{Y},t)\, \mathcal{G}(\mathbf{Y},t|\mathbf{Y}',t')\, \delta\mathcal{K}^\alpha(\mathbf{Y}',t')\rangle \ F^\alpha(\mathbf{Y},\mathbf{v},t') \right.$$

$$\left. - \langle \delta\mathcal{K}^\alpha(\mathbf{Y},t)\, \mathcal{G}(\mathbf{Y},t|\mathbf{Y}',0)\, \delta f^\alpha(\mathbf{Y}',\mathbf{v},0)\rangle \right\}. \qquad (6.28)$$

---

[3] It suffices at this point to think qualitatively of the correlation length as the range over which the correlation function of the fluctuations is effectively non-zero. A more precise definition will be given in Sec. 6.3.

This is the final form of the **kinetic equation for the average distribution function**. This form is quite familiar from non-equilibrium statistical mechanics (BALESCU 1975, 1997): it is a closed equation expressing the rate of change of the distribution function as resulting from two contributions. Its most striking property is its *non-Markovian character*: the rate of change of the distribution function at time t depends on all its values taken at previous times $0 \leq t' \leq t$. The second term in the right hand side, called the *"destruction term"* in statistical mechanics, is a source term depending on the initial value of the fluctuations. The first derivation of a non-Markovian equation for the average distribution function of a plasma in a fluctuating electromagnetic field (starting from the Vlasov, rather than the gyrokinetic equation) appears to be due to GUREVICH et al. 1983, 1987. A similar, more recent work is ZAGORODNY & WEILAND 1999.

The turbulent plasma thus displays a *finite memory effect*. In many cases it may appear that the memory range is sufficiently short to justify a Markovian approximation. But we shall keep the equation in its non-Markovian form as long as possible.

Eq. (6.28) can be rewritten in the following explicit form:

$$(\partial_t + v_\parallel \nabla_Z) F^\alpha(\mathbf{Y}, \mathbf{v}, t)$$

$$= \int_0^t dt' \; \{ \nabla_j L_{jn}^{VV}(\mathbf{Y}, t|t') \nabla_n \; + \; \nabla_j L_j^{Va}(\mathbf{Y}, t|t') \partial_\parallel$$

$$+ \; \partial_\parallel L_n^{aV}(\mathbf{Y}, t|t') \nabla_n \; + \; \partial_\parallel L^{aa}(\mathbf{Y}, t|t') \partial_\parallel \} \; F^\alpha(\mathbf{Y}, \mathbf{v}, t') - \Delta^\alpha(\mathbf{Y}, t|0)$$

$$(6.29)$$

with the *memory kernels*:

$$L_{jn}^{VV}(\mathbf{Y}, t|t') = \int d\mathbf{Y}' \; \langle \, \delta V_j^E(\mathbf{Y}, t) \, \mathcal{G}(\mathbf{Y}, t|\mathbf{Y}', t') \, \delta V_n^E(\mathbf{Y}', t') \, \rangle, \quad (6.30)$$

$$L_j^{Va}(\mathbf{Y}, t|t') = \int d\mathbf{Y}' \; \langle \, \delta V_j^E(\mathbf{Y}, t) \, \mathcal{G}(\mathbf{Y}, t|\mathbf{Y}', t') \, \delta a^\alpha(\mathbf{Y}', t') \, \rangle, \quad (6.31)$$

$$L_n^{aV}(\mathbf{Y}, t|t') = \int d\mathbf{Y}' \; \langle \, \delta a^\alpha(\mathbf{Y}, t) \, \mathcal{G}(\mathbf{Y}, t|\mathbf{Y}', t') \, \delta V_n^E(\mathbf{Y}', t') \, \rangle, \quad (6.32)$$

$$L^{aa}(\mathbf{Y}, t|t') = \int d\mathbf{Y}' \; \langle \, \delta a^\alpha(\mathbf{Y}, t) \, \mathcal{G}(\mathbf{Y}, t|\mathbf{Y}', t') \, \delta a^\alpha(\mathbf{Y}', t') \, \rangle. \quad (6.33)$$

These expressions can also be written in the following alternative form:

$$L_{jn}^{VV}(\mathbf{Y}, t|t') = \big\langle\, \delta V_j^E(\mathbf{Y}, t)\, \delta V_n^E[\mathbf{Y}(t'|\mathbf{Y}, t), t')\,\big\rangle, \qquad (6.34)$$

and similar expression for the other coefficients. The memory kernels are thus *Lagrangian correlation tensors* of various kinds, to be defined and further discussed in Secs. 6.4 and 6.5.

The last term in Eq. (6.2) is written as:

$$\Delta^\alpha(\mathbf{Y}, t|0) = -\int d\mathbf{Y}'\, \langle \delta\mathcal{K}^\alpha(\mathbf{Y}, t)\, \mathcal{G}(\mathbf{Y}, t|\mathbf{Y}', 0)\, \delta f^\alpha(\mathbf{Y}', \mathbf{v}, 0)\rangle. \qquad (6.35)$$

This "destruction term" in the kinetic equation has a very peculiar status. Clearly, we have no information about the fluctuation of the distribution function at time 0. We cannot even make reasonable assumptions about its "value".[4] In the analogous problem of nonequilibrium statistical mechanics it is shown that the destruction term tends to zero after a relatively short time (the duration of a collision). It may therefore be reasonable to assume that in the present case, whenever the initial state is not too "exotic", the destruction term tends asymptotically to zero and may be neglected after a time of the order of the correlation time (see Sec. 6.3). In practice, this approximation is implemented by setting simply $\delta f^\alpha(\mathbf{Y}, \mathbf{v}; 0) \equiv 0$. The resulting simplified kinetic equation then reduces to:

$$(\partial_t + v_\parallel \nabla_Z)\, F^\alpha(\mathbf{Y}, \mathbf{v}, t) = \int_0^t dt'\, L^{KK}(\mathbf{Y}, t|t')\, F^\alpha(\mathbf{Y}, \mathbf{v}, t'), \qquad (6.36)$$

where the operator L^{KK} is defined as:

$$L^{KK}(\mathbf{Y}, t|t') = \int d\mathbf{Y}'\, \big\langle\, \delta\mathcal{K}^\alpha(\mathbf{Y}, t)\, \mathcal{G}(\mathbf{Y}, t|\mathbf{Y}', t')\, \delta\mathcal{K}^\alpha(\mathbf{Y}', t')\,\big\rangle. \qquad (6.37)$$

6.3 Eulerian Correlations

In most of the previous expressions we see the appearance of averages of products of fluctuating quantities. An average of a product of n such functions is called a *moment of order n* of the probability distribution associated with the turbulent ensemble. In particular, the average of a product of two fluctuating quantities is called a *correlation function* (or, briefly, *correlation*). Because of their utmost importance in turbulence theory, the

---

[4] Being a random function, the numerical value of δf^α is different in different realizations.

correlations deserve a special analysis. More extensive discussions of the correlations can be found in the classical books on turbulence theory (*e.g.*, BATCHELOR 1956, MONIN & YAGLOM 1971-1975, McCOMB 1992, POPE 2000).

Let $\delta A(\mathbf{Y},t)$ and $\delta B(\mathbf{Y},t)$ be two fluctuation *fields*, depending on the spatial coordinates $\mathbf{Y}$ and on time t. Various types of correlation functions can be defined, which must be carefully distinguished.

The most general *Eulerian correlation* of the two fields is defined as:

$$E^{AB}(\mathbf{Y},t|\mathbf{Y}',t') = \langle\, \delta A(\mathbf{Y},t)\,\delta B(\mathbf{Y}',t')\,\rangle. \tag{6.38}$$

This is also appropriately called the two-point, two-time correlation of δA and δB. Here t and t' are two (generally different) values of the time, and $\mathbf{Y}$ and $\mathbf{Y}'$ are the coordinates of two fixed points in space. The importance of the Eulerian correlation lies in the fact that it *defines* (at least partially) the stochastic process represented by the fluctuating ensemble. The statistical definition of the process requires the knowledge of all moments. In the particular case of a *Gaussian stochastic process*, however, the knowledge of the second moments, *i.e.*, of the Eulerian correlations E^{AA}, E^{AB}, E^{BA}, E^{BB} completely defines the process. Indeed, the higher-order moments are functionals of the second-order ones (GARDINER 1985). Various special cases are of great interest because of their simpler properties, due to special symmetries.

- The *autocorrelation function of A* is obtained when $A = B$, thus:

$$E^{AA}(\mathbf{Y},t|\mathbf{Y}',t') = \langle\, \delta A(\mathbf{Y},t)\,\delta A(\mathbf{Y}',t')\,\rangle. \tag{6.39}$$

It expresses the statistical dependence of the fluctuating quantity δA evaluated at two different (fixed) points and two different times. In the forthcoming formulae we shall consider autocorrelation functions of varous types (which will be simply called "correlations" whenever no confusion is possible); the concepts are trivially extended to correlations of different functions.

- The *one-point, two-time correlation* is a purely temporal correlation, obtained when $\mathbf{Y} = \mathbf{Y}'$:

$$E^{AA}(\mathbf{Y},t|t') = \langle\, \delta A(\mathbf{Y},t)\,\delta A(\mathbf{Y},t')\,\rangle. \tag{6.40}$$

- The *two-point, one-time correlation* corresponds to a purely spatial correlation, for $t = t'$:

$$E^{AA}(\mathbf{Y}|\mathbf{Y}',t) = \langle\, \delta A(\mathbf{Y},t)\,\delta A(\mathbf{Y}',t)\,\rangle. \tag{6.41}$$

- The turbulence is called *stationary* when the Eulerian correlation does not depend separately on the two times t, t' but only on their difference $t' - t$:

$$E^{AA}(\mathbf{Y}, t | \mathbf{Y}', t') = E^{AA}(\mathbf{Y} | \mathbf{Y}', t' - t). \quad [S] \qquad (6.42)$$

- The turbulence is called *homogeneous* when the Eulerian correlation does not depend separately on the two coordinates $\mathbf{Y}, \mathbf{Y}'$ but only on their difference $\mathbf{Y}' - \mathbf{Y}$:

$$E^{AA}(\mathbf{Y}, t | \mathbf{Y}', t') = E^{AA}(\mathbf{Y}' - \mathbf{Y}, t | t'). \quad [H] \qquad (6.43)$$

- The turbulence is called *homogeneous and isotropic* when the Eulerian correlation depends only on the absolute value of the distance between the two points:

$$E^{AA}(\mathbf{Y}, t | \mathbf{Y}', t') = E^{AA}(|\mathbf{Y}' - \mathbf{Y}|, t | t'). \quad [H, I] \qquad (6.44)$$

It is often useful to express the Eulerian correlations in terms of the variables $\mathbf{Y}$, t, $\mathbf{R} = \mathbf{Y}' - \mathbf{Y}$, $\tau = t' - t$. We then adopt the notation [for simplicity, we no longer write explicitly the superscripts AA]:

$$E(\mathbf{Y}, t | \mathbf{Y} + \mathbf{R}, t + \tau) \equiv \mathcal{E}(\mathbf{R}, \tau; \mathbf{Y}, t). \qquad (6.45)$$

For a homogeneous and stationary turbulence this function is independent of $\mathbf{Y}$ and t:

$$\mathcal{E}(\mathbf{R}, \tau; \mathbf{Y}, t) = \mathcal{E}(\mathbf{R}, \tau). \quad [H, S] \qquad (6.46)$$

Because of this property of translation invariance, the Eulerian correlation can always be written in this case as:

$$\mathcal{E}(\mathbf{R}, \tau) = \langle \delta A(0, 0) \, \delta A(\mathbf{R}, \tau) \rangle. \quad [H, S] \qquad (6.47)$$

It is physically obvious that the mutual statistical influence of quantities located at different points and different times is a decreasing function of the distance between these points and of the difference between the corresponding times. We may thus estimate qualitatively the range of the correlations by introducing a *correlation length* λ_c and a *correlation time* τ_c by the following conditions:

$$\mathcal{E}(\mathbf{R}, \tau) \approx 0, \qquad for \ \ |\mathbf{R}| \gg \lambda_c, \qquad (6.48)$$

$$\mathcal{E}(\mathbf{R}, \tau) \approx 0, \qquad for \ \ \tau \gg \tau_c, \ \ (\tau > 0). \qquad (6.49)$$

The *one-point, two-time correlation* of the fluctuations plays a specially important role in the forthcoming developments; in particular, the correlations appearing in the average kinetic equation (6.3) and in the expressions of the fluxes (5.12), (5.22) are of this type. It is, clearly, the limit of the two-point correlation as the distance between the points tends to zero. We denote it by:

$$E(\mathbf{Y}, t | \mathbf{Y}, t + \tau) \equiv \mathcal{E}(\tau; \mathbf{Y}, t) = \mathcal{E}(\mathbf{R}, \tau; \mathbf{Y}, t)|_{\mathbf{R}=0}. \qquad (6.50)$$

Its *spatial Fourier transform* with respect to $\mathbf{Y}$ is defined as follows:[5]

$$\mathcal{E}(\tau; \mathbf{Y}, t) = \int d\mathbf{k}\, e^{i\mathbf{k}\cdot\mathbf{Y}}\, \mathcal{E}_{\mathbf{k}}(\tau; t), \qquad (6.51)$$

with the inversion formula:

$$\mathcal{E}_{\mathbf{k}}(\tau; t) = (2\pi)^{-3} \int d\mathbf{Y}\, e^{-i\mathbf{k}\cdot\mathbf{Y}}\, \mathcal{E}(\tau; \mathbf{Y}, t). \qquad (6.52)$$

One may further consider the *temporal Fourier transform* of the correlation with respect to the *relative* time τ

$$\mathcal{E}_{\mathbf{k}}(\tau; t) = (2\pi)^{-1} \int_{-\infty}^{\infty} d\omega\, e^{-i\omega\tau}\, \widetilde{\mathcal{E}}_{\mathbf{k}}^{F}(\omega; t), \qquad (6.53)$$

with the inverse:

$$\widetilde{\mathcal{E}}_{\mathbf{k}}^{F}(\omega; t) = \int_{-\infty}^{\infty} d\tau\, e^{i\omega\tau}\, \mathcal{E}_{\mathbf{k}}(\tau; t). \qquad (6.54)$$

We now relate the Fourier transform of the one-point Eulerian correlation to the correlation of the Fourier transforms of the fluctuations $\delta A_{\mathbf{k}}(\omega)$, defined as in Eqs. (4.64), (4.65):

$$\mathcal{E}(\tau; \mathbf{Y}, t)$$

$$= \int d\mathbf{k}_1 d\mathbf{k}_2\, e^{i(\mathbf{k}_1+\mathbf{k}_2)\cdot\mathbf{Y}} (2\pi)^{-2} \int d\omega_1 d\omega_2\, e^{-i(\omega_1+\omega_2)t - i\omega_2\tau}$$

$$\times \langle \delta A_{\mathbf{k}_1}(\omega_1)\, \delta B_{\mathbf{k}_2}(\omega_2) \rangle. \qquad (6.55)$$

If the turbulence is homogeneous and stationary, this function must be independent of $\mathbf{Y}$ and of t. This requires the following relation:

$$\langle \delta A_{\mathbf{k}_1}(\omega_1)\, \delta B_{\mathbf{k}}(\omega) \rangle = 2\pi\, \delta(\omega_1 + \omega)\, \delta(\mathbf{k}_1 + \mathbf{k})\, S_{\mathbf{k}}^{F}(\omega). \qquad (6.56)$$

---

[5] We do not use a separate symbol for the Fourier transform; the suffix $_{\mathbf{k}}$ clearly indicates that we are considering a Fourier representation.

The function $S_{\mathbf{k}}^{F}(\omega)$ will be called the Fourier $\mathbf{k} - \omega$ *spectral function* (or, simply, the *spectrum*) of the fluctuations. Its relation to the Eulerian one-point, two-time correlation in a homogeneous and stationary turbulence is:

$$\mathcal{E}(\tau) = \int d\mathbf{k} \, (2\pi)^{-1} \int d\omega \, e^{-i\omega\tau} \, S_{\mathbf{k}}^{F}(\omega). \qquad (6.57)$$

The Fourier transformation with respect to the relative time τ is, however, inconvenient in some problems, as will be seen later. The *Laplace transformation* is then much more useful; it is defined as follows (DOETSCH 1943, WIDDER 1946, KORN & KORN 1968, CHURCHILL 1972, LAVRENTIEV & SHABAT 1972):

$$\widetilde{\mathcal{E}}_{\mathbf{k}}^{L}(w;t) = \int_{0}^{\infty} d\tau \, e^{iw\tau} \, \mathcal{E}_{\mathbf{k}}(\tau;t). \qquad (6.58)$$

The difference with the Fourier transformation is, *apparently*, very slight: the lower limit of integration over time is 0 (rather than $-\infty$): thus, only *positive times* τ contribute to the transform. Most important, instead of the real frequency ω we introduce the *complex* variable[6]

$$w = \omega + i\gamma. \qquad (6.59)$$

The inversion formula is defined as follows *for positive time,* $\tau > 0$:

$$\mathcal{E}_{\mathbf{k}}(\tau;t) = (2\pi)^{-1} \int_{\Gamma} dw \, e^{-iw\tau} \, \widetilde{\mathcal{E}}_{\mathbf{k}}^{L}(w;t), \qquad (6.60)$$

The contour of integration Γ in the complex plane is a straight line parallel to the real axis lying above all singularities of the integrand (Bromwich contour, Fig. 6.1):

It is shown in the books cited above that any function of a complex variable w: $\widetilde{\mathcal{E}}_{\mathbf{k}}^{L}(w)$, that is the Laplace transform of a (sufficiently smooth) time-dependent function is *holomorphic* (*i.e.*, has no singularities) in the half-plane lying above the contour Γ.

The relation to the correlation of the Fourier-Laplace transforms of the fluctuations is more complicated in the present case. As there exists no distribution $\delta(w - w_1)$ for *complex* w, w_1, the integrations in the complex plane must be done carefully. We now show that for a *spatially homogeneous* turbulence the relation is:

$$\langle \delta A_{\mathbf{k}_1}(w_1) \, \delta B_{\mathbf{k}}(w) \rangle = \delta(\mathbf{k} + \mathbf{k}_1) \, S_{\mathbf{k}}^{L}(w_1, w) \quad [H]. \qquad (6.61)$$

---

[6] The usual notation for the complex Laplace variable is $s = -iw$. We prefer to use here the present notation which is closer to the corresponding Fourier transformation. The passage from one notation to the other is trivial. Eq. (6.58) is sometimes called the *one-sided Fourier transformation. The variable s* will be used in Chaps. 12 and 17.

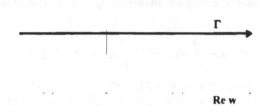

Figure 6.1. The Bromwich contour.

If the turbulence is both homogeneous and *stationary*, then:

$$\langle \delta A_{\mathbf{k}_1}(w_1)\,\delta B_{\mathbf{k}}(w)\rangle = \delta(\mathbf{k}+\mathbf{k}_1)\,\frac{1}{-i(w_1+w)}\,S_{\mathbf{k}}^{L}(w) \quad [HS]. \qquad (6.62)$$

In order to prove this, we calculate the Laplace transform of the correlation function as follows:

$$\int_0^\infty d\tau\, e^{iw\tau}\,\langle \delta A_{-\mathbf{k}}(t)\,\delta B_{\mathbf{k}}(t+\tau)\rangle$$

$$= \int_0^\infty d\tau\, e^{iw\tau}(2\pi)^{-2}\int_{\Gamma_1} dw_1 \int_{\Gamma_2} dw_2\, e^{-iw_1 t}\, e^{-iw_2 t - iw_2 \tau}$$

$$\times \langle \delta A_{-\mathbf{k}}(w_1)\,\delta B_{\mathbf{k}}(w_2)\rangle$$

$$= (2\pi)^{-2}\int_{\Gamma_2} dw_2 \int_{\Gamma_1} dw_1\, \frac{e^{-i(w_1+w_2)t}}{-i(w-w_2)}\,\frac{1}{-i(w_1+w_2)}\,S_{\mathbf{k}}^{L}(w_1),$$

provided $\mathcal{I}m(w-w_2)>0$. With the correlation of the form (6.61) the integration over w_1 can be performed (by the method of residues), provided $\mathcal{I}m(w_1+w_2)>0$, yielding a result independent of t (thus expressing stationarity):

$$\int_0^\infty d\tau\, e^{iw\tau}\,\langle \delta A_{-\mathbf{k}}(t)\,\delta B_{\mathbf{k}}(t+\tau)\rangle = (2\pi)^{-1}\int_{\Gamma_2} dw_2\,\frac{1}{-i(w-w_2)}\,S_{\mathbf{k}}^{L}(w_2).$$

Finally, if w lies above all singularities of $S_{\mathbf{k}}^{L}(w_2)$, the contour Γ_2 must be chosen to lie above all singularities of $S_{\mathbf{k}}^L(w_2)$ but below w. It can then be closed by a semi-circle in the upper half-plane, thus yielding

$$(2\pi)^{-1} \int_{\Gamma_2} dw_2 \frac{1}{-i(w - w_2)} S_{\mathbf{k}}^L(w_2) = S_{\mathbf{k}}^L(w), \qquad (6.63)$$

which proves our assertion. It may be said that the function $[-i(w-w_2)]^{-1}$ plays in the complex plane a role analogous to the Dirac delta function $\delta(\omega - \omega_2)$ on the real axis. But this function must always be handled with care (in particular, in the correct choice of integration contours). The function $S_{\mathbf{k}}^L(w)$ is called the Fourier-Laplace *spectrum of the fluctuations*.

6.4 Lagrangian Correlations

A quite different, much more complex but very important type of correlations will be introduced now. In any problem of stochastic dynamics, such as the kinetic equations under discussion here, the *Eulerian correlations* (6.38) of the various fluctuations are *given* (or are derived separately): they are part of the statistical definition of the problem. They describe the mutual statistical dependence of the fluctuations of A and B at two *fixed* points and two times. We now direct our attention on a small element of fluid or - microscopically - on a specific ("test") particle, moving according to the dynamical law of motion. We may calculate the correlation between the fluctuations evaluated at two *instantaneous positions* of the guiding centre, $\mathbf{Y}$ and $\mathbf{Y}'$. Here $\mathbf{Y} \equiv \mathbf{Y}(t|\mathbf{Y}, t)$ is the position of the particle at time t, and $\mathbf{Y}' \equiv \mathbf{Y}(t'|\mathbf{Y}, t)$ is its position at time t', given that it reaches the position $\mathbf{Y}$ at time t, by moving along its orbit [see Eq. (6.22)]. It should not be forgotten that this correlation also depends parametrically on $\mathbf{v}$ [through $\mathbf{Y}(t'|\mathbf{Y}, t)$]. The resulting expression is called the *Lagrangian correlation* of A and B:

$$L^{AB}(\mathbf{Y}, t|t') = \langle \delta A(\mathbf{Y}, t) \; \delta B[\mathbf{Y}(t'|\mathbf{Y}, t), t'] \rangle. \qquad (6.64)$$

Clearly, $\mathbf{Y}(t'|\mathbf{Y}, t)$ is determined by solving the underlying *Langevin equations* (6.8), as explained in Sec. 6.2. This is a highly non-trivial problem, which makes the calculation of Lagrangian correlations a very difficult dynamical problem.

Having explained the meaning of the symbols, we will henceforth use, whenever no confusion is possible, a simpler notation:

$$L^{AB}(\mathbf{Y}, t|t') = \langle \delta A(\mathbf{Y}, t) \; \delta B[\mathbf{Y}(t'), t'] \rangle, \qquad (6.65)$$

where:

$$\mathbf{Y}(t') = \mathbf{Y}(t'|\mathbf{Y}, t). \qquad (6.66)$$

Using the solution of Eq. (6.9), Eq. (6.64) can be written more explicitly as:

$$L^{AB}(\mathbf{Y}_\perp, Z, t|t') = \langle\, \delta A(\mathbf{Y}_\perp, Z, t)\ \delta B[\mathbf{Y}_\perp(t'), Z - v_\parallel(t' - t), t'] \,\rangle. \quad (6.67)$$

The Lagrangian correlations are a crucial element in the theory of transport, as will be seen below. As a result, the problem of finding an exact or an approximate relationship between Lagrangian correlations and the corresponding (known) Eulerian correlations is one of the most important problems in turbulence theory. This problem is clearly discussed in the book of McCOMB 1992.

Let us consider, for instance, the *Lagrangian autocorrelation of the fluctuating potential* $\delta\phi(\mathbf{Y}, t)$. For simplicity, we henceforth drop the subscripts $\phi\phi$ in the notations of the (Eulerian and Lagrangian) potential autocorrelations; thus $L^{\phi\phi} \equiv L$, $E^{\phi\phi} \equiv E$.

Some important symmetry properties of Lagrangian velocity correlations have been derived by LUMLEY 1962 for the case of Navier-Stokes turbulence (see also McCOMB 1992). We quote them here without proof. If the flow is incompressible (Here: $\nabla\cdot\delta\mathbf{V}^E = 0$) and the Eulerian correlations are homogeneous, the Lagrangian correlations are also homogeneous, hence independent of $\mathbf{Y}$. If the flow is moreover stationary, the Lagrangian correlations are also stationary, hence depend only on $(t' - t)$:

$$L(\mathbf{Y}, t|t') = L(t' - t), \quad [H, S] \quad\quad (6.68)$$

Without loss of generality, we may take in this case $\mathbf{Y} = 0$, $t = 0$, $\tau = t' - t$ and write the Lagrangian correlation in the form [see Eq. (6.47)]:

$$L(\tau) = \langle \delta\phi(0, 0)\ \delta\phi[\mathbf{Y}(\tau), \tau] \rangle. \quad [H, S] \quad\quad (6.69)$$

The *exact* evaluation of the Lagrangian correlation (6.67) requires the following steps. First, the Langevin equation (6.8) must be solved for a given realization of $\delta\phi(\mathbf{Y}_\perp, Z, t)$, yielding the trajectory $\mathbf{Y}_\perp(t'|\mathbf{Y}_\perp, Z, t)$: this is a nonlinear functional of the potential fluctuation. Upon substitution into (6.67) the Lagrangian correlation involves $\delta\phi$ in two ways: on one hand through the explicit factors $\delta\phi(\mathbf{Y}_\perp, Z, t)$, $\delta\phi[\mathbf{Y}_\perp(t'), Z(t'), t']$, on the other hand implicitly through $\mathbf{Y}_\perp(t')$. This highly nonlinear functional of $\delta\phi$ is to be averaged over all the realizations of the potential fluctuations. Clearly, this programme cannot be accomplished exactly.

In most existing works a very simple assumption is used for the calculation of the Lagrangian potential autocorrelation. It was introduced (for Navier-Stokes turbulence) by CORRSIN 1959, 1962 as a possible asymptotic approximation for long times. It was further developed by WEINSTOCK 1976; a clear and concise presentation is given in the book of McCOMB 1992.

In order to understand the basic idea, we start from an expression of the Lagrangian correlation in terms of the propagator [see Eq. (6.23)]:

$$L(\mathbf{Y}, t | t') = \int d\mathbf{Y}' \; \langle \delta\phi(\mathbf{Y}, t) \; \mathcal{G}(\mathbf{Y}, t | \mathbf{Y}', t') \; \delta\phi(\mathbf{Y}', t') \rangle. \qquad (6.70)$$

If Eq. (6.8) were deterministic, $\mathcal{G}(\mathbf{Y}, t | \mathbf{Y}', t')$ would be simply the propagator determining the dynamics of the guiding centre motion in a given potential. As the Langevin equation (6.8) is stochastic, this propagator will be different in different realizations of $\delta\phi$. *Corrsin's approximation* consists of replacing the fluctuating propagator inside Eq. (6.70) by its ensemble average:

$$\langle \mathcal{G}(\mathbf{Y}, t | \mathbf{Y}', t') \rangle = \langle \; \delta[\mathbf{Y}_\perp - \mathbf{Y}_\perp(t | \mathbf{Y}'_\perp, Z', t')] \; \rangle \;\; \delta[Z - Z' - v_\parallel (t' - t)]. \quad (6.71)$$

Thus, the fluctuating dynamics of $\mathbf{Y}(t)$ is replaced by an average, deterministic dynamics. This process is performed independently of the total average of Eq. (6.70): for this reason, Corrsin's conjecture is sometimes called the "independence hypothesis". As a result, the average in Eq. (6.70) is approximated by the following *factorized* form:

$$L(\mathbf{Y}, t | t') = \int d\mathbf{Y}' \; \langle \delta\phi(\mathbf{Y}, t) \; \delta\phi(\mathbf{Y}', t') \rangle \; \langle \mathcal{G}(\mathbf{Y}, t | \mathbf{Y}', t') \rangle. \quad \text{(Corrsin)}$$
$$(6.72)$$

The first factor in the integrand is simply the *Eulerian* potential autocorrelation. The average propagator must be further approximated in an appropriate way, depending on the specific problem under consideration. Thus, within the Corrsin approximation, the following simple relation holds between Lagrangian and Eulerian correlations:

$$L(\mathbf{Y}, t | t') = \int d\mathbf{Y}' \; \langle \mathcal{G}(\mathbf{Y}, t | \mathbf{Y}', t') \rangle \; E(\mathbf{Y}, t | \mathbf{Y}', t'). \qquad (6.73)$$

The Corrsin independence conjecture appears to be a rather strong approximation. Its domain of validity and its limitations will be discussed in Chap. 14 (see also WEINSTOCK 1976). It will appear to be valid in the case of sufficiently weak turbulence. A possible way of going beyond this approximation will be developed in the last chapters of this book.

6.5 Fluxes and Lagrangian Correlations

Coming back to Eq. (6.36), it is immediately seen that *all the coefficients of the kinetic equation are exactly expressed in terms of Lagrangian correlation functions.* Thus, the memory kernel is the Lagrangian autocorrelation of the fluctuating operator $\delta \mathcal{K}^{\alpha}$. More explicitly, as appears in Eqs. (6.2) - (6.33), this is a combination of four Lagrangian correlations involving the drift velocity components and the parallel acceleration.

Upon integration over the velocities $v_{\parallel}, v_{\perp}$ [remember the notation (4.62)], the kinetic equation (6.36) is transformed into an equation for the density profile (see Sec. 5.2). Given that the terms starting with $\partial_{\parallel}$ in the right hand side of Eq. (6.2) integrate out to zero, the result has the correct form of the conservation equation $\partial_t n_{\alpha 0} = -\nabla \cdot \Gamma^{\alpha}$, from which we derive the expression of the j-th component of the *anomalous radial particle flux*:

$$\Gamma_j^{\alpha}(\mathbf{x}, t) = -\int_0^t dt' \int d\mathbf{v} \, L_j^{VK}(\mathbf{x}, t|t') \, F^{\alpha}(\mathbf{x}, \mathbf{v}, t'), \qquad (6.74)$$

where $L_j^{VK}(t|t')$ is the Lagrangian correlation of the drift velocity and of the operator $\delta \mathcal{K}^{\alpha}$:

$$\begin{aligned} L_j^{VK}(\mathbf{x}, t|t') &= \int d\mathbf{Y}' \, \langle \delta V_j^E(\mathbf{x}, t) \, \mathcal{G}(\mathbf{x}, t|\mathbf{Y}', t') \, \delta \mathcal{K}^{\alpha}(\mathbf{Y}', t') \rangle \\ &= \mathcal{L}_{jn}^{VV}(\mathbf{x}, t|t') \, \nabla_n + \mathcal{L}_j^{Va}(\mathbf{x}, t|t') \, \partial_{\parallel}. \end{aligned} \qquad (6.75)$$

From Eqs. (6.6) and (6.18) [with $\mathcal{C}^{\alpha} = 0$ and $\delta f^{\alpha}(\mathbf{Y}, \mathbf{v}, 0) = 0$] we see that, (in the local approximation of the kinetic equation):

$$\Gamma_j^{\alpha}(\mathbf{x}, t)$$

$$= -\int d\mathbf{v} \int_0^t dt' \int d\mathbf{Y}' \, \langle \delta V_j^E(\mathbf{x}, t) \, \mathcal{G}(\mathbf{x}, t|\mathbf{Y}', t') \, \delta \mathcal{K}^{\alpha}(\mathbf{Y}', t') \rangle \, F^{\alpha}(\mathbf{x}, \mathbf{v}, t')$$

$$= \int d\mathbf{v} \, \langle \delta V_j^E(\mathbf{x}, t) \, \delta f^{\alpha}(\mathbf{x}, \mathbf{v}, t) \rangle. \qquad (6.76)$$

Hence, Eq. (6.74) is equivalent to the definition (5.11) of the particle flux.

Similarly, we obtain as in Sec. 5.3 the internal energy balance equation (5.19), with the *anomalous radial energy flux* given by:

$$Q_j^{\alpha}(\mathbf{x}; t) = -\frac{m_{\alpha}}{2} \int d\mathbf{v} \, (v_{\parallel}^2 + v_{\perp}^2) \int_0^t dt' \, L_j^{VK}(\mathbf{x}, t|t') \, F^{\alpha}(\mathbf{x}, \mathbf{v}, t') \qquad (6.77)$$

The equivalence of this expression with Eq. (5.21) is proved in the same way as above. Note that the terms of Eq. (6.2) starting with an operator $\partial_{\parallel}$ at the left only contribute to the source term in the macroscopic energy balance equation, as in Eq. (5.23); they are not further discussed here.

We thus obtained expressions of the anomalous fluxes that are more explicit than Eqs. (5.11) and (5.21). These expressions of the fluxes at time t contain three ingredients:

- *The Lagrangian drift velocity autocorrelation $L_{jn}^{VV}(t|t')$,*
- *The Lagrangian correlation of drift velocity and parallel acceleration $L_j^{Va}(t|t')$,*
- *The nonequilibrium average distribution function $F^\alpha(\mathbf{x}, \mathbf{v}, t')$, whose value has to be known at all times t' between 0 and $t > 0$.*

The determination of these quantities poses extremely difficult mathematical problems, that were already mentioned above. The first two problems require a relation between the Lagrangian correlation and the corresponding Eulerian correlation. The third one requires the solution of the kinetic equation (6.2) (which can only be done when the first two problems are solved). These three problems exhibit in a nutshell why the problem of anomalous transport is so difficult!

6.6 The Local Equilibrium Approximation

There is a problem with Eqs. (6.74) and (6.77): they are not in the form of macroscopic transport equations. They provide us with a *microscopic* definition of the anomalous fluxes, but they do not tell us how these fluxes are related to the *macroscopic* thermodynamic forces that are driving them. In the simple case considered here, these are the gradients of density and of temperature. If a macroscopic description of the plasma (as in Chaps. 2 and 5) is at all possible, the anomalous fluxes should be expressible in terms of $n_{\alpha 0}$, $P_{\alpha 0}$ and $T_{\alpha 0}$ in order to close the plasmadynamical equations. These quantities are "buried" in the distribution function F^α, which requires the solution of the kinetic equation. Moreover, whereas $n_{\alpha 0}$ and $P_{\alpha 0}$ are well-defined averages of dynamical quantities (5.4), (5.15), the temperature $T_{\alpha 0}$ is a concept related to the thermal equilibrium state. It is a *parameter* of the Maxwellian distribution function. Its relation to the density and the pressure is determined by an *equation of state*, which has the well-known form $P_{\alpha 0} = n_{\alpha 0} T_{\alpha 0}$ [Eq. (2.1)] only in equilibrium (and for sufficiently small density).

In order to cope with this difficulty, most authors dealing with plasma turbulence introduce more or less explicitly a strong assumption, which is inspired from "ordinary" collisional kinetic theory of gases. It is assumed that the plasma is in a state close to *local equilibrium*. This implies that the distribution function of both species has the same form (Maxwellian) as in equilibrium, but with parameters $n_{\alpha 0}$ and $T_{\alpha 0}$ depending (slowly) on space and time.

It is moreover assumed that the unperturbed macroscopic quantities depend on the single coordinate X [see Eq. (2.12)]. This implies that all gradients of these quantities are oriented in the same direction. As a result, the average distribution function also depends on the single coordinate X: $F^\alpha = F^\alpha(X, \mathbf{v}, t)$. Note that in the case of a stationary and homogeneous turbulent state, the Lagrangian corelations appearing in Eqs. (6.30) - (6.33) are independent of $\mathbf{Y}$, as mentioned in Sec. 6.4.[7] Hence a solution of the form $F^\alpha(X, \mathbf{v}, t)$ is compatible with the kinetic equation (6.36), which reduces to:

$$\partial_t F^\alpha(X, \mathbf{v}, t) = \int_0^t dt' \, L^{KK}(t|t') \, F^\alpha(X, \mathbf{v}, t'). \tag{6.78}$$

The local equilibrium distribution is now defined as:

$$F^\alpha = F_M^\alpha(X, \mathbf{v}, t) = \pi^{-3/2} V_{T\alpha}^{-3}(X, t) \, n_0(X, t) \exp\left(-\frac{v_\parallel^2 + v_\perp^2}{V_{T\alpha}^2(X, t)}\right)$$

$$\equiv n_0(X, t) \, \varphi_M^\alpha(X, \mathbf{v}, t), \tag{6.79}$$

where $V_{T\alpha}(X, t)$ is the *thermal velocity* of species α:

$$V_{T\alpha}(X, t) = \left(\frac{2T_{\alpha0}(X, t)}{m_\alpha}\right)^{1/2}. \tag{6.80}$$

It should be recalled that $n_{e0} = n_{i0} = n_0$ by the global electroneutrality constraint. It is sometimes useful to introduce the local Maxwellian velocity distribution $\varphi_M^\alpha(X, \mathbf{v}, t)$ in order to simplify the form of certain formulae: it depends on X and t only through the local temperature.

It is not easy to really justify this assumption for a "collisionless" (or weakly collisional) plasma. The least questionable argument could be as follows. The plasma starts in the remote past at a low temperature, when collisions are dominant: these drive the system close to a local equilibrium state. This velocity distribution is (perhaps!) not too much distorted as the plasma is slowly heated to a high temperature, when collisions become very rare. The argument is not very convincing, but at present one can hardly do better. The only test is to judge the resulting theory *a posteriori* from its results.

Substitution of Eq. (6.79) into the kinetic equation (6.2) (with $\Delta^\alpha = 0$) essentially transforms the latter into an equation for the density and for the temperature:

---

[7] In the general case of a non-stationary turbulence, the kinetic equation must first be integrated over Y and Z. In toroidal geometry, this amounts to a surface-average.

$$\left[\frac{\partial_t n_0(X,t)}{n_0(X,t)} + \left(\frac{m_\alpha}{2}\frac{v_\parallel^2 + v_\perp^2}{T_{\alpha 0}(X,t)} - \frac{3}{2}\right)\frac{\partial_t T_{\alpha 0}(X,t)}{T_{\alpha 0}(X,t)}\right] F_M^\alpha(X,\mathbf{v},t)$$

$$= \int_0^t dt' \, L^{KK}(t|t') \, F_M^\alpha(X,\mathbf{v},t'). \tag{6.81}$$

The action of the operator $L_{KK}^\alpha(t|t')$ on the *known* function $F_M^\alpha(X,\mathbf{v},t')$ can now be evaluated explicitly. This elementary calculation will not be done at the present stage, in order not to burden the text.

The third difficulty mentioned at the end of Sec. 6.5 is thus (somewhat arbitrarily) eliminated. The problem of calculating the anomalous fluxes now rests entirely on the determination of the *Lagrangian correlations* L_{jn}^{VV} and L_j^{Va} or, basically, of the *Lagrangian autocorrelation* $L(t|t')$ *of the fluctuating gyro-averaged electrostatic potential* $\overline{\delta\phi}[\mathbf{Y}(t),t]$. If that quantity is known, the transport coefficients are obtained by simple quadratures.

Chapter 7

Quasilinear Transport Theory

7.1 The Quasilinear Approximation

The evaluation of the Lagrangian velocity correlations, hence of the anomalous fluxes, is a very difficult problem, because of its high nonlinearity. It can only be solved by introducing additional approximations. The crudest of these is called the *Quasilinear approximation*: it has been extensively studied because of its simplicity (*e.g.*, SAGDEEV & GALEEV 1969, DAVIDSON 1972, NICHOLSON 1983, KRALL & TRIVELPIECE 1986, ICHIMARU 1992). Although (in its strict form) it will turn out to yield an incomplete transport theory, it is an important starting point for further analysis.

The idea of the quasilinear approximation is in the spirit of perturbation theory. It is applicable to *weak turbulence*, defined as follows [see Eq. (4.52)]:

$$\frac{\delta f^\alpha}{F^\alpha} \sim \frac{e_\alpha \delta \phi}{T_\alpha} \sim \frac{\delta V^E}{V_{T\alpha}} \ll 1. \tag{7.1}$$

These relations should hold for all realizations of the turbulent ensemble, in all points of space and for all times. This approximation is implemented by neglecting in the gyrokinetic equation (6.4) all terms that are quadratic in the fluctuations. The gyrokinetic equation then reduces (in shearless slab geometry) to:

$$\left(\partial_t + v_\parallel \nabla_Z\right) \delta f^\alpha = -\delta \mathcal{K}^\alpha F^\alpha, \tag{7.2}$$

where the operator $\delta \mathcal{K}^\alpha$ was defined in Eq. (6.2). This is a very simple linear equation for δf^α which can be solved exactly. The characteristic equations (6.8) - (6.11) reduce to a single non-trivial one (for Z):

$$\frac{dX(t|\mathbf{Y}',t')}{dt} = 0, \quad X(t|\mathbf{Y}',t) = X',$$

$$\frac{dY(t|\mathbf{Y}',t')}{dt} = 0, \quad Y(t|\mathbf{Y}',t) = Y',$$

$$\frac{dZ(t|\mathbf{Y}',t')}{dt} = v_{\|}, \quad Z(t|\mathbf{Y}',t) = Z',$$

$$\frac{d\mathbf{v}(t|\mathbf{Y}',t')}{dt} = 0, \quad \mathbf{v}(t|\mathbf{Y}',t') = \mathbf{v}'. \qquad (7.3)$$

The right hand sides are independent of the fluctuating velocity δV^E and of the fluctuating acceleration δa^α; hence these "Langevin" equations are no longer stochastic equations, but rather ordinary, *deterministic* equations. Their solution is:

$$X(t|\mathbf{Y}',t') = X', \quad Y(t|\mathbf{Y}',t') = Y', \quad Z(t|\mathbf{Y}',t') = Z' + v_{\|}(t-t'),$$
$$v_{\|}(t|\mathbf{Y}',t') = v_{\|}', \quad v_{\perp}(t|\mathbf{Y}',t') = v_{\perp}'. \qquad (7.4)$$

The propagator g, Eq. (6.14) reduces to the approximate propagator $g_0^{\alpha\beta}$:[1]

$$g_0^{\alpha\beta}(\mathbf{Y},\mathbf{v},t|\mathbf{Y}',\mathbf{v}',t') = \delta^{\alpha\beta} \, \mathcal{G}_0(\mathbf{Y},t|\mathbf{Y}',t') \, \delta(\mathbf{v}-\mathbf{v}'), \qquad (7.5)$$

with the spatial part:

$$\mathcal{G}_0(\mathbf{Y},t|\mathbf{Y}',t') = \delta(\mathbf{Y}_{\perp}-\mathbf{Y}_{\perp}') \, \delta[Z - Z' - v_{\|}(t-t')] . \qquad (7.6)$$

The difficult problem mentioned after Eq. (6.20), *viz.*, the inversion of the equation $\mathbf{Y} = \mathbf{Y}(t|\mathbf{Y}',t')$ is now trivial $[Z' = Z - v_{\|}(t-t')]$. The integration over $\mathbf{Y}'$ in the general formula (6.18) can thus be done explicitly, yielding the solution of the linearized gyrokinetic equation, Eq. (6.18) (with $C^\alpha = 0$):

$$\delta f^\alpha(X,Y,Z,\mathbf{v},t)$$

$$= \sum_\beta \int d\mathbf{Y}' \int d\mathbf{v}' \left\{ \int_0^t dt' g_0^{\alpha\beta}(\mathbf{Y},\mathbf{v},t|\mathbf{Y}',\mathbf{v}',t') \, S^\beta(\mathbf{Y}',\mathbf{v}',t') \right.$$

$$\left. + g_0^{\alpha\beta}(\mathbf{Y},\mathbf{v},t|\mathbf{Y}',\mathbf{v}',0) \, \delta f^\beta(\mathbf{Y}',\mathbf{v}',0) \right\} . \qquad (7.7)$$

[1] We introduce here explicitly a dependence on the species indices α, β. This dependence is trivial in the quasilinear case, but it will become non-trivial in the nonlinear situation treated in Chap. 8.

Using Eq. (7.6), this expression is rewritten as:

$$\delta f^\alpha(X, Y, Z, \mathbf{v}, t)$$

$$= - \int_0^t dt' \delta \mathcal{K}^\alpha[X, Y, Z - v_\parallel(t - t'), \ t'] \ F_M^\alpha(X, \mathbf{v}, t')$$

$$+ \delta f^\alpha(X, Y, Z - v_\parallel t, \mathbf{v}, 0) \tag{7.8}$$

The average kinetic equation is obtained by substituting this result into Eq. (6.3). In particular, the memory kernel, *i.e.*, the Lagrangian autocorrelation $\mathcal{L}^{KK}(t|t')$, is obtained as follows:

$$\mathcal{L}^{KK}(t|t') = \left\langle \delta \mathcal{K}^\alpha[X, Y, Z, t] \, \delta \mathcal{K}^\alpha[X, Y, Z - v_\parallel(t - t'), t'] \right\rangle. \tag{7.9}$$

This follows from the general form:

$$\mathcal{L}^{KK}(t|t') = \int d\mathbf{Y}' \left\langle \delta \mathcal{K}^\alpha(\mathbf{Y}, t) \, \delta \mathcal{K}^\alpha(\mathbf{Y}', t') \, \mathcal{G}_0(\mathbf{Y}, t|\mathbf{Y}', t') \right\rangle$$

$$= \int d\mathbf{Y}' \left\langle \delta \mathcal{K}^\alpha(\mathbf{Y}, t) \, \delta \mathcal{K}^\alpha(\mathbf{Y}', t') \, \delta[\mathbf{Y} - \mathbf{Y}(t|\mathbf{Y}', t')] \right\rangle. \tag{7.10}$$

This equation has the same form as Eq. (6.70). But in the present case, the orbit $\mathbf{Y}(t|\mathbf{Y}', t')$, given by Eq. (7.4) is *deterministic*: it does not depend on the fluctuations. Hence, the delta-function in (7.10) can be taken out of the average. The Lagrangian autocorrelation can then be written as in Eq. (6.73):

$$\mathcal{L}^{KK}(t|t') = \int d\mathbf{Y}' \ E^{KK}(\mathbf{Y}, t|\mathbf{Y}', t') \, \mathcal{G}_0(\mathbf{Y}, t|\mathbf{Y}', t'). \tag{7.11}$$

Thus:

$$\left\langle \mathcal{G}_0(\mathbf{Y}, t|\mathbf{Y}', t') \right\rangle = \mathcal{G}_0(\mathbf{Y}, t|\mathbf{Y}', t') \qquad (Q - lin.). \tag{7.12}$$

Several interesting features are expressed by this result.

- Eq. (7.12) has exactly the same form as (6.73). Thus, *the Corrsin independence hypothesis is exactly fulfilled in the quasilinear approximation.*
- The orbits (7.4) are deterministic.
- As a result of the deterministic character of the orbits, the propagator $g_0^{\alpha\beta}(\mathbf{Y}, \mathbf{v}, t|\mathbf{Y}', \mathbf{v}', t')$ is a product of *singular delta-functions*.

The previous results are now incorporated into the average kinetic equation (6.28) which becomes:

$$\partial_t F^\alpha(\mathbf{Y}, \mathbf{v}, t)$$

$$= \int_0^t dt' \, \big\langle \delta \mathcal{K}^\alpha(X, Y, Z, t) \, \delta \mathcal{K}^\alpha[X, Y, Z - v_\parallel(t - t'), t'] \big\rangle \, F^\alpha(\mathbf{Y}, \mathbf{v}, t')$$

$$- \big\langle \delta \mathcal{K}^\alpha(X, Y, Z, t) \, \delta f^\alpha[X, Y, Z - v_\parallel t, 0] \big\rangle. \tag{7.13}$$

We thus see that even in this very simple quasilinear approximation the memory effects are present. Unfortunately, their explicit treatment is in general difficult.[2] For this reason, the memory effects will be neglected in the present chapter. The kinetic equation will be Markovianized in the usual way, assuming that the time scale of the average distribution is very long compared to the variation of the Lagrangian autocorrelation. This assumption is quite well realized in the weakly turbulent case.

7.2 The Fourier Representation

Kinetic theory in the quasilinear approximation is conveniently formulated in the spatial Fourier picture. We first assume as in Sec. 6.6 that the average distribution function $F^\alpha(X, \mathbf{v}; t)$ depends *slowly* (*i.e.* on the scale L_M) on the single spatial coordinate X, and also *slowly* on time. It will moreover be approximated by a local equilibrium distribution (6.79). As a result, this function is treated in Eq. (7.2) as a "constant" in space and time, in the sense that it is not affected by the Fourier transformation ("local approximation"). The latter equation then becomes a linear differential equation with constant coefficients, obtained by neglecting the nonlinear terms in Eq. (4.83) and using Eq. (4.76):

$$(\partial_t + i k_\parallel v_\parallel) \, \delta f_\mathbf{k}^\alpha(\mathbf{v}, t)$$

$$= i \left[\frac{c}{B} k_y \nabla_X + \frac{e_\alpha}{m_\alpha} k_\parallel \partial_\parallel \right] F_M^\alpha(X, \mathbf{v}) \, J_0(k_\perp \hat{\rho}_{L\alpha}) \, \delta \phi_\mathbf{k}(t)$$

$$\equiv s_\mathbf{k}^\alpha(\mathbf{v}; t). \tag{7.14}$$

We introduce several notations which make this equation more transparent. Evaluating explicitly the derivatives of the Maxwelian function (6.79), we obtain the following useful relations:[3]

$$\frac{c}{B} k_y \, \nabla_X F_M^\alpha = \frac{e_\alpha}{T_\alpha} \, \omega_{*\mathbf{k}}^{\alpha T}(\mathbf{v}) \, F_M^\alpha(\mathbf{v}), \tag{7.15}$$

[2] A specific case where these memory effects can be handled explicitly will be treated in Chap. 14.

[3] The variables X and t on which the distribution function F_M^α (hence also $\delta f_\mathbf{k}^\alpha$) depends slowly [through the density $n_0(X; t)$ and the temperatures $T_{\alpha 0}(X; t)$] will not be written down explicitly whenever no confusion is possible.

$$\frac{e_\alpha}{m_\alpha} k_\parallel \partial_\parallel F_M^\alpha = -\frac{e_\alpha}{T_\alpha} k_\parallel v_\parallel F_M^\alpha(\mathbf{v}), \qquad (7.16)$$

with:

$$\omega_{*\mathbf{k}}^{\alpha T}(\mathbf{v}) = \left[1 + \left(\frac{v_\parallel^2 + v_\perp^2}{V_{T\alpha}^2} - \frac{3}{2}\right)\eta_\alpha\right]\omega_{*\mathbf{k}}^\alpha. \qquad (7.17)$$

Here $\omega_{*\mathbf{k}}^\alpha$ is the *characteristic drift wave frequency of species* α:

$$\omega_{*\mathbf{k}}^\alpha = k_y V_{d\alpha}, \qquad (7.18)$$

where the *diamagnetic drift velocity* $V_{d\alpha}$ of species α is defined in Eqs. (2.27), (2.28) and the parameter η_α is defined in Eq. (2.19). We also use the convenient abbreviation introducing the microscopic Larmor radius (4.17):

$$J_{\mathbf{k}}^\alpha(\mathbf{v}) = J_0(k_\perp \widehat{\rho}_{L\alpha}) = J_0(|\Omega_\alpha|^{-1} k_\perp v_\perp). \qquad (7.19)$$

We introduce the concept of the *unperturbed vertex operator* (which reduces to a function in the Fourier picture):

$$\Delta_{0\mathbf{k}}^\alpha(\mathbf{v}) = i\frac{e_\alpha}{T_\alpha}\left[k_\parallel v_\parallel - \omega_{*\mathbf{k}}^{\alpha T}(\mathbf{v})\right] J_{\mathbf{k}}^\alpha(\mathbf{v}) F_M^\alpha(\mathbf{v}). \qquad (7.20)$$

Using all these notations, the gyrokinetic equation (7.14) is rewritten as:

$$(\partial_t + i k_\parallel v_\parallel)\,\delta f_{\mathbf{k}}^\alpha(\mathbf{v}, t) = -\Delta_{0\mathbf{k}}^\alpha(\mathbf{v})\,\delta\phi_{\mathbf{k}}(t). \qquad (7.21)$$

The solution of this linear, inhomogeneous equation is:

$$\delta f_{\mathbf{k}}^\alpha(\mathbf{v}, t) = -\int_0^t dt'\,\mathcal{G}_{0\mathbf{k}}(\mathbf{v}; t - t')\,\Delta_{0\mathbf{k}}^\alpha(\mathbf{v})\,\delta\phi_{\mathbf{k}}(t')$$
$$+ \mathcal{G}_{0\mathbf{k}}(\mathbf{v}; t)\,\delta f_{\mathbf{k}}^\alpha(\mathbf{v}; 0), \qquad (7.22)$$

where:

$$\mathcal{G}_{0\mathbf{k}}(\mathbf{v}; t - t') = \exp[-i k_\parallel v_\parallel (t - t')]. \qquad (7.23)$$

This is just the Fourier representation of the propagator (7.6). Its full matrix form, corresponding to Eq. (7.5), is:

$$g_0^{\alpha\beta}(\mathbf{k}, \mathbf{v}, t | \mathbf{k}', \mathbf{v}', t') = \delta^{\alpha\beta}\,\delta(\mathbf{k} - \mathbf{k}')\delta(\mathbf{v} - \mathbf{v}')\,\mathcal{G}_{0\mathbf{k}}(\mathbf{v}; t - t'). \qquad (7.24)$$

The velocity delta function (in cylindrical coordinates) is defined as:

$$\delta(\mathbf{v} - \mathbf{v}') = \frac{1}{2\pi v'_\perp}\,\delta(v_\perp - v'_\perp)\delta(v_\parallel - v'_\parallel). \qquad (7.25)$$

This propagator obeys the following differential equation:

$$(\partial_t + ik_\parallel v_\parallel)\,g_0^{\alpha\beta}(\mathbf{k}, \mathbf{v}, t|\mathbf{k}', \mathbf{v}', t') = 0, \qquad (7.26)$$

with the initial (= equal time) condition:

$$g_0^{\alpha\beta}(\mathbf{k}, \mathbf{v}, t'|\mathbf{k}', \mathbf{v}', t') = \delta^{\alpha\beta}\,\delta(\mathbf{k} - \mathbf{k}')\,\delta(\mathbf{v} - \mathbf{v}'). \qquad (7.27)$$

7.3   The Fourier-Laplace Representation

The quasilinear gyrokinetic equation could be further Fourier-transformed in time; this is, however, not always appropriate for solving certain problems, such as the initial value problem of the original equation. The standard method for this problem makes use of the *Laplace transformation*, which was discussed in detail in Sec. 6.3. The Fourier-Laplace transform of the fluctuating potential is:

$$\delta\widetilde{\phi}_\mathbf{k}(w) = \int_0^\infty dt\, e^{iwt}\,\delta\phi_\mathbf{k}(t), \qquad (7.28)$$

where w is a *complex* variable. The inversion formula is:

$$\delta\phi_\mathbf{k}(t) = (2\pi)^{-1}\int_\Gamma dw\, e^{-iwt}\,\delta\widetilde{\phi}_\mathbf{k}(w), \quad t > 0, \qquad (7.29)$$

where the Bromwich contour Γ is a straight line in the upper complex plane, as defined in Fig. 6.1. A well-known property of the Laplace transformation should be stressed here: *the integral in the right hand side of Eq. (7.29) defines the function $\delta\phi_\mathbf{k}(t)$ only for positive times. For $t < 0$ this integral is zero.*[4] The consequences of this important point are further developed below.

Similar formulae define $\delta\widetilde{f}_\mathbf{k}^\alpha(\mathbf{v}, w)$. The following relation for the Laplace transform of a derivative must be kept in mind: $\partial_t f(t) \Rightarrow -iw\widetilde{f}(w) - f(t = 0)$. The Fourier-Laplace transformed gyrokinetic equation is thus:

$$-i(w - k_\parallel v_\parallel)\,\delta\widetilde{f}_\mathbf{k}^\alpha(\mathbf{v}, w) = -\Delta_{0\mathbf{k}}^\alpha(\mathbf{v})\,\delta\widetilde{\phi}_\mathbf{k}(w) + \delta f_\mathbf{k}^\alpha(\mathbf{v}, 0). \qquad (7.30)$$

Its solution is immediate:

---

[4] For negative times, the Bromwich contour can be supplemented by a half-circle at infinity in the upper half-plane. By definition of Γ, there are no singularities in the resulting closed region, hence the integral is zero.

$$\delta \tilde{f}_{\mathbf{k}}^{\alpha}(\mathbf{v}, w) = \tilde{\mathcal{G}}_{0\mathbf{k}}(\mathbf{v}, w) \left\{ -\Delta_{0\mathbf{k}}^{\alpha}(\mathbf{v}) \, \delta \tilde{\phi}_{\mathbf{k}}(w) + \delta f_{\mathbf{k}}^{\alpha}(\mathbf{v}, 0) \right\}. \tag{7.31}$$

We note here the appearance of the (Fourier-Laplace) **unperturbed propagator** of the gyrokinetic equation:

$$\tilde{\mathcal{G}}_{0\mathbf{k}}(\mathbf{v}, w) = \frac{1}{-i(w - k_{\parallel} v_{\parallel})}; \tag{7.32}$$

its properties are further discussed below. For comparison with forthcoming results, Eq. (7.31) can be rewritten in matrix form:

$$\delta \tilde{f}_{\mathbf{k}}^{\alpha}(\mathbf{v}, w)$$
$$= \sum_{\beta} \int d\mathbf{k}' d\mathbf{v}' \, \tilde{g}_{0\mathbf{k}}^{\alpha\beta}(\mathbf{k}, \mathbf{v} | \mathbf{k}', \mathbf{v}', w) \left\{ -\Delta_{0\mathbf{k}'}^{\beta}(\mathbf{v}') \, \delta \tilde{\phi}_{\mathbf{k}'}(w) + \delta f_{\mathbf{k}'}^{\beta}(\mathbf{v}', 0) \right\}$$

$$\tag{7.33}$$

with:

$$\tilde{g}_{0\mathbf{k}}^{\alpha\beta}(\mathbf{k}, \mathbf{v} | \mathbf{k}', \mathbf{v}', w) = \delta^{\alpha\beta} \, \delta(\mathbf{k} - \mathbf{k}') \, \delta(\mathbf{v} - \mathbf{v}') \, \tilde{\mathcal{G}}_{0\mathbf{k}}(\mathbf{v}, w). \tag{7.34}$$

It is important to recall that w is a *complex variable located in the upper half plane* S_+, i.e., $\mathcal{I}m \, w > 0$. As a result, whenever $\delta \tilde{\phi}_{\mathbf{k}}(w)$ has no poles in S_+, the function $\delta \tilde{f}_{\mathbf{k}}^{\alpha}(\mathbf{v}, w)$ is holomorphic in S_+, but has a simple pole on the real axis, for $w = k_{\parallel} v_{\parallel}$. Next, we note that just because of this property, the inverse Laplace transform (7.29) equals $\delta f_{\mathbf{k}}^{\alpha}(\mathbf{v}, t)$ for $t > 0$, but equals zero for $t < 0$.

Note that the function $\delta f_{\mathbf{k}}^{\alpha}(\mathbf{v}, t)$ can always be represented as a sum of two terms:

$$\delta f_{\mathbf{k}}^{\alpha}(\mathbf{v}, t) = \delta f_{\mathbf{k}+}^{\alpha}(\mathbf{v}, t) + \delta f_{\mathbf{k}-}^{\alpha}(\mathbf{v}, t), \tag{7.35}$$

where $\delta f_{\mathbf{k}+}^{\alpha}(\mathbf{v}, t) \, [\delta f_{\mathbf{k}-}^{\alpha}(\mathbf{v}, t)]$ represents the solution of the gyrokinetic equation (7.21) for positive [negative] time and equals 0 for negative [positive] time. The two terms can therefore be written as:

$$\delta f_{\mathbf{k}+}^{\alpha}(\mathbf{v}, t) = H(t) \, \delta f_{\mathbf{k}}^{\alpha}(\mathbf{v}, t), \quad \delta f_{\mathbf{k}-}^{\alpha}(\mathbf{v}, t) = [1 - H(t)] \, \delta f_{\mathbf{k}}^{\alpha}(\mathbf{v}, t), \tag{7.36}$$

where $H(t)$ is the discontinuous Heaviside function:

$$H(t) = \begin{cases} 1 & for \ t > 0 \\ 0 & for \ t < 0. \end{cases} \tag{7.37}$$

The Heaviside function acts like a *projection operator*, projecting the function $\delta f_{\mathbf{k}}^{\alpha}(\mathbf{v},t)$ on the subspace of functions defined only for $t > 0$. As explained after Eq. (7.29), the inverse Laplace transformation of any function that is holomorphic in the upper half-plane of w necessarily yields only functions of this subspace ("plus-functions"), thus:

$$\frac{1}{2\pi}\int_\Gamma dw\, e^{-iwt}\, \delta\widetilde{f}_{\mathbf{k}}^{\alpha}(\mathbf{v},w) = \delta f_{\mathbf{k}+}^{\alpha}(\mathbf{v},t). \tag{7.38}$$

We now consider the unperturbed propagator, Eq. (7.32). The inverse Laplace transformation of this function yields the time propagator (7.23) *for positive time difference* and zero otherwise:

$$\frac{1}{2\pi}\int_\Gamma dw\, e^{-iw(t-t')}\,\widetilde{\mathcal{G}}_{0\mathbf{k}}(\mathbf{v},w) = H(t-t')\,\mathcal{G}_{0\,\mathbf{k}}(\mathbf{v},t-t')$$

$$\equiv \mathcal{G}_{0+\,\mathbf{k}}(\mathbf{v},t-t'). \tag{7.39}$$

This is called a *causal propagator*: it propagates the effect of the initial condition towards the future. Its complete matrix form is obtained from Eq. (7.24):

$$g_{0+}^{\alpha\beta}(\mathbf{k},\mathbf{v},t|\mathbf{k}',\mathbf{v}',t')$$

$$= \delta^{\alpha\beta}\,\delta(\mathbf{k}-\mathbf{k}')\,\delta(\mathbf{v}-\mathbf{v}')\,H(t-t')\,\exp[-ik_{\|}v_{\|}(t-t')]. \tag{7.40}$$

Because of the extra singularity at $t = t'$, the causal propagator obeys an equation different from (7.26):

$$(\partial_t + ik_{\|}v_{\|})\,g_{0+}^{\alpha\beta}(\mathbf{k},\mathbf{v},t|\mathbf{k}',\mathbf{v}',t')$$

$$= \delta^{\alpha\beta}\,\delta(\mathbf{k}-\mathbf{k}')\,\delta(\mathbf{v}-\mathbf{v}')\,\delta(t-t'). \tag{7.41}$$

The Laplace version of this equation is a trivial rewriting of Eq. (7.32):

$$-i(w - k_{\|}v_{\|})\,\widetilde{\mathcal{G}}_{0\mathbf{k}}^{\alpha\beta}(\mathbf{v}|\mathbf{v}',w) = \delta^{\alpha\beta}\delta(\mathbf{v}-\mathbf{v}'). \tag{7.42}$$

The initial condition must be understood as a limiting value:

$$g_{0+}^{\alpha\beta}(\mathbf{k},\mathbf{v},t'+\epsilon|\mathbf{k}',\mathbf{v}',t') \Rightarrow \delta^{\alpha\beta}\,\delta(\mathbf{k}-\mathbf{k}')\delta(\mathbf{v}-\mathbf{v}'),\ \epsilon \to +0. \tag{7.43}$$

Thus, when working with Laplace transforms, one always deals with "plus-functions" of time, $f_+(t)$, defined for $t > 0$, and with causal propagators. It may be noted that this produces the correct form of the convolution of Eq. (7.22) by inverse transformation of (7.31). Indeed, using the convolution theorem, we have:

$$\frac{1}{2\pi} \int_\Gamma e^{-iwt} \widetilde{\mathcal{G}}_{0\mathbf{k}}(w) \delta\widetilde{\phi}_\mathbf{k}(w) = \int_{-\infty}^\infty dt'\, H(t-t') \mathcal{G}_{0\mathbf{k}}(t-t')\, H(t') \delta\phi_\mathbf{k}(t')$$

We now note the simple, but useful identity:

$$\int_{-\infty}^\infty dt'\, H(t-t')\, H(t')\, f(t') = H(t) \int_0^t dt'\, f(t') \qquad (7.44)$$

[indeed, $H(t')$ requires $t' > 0$, and $H(t-t')$ requires $t > t'$, hence also $t > 0$.] We thus obtain from (7.31) the transformed equation:

$$\delta f^\alpha_{\mathbf{k}+}(\mathbf{v},t)$$
$$= H(t) \left\{ \int_0^t dt'\, \mathcal{G}_{0\mathbf{k}}(\mathbf{v},t-t')\, \Delta^\alpha_{0\mathbf{k}}(\mathbf{v})\, \delta\phi_\mathbf{k}(t') + \mathcal{G}_{0\mathbf{k}}(\mathbf{v},t)\, \delta f^\alpha_\mathbf{k}(\mathbf{v},0) \right\},$$

which is identical to Eq. (7.22) for positive time.

The result (7.31) (in the Markovian approximation) can be used for calculating the correlation function appearing in the right hand side of the average equation (6.3). But we rather skip this step here and go over directly to the calculation of the transport autocorrelation operator $\mathcal{L}_j^{VK}(t|t')$ that will lead us to the expression of the anomalous fluxes.

$$\mathcal{L}_j^{VK}(\mathbf{Y},t|t')$$
$$= \left\langle \delta V_j^E(X,Y,Z;t)\, \delta\mathcal{K}^\alpha[X,Y,Z-v_\parallel(t-t'),\mathbf{v};t'] \right\rangle$$
$$= (2\pi)^{-1} \int d\mathbf{k}_1 \int_{\Gamma_1} dw_1 e^{-iw_1 t + i\mathbf{k}_{1\perp}\cdot\mathbf{Y}_\perp + ik_{1\parallel}Z} \left(-ik_{1y}\frac{c}{B} \right) J^\alpha_{\mathbf{k}_1}(\mathbf{v})$$
$$\times (2\pi)^{-1} \int d\mathbf{k}_2 \int_{\Gamma_2} dw_2 e^{-iw_2 t' + i\mathbf{k}_{2\perp}\cdot\mathbf{Y}_\perp + ik_{2\parallel}Z - ik_{2\parallel}v_\parallel(t-t')}$$
$$\times \left(-i\frac{e_\alpha}{T_\alpha}\left[k_{2\parallel}v_\parallel - \omega^{\alpha T}_{*\mathbf{k}_2}(\mathbf{v}) \right] \right) J^\alpha_{\mathbf{k}_2}(\mathbf{v}) \left\langle \delta\widetilde{\phi}_{\mathbf{k}_1}(w_1)\, \delta\widetilde{\phi}_{\mathbf{k}_2}(w_2) \right\rangle.$$

$$(7.45)$$

Both contours Γ_1, Γ_2 are straight lines lying in the upper half plane of their respective variables, and above the highest singularity of $\delta\widetilde{\phi}_\mathbf{k}(w)$. We assume that the turbulence is spatially homogeneous, but not necessarily stationary: the potential Eulerian correlation function then has the form (6.61):

$$\left\langle \delta\widetilde{\phi}_{\mathbf{k}_1}(w_1)\, \delta\widetilde{\phi}_{\mathbf{k}_2}(w_2) \right\rangle = \delta(\mathbf{k}_1 + \mathbf{k}_2)\, S_{\mathbf{k}_2}(w_1, w_2). \qquad (7.46)$$

The integration over $\mathbf{k}_1$ is trivial, and we obtain:

$$\mathcal{L}_j^{VK}(t|t')$$

$$= \int d\mathbf{k}\,(2\pi)^{-2} \int dw_1 dw_2\, e^{-i(w_1+w_2)t}\, e^{i(w_2-k_\parallel v_\parallel)(t-t')}\, S_{\mathbf{k}}(w_1,w_2)$$

$$\times \left(\frac{c}{B}k_y\right)\left(\frac{e_\alpha}{T_\alpha}\left[k_\parallel v_\parallel - \omega_{\bullet\mathbf{k}}^{\alpha T}(\mathbf{v})\right]\right)[J_{\mathbf{k}}^\alpha(\mathbf{v})]^2. \tag{7.47}$$

7.4 Quasilinear Anomalous Fluxes (I)

We now use the results obtained so far in conjunction with Eq. (6.74) in order to obtain a first expression of the quasilinear anomalous particle fluxes. As explained at the end of Sec. 7.1, we limit ourselves to the Markovian approximation, combined with the local equilibrium approximation (Sec. 6.6). The average distribution function is then replaced by a Maxwellian that is quasi-constant in space and time $F_M^\alpha(\mathbf{Y},\mathbf{v};t-t') \to F_M^\alpha(X,\mathbf{v};t) \equiv F_M^\alpha(\mathbf{v})$; moreover, the t'-integration in Eq. (6.74) is extended to infinity. Performing the latter integration we thus obtain:

$$\Gamma_x^\alpha = -\frac{e_\alpha}{T_\alpha}\frac{c}{B}$$

$$\times \int d\mathbf{v} \int d\mathbf{k}\,(2\pi)^{-2} \int dw_1 dw_2\, e^{-i(w_1+w_2)t}\, k_y\, \frac{k_\parallel v_\parallel - \omega_{\bullet\mathbf{k}}^{\alpha T}(\mathbf{v})}{-i(k_\parallel v_\parallel - w_2)}$$

$$\times [J_{\mathbf{k}}^\alpha(\mathbf{v})]^2\, S_{\mathbf{k}}(w_1,w_2)\, F_M^\alpha(\mathbf{v}). \tag{7.48}$$

The fraction in the integrand can be transformed as follows:

$$\frac{k_\parallel v_\parallel - \omega_{\bullet\mathbf{k}}^{\alpha T}(\mathbf{v})}{k_\parallel v_\parallel - w_2} = 1 + \frac{w_2 - \omega_{\bullet\mathbf{k}}^{\alpha T}(\mathbf{v})}{k_\parallel v_\parallel - w_2}, \tag{7.49}$$

and we note that the term 1 contributes zero to the integral. Indeed, the resulting integrand involves $[J_{\mathbf{k}}^\alpha(\mathbf{v})]^2 S_{\mathbf{k}}(w_1,w_2)$, an even function of $\mathbf{k}$, multiplied by the odd function k_y. We thus rewrite (7.48) as:

$$\Gamma_x^\alpha = -\frac{e_\alpha n}{T_\alpha}\frac{c}{B}$$

$$\times (2\pi)^{-2} \int d\mathbf{k}\, ik_y \int dw_1 dw_2\, e^{-i(w_1+w_2)t} S_{\mathbf{k}}(w_1,w_2)\, M_{\mathbf{k}}^{\alpha(n)}(w_2), \tag{7.50}$$

where:

$$M_{\mathbf{k}}^{\alpha(n)}(w) = \frac{2}{\sqrt{\pi}} \int_{-\infty}^{\infty} du_{\parallel} \int_{0}^{\infty} du_{\perp} u_{\perp} \left[J_0(k_{\perp} \rho_{L\alpha} u_{\perp}) \right]^2$$

$$\times \frac{w - \omega_{*\mathbf{k}}^{\alpha} \left[1 + \left(u^2 - \frac{3}{2} \right) \eta_{\alpha} \right]}{k_{\parallel} V_{T\alpha} u_{\parallel} - w} e^{-u_{\perp}^2 - u_{\parallel}^2}. \tag{7.51}$$

We introduced here dimensionless velocity variables, scaled with the thermal velocity:

$$u_{\parallel} = \frac{v_{\parallel}}{V_{T\alpha}}, \quad u_{\perp} = \frac{v_{\perp}}{V_{T\alpha}}, \quad u^2 = u_{\perp}^2 + u_{\parallel}^2, \tag{7.52}$$

and used the thermal Larmor radius: $\rho_{L\alpha} = V_{T\alpha}/|\Omega_{\alpha}|$.

A similar calculation starting from Eq. (6.77) provides us with the expression of the anomalous energy flux:

$$Q_x^{\alpha} = -\frac{e_{\alpha} n}{T_{\alpha}} V_{T\alpha}^2 \frac{c}{B}$$

$$\times (2\pi)^{-2} \int d\mathbf{k} \, ik_y \int dw_1 dw_2 \, e^{-i(w_1 + w_2)t} S_{\mathbf{k}}(w_1, w_2) M_{\mathbf{k}}^{\alpha(P)}(w_2). \tag{7.53}$$

The function $M_{\mathbf{k}}^{\alpha(P)}(w_2)$ differs from $M_{\mathbf{k}}^{\alpha(n)}(w_2)$ only by an additional factor u^2 in the integrand; we therefore write both functions as:

$$M_{\mathbf{k}}^{\alpha(B)}(w) = \frac{2}{\sqrt{\pi}} \int_{-\infty}^{\infty} du_{\parallel} \int_{0}^{\infty} du_{\perp} u_{\perp} \, \psi^{(B)}(u) \left[J_0(k_{\perp} \rho_{L\alpha} u_{\perp}) \right]^2$$

$$\times \frac{w - \omega_{*\mathbf{k}}^{\alpha} \left[1 + \left(u^2 - \frac{3}{2} \right) \eta_{\alpha} \right]}{k_{\parallel} V_{T\alpha} u_{\parallel} - w} e^{-u_{\parallel}^2 - u_{\perp}^2}, \quad B = n, P \tag{7.54}$$

with:

$$\psi^{(n)}(u) = 1, \quad \psi^{(P)}(u) = u^2. \tag{7.55}$$

The integrals in Eq. (7.54) can be evaluated analytically in the form:

$$M_{\mathbf{k}}^{\alpha(n)}(w) = (w - d_1^{\alpha}) K_0^{\alpha} - d_3^{\alpha} K_1^{\alpha},$$

$$M_{\mathbf{k}}^{\alpha(P)}(w) = (w - d_1^{\alpha}) K_1^{\alpha} - d_3^{\alpha} K_2^{\alpha}, \tag{7.56}$$

where:

$$d_1^{\alpha} = \omega_{*\mathbf{k}}^{\alpha} \left(1 - \frac{3}{2} \eta_{\alpha} \right), \quad d_3^{\alpha} = \omega_{*\mathbf{k}}^{\alpha} \eta_{\alpha}, \tag{7.57}$$

and the integrals K_n^{α} are defined as follows:

$$K_n^{\alpha} = 2\pi^{-1/2} \int_{0}^{\infty} du_{\perp} u_{\perp} \int_{-\infty}^{\infty} du_{\parallel} \frac{J_0^2(k_{\perp} \rho_{L\alpha} u_{\perp})}{k_{\parallel} V_{T\alpha} u_{\parallel} - w} u^{2n} e^{-u_{\perp}^2 - u_{\parallel}^2}. \tag{7.58}$$

By expanding $u^2 = u_\perp^2 + u_\parallel^2$, it is easily seen that these integrals can be written as sums of products of a $u_\perp$-integral and a $u_\parallel$-integral:

$$K_n^\alpha = \frac{1}{|k_\parallel| V_{T\alpha}} \sum_{m=0}^{n} \frac{n!}{m!(n-m)!} G_m(b_\alpha) H_{n-m}(\zeta_\alpha), \quad n = 0, 1, 2, \ldots$$

(7.59)

where:

$$G_m(b_\alpha) = 2 \int_0^\infty du_\perp \, u_\perp^{2m+1} J_0^2(k_\perp \rho_{L\alpha} u_\perp) \exp(-u_\perp^2),$$

(7.60)

$$H_p(z_\alpha) = \pi^{-1/2} \int du_\parallel \, u_\parallel^{2p} \frac{1}{u_\parallel - z_\alpha} \exp(-u_\parallel^2).$$

(7.61)

We note that the $u_\perp$-integrals depend on the following dimensionless variable, involving the thermal Larmor radius and the perpendicular wave vector:

$$b_\alpha = \frac{1}{2} k_\perp^2 \rho_{L\alpha}^2.$$

(7.62)

The $u_\parallel$-integrals depend on a different dimensionless variable, involving the complex frequency w and the parallel wave vector:[5]

$$z_\alpha = \frac{w}{|k_\parallel| V_{T\alpha}}.$$

(7.63)

The integrals are rather easily evaluated. For the $u_\perp$-integrals one uses basically the following formula (GRADSHTEIN & RYZHIK 1980):

$$\int_0^\infty dx \, x \, e^{-\alpha^2 x^2} J_m^2(\beta x) = \frac{1}{2\alpha^2} e^{-\beta^2/2\alpha^2} I_m(\beta^2/2\alpha^2),$$

(7.64)

where $I_m(x)$ is the modified Bessel function of order m. When applied to (7.60), some standard algebra, involving the identities obeyed by the functions $I_m(x)$, leads to:

$$\begin{aligned}
G_0(b) &= \Gamma_0(b), \\
G_1(b) &= \Gamma_0(b) \left[1 - M(b)\right], \\
G_2(b) &= \Gamma_0(b) \left\{2 - 3M(b) - b\left[1 - 2M(b)\right]\right\},
\end{aligned}$$

(7.65)

where:

$$\Gamma_0(b) = e^{-b} I_0(b), \quad M(b) = b\left\{1 - \left[I_1(b)/I_0(b)\right]\right\}.$$

(7.66)

[5] The difference between z_α and ζ_α will be explained below.

These two functions are important, because all the finite Larmor radius (FLR) effects can be described in terms of their combinations. They are represented in Figs. 7.1 and 7.2. We note their limiting values:

$$\Gamma_0(\infty) = 0, \quad M(\infty) = 0,$$
$$\Gamma_0(0) = 1, \quad M(0) = 0. \tag{7.67}$$

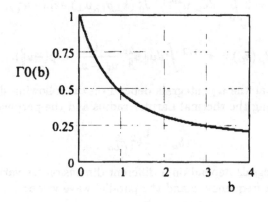

Figure 7.1. The Finite Larmor Radius (FLR) function $\Gamma_0(b)$.

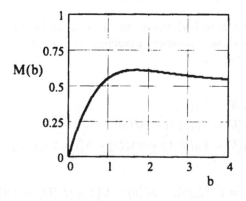

Figure 7.2. The finite Larmor Radius (FLR) function $M(b)$.

Until now we have always treated the FLR effects symmetrically for the ions and the electrons, by introducing the functions $J_0(k_\perp \rho_{L\alpha} u_\perp)$ with the appropriate value of the index α. It should be clear, however, that the Larmor radius of the electrons is smaller than of the ions, because of the large difference in mass. For the data of Table 2.1, we find: $b_e/b_i = m_e/m_i = 5.45 \cdot 10^{-4}$. It is therefore a good approximation to neglect the FLR for the electrons. Thus, for the electrons, in the zero Larmor radius approximation, the integrals G_m reduce to simple constants:

$$b_e = 0: \quad G_0(b_e) = G_1(b_e) = 1, \quad G_2(b_e) = 2. \qquad (7.68)$$

Hence, only the ion quantities are affected by a finite Larmor radius effect.

The integral over $u_\parallel$ requires a detailed discussion. This object is a *Cauchy integral* of the form:

$$\Phi(z) = \int_{-\infty}^{\infty} dx \, \frac{\varphi(x)}{x - z}. \qquad (7.69)$$

Its properties are described in detail in Appendix B. It is a well-defined analytical function of z whenever this variable lies in its "natural" domain S_+ (= upper half-plane, $Im\ z > 0$), as required by the Laplace integral. Great care must be taken as z approaches the real axis from above, as shown in Appendix B.

The integrals $H_p(z_\alpha)$ introduce a class of Cauchy integrals called the *generalized plasma dispersion functions* $Z_{2p}(\zeta_\alpha)$. They must be evaluated in the limit of the complex variable z_α tending to a real value from above: $z_\alpha = \zeta_\alpha + i\varepsilon$:

$$H_p(\zeta_\alpha) \equiv Z_{2p}(\zeta_\alpha)$$

$$= \frac{1}{\sqrt{\pi}} \int_{-\infty}^{\infty} du \, \frac{1}{u - \zeta_\alpha - i\varepsilon} u^{2p} e^{-u^2}, \quad \zeta_\alpha : \text{real}; \quad \varepsilon \to +0. \qquad (7.70)$$

The particular function $Z_0(\zeta) \equiv Z(\zeta)$ is simply called the *plasma dispersion function*. The functions $Z_{2p}(\zeta)$ play an important role in many parts of plasma physics. Their main properties are described in Appendix C.

Using now these results in Eqs.(7.59), (7.56) and (7.54), we find the following expressions for the functions $M_\mathbf{k}^{\alpha(B)}(\zeta_\alpha)$:

$$M_\mathbf{k}^{\alpha(n)}(\zeta_\alpha) = \frac{\Gamma_0(b_\alpha)}{|k_\parallel| V_{T\alpha}} \left\{ \left[\zeta_\alpha - \omega_{*k}^\alpha \left(1 - \tfrac{3}{2}\eta_\alpha \right) \right] Z(\zeta_\alpha) \right.$$

$$\left. - \omega_{*k}^\alpha \eta_\alpha \left[Z_2(\zeta_\alpha) + (1 - M(b_\alpha)) Z(\zeta_\alpha) \right] \right\} \qquad (7.71)$$

and:

$$M_{\mathbf{k}}^{\alpha(P)}(\zeta_\alpha)$$

$$= \frac{\Gamma_0(b_\alpha)}{|k_\parallel| V_{T\alpha}} \left\{ \left[\zeta_\alpha - \omega_{*\mathbf{k}}^\alpha \left(1 - \tfrac{3}{2}\eta_\alpha\right) \right] \left[Z_2(\zeta_\alpha) + (1 - M(b_\alpha)) \, Z(\zeta_\alpha) \right] \right.$$

$$-\omega_{*\mathbf{k}}^\alpha \eta_\alpha \left[Z_4(\zeta_\alpha) + 2(1 - M(b_\alpha)) Z_2(\zeta_\alpha) \right.$$

$$\left. + (2 - 3M(b_\alpha) - b_\alpha(1 - 2M(b_\alpha)) Z(\zeta_\alpha)] \right\}. \tag{7.72}$$

Eqs. (7.50), (7.71) - (7.72) provide us with explicit expressions of the anomalous fluxes. A crucial element is, however, still lacking: the spectrum of potential fluctuations $S_{\mathbf{k}}(w_1, w_2)$. In order to determine this quantity we turn now to the last unexploited tool: the Poisson equation. This should, in principle, close the system of equations and make the theory self-consistent.

7.5   The Poisson Equation

The Poisson equation in Fourier-Laplace representation is obtained from Eq. (4.88):

$$k^2 \, \delta\tilde{\phi}_{\mathbf{k}}(w) = 4\pi \sum_\alpha e_\alpha \int d\mathbf{v} \, J_{\mathbf{k}}^\alpha(\mathbf{v}) \, \delta\tilde{f}_{\mathbf{k}}^\alpha(\mathbf{v}, w). \tag{7.73}$$

This equation establishes a linear relationship between the fluctuating potential $\delta\tilde{\phi}_{\mathbf{k}}(w)$ and the fluctuation $\delta\tilde{f}_{\mathbf{k}}^\alpha(\mathbf{v}, w)$.

We now substitute the solution of the quasilinear gyrokinetic equation (7.31) [with $\delta f_{\mathbf{k}}^\alpha(\mathbf{v}, 0) = 0$, as usual] into the Poisson equation (7.73); the result is written in the form:

$$\hat{\varepsilon}_{0\mathbf{k}}(w) \, \delta\tilde{\phi}_{\mathbf{k}}(w) = 0, \tag{7.74}$$

with:

$$\tilde{\varepsilon}_{0\mathbf{k}}(w) = 1 + \frac{4\pi}{k^2} \sum_\alpha e_\alpha \int d\mathbf{v} \, J_{\mathbf{k}}^\alpha(\mathbf{v}) \, \tilde{\mathcal{G}}_{0\mathbf{k}}(\mathbf{v}, w) \, \Delta_{0\mathbf{k}}^\alpha(\mathbf{v}). \tag{7.75}$$

$\tilde{\varepsilon}_{0\mathbf{k}}(w)$ will be called the *unperturbed dielectric function* (or simply the *dielectric function*).

There appears a serious difficulty with Eq. (7.74). Being a *homogeneous linear equation*, it possesses generically the trivial solution: $\delta\tilde{\phi}_{\mathbf{k}}(w) = 0$. The only way of satisfying Eq. (7.74) with a non-trivial $\delta\tilde{\phi}_{\mathbf{k}}(w)$ is to annul the dielectric function: this problem thus amounts to finding the root(s) $w_{\mathbf{k}}$ of the dielectric function, *i.e.*, solving the equation:

$$\widetilde{\varepsilon}_{0\mathbf{k}}(w_{\mathbf{k}}) = 0. \tag{7.76}$$

This is called the *microscopic drift wave dispersion equation.* The dielectric function is a complex function, and its roots $w_{\mathbf{k}}$ will, in general, also be complex functions of the real wave vector $\mathbf{k}$. It will be shown that there exists one root with a small imaginary part (compared to its real part), and possibly other roots lying far down in the complex plane, which are irrelevant for transport theory. It follows that the solution of the dispersion equation (neglecting the contribution of all possible roots other than the dominant one) is:

$$\delta\widetilde{\phi}_{\mathbf{k}}(w) = \frac{1}{-i(w - w_{\mathbf{k}})}\,\delta\phi_{\mathbf{k}}. \tag{7.77}$$

It was shown, indeed, that the function $[-i(w - w_{\mathbf{k}})]$ plays the role of the Dirac delta function in the complex plane.[6] This random function is used in Eq. (7.46) for calculating the Fourier-Laplace spectrum of the potential fluctuations:

$$S_{\mathbf{k}}(w_1, w_2) = \frac{1}{-i(w_1 - w_{-\mathbf{k}})}\,\frac{1}{-i(w_2 - w_{\mathbf{k}})}\,S_{\mathbf{k}}. \tag{7.78}$$

Two features characterize this solution:

- The Fourier-Laplace spectrum of potential fluctuations $S_{\mathbf{k}}(w_1, w_2)$ is a *singular function of the (complex) frequencies* w_1, w_2. It has simple poles whose location is determined by the roots of the dispersion equation. Note that when $w_{\mathbf{k}} = \omega_{\mathbf{k}}$ is real, Eq. (7.78) becomes:

$$S_{\mathbf{k}}(\omega_1, \omega_2) = \delta(\omega - \omega_{-\mathbf{k}})\,\delta(\omega - \omega_{\mathbf{k}})\,S_{\mathbf{k}}. \tag{7.79}$$

- The k-spectrum $S_{\mathbf{k}}$ *cannot be determined from the Poisson equation,* because the latter is a *homogeneous linear equation.* If the relation between $\delta\widetilde{f}_{\mathbf{k}}^{\alpha}(\mathbf{v}; w)$ and $\delta\widetilde{\phi}_{\mathbf{k}}(w)$ were nonlinear [*i.e.*, by solving the full gyrokinetic equation (4.83)], the possibility of an intrinsic non-trivial solution of the Poisson equation exists.[7]

The consequences of these facts on transport theory will be further discussed in Sec. 7.10.

[6] In order to understand in what sense Eq. (7.77) is a solution of the dispersion equation, we approximate (as stated above) the dielectric function by a function having a single zero, thus disregarding the effect of the roots lying far down in the complex plane: $\widetilde{\varepsilon}_{0\mathbf{k}}(w) \approx (w - w_{\mathbf{k}})e_{\mathbf{k}}$. Then, using (7.77), we find that $\widetilde{\varepsilon}_{0\mathbf{k}}(w)\delta\widetilde{\phi}_{\mathbf{k}}(w) = a_{\mathbf{k}}$, a constant independent of w. The inverse Laplace transform of a constant is zero, as can be seen from Eq. (7.29).

[7] This statement can be illustrated by a trivial mathematical example. Compare the linear homogeneous equation $\varepsilon(w)\,\delta\phi(w) = 0$ with the nonlinear - but also homogeneous - equation $\varepsilon(w)\,[a(w) - \delta\phi(w)]\,\delta\phi(w) = 0$.

7.6 Self-consistent Quasilinear Theory

Before going further into the determination of the anomalous fluxes, we use the results obtained so far for obtaining a completely self-consistent solution of the quasilinear gyrokinetic equation. By a *"self-consistent solution"* is meant one which incorporates the solution of the Poisson equation into the gyrokinetic equation.

We consider a slightly more general equation than the (Markovian) quasilinear equation (7.30): not only do we retain a non-zero initial condition, but we add an arbitrary source term $\hat{\sigma}_k^\alpha(\mathbf{v}; w)$:

$$-i(w - k_\parallel v_\parallel)\,\delta \widetilde{f}_k^\alpha(\mathbf{v}, w)$$
$$= -\Delta_{0k}^\alpha(\mathbf{v})\,\delta\widetilde{\phi}_k(w) + \widetilde{\sigma}_k^\alpha(\mathbf{v}, w) + \delta f_k^\alpha(\mathbf{v}, 0), \qquad (7.80)$$

Using the expression of the unperturbed propagator defined in Eq. (7.32), the solution of the gyrokinetic equation is written in the following compact form:

$$\delta \widetilde{f}_k^\alpha(\mathbf{v}, w) = \widetilde{\mathcal{G}}_{0k}(\mathbf{v}; w)\left\{-\Delta_{0k}^\alpha(\mathbf{v})\,\delta\widetilde{\phi}_k(w) + \widetilde{\sigma}_k^\alpha(\mathbf{v}, w) + \delta f_k^\alpha(\mathbf{v}, 0)\right\}.$$
$$(7.81)$$

This expression is substituted into the Poisson equation (7.73):

$$\widetilde{\varepsilon}_{0k}(w)\delta\widetilde{\phi}_k(w)$$
$$= \frac{4\pi}{k^2}\sum_\alpha e_\alpha \int d\mathbf{v}\, J_k^\alpha(\mathbf{v})\widetilde{\mathcal{G}}_{0k}(\mathbf{v}, w)\left[\widetilde{\sigma}_k^\alpha(\mathbf{v}, w) + \delta f_k^\alpha(\mathbf{v}, 0)\right]. \qquad (7.82)$$

The integral in the expression of the dielectric function (7.75) was calculated above. Indeed, it is easily related to the function $M_k^{\alpha(n)}(w)$ defined in Eq. (7.51):

$$4\pi e_\alpha \int d\mathbf{v}\, J_k^\alpha(\mathbf{v})\,\widetilde{\mathcal{G}}_{0k}(\mathbf{v}, w)\,\Delta_{0k}^\alpha(\mathbf{v}) = k_{D\alpha}^2\left[\Gamma_0(b_\alpha) + M_k^{\alpha(n)}(w)\right], \quad (7.83)$$

where the Debye wave vector of species α, $k_{D\alpha}$ is defined by Eq. (2.31). The dielectric function (7.75) is thus:

$$\widetilde{\varepsilon}_{0k}(w) = 1 + \sum_\alpha \frac{k_{D\alpha}^2}{k^2}\left[\Gamma_0(b_\alpha) + M_k^{\alpha(n)}(w)\right]. \qquad (7.84)$$

We now solve Eq. (7.82) for $\delta\widetilde{\phi}_k(w)$ and substitute the result into Eq. (7.81), thus obtaining the final form of the self-consistent solution of the initial value problem of the quasilinear gyrokinetic equation:

$$\delta \tilde{f}_{\mathbf{k}}^{\alpha}(\mathbf{v}, w)$$

$$= \sum_{\beta} \int d\mathbf{k}' \int d\mathbf{v}' \, \tilde{R}_{0}^{\alpha\beta}(\mathbf{k}, \mathbf{v}|\mathbf{k}', \mathbf{v}', w) \, \left\{ \tilde{\sigma}_{\mathbf{k}'}^{\beta}(\mathbf{v}', w) + \delta f_{\mathbf{k}'}^{\beta}(\mathbf{v}', 0) \right\}, \quad (7.85)$$

with:

$$\tilde{R}_{0}^{\alpha\beta}(\mathbf{k}, \mathbf{v}|\mathbf{k}', \mathbf{v}', w) = \delta(\mathbf{k} - \mathbf{k}') \, \tilde{R}_{0\mathbf{k}}^{\alpha\beta}(\mathbf{v}|\mathbf{v}', w), \quad (7.86)$$

and:

$$\tilde{\mathcal{R}}_{0\mathbf{k}}^{\alpha\beta}(\mathbf{v}|\mathbf{v}', w) = \delta^{\alpha\beta} \, \delta(\mathbf{v} - \mathbf{v}') \, \tilde{\mathcal{G}}_{0\mathbf{k}}(\mathbf{v}, w)$$

$$-\tilde{\mathcal{G}}_{0\mathbf{k}}(\mathbf{v}, w) \, \Delta_{0\mathbf{k}}^{\alpha}(\mathbf{v}) \, \frac{1}{\tilde{\varepsilon}_{0\mathbf{k}}(w)} \, \frac{4\pi}{k^2} \, e_{\beta} \, J_{\mathbf{k}}^{\beta}(\mathbf{v}) \, \tilde{\mathcal{G}}_{0\mathbf{k}}(\mathbf{v}', w). \quad (7.87)$$

The integral kernel $\tilde{\mathcal{R}}_{0\mathbf{k}}^{\alpha\beta}(\mathbf{v}|\mathbf{v}', w)$ defines the *unperturbed resolvent operator*: it relates the solution of the (Fourier-Laplace) gyrokinetic equation to the initial condition and to the source term.[8] The solution (7.85) was first obtained by LANDAU 1946 [see also BALESCU 1963].

It is easily checked that the resolvent operator is a solution of the following inhomogeneous equation:

$$\tilde{\mathcal{G}}_{0\mathbf{k}}^{-1}(\mathbf{v}; w) \, \tilde{\mathcal{R}}_{0\mathbf{k}}^{\alpha\beta}(\mathbf{v}|\mathbf{v}'; w)$$

$$+\Delta_{0\mathbf{k}}^{\alpha}(\mathbf{v}) \frac{4\pi}{k^2} \sum_{\gamma} e_{\gamma} \int d\mathbf{v}'' J_{\mathbf{k}}^{\gamma}(\mathbf{v}'') \, \tilde{\mathcal{R}}_{0\mathbf{k}}^{\gamma\beta}(\mathbf{v}''|\mathbf{v}'; w)$$

$$= \delta^{\alpha\beta} \, \delta(\mathbf{v} - \mathbf{v}'). \quad (7.88)$$

Eq. (7.85) should be compared to (7.31): it expresses the solution of the gyrokinetic equation (7.80), completed by the Poisson equation (7.73) in terms of the source term and of the initial condition. This link is performed by the propagator $\tilde{\mathcal{G}}_{0\mathbf{k}}$ in (7.31), and by the resolvent $\tilde{\mathcal{R}}_{0\mathbf{k}}$ in (7.85). The

---

[8] In the plasma turbulence literature [DuBOIS 1976, DuBOIS & ESPEDAL 1978, KROMMES 1984] the kernel $\hat{\mathcal{R}}_{0\mathbf{k}}^{\alpha\beta}(\mathbf{v}|\mathbf{v}', w)$ [and the related quantity $\hat{\mathcal{R}}_{\mathbf{k}}^{\alpha\beta}(\mathbf{v}|\mathbf{v}', w)$ to be introduced later] is called the *infinitesimal response function*, a name introduced by KRAICHNAN 1959 in his classical work on hydrodynamic turbulence. We believe that this name is doubly ambiguous. First, because $\mathcal{R}_0$ is not a function but an *operator* which, in more general cases, is used as such, without its representation as a kernel (or, equivalently, as a matrix). Next, the term "response" is ambiguous in the same literature: it also (more properly) denotes the effect of a perturbation of the plasma quantities, *e.g.* $\delta \tilde{f}_{\mathbf{k}}^{\alpha}(\mathbf{v}, w)$. For these reasons, in spite of the popularity of the name "response function" we prefer the mathematically well-defined name of "resolvent operator" [RIESZ & SZ.NAGY 1955, BALESCU 1963].

difference between these equations should be clearly understood. In Eq. (7.30) the vertex term appears as a source term, because $\delta\phi$ is treated as a given function, independent of $\delta\tilde{f}^\alpha$. In (7.80) the potential is expressed self-consistently in terms of $\delta\tilde{f}^\alpha$: the operator acting on the latter function is thus a new linear operator including the effect of the Poisson equation. In other words, the potential has been eliminated between the gyrokinetic equation and the Poisson equation. The resolvent obeys Eq. (7.88) which is the analog of (7.42). It shows that the resolvent $\tilde{\mathcal{R}}_{0k}^{\alpha\beta}(\mathbf{v}|\mathbf{v}';w)$ is, indeed, the Laplace transform of a *causal* propagator - or *causal Green's operator* G_{0+} - in time. The latter obeys the following equation, obtained from (7.88):

$$\sum_\gamma \int d\mathbf{v}'' \left\{ (\partial_t + ik_\parallel v_\parallel)\, \delta(\mathbf{v} - \mathbf{v}'')\, \delta^{\alpha\gamma} \right.$$

$$\left. +\Delta_{0k}^\alpha(\mathbf{v})\, \frac{4\pi}{k^2}\, e_\gamma J_\mathbf{k}^\gamma(\mathbf{v}'') \right\} G_{0+k}^{\gamma\beta}(\mathbf{v}''|\mathbf{v}';t-t')$$

$$= \delta^{\alpha\beta}\, \delta(\mathbf{v} - \mathbf{v}')\, \delta(t - t'). \tag{7.89}$$

In the mathematical literature, the Laplace transform of a causal Green's operator is, indeed, called a "resolvent". We already noted earlier that the propagator $\tilde{\mathcal{G}}_{0+k}^{\alpha\beta}(\mathbf{v}|\mathbf{v}';w)$, Eq. (7.34), is technically speaking, also a resolvent operator, corresponding to the gyrokinetic equation (7.81) in which the potential $\delta\tilde{\phi}_\mathbf{k}(w)$ is set equal to zero: it is simply the first term in the right hand side of Eq. (7.87). Given the importance of $\tilde{\mathcal{G}}_{0k}(\mathbf{v};w)$ and of its derived quantities in the theory, we choose (in agreement here with common practice) to call it specifically by the name "*propagator*", even in the Laplace representation.

Another important property of the resolvent operator results from the fact that the *temporal* behaviour of the fluctuation $\delta f_\mathbf{k}^\alpha(\mathbf{v};t)$ can be inferred from the study of the analytical behaviour of the resolvent as a function of the complex frequency w. Indeed, the inverse Laplace transform of the fluctuation [see Eq. (7.29)] yields, for the homogeneous gyrokinetic equation [Eq. (7.85) with $\sigma_\mathbf{k}^\beta(\mathbf{v}';w) = 0$]:

$$\delta f_\mathbf{k}^\alpha(\mathbf{v};t) = \frac{1}{2\pi} \int_\Gamma dw\, e^{-iwt} \sum_\beta \int d\mathbf{v}'\, \tilde{\mathcal{R}}_{0k}^{\alpha\beta}(\mathbf{v}|\mathbf{v}';w)\, \delta f_\mathbf{k}^\beta(\mathbf{v}';0). \tag{7.90}$$

From the theory of functions of a complex variable we know that if the resolvent has (for instance) simple poles at $w = w_n$, $(n = 1, 2, ...)$, the time-dependent response is of the form:

$$\delta f_\mathbf{k}^\alpha(\mathbf{v};t) = -i \sum_n a_{n\,\mathbf{k}}^\alpha(\mathbf{v})\, e^{-iw_n t}, \tag{7.91}$$

where $a_{n\,k}^{\alpha}(\mathbf{v})$ is the residue of the integrand at $w = w_n$. In particular, we see that any pole in the upper complex half-plane, *i.e.*, $w_n = \omega_n + i\gamma_n$, $\gamma_n > 0$, gives rise to an exponentially amplified contribution. Looking at the form of the resolvent (7.87), it appears that such an unstable pole can only possibly arise from a zero of the dielectric function $\widetilde{\varepsilon}_{0k}(w)$. Eq. (7.91) thus clearly establishes the relation between the location of the zeroes of the dielectric function and the temporal behaviour of the response.

7.7   Solution of the Dispersion Equation

We now return to the dispersion equation (7.76) and determine explicitly its root(s). Using the form (7.84) of the dielectric function, multiplied by k^2, we find:

$$k^2 + \sum_{\alpha} k_{D\alpha}^2 \left[\Gamma_0(b_\alpha) + M_{\mathbf{k}}^{\alpha(n)}(w_{\mathbf{k}}) \right] = 0. \qquad (7.92)$$

We note that drift waves are characterized by very small wave vectors, compared to $k_{D\alpha}$ (for both species): $k^2 \ll k_{D\alpha}^2$. (This can be checked from the data given in Table 2.1.) We are then allowed to neglect the first term in Eq. (7.92). Moreover, as explained in Sec. 7.4, the Larmor radius effects are negligible for the electrons, and Eq. (7.67) is applicable for $b_e = 0$. Noting also that $k_{Di}^2/k_{De}^2 = \theta_i^{-1}$, with [see (2.65)]:

$$\theta_i = \frac{T_i}{T_e}, \qquad (7.93)$$

the dispersion equation is rewritten as:

$$1 + M_{\mathbf{k}}^{e(n)}(w_{\mathbf{k}}) + \theta_i^{-1} \left[\Gamma_0 + M_{\mathbf{k}}^{i(n)}(w_{\mathbf{k}}) \right] = 0. \qquad (7.94)$$

Here and below, we no longer write explicitly the argument b_i in the ion Larmor radius functions; thus: $\Gamma_0 \equiv \Gamma_0(b_i)$, $M \equiv M(b_i)$. Any root, $w_{\mathbf{k}}$, of Eq. (7.94) will be called an *eigenfrequency* of the plasma. From Eq. (7.91) we see that this quantity determines indeed the frequency, as well as the damping (or growth) rate of the possible "eigenmodes" of the plasma (at least in the linear approximation). In general, the eigenfrequencies are complex:

$$w_{\mathbf{k}} = \omega_{\mathbf{k}} + i\gamma_{\mathbf{k}}. \qquad (7.95)$$

Clearly, the solutions for which $|\gamma_{\mathbf{k}}| > |\omega_{\mathbf{k}}|$ cannot really qualify as waves, because they are damped out before a single oscillation period. We thus only consider modes with $|\gamma_{\mathbf{k}}| \ll |\omega_{\mathbf{k}}|$. It appears (as will be confirmed *a posteriori*) that these modes exist in a doubly bounded range of frequencies:

$$|k_\parallel| V_{Ti} \ll \omega \ll |k_\parallel| V_{Te}. \tag{7.96}$$

Given the complicated transcendental form of the functions $M_k^{\alpha(n)}(w_k)$, Eq. (7.71) it is impossible to solve exactly the dispersion equation (7.94). The restriction (7.96) allows us to define two small quantities, thus justifying an approximate solution by using perturbation theory. We thus define the following quantities:

$$\lambda f = \frac{\omega_{*k}^e}{|k_\parallel| V_{Te}}, \qquad \lambda h = \frac{|k_\parallel| V_{Ti}}{\omega_{*k}^e}, \tag{7.97}$$

with $\lambda \ll 1$. The eigenfrequency will also be scaled with the drift wave frequency:

$$\overline{\omega} = \frac{w_k}{\omega_{*k}^e}. \tag{7.98}$$

The double inequality (7.96) shows that the electrons and the ions must be treated differently. The plasma dispersion functions $Z_n(\zeta_\alpha)$ appearing in Eq. (7.71) for the electrons will be expanded for small values of ζ_e, whereas the corresponding functions for the ions will be expanded for large values of ζ_i. We will limit ourselves here to the dominant order of approximation. This choice is made not only for pedagogical reasons, as will be discussed below. Using Eq. (C.11) in Appendix C, a somewhat tedious calculation leads to:

$$M_k^{e(n)}(\overline{\omega}) = \lambda f \left(\overline{\omega} - 1 + \tfrac{1}{2}\eta_e\right) i\sqrt{\pi} + O(\lambda^2). \tag{7.99}$$

The corresponding function for the ions is expanded according to Eq. (C.12):

$$M_k^{i(n)}(\overline{\omega}) = -\Gamma_0 - \theta_i \frac{\Gamma_0(1 - M\eta_i)}{\overline{\omega}} + O(\lambda^2). \tag{7.100}$$

These results are substituted into the dispersion equation (7.94). We then expand the root as:

$$\overline{\omega} = a_0 + \lambda i a_1 + O(\lambda^2). \tag{7.101}$$

A standard calculation results in the following values:

$$a_0 = \Gamma_0(1 - M\eta_i),$$
$$a_1 = \sqrt{\pi} f \Gamma_0 \left[1 - \Gamma_0(1 - M\eta_i) - \tfrac{1}{2}\eta_e\right](1 - M\eta_i). \tag{7.102}$$

Reverting to dimensional quantities (7.95), (7.98), we find the real frequency:

$$\omega_k = \Gamma_0 (1 - M\eta_i) \omega_{*k}^e. \tag{7.103}$$

This corresponds to the value found in all textbooks, where, however, only the case $\eta_e = \eta_i = 0$ is considered [KADOMTSEV 1965, KRALL & TRIVELPIECE 1986, NISHIKAWA & WAKATANI 1990, GOLDSTON & RUTHERFORD 1995]. For a very strong magnetic field, such that the ion Larmor radius tends to zero, and also $\rho_s \to 0$, we use the relations (7.67) and find:

$$\omega_{\mathbf{k}} = \omega^e_{*\mathbf{k}}, \quad [b_i \to 0]. \tag{7.104}$$

This is the macroscopic result for a pure drift wave (2.29). Clearly, the FLR effects appearing in Eq. (7.103) are missed in the macroscopic plasmadynamical picture.

We now consider the imaginary part of the eigenfrequency. In the case $\eta_e = \eta_i = 0$, one finds the following result, which is essentially[9] the one found in the textbooks [KRALL & TRIVELPIECE 1986, CAP 1976]:

$$\gamma_{\mathbf{k}} = \sqrt{\pi} \lambda f \Gamma_0 \left(1 - \Gamma_0\right) \omega^e_{*\mathbf{k}}. \tag{7.105}$$

It is important to note that:

$$1 - \Gamma_0 \geq 0. \tag{7.106}$$

This function is zero for $b_i = 0$: thus, for $\eta_e = \eta_i = 0$, there is no imaginary part in the eigenfrequency: we find again an agreement with the macroscopic result. But for any $b_i \neq 0$, $1 - \Gamma_0 > 0$ [see Fig. 7.1], hence:

$$\gamma_{\mathbf{k}} > 0, \quad [\eta_e = \eta_i = 0]. \tag{7.107}$$

Thus, the drift waves (for $\eta_e = \eta_i = 0$) in shearless slab geometry, in the range of frequencies (7.96), are always *linearly unstable* (GALEEV et al. 1963, GARY & SANDERSON 1978, 1979, GARY & ABRAHAM-SHRAUNER 1981, BALESCU 1992). This phenomenon has been called the *universal instability*. The phenomenon is, however, not as universal as it seems at first sight. Indeed, the presence of an electron temperature gradient ($\eta_e \neq 0$) is a stabilizing factor: in the limit $b_i \to 0$ there remains a pure damping, proportional to η_e. In presence of all three thermodynamic forces (∇n, ∇T_e, ∇T_i) (as in any tokamak situation), nothing general can be said *a priori* about the sign of $\gamma_{\mathbf{k}}$.

We also note that the non-vanishing imaginary part of the eigenfrequency can be traced back (through the factor $\sqrt{\pi}$, acting as a "signature") to the imaginary part of the dielectric function (7.84), (7.71), hence

---

[9] There is an obvious misprint in Eq. (8.6.11) of KRALL & TRIVELPIECE 1986: the denominator is written there as $\sqrt{2\kappa T_e / m_e} I_0 e^{-b}$. The square root sign should extend only up to "m_e", the factor $I_0 e^{-b}$ being outside. Moreover, an unfortunate mistake leads to their adoption of an additional factor $(1 + \theta_i^{-1})$ in Eq. (7.105). Finally, for consistency, their exponential factor $exp(-\lambda^2 f^2) \approx 1 + O(\lambda^2)$ should be expanded. Unfortunately these mistakes have been inherited in CAP 1976.

to the imaginary part of the plasma dispersion function (C.4), hence to the delta-function part of the propagator, Eq. (7.32). The latter contribution, $\delta\left(\omega - k_\parallel v_\parallel\right)$, shows that the instability (or damping) is entirely due to a particular population of the electrons, the *resonant particles*, whose parallel velocity equals the parallel phase velocity of the drift waves. The growth rate $\gamma_\mathbf{k}$ has the same origin as the well-known Landau damping in the elementary theory of plasma waves, based on the Vlasov equation (see, *e.g.*, KRALL & TRIVELPIECE 1986, NISHIKAWA & WAKATANI 1990, GOLDSTON & RUTHERFORD 1995).

It is important to stress the parity properties of the drift wave frequency. From the definition (7.18) it is clear that the real part $\omega_\mathbf{k}$ is an *odd* function of $\mathbf{k}$, whereas the imaginary part $\gamma_\mathbf{k}$ is an even function:

$$\omega_{-\mathbf{k}} = -\omega_\mathbf{k}, \qquad \gamma_{-\mathbf{k}} = \gamma_\mathbf{k}. \tag{7.108}$$

A word of caution is necessary at this point. One should not think that the simple result obtained here is general. A change in the assumptions defining the plasma model can change radically the conclusions reached here. If higher order corrections to the eigenfrequencies are included, some strange phenomena seem to appear. For certain relative values of the density gradient ($\propto f$) and of the ion temperature gradient ($\propto f\eta_i$), the perturbative solution (through order λ^3) becomes singular (BALESCU 1992). Actually a careful non-perturbative analysis of the dispersion equation (BALESCU et al. 1993) shows that there is no such singularity, but that the perturbation theory breaks down (the solution is not analytic in λ) in a certain domain of the parameter space. The presence of an (unperturbed) electric current, a different geometry of the magnetic field, *e.g.*, the presence of shear, or of a toroidal geometry as in a tokamak, may also strongly modify the nature of the eigenmodes. The study of the (linear) plasma instabilities is thus a very sensitive subject that must be handled with great care. This justifies the enormous volume of literature existing on the subject.

7.8   DG Drift-Ion Acoustic Modes

In order to check the complete compatibility of the microscopic result with the macroscopic theory of Chap. 3, in particular with the expressions obtained in Sec. 3.5, we must consider an identical situation. This requires consideration of constant temperatures, thus $\eta_e = \eta_i = 0$. Moreover the limit $m_e \to 0$ for the electrons, and the limit $T_i \to 0$ for the ions must be taken. The latter operation is especially tricky: the form of the dispersion equation (7.92) seems to produce a singularity, because of the factor $k_{Di}^2 \to \infty$; moreover it is not obvious that k^2 can be neglected compared to k_{Di}^2.

We therefore reconsider the dispersion equation *ab initio*, writing it in a slightly different form:

$$k^2 + 4\pi e^2 n_0 \left[N_{\mathbf{k}}^e(w) + N_{\mathbf{k}}^i(w)\right] = 0, \tag{7.109}$$

with:

$$N_{\mathbf{k}}^e(w) = \frac{1}{T_{e0}}$$

$$\times \int dv_{\parallel} \frac{e^{-v_{\parallel}^2/V_{Te}^2}}{\pi^{1/2} V_{Te}} \frac{k_{\parallel} v_{\parallel} - \omega_{*\mathbf{k}}^e}{k_{\parallel} v_{\parallel} - w} \int 2 dv_{\perp} v_{\perp} J_0^2 \left(\frac{k_{\perp} v_{\perp}}{\Omega_e}\right) \frac{e^{-v_{\perp}^2/V_{Te}^2}}{\pi V_{Te}^2}, \tag{7.110}$$

and:

$$N_{\mathbf{k}}^i(w) = \frac{1}{T_{i0}}$$

$$\times \int dv_{\parallel} \frac{e^{-v_{\parallel}^2/V_{Ti}^2}}{\pi^{1/2} V_{Ti}} \frac{k_{\parallel} v_{\parallel} + \theta_i \, \omega_{*\mathbf{k}}^e}{k_{\parallel} v_{\parallel} - w} \int 2 dv_{\perp} v_{\perp} J_0^2 \left(\frac{k_{\perp} v_{\perp}}{\Omega_i}\right) \frac{e^{-v_{\perp}^2/V_{Ti}^2}}{\pi V_{Ti}^2}, \tag{7.111}$$

[we used Eqs. (7.48) and (2.28)].

Consider first the electron contribution. The integrals can be done as in Sec. 7.4, with the result:

$$4\pi e^2 n_0 N_{\mathbf{k}}^e(w)$$

$$= k_{De}^2 \Gamma_0(b_e) \left[1 + \frac{1}{|k_{\parallel}| V_{Te}} (w - \omega_{*\mathbf{k}}^e) \left(-2 \frac{w}{|k_{\parallel}| V_{Te}} + i\sqrt{\pi}\right)\right] \tag{7.112}$$

At this point we can take the limit $m_e \to 0$, which implies:

$$V_{Te} = \left(\frac{2T_{e0}}{m_e}\right) \to \infty, \quad b_e \to 0, \quad \Gamma_0(b_e) \to 1$$

Hence, Eq. (7.8) reduces to:

$$4\pi e^2 n_0 N_{\mathbf{k}}^e(w) = k_{De}^2. \tag{7.113}$$

It is easily checked that this result corresponds to an adiabatic density response: $\delta n_e = (e/T_{e0})\delta\phi$.

We now turn to the ion contribution, Eq. (7.8). The integral over $v_\perp$ is performed as above, and yields the familiar factor $\Gamma_0(b_i)$. But in the limit $T_{i0} \to 0$,

$$b_i = \frac{1}{2} k_\perp^2 \rho_{Li}^2 = \frac{1}{2} k_\perp^2 \frac{T_{i0}}{m_i \Omega_i^2} \to 0.$$

Hence $\Gamma_0(b_i) = 1$ in this limit. The remaining expression is written as a sum of two integrals:

$$N_{\mathbf{k}}^i(w) \equiv J_1 + J_2. \tag{7.114}$$

The first integral is:

$$J_1 = \frac{1}{T_{i0}} \int dv_\parallel \frac{e^{-v_\parallel^2/V_{Ti}^2}}{\pi^{1/2} V_{Ti}} \frac{k_\parallel v_\parallel}{k_\parallel v_\parallel - w}. \tag{7.115}$$

In the limit $V_{Ti} \to 0$, the Gaussian factor is extremely narrow, hence the integral effectively extends only over a small range around $v_\parallel = 0$. The second factor in the integrand can therefore be expanded for $k_\parallel v_\parallel \ll w$; thus:

$$J_1 = -\frac{1}{T_{i0}} \int dv_\parallel \frac{e^{-v_\parallel^2/V_{Ti}^2}}{\pi^{1/2} V_{Ti}} \frac{k_\parallel}{w} v_\parallel \left[1 + \frac{k_\parallel v_\parallel}{w} + \left(\frac{k_\parallel v_\parallel}{w} \right)^2 + \dots \right]; \tag{7.116}$$

only the second bracketed term yields a non-zero integral, hence:

$$J_1 = -\frac{1}{T_{i0}} \frac{k_\parallel^2}{w^2} \frac{1}{2} V_{Ti}^2 = -\frac{1}{T_{i0}} \frac{k_\parallel^2}{w^2} \frac{T_{i0}}{m_i} = -\frac{k_\parallel^2}{w^2 m_i}. \tag{7.117}$$

This important result shows that the apparent singularity of Eq. (7.115) is spurious: J_1 tends to a finite limit as $T_{i0} \to 0$.

The second integral in Eq. (7.114) is:

$$J_2 = \frac{1}{T_{i0}} \frac{T_{i0}}{T_{e0}} \omega_{*\mathbf{k}}^e \int dv_\parallel \frac{e^{-v_\parallel^2/V_{Ti}^2}}{\pi^{1/2} V_{Ti}} \frac{1}{k_\parallel v_\parallel - w}. \tag{7.118}$$

Here there is no spurious singularity. The integral is expanded as above and yields, in the limit $T_{i0} \to 0$:

$$J_2 = -\frac{1}{T_{e0}} \frac{\omega_{*\mathbf{k}}^e}{w}. \tag{7.119}$$

Substituting now the partial results (7.113), (7.114), (7.117), (7.119) into the dispersion equation (7.109), the latter becomes:

$$k^2 + k_{De}^2 \left(1 - \frac{k_\parallel^2 T_{e0}}{m_i w^2} - \frac{\omega_{*k}^e}{w} \right) = 0.$$

At this stage, k^2 can clearly be neglected against k_{De}^2, and the dispersion equation reduces to:

$$w^2 - \omega_{*k}^e \, w - k_\parallel^2 c_s^2 = 0. \qquad (7.120)$$

This is the DG drift-ion acoustic dispersion equation, first obtained by KADOMTSEV 1965 (see also NISHIKAWA & WAKATANI 1990). It differs from Eq. (3.12) by a factor $(1 + \rho_s^2 k_\perp^2)$. The same factor is missing in Eq. (7.104) Thus, the microscopic theory developed here agrees in the present limit with the macroscopic one of Sec. 3.5 in the range of perpendicular wave vectors $k_\perp \ll \rho_s^{-1}$. The term $\rho_s^2 k_\perp^2$ poses a problem. It appears as a kind of finite Larmor radius effect: in the double limit $m_e \to 0$, $T_{i0} \to 0$, both the electron and the ion Larmor radius are zero; but the ion-acoustic Larmor radius, evaluated with the finite electron temperature T_{e0} is finite in this limit. A careful analysis of the derivation of the Hasegawa-Wakatani (hence of the Hasegawa-Mima) equations in Chap. 2 shows that the term $\rho_s^2 k_\perp^2$ arises from the ion polarization drift [see Eq. (2.5)]. This term is formally of second order in Ω_i^{-1} and can usually be neglected (as in KADOMTSEV 1965). But for large perpendicular wave vectors it yields a finite correction, even in the limit $T_{i0} = 0$. A thorough discussion of this point is found in GOLDSTON & RUTHERFORD 1995.

7.9 Quasilinear Anomalous Fluxes (II)

We now possess all the tools offered by the quasilinear theory for the calculation of the anomalous particle and energy fluxes. We start with the *anomalous electron flux*. The necessary ingredients are the function $M_k^{e(n)}(w)$ provided by Eq. (7.71), the spectrum $S_k(w_1, w_2)$ provided by Eq. (7.78) and the drift wave eigenfrequency provided by Eqs. (7.103), (7.105). Eq. (7.50) then yields:

$$\Gamma_x^e = -\frac{e_e n}{T_e} \frac{c}{B} (2\pi)^{-2} \int dk \, ik_y \int dw_1 dw_2 \, e^{-i(w_1+w_2)t}$$

$$\times \frac{1}{-i(w_1 - w_{-k})} \frac{1}{-i(w_2 - w_k)} S_k \, M_k^{e(n)}(w_2). \qquad (7.121)$$

The integrals over w_1, w_2 are immediately calculated; using the parity property (7.108), we find:

$$\Gamma_x^e = -\frac{e_e n}{T_e} \frac{c}{B} \int dk \, ik_y \, e^{2\gamma_k t} S_k \, M_k^{e(n)}(w_k). \qquad (7.122)$$

Eq. (7.71) yields:

$$M_{\mathbf{k}}^{e(n)}(w_{\mathbf{k}}) = \frac{\omega_{*\mathbf{k}}^e}{|k_\parallel| V_{Te}} \left\{ \left[\frac{w_{\mathbf{k}}}{\omega_{*\mathbf{k}}^e} - \left(1 - \frac{1}{2}\eta_e\right) \right] Z\left(\frac{w_{\mathbf{k}}}{|k_\parallel| V_{Te}}\right) \right.$$

$$\left. - \eta_e \, Z_2\left(\frac{w_{\mathbf{k}}}{|k_\parallel| V_{Te}}\right) \right\}. \tag{7.123}$$

Because of its first factor, this quantity is of order λ [see Eq. (7.97)], hence we calculate only the zeroth order term in the bracketed expression: From Eq. (C.11) we see that it comes from the imaginary part of $Z(\zeta) = i\sqrt{\pi} + O(\lambda)$. Hence:

$$\Gamma_x^e = \sqrt{\pi}\frac{e_e n}{T_e}\frac{c}{B} \int dk \, k_y \, S_{\mathbf{k}} \, e^{2\gamma_{\mathbf{k}} t} \frac{\omega_{*\mathbf{k}}^e}{|k_\parallel| V_{Te}} \left[1 - \Gamma_0(1 - M\eta_i) + \tfrac{1}{2}\eta_e\right].$$

$$\tag{7.124}$$

A more explicit expression is obtained by using Eqs. (7.17) - (7.18):

$$\Gamma_x^e = \sqrt{\pi}n \frac{c^2}{B^2} \int dk \, k_y^2 \, S_{\mathbf{k}} \, e^{2\gamma_{\mathbf{k}} t} \frac{1}{|k_\parallel| V_{Te}}$$

$$\times \left[(1 - \Gamma_0)\left(-\frac{d\ln n}{dx}\right) + \frac{1}{2}\left(-\frac{d\ln T_e}{dx}\right) + \Gamma_0 M\left(-\frac{d\ln T_i}{dx}\right) \right].$$

$$\tag{7.125}$$

This result exhibits, in particular the *Quasilinear diffusion coefficient*, which is a reference point in transport theory:

$$D_{QL} = \sqrt{\pi}n \frac{c^2}{B^2} \int dk \, k_y^2 \, S_{\mathbf{k}} \, e^{2\gamma_{\mathbf{k}} t} (1 - \Gamma_0) \frac{1}{|k_\parallel| V_{Te}}. \tag{7.126}$$

We shall comment about this quantity in the next section. For the *anomalous ion flux* the task is very easy. We note that in Eq. (7.122) we may make the replacement: $M_{\mathbf{k}}^{e(n)}(w_{\mathbf{k}}) \to \Gamma_0 + M_{\mathbf{k}}^{e(n)}(w_{\mathbf{k}})$ without changing the value of the integral [see Eq. (7.49) and the comment thereafter]. We then calculate the difference between electron and ion fluxes:

$$\Gamma_x^e - \Gamma_x^i = -\frac{e_e n}{T_e}\frac{c}{B} \int dk \, i k_y \, e^{2\gamma_{\mathbf{k}} t} \, S_{\mathbf{k}}$$

$$\times \left\{ 1 + M_{\mathbf{k}}^{e(n)}(w_{\mathbf{k}}) + \theta_i^{-1}\left[\Gamma_0 + M_{\mathbf{k}}^{i(n)}(w_{\mathbf{k}})\right] \right\}$$

$$= 0. \tag{7.127}$$

This quantity is zero because w_k is a solution of the dispersion equation (7.94). Thus:

$$\Gamma_x^e = \Gamma_x^i. \tag{7.128}$$

This equality shows that the particle fluxes are *ambipolar*: the electrons and the ions (more precisely, their guiding centres) move together in the x-direction. In other words, the electric current is zero in the "radial" direction. This should not be surprising, as the motion of the individual guiding centres is determined by the electrostatic drift velocity, which is the same for electrons and for ions [see Eq. (4.34)].

The *anomalous energy fluxes* are calculated in the same way; we do not write down their explicit expressions.

7.10   Discussion of the Quasilinear Fluxes

We found relatively simple expressions for the anomalous fluxes in the quasilinear approximation. These expressions are, however, useless as they stand: they demonstrate that *the quasilinear approximation cannot yield a complete transport theory.*

The main failure was already discussed earlier: the k-spectrum of the potential fluctuations S_k cannot be determined *within the theory*. This is due to the Poisson equation being a linear homogeneous equation. It is therefore impossible to calculate the integrals appearing in the anomalous fluxes.

The next failure is in the appearance of the exponential $e^{2\gamma_k t}$: it is due to the same origin, *i.e.* the linear homogeneous dispersion equation. The latter yields a *"line spectrum"* in frequency, *i.e.* a singular function having simple poles in the complex plane (or delta-functions if the eigenfrequency is real). Only the imaginary part of the frequency appears in the exponential. As a result, if γ_k is negative, the fluxes decay in time. Hence the fluxes tend asymptotically to zero: the behaviour is called *subdiffusive*. On the other hand, if the plasma is unstable, the fluxes grow exponentially (the behaviour is *superdiffusive*). Such behaviour is unphysical. As the fluctuations are amplified, the quasilinear approximation breaks down [Eqs. (7.1) are violated]. At this point, the nonlinear effects become important and are expected to lead to a saturation. Thus, *there is a non-trivial anomalous transport only when the plasma is linearly unstable, provided one can show that the nonlinearity produces a suitable saturation mechanism.*

We note that the quasilinear fluxes, such as Γ_x^e, Eq. (7.9), depend on the magnetic field as B^{-2}. This is the same dependence as the classical and the neoclassical fluxes, and is known to disagree with experiments.

In conclusion, anomalous transport theory cannot be constructed on the basis of the quasilinear approximation. A nonlinear theory is indis-

pensable for a reasonable transport model. Many attempts appear in the
literature for "patching up" the quasilinear transport coefficients. A very
common estimate of this kind is the so-called *mixing length* approximation,
that will be discussed in Chap. 11. But these are at best semi-empirical
procedures, which require a serious theoretical justification. Nevertheless,
as will appear in forthcoming chapters, quasilinear theory constitutes an
important starting point for further progress.

Chapter 8

Closure and Renormalization

8.1 The Basic Equations of Drift Wave Turbulence

We now attack for the first time the fully nonlinear gyrokinetic equation (GKE) (4.83).[1] We use here the Fourier representation for the spatial dependence of the fluctuations, as defined in Eqs. (4.63) - (4.65). Occasionally, we may use an additional Laplace transformation for the time dependence, but only in the final results; the general theory is more simply developed in the time-representation. The resulting GKE is rewritten here by combining the notations of Eq. (4.83) with those of (7.21) and (7.23) as follows:

$$\mathcal{G}_{0+\mathbf{k}}^{-1}(\mathbf{v},t)\,\delta f_{\mathbf{k}}^{\alpha}(\mathbf{v},t) + \Delta_{0\mathbf{k}}^{\alpha}(\mathbf{v})\,\delta\phi_{\mathbf{k}}(t)$$

$$+\frac{c}{B}\frac{1}{2\pi}\int d\mathbf{k}'\, J_{\mathbf{k}'}^{\alpha}(\mathbf{v})\,\mathbf{k}\cdot(\mathbf{k}'\times\mathbf{b})\,\mathcal{Q}\,\delta\phi_{\mathbf{k}'}(t)\,\delta f_{\mathbf{k}-\mathbf{k}'}^{\alpha}(\mathbf{v},t)$$

$$= 0. \tag{8.1}$$

The unperturbed causal propagator $\mathcal{G}_{0+\mathbf{k}}(\mathbf{v},t)$, the unperturbed vertex operator $\Delta_{0\mathbf{k}}^{\alpha}(\mathbf{v})$, the Bessel function $J_{\mathbf{k}}^{\alpha}(\mathbf{v})$ and the subtraction operator $\mathcal{Q}$ were defined, respectively, in Eqs. (7.39), (7.20), (7.19) and (4.45). The inverse propagator is the differential *operator* appearing in the linear GKE [see Eq. (7.41):[2]

$$\mathcal{G}_{0+\mathbf{k}}^{-1}(\mathbf{v};t) = \partial_t + ik_{\parallel}v_{\parallel}. \tag{8.2}$$

---

[1] The presentation given here has benefited from many discussions with H. Wang and E. Vanden Eijnden.
[2] It may be noted that the two-sided propagator $g_0(\mathbf{k},\mathbf{v};t|\mathbf{k}',\mathbf{v}';t')$, which obeys Eq. (7.26), has no inverse!

The two equations (8.1) (for $\alpha = e, i$) are to be completed by the Poisson equation (4.88):

$$\delta\phi_{\mathbf{k}}(t) = \frac{4\pi}{k^2} \sum_{\alpha} e_{\alpha} \int d\mathbf{v} \; J_{\mathbf{k}}^{\alpha}(\mathbf{v}) \, \delta f_{\mathbf{k}}^{\alpha}(\mathbf{v}, t). \tag{8.3}$$

The innocently looking last term in the left hand side of Eq. (8.1) is the source of the enormous complication of the nonlinear plasma problem. The reason can immediately be seen from the fact that it contains a *product* of the fluctuations: $\delta\phi_{\mathbf{k}'}(t) \, \delta f_{\mathbf{k}-\mathbf{k}'}^{\alpha}(\mathbf{v}, t)$ [or, what amounts to the same by (8.3), $\delta f_{\mathbf{k}'}^{\beta}(\mathbf{v}, t) \, \delta f_{\mathbf{k}-\mathbf{k}'}^{\alpha}(\mathbf{v}, t)$]. The linear GKE (7.21) could also be considered as an infinite set of *uncoupled equations*, one for each Fourier component $\delta f_{\mathbf{k}}^{\alpha}(\mathbf{v}, t)$; on the contrary, (8.1) is an infinite set of *coupled equations* for these components.

We now set up a system of notations that will greatly simplify the forthcoming calculations. [A similar, but not identical system was introduced by DuBois & Espedal 1978, and by Krommes 1984].

The unknowns of our problem are the *fluctuations of the distribution function* $\delta f_{\mathbf{k}_1}^{\alpha_1}(\mathbf{v}_1, t_1)$ *i.e.* a *set* of quantities depending on a species index α_1, a wave vector $\mathbf{k}_1$, a velocity vector $\mathbf{v}_1 (\equiv v_{1\parallel}, v_{1\perp})$ [without the gyrophase φ_1 which is averaged out] and a time t_1. We shall henceforth use the shorter term *"response"* for the "fluctuation of the distribution function" $\delta f_{\mathbf{k}}^{\alpha}(\mathbf{v}, t)$: this widely used terminology is quite intuitive for denoting the response of the plasma (in phase space) to the potential fluctuations $\delta\phi_{\mathbf{k}}(t)$. We further note that the calculations in the time formalism naturally deal with "plus-functions" of time, *i.e.*, functions defined for *positive times* t; thus: $\delta f_{\mathbf{k}}^{\alpha}(\mathbf{v}, t) \to \delta f_{+\mathbf{k}}^{\alpha}(\mathbf{v}, t)$ [see the discussions of Sec. 7.3, Eqs. (7.35), (7.36)].[3] These quantities will be grouped together in a "vector" depending on the *set of variables* denoted by the numeral 1:

$$\delta f(1) \to \delta f_{+\mathbf{k}_1}^{\alpha_1}(\mathbf{v}_1, t_1) \equiv H(t_1) \, \delta f_{\mathbf{k}_1}^{\alpha_1}(\mathbf{v}_1, t_1). \tag{8.4}$$

The *same* convention holds for quantities independent of one or two of the variables of the set 1. In particular, the potential fluctuation, which is independent of α and of $\mathbf{v}$, is represented as follows:

$$\delta\phi(1) \to H(t_1) \, \delta\phi_{\mathbf{k}_1}(t_1). \tag{8.5}$$

The next type of objects entering the theory are "matrices" depending on two or three sets of variables $(\alpha, \mathbf{k}, \mathbf{v}; t)$:

$$A(1\,2) \to A_{\mathbf{k}_1, \mathbf{k}_2}^{\alpha_1, \alpha_2}(\mathbf{v}_1, \mathbf{v}_2, t_1, t_2),$$
$$B(1\,2\,3) \to B_{\mathbf{k}_1, \mathbf{k}_2, \mathbf{k}_3}^{\alpha_1, \alpha_2, \alpha_3}(\mathbf{v}_1, \mathbf{v}_2, \mathbf{v}_3, t_1, t_2, t_3). \tag{8.6}$$

---

[3] When no confusion arises, the subscript "+" will not be written explicitly.

The most important operation with these objects is the *contraction*, or *scalar product*, or *matrix product*. It will be abbreviated by introducing a convention of *summation over repeated arguments*, which involves a summation over species indices α, an integration over wave vectors $\mathbf{k}$, an integration over velocities $\mathbf{v}$ and an integration *over the whole range of times* t (from $-\infty$ to $+\infty$). The rule will be clear from the following example:

$$A(1\,2)\,B(2\,3) = C(1\,3)$$

$$\rightarrow \sum_{\alpha_2} \int d\mathbf{k}_2 \int d\mathbf{v}_2 \int_{-\infty}^{\infty} dt_2 \; A_{\mathbf{k}_1,\mathbf{k}_2}^{\alpha_1,\alpha_2}(\mathbf{v}_1,\mathbf{v}_2,t_1,t_2)\, B_{\mathbf{k}_2,\mathbf{k}_3}^{\alpha_2,\alpha_3}(\mathbf{v}_2,\mathbf{v}_3,t_2,t_3)$$

$$= C_{\mathbf{k}_1\mathbf{k}_3}^{\alpha_1\alpha_3}(\mathbf{v}_1,\mathbf{v}_3,t_1,t_3). \tag{8.7}$$

We underscore the following feature of this equation: 1 and 3 are *fixed indices*, whereas the (repeated) index 2 is a *dummy summation variable*. The latter may be arbitrarily renamed as $4, 5, 8, ...$ (provided their pairing is respected). This renaming process will be repeatedly used below. In order to make the difference between fixed and repeated variables more visible, we will often (but not always) use **bold** numerals for the fixed indices in the expressions involving matrix products.

We now define a set of operators that allow us to write the GKE in a compact way. We begin with the Poisson equation, which is a bit tricky. We show that it can be written as:

$$\delta\phi(1) = \Phi(1\,2)\,\delta f(2). \tag{8.8}$$

The operator $\Phi(1\,2)$ that does the job is:

$$\Phi(1\,2) \rightarrow \Phi_{\mathbf{k}_1\mathbf{k}_2}^{\alpha_1\alpha_2}(\mathbf{v}_1,\mathbf{v}_2,t_1,t_2)$$

$$= \frac{4\pi}{k_1^2}\, e_{\alpha_2}\, J_{\mathbf{k}_2}^{\alpha_2}(\mathbf{v}_2)\,\delta(\mathbf{k}_1 - \mathbf{k}_2)\,\delta(t_1 - t_2). \tag{8.9}$$

Note that this operator is independent of α_1 and of $\mathbf{v}_1$. By our rule, the right hand side of Eq. (8.8) is realized as follows:

$$\Phi(1\,2)\,\delta f(2)$$

$$\rightarrow \sum_{\alpha_2} \int d\mathbf{k}_2 \int d\mathbf{v}_2 \int_{-\infty}^{\infty} dt_2\; \Phi_{\mathbf{k}_1,\mathbf{k}_2}^{\alpha_1;\alpha_2}(\mathbf{v}_1,\mathbf{v}_2,t_1,t_2)\, H(t_2)\,\delta f_{\mathbf{k}_2}^{\alpha_2}(\mathbf{v}_2,t_2)$$

$$= \frac{4\pi}{k_1^2} \sum_{\alpha_2} \int d\mathbf{k}_2 \int d\mathbf{v}_2 \int_{-\infty}^{\infty} dt_2\; e_{\alpha_2} J_{\mathbf{k}_2}^{\alpha_2}(\mathbf{v}_2)$$

$$\times\, \delta(\mathbf{k}_1 - \mathbf{k}_2)\,\delta(t_1 - t_2)\, H(t_2)\,\delta f_{\mathbf{k}_2}^{\alpha_2}(\mathbf{v}_2,t_2)$$

$$= \frac{4\pi}{k_1^2}\, H(t_1) \sum_{\alpha_2} e_{\alpha_2} \int d\mathbf{v}_2\; J_{\mathbf{k}_1}^{\alpha_2}(\mathbf{v}_2)\delta f_{\mathbf{k}_1}^{\alpha_2}(\mathbf{v}_2,t_1) = \delta\phi_{+\mathbf{k}_1}(t_1),$$

which is, indeed, identical to $\delta\phi_{\mathbf{k}_1}(t_1)$ as given by the Poisson equation (8.3).

The coefficients of the linear part of the GKE are easily expressed in terms of diagonal operators. Thus, the *inverse unperturbed propagator* is defined as:

$$g_0^{-1}(1\,2) \rightarrow \left(g_{+0}^{-1}\right)_{\mathbf{k}_1,\mathbf{k}_2}^{\alpha_1,\alpha_2}(\mathbf{v}_1,\mathbf{v}_2,t_1,t_2)$$

$$\equiv \left(\partial_{t_1} + ik_{1\parallel}v_{1\parallel}\right) \delta^{\alpha_1\,\alpha_2} \delta(\mathbf{v}_1 - \mathbf{v}_2) \delta(\mathbf{k}_1 - \mathbf{k}_2) \delta(t_1 - t_2), \qquad (8.10)$$

and the *unperturbed vertex operator* is given by:

$$\Delta_0(1\,2) \rightarrow (\Delta_0)_{\mathbf{k}_1,\mathbf{k}_2}^{\alpha_1,\alpha_2}(\mathbf{v}_1,\mathbf{v}_2,t_1,t_2)$$

$$\equiv \Delta_{0\,\mathbf{k}_1}^{\alpha_1}(\mathbf{v}_1) \delta^{\alpha_1\,\alpha_2} \delta(\mathbf{v}_1 - \mathbf{v}_2) \delta(\mathbf{k}_1 - \mathbf{k}_2) \delta(t_1 - t_2)$$

$$= i\frac{e_{\alpha_1}}{T_{\alpha_1}} J_{\mathbf{k}_1}^{\alpha_1}(\mathbf{v}_1) F_M^{\alpha_1}(\mathbf{v}_1) \left[k_{1\parallel}v_{1\parallel} - \omega_{*\mathbf{k}_1}^{\alpha_1 T}(\mathbf{v}_1)\right]$$

$$\times \delta^{\alpha_1\,\alpha_2} \delta(\mathbf{k}_1 - \mathbf{k}_2) \, \delta(t_1 - t_2) \delta(\mathbf{v}_1 - \mathbf{v}_2), \qquad (8.11)$$

where:

$$\omega_{*\mathbf{k}_1}^{\alpha_1 T}(\mathbf{v}_1) = -\frac{cT_{\alpha_1}}{e_{\alpha_1}BL_n} k_{1y} \left[1 + \left(\frac{v_1^2}{V_{T\alpha_1}^2} - \frac{3}{2}\right)\eta_{\alpha_1}\right]. \qquad (8.12)$$

We recall that the *unperturbed causal propagator*, i.e., the inverse of the operator $g_0(1\,2)$, is:

$$g_0(1\,2) \rightarrow (g_{0+})_{\mathbf{k}_1,\mathbf{k}_2}^{\alpha_1,\alpha_2}(\mathbf{v}_1,\mathbf{v}_2,t_1,t_2)$$

$$= \delta^{\alpha_1,\alpha_2} \delta(\mathbf{v}_1 - \mathbf{v}_2) \delta(\mathbf{k}_1 - \mathbf{k}_2)$$

$$\times H(t_1 - t_2) \exp[-ik_{1\parallel}v_{1\parallel}(t_1 - t_2)]. \qquad (8.13)$$

For the nonlinear term we introduce the operator M:

$$M(1\,2\,3) \rightarrow M_{\mathbf{k}_1,\mathbf{k}_2,\mathbf{k}_3}^{\alpha_1,\alpha_2,\alpha_3}(\mathbf{v}_1,\mathbf{v}_2,\mathbf{v}_3,t_1,t_2,t_3)$$

$$\equiv \delta^{\alpha_1\alpha_2}\delta^{\alpha_1\alpha_3} \delta(\mathbf{v}_1 - \mathbf{v}_2) \delta(\mathbf{v}_1 - \mathbf{v}_3) \delta(\mathbf{k}_1 - \mathbf{k}_2 - \mathbf{k}_3)$$

$$\times \delta(t_1 - t_2) \delta(t_1 - t_3) \frac{c}{B} \mathbf{k}_1 \cdot (\mathbf{k}_2 \times \mathbf{b}) J_{\mathbf{k}_2}^{\alpha_2}(\mathbf{v}_2) ; \qquad (8.14)$$

and finally, we define the operator N:

$$N(1\,2\,3) = M(1\,4\,3)\,\Phi(4\,2) + M(1\,4\,2)\,\Phi(4\,3). \qquad (8.15)$$

We note the important symmetry property:

$$N(1\,2\,3) = N(1\,3\,2). \tag{8.16}$$

In terms of all these quantities, the GKE takes the following compact form:

$$g_0^{-1}(1\,2)\,\delta f(2) + \Delta_0(1\,2)\,\delta\phi(2) + \tfrac{1}{2}N(1\,2\,3)\,\mathcal{Q}\,\delta f(2)\,\delta f(3) = 0. \tag{8.17}$$

It is useful to write down two additional, equivalent forms. Explicit reference to the Poisson equation (8.8) results in:

$$\left[g_0^{-1}(1\,2) + \Delta_0(1\,3)\,\Phi(3\,2)\right]\delta f(2) + \tfrac{1}{2}N(1\,2\,3)\,\mathcal{Q}\,\delta f(2)\,\delta f(3) = 0. \tag{8.18}$$

In this form, the potential no longer appears explicitly and the Poisson equation is incorporated into the GKE: this represents the completely *self-consistent* form of the GKE.

Finally, we may define an operator L:

$$L(1\,2) = g_0^{-1}(1\,2) + \Delta_0(1\,3)\,\Phi(3\,2), \tag{8.19}$$

and write the GKE in the form:

$$L(1\,2)\,\delta f(2) + \tfrac{1}{2}\,N(1\,2\,3)\,\mathcal{Q}\,\delta f(2)\,\delta f(3) = 0. \tag{8.20}$$

In this form, the stress is laid on the structure of the GKE operator, consisting of a linear part L and a quadratically non-linear part N [see (4.90)].

Before going further, we may make a few general remarks on a difference between linear and nonlinear GKE. As appears clearly from its derivation in Chap. 4, the GKE must be interpreted as a *stochastic equation*: the signature of this feature is the appearance of the ensemble-averaging operation $\langle...\rangle$ in the equation.

In the *linearized GKE*, however, this averaging operator is absent; hence, this equation can *also* be interpreted deterministically. It may describe the evolution of a small perturbation δf from a reference state F, the propagation of (linear) waves, their dispersion, damping or amplification, etc. Thus, all the properties discussed in Chap. 7 can also be interpreted deterministically (*i.e.*, the symbols δf denote functions which have numerical values). It is only in applying these results to transport theory (in the quasilinear approximation) that we superimposed from outside (so to say) the assumptions that δf, $\delta\phi$ were fluctuating quantities and that only ensemble-averages of their products may have definite numerical values.

Whenever the nonlinear processes can no longer be neglected in the evolution equation, because the amplitude of the perturbation becomes sufficiently large, the situation changes radically. The distribution functions undergo important fluctuations due, on one hand, to the intrinsic dynamical instability of the particle motions (non-integrable Hamiltonian) and, on the other hand, to the incessant transformations of the collective degrees of freedom through decay, amplification or change of nature produced by the non-linear coupling. Starting from any "quiet" state, these phenomena progressively lead the plasma toward a very complex state of motion, which is practically irreproducible, because of the extreme sensitivity to initial conditions which characterizes the chaotic dynamics. In this state of *"deterministic chaos"* the plasma is said to be **turbulent**.

The transition from quiet to turbulent motion is an extremely difficult and poorly understood problem. On the other hand, the final, *fully developed turbulence* can be adequately described by a *statistical formulation* (see Sec. 1.6). In this case, the complex turbulent state is modelled by interpreting the nonlinear GKE (8.17) or (8.20) as a *stochastic equation*. We now develop this point of view systematically. For general expositions of the theory of stochastic processes, that are quite clear and easily accessible to non-professional mathematicians, we quote the monographs of RESIBOIS & DE LEENER 1977, VAN KAMPEN 1981 and GARDINER 1985.

8.2   Potential and Response Correlations

The new point of view implies that numerical values of the functions $\delta f(1)$ or $\delta\phi(1)$ have no meaning. The equations (8.17), (8.18) can only be used as a starting point for the computation of *average values* taken over a large number of - individually irreproducible - realizations. By definition, we have $\langle \delta f_{\mathbf{k}}^{\alpha}(\mathbf{v},t) \rangle = 0$, $\langle \delta\phi_{\mathbf{k}}(t) \rangle = 0$ [see (1.33)].

The next most important average quantities are the *Eulerian correlation functions* (in the present chapter, they will be simply called: *correlations*) of various kinds. The basic quantity of this type is the correlation of two responses $\delta f_{\mathbf{k}_1}^{\alpha_1}(\mathbf{v}_1, t_1)$ and $\delta f_{\mathbf{k}_2}^{\alpha_2}(\mathbf{v}_2, t_2)$: it is a function of α_1, α_2, $\mathbf{v}_1$, $\mathbf{v}_2$, $\mathbf{k}_1$, $\mathbf{k}_2$, t_1, t_2. The correlations are readily incoporated into our system of compact notations:

$$\langle \delta f(1)\, \delta f(2) \rangle \equiv C(1\,2) \rightarrow \left\langle \delta f_{+\mathbf{k}_1}^{\alpha_1}(\mathbf{v}_1, t_1)\, \delta f_{+\mathbf{k}_2}^{\alpha_2}(\mathbf{v}_2, t_2) \right\rangle. \qquad (8.21)$$

This quantity is also called the *second moment* of the random function $\delta f(1)$. A discussion of the properties of Eulerian correlations was given in Sec. 6.3, together with a list of further references.

A major role will be played by the correlations involving the potential [they are already present in Eq. (8.17), through the operator $\mathcal{Q}$, Eq. (4.45)]:

$$\langle \delta f(1)\,\delta\phi(2)\rangle \equiv \mathcal{T}(1\,2) \rightarrow \big\langle \delta f^{\alpha_1}_{+\mathbf{k}_1}(\mathbf{v}_1, t_1)\,\delta\phi_{+\mathbf{k}_2}(t_2)\big\rangle,$$

$$\langle \delta\phi(1)\,\delta f(2)\rangle \equiv \mathcal{T}^\dagger(1\,2) \rightarrow \big\langle \delta\phi_{+\mathbf{k}_1}(t_1)\,\delta f^{\alpha_2}_{+\mathbf{k}_2}(\mathbf{v}_2, t_2)\big\rangle, \qquad (8.22)$$

$$\langle \delta\phi(1)\,\delta\phi(2)\rangle \equiv \mathcal{S}(1\,2) \rightarrow \big\langle \delta\phi_{+\mathbf{k}_1}(t_1)\,\delta\phi_{+\mathbf{k}_2}(t_2)\big\rangle. \qquad (8.23)$$

We underscore the convention defining the mixed correlation $\mathcal{T}(1\,2)$: the first label corresponds to the variable associated with $\delta f(1)$; in $\mathcal{T}^\dagger(1\,2)$ the first variable is associated with the potential fluctuation, $\delta\phi(1)$. Of course, these correlations are not independent of $C(1\,2)$: from Eq. (8.8) we have:

$$\langle \delta f(1)\,\delta\phi(2)\rangle = \Phi(2\,3)\,\langle \delta f(1)\,\delta f(3)\rangle,$$

$$\langle \delta\phi(1)\,\delta\phi(2)\rangle = \Phi(1\,3)\,\Phi(2\,4)\,\langle \delta f(3)\,\delta f(4)\rangle,$$

i.e.,

$$\mathcal{T}(1\,2) = \Phi(2\,3)\,C(1\,3), \quad \mathcal{S}(1\,2) = \Phi(1\,3)\,\Phi(2\,4)\,C(3\,4). \qquad (8.24)$$

In most problems to be discussed here, it will be assumed that the turbulent state of the plasma is spatially quasi-homogeneous or else, *weakly inhomogeneous*. These properties will be expressed by the following relations:[4]

$$\big\langle \delta f^{\alpha_1}_{+\mathbf{k}_1}(\mathbf{v}_1, t_1)\,\delta f^{\alpha_2}_{+\mathbf{k}_2}(\mathbf{v}_2, t_2)\big\rangle = \delta(\mathbf{k}_1 + \mathbf{k}_2)\,C^{\alpha_1\alpha_2}_{\mathbf{k}_1}(\mathbf{v}_1, \mathbf{v}_2, t_1, t_2),$$

$$\big\langle \delta f^{\alpha_1}_{+\mathbf{k}_1}(\mathbf{v}_1, t_1)\,\delta\phi_{+\mathbf{k}_2}(t_2)\big\rangle = \delta(\mathbf{k}_1 + \mathbf{k}_2)\,T^{\alpha_1}_{\mathbf{k}_1}(\mathbf{v}_1, t_1, t_2),$$

$$\big\langle \delta\phi_{+\mathbf{k}_1}(t_1)\,\delta\phi_{+\mathbf{k}_2}(t_2)\big\rangle = \delta(\mathbf{k}_1 + \mathbf{k}_2)\,S_{\mathbf{k}_1}(t_1, t_2). \qquad (8.25)$$

The functions $C^{\alpha_1\alpha_2}_{\mathbf{k}_1}(\mathbf{v}_1, \mathbf{v}_2, t_1, t_2)$, $T^{\alpha_1}_{\mathbf{k}_1}(\mathbf{v}_1, t_1, t_2)$, $S_{\mathbf{k}_1}(t_1, t_2)$ also depend weakly (in the standard shearless slab geometry) on the space coordinate x; but this dependence will usually not be written explicitly. Note that no relative ordering is required for the time variables in these formulae, except that $t_1 > 0$, $t_2 > 0$.

The stochastic equation (8.17) is used as a starting point for the derivation of equations of evolution for the correlations. Thus, multiplying throughout this equation by $\delta f(2)$ (with a *fixed* index **2**) and averaging the result, we obtain:[5]

---

[4] See Eq. (6.56), without the temporal Fourier transformation.
[5] We underscore again the structure of this equation: **1** and **2** are *fixed variables*, whereas the repeated indices **3** and **4** are *dummy summation variables*.

$$g_0^{-1}(1\,3)\,\langle \delta f(3)\,\delta f(2)\rangle + \Delta_0(1\,3)\,\langle \delta\phi(3)\,\delta f(2)\rangle$$
$$+\tfrac{1}{2}\,N(1\,3\,4)\,\langle \delta f(3)\,\delta f(4)\,\delta f(2)\rangle = 0. \qquad (8.26)$$

In obtaining this result, we used the properties of the subtraction operator as follows:

$$\langle [\mathcal{Q}\,\delta f(3)\,\delta f(4)]\,\delta f(2)\rangle$$
$$= \langle \delta f(3)\,\delta f(4)\,\delta f(2)\rangle - \langle \langle \delta f(3)\,\delta f(4)\rangle\,\delta f(2)\rangle$$
$$= \langle \delta f(3)\,\delta f(4)\,\delta f(2)\rangle - \langle \delta f(3)\,\delta f(4)\rangle\,\langle \delta f(2)\rangle,$$

where the last term is zero, by Eq. (1.33). Equivalently, we get from Eq. (8.20)

$$L(1\,3)\,\langle \delta f(3)\,\delta f(2)\rangle + \tfrac{1}{2}\,N(1\,3\,4)\,\langle \delta f(3)\,\delta f(4)\,\delta f(2)\rangle = 0. \qquad (8.27)$$

We are thus faced with the ubiquitous plague of all the statistical theories of interacting (hence, nonlinear) systems (BALESCU 1975, 1997). The determination of the two-body correlations requires the knowledge of the three-body correlations. These, in turn are determined by an equation involving the four-body correlations:

$$L(1\,4)\,\langle \delta f(4)\,\delta f(2)\,\delta f(3)\rangle$$
$$+\tfrac{1}{2}N(1\,4\,5)\{\langle \delta f(4)\,\delta f(5)\,\delta f(2)\,\delta f(3)\rangle$$
$$-\langle \delta f(4)\,\delta f(5)\rangle\,\langle \delta f(2)\,\delta f(3)\rangle\} = 0, \qquad (8.28)$$

and so on. The form of the nonlinear term is again obtained by using the properties of the subtraction operator:

$$\langle \mathcal{Q}\,[\delta f(4)\,\delta f(5)]\,\delta f(2)\,\delta f(3)\rangle$$
$$= \langle \delta f(4)\,\delta f(5)\,\delta f(2)\,\delta f(3)\rangle - \langle \langle \delta f(4)\,\delta f(5)\rangle\,\delta f(2)\,\delta f(3)\rangle$$

In order to pursue this discussion it is necessary to introduce the following refinement, well known in nonequilibrium statistical mechanics (BALESCU 1975, Sec. 3.5; BALESCU 1997, Sec. 4.4). [For simplicity, we use the provisional abbreviation $\langle 1\,2...\rangle$ for $\langle \delta f(1)\,\delta f(2)...\rangle$]. A correlation such as $\langle 1\,2...s\rangle$ represents the global deviation of the microscopic state from one in which all particles are statistically independent; in many cases this representation does not yield sufficient information. For $s = 2$, $\langle 1\,2\rangle$ cannot be further analyzed. But for $s = 3$ we can imagine situations where particles 1 and 2 are correlated, the third one being independent (*e.g.*, because it is far away), or particles $1, 2, 3$ are altogether connected. Both cases are included in $\langle 1\,2\,3 \rangle$, but the type of correlation is different in the two cases. In order to analyze all situations, we consider all possible *partitions of the set* $(1\,2...s)$ into disjoint subsets containing at least one particle. Interpreting each subset, or *cluster*, as representing a group of irreducibly correlated particles, statistically independent of the other subsets, we obtain the following *cluster representation* of the correlations:

$$\langle 1 \rangle = \langle\!\langle 1 \rangle\!\rangle,$$
$$\langle 1\,2 \rangle = \langle\!\langle 1 \rangle\!\rangle \langle\!\langle 2 \rangle\!\rangle + \langle\!\langle 1\,2 \rangle\!\rangle,$$
$$\langle 1\,2\,3 \rangle = \langle\!\langle 1 \rangle\!\rangle \langle\!\langle 2 \rangle\!\rangle \langle\!\langle 3 \rangle\!\rangle + [\langle\!\langle 1 \rangle\!\rangle \langle\!\langle 2\,3 \rangle\!\rangle + perm.] + \langle\!\langle 1\,2\,3 \rangle\!\rangle,$$
$$\langle 1\,2\,3\,4 \rangle = \langle\!\langle 1 \rangle\!\rangle \langle\!\langle 2 \rangle\!\rangle \langle\!\langle 3 \rangle\!\rangle \langle\!\langle 4 \rangle\!\rangle + [\langle\!\langle 1 \rangle\!\rangle \langle\!\langle 2 \rangle\!\rangle \langle\!\langle 3\,4 \rangle\!\rangle + perm.]$$
$$+ [\langle\!\langle 1 \rangle\!\rangle \langle\!\langle 2\,3\,4 \rangle\!\rangle + perm.] + [\langle\!\langle 1\,2 \rangle\!\rangle \langle\!\langle 3\,4 \rangle\!\rangle + perm.] + \langle\!\langle 1\,2\,3\,4 \rangle\!\rangle.$$
$$(8.29)$$

These relations introduce the irreducible correlations $\langle\!\langle\, \delta f \delta f... \rangle\!\rangle$, called *cumulants* in the literature of probability theory (see GARDINER 1985, Sec. 2.7). When the moments are centred, *i.e.* when $\langle \delta f(1) \rangle = 0$ holds, these relations (reverting to the complete notation) reduce to:

$$\langle \delta f(1) \rangle = \langle\!\langle \delta f(1) \rangle\!\rangle = 0,$$
$$\langle \delta f(1)\, \delta f(2) \rangle = \langle\!\langle \delta f(1)\, \delta f(2) \rangle\!\rangle,$$
$$\langle \delta f(1)\, \delta f(2)\, \delta f(3) \rangle = \langle\!\langle \delta f(1)\, \delta f(2)\, \delta f(3) \rangle\!\rangle,$$
$$\langle \delta f(1)\, \delta f(2)\, \delta f(3)\, \delta f(4) \rangle = \langle\!\langle \delta f(1)\, \delta f(2) \rangle\!\rangle \langle\!\langle \delta f(3)\, \delta f(4) \rangle\!\rangle$$
$$+ \langle\!\langle \delta f(1)\, \delta f(3) \rangle\!\rangle \langle\!\langle \delta f(2)\, \delta f(4) \rangle\!\rangle$$
$$+ \langle\!\langle \delta f(1)\, \delta f(4) \rangle\!\rangle \langle\!\langle \delta f(2)\, \delta f(3) \rangle\!\rangle$$
$$+ \langle\!\langle \delta f(1)\, \delta f(2)\, \delta f(3)\, \delta f(4) \rangle\!\rangle.$$
$$(8.30)$$

We now express Eqs. (8.27) and (8.28) in terms of cumulants. Given, however, that for centred moments the second and the third cumulants equal the respective moments, Eq. (8.27) remains unchanged; Eq. (8.28) becomes:

$$L(1\,4)\ \langle \delta f(4)\,\delta f(2)\,\delta f(3)\rangle$$
$$+\tfrac{1}{2}N(1\,4\,5)\ \{\ [\langle \delta f(4)\,\delta f(5)\rangle\ \langle \delta f(2)\,\delta f(3)\rangle$$
$$+\langle \delta f(2)\,\delta f(5)\rangle\ \langle \delta f(3)\,\delta f(4)\rangle + \langle \delta f(2)\,\delta f(4)\rangle\ \langle \delta f(3)\,\delta f(5)\rangle$$
$$+\langle\langle\,\delta f(4)\delta f(5)\delta f(2)\delta f(3)\,\rangle\rangle\,]$$
$$-\ \langle \delta f(2)\,\delta f(3)\rangle\ \langle \delta f(4)\,\delta f(5)\rangle\ \} = 0. \qquad (8.31)$$

The structure of this equation, generated from the moment equation (8.28), is best visualized in terms of the graph of Fig. 8.1. The coupling of particles 4 and 5 yielding 1 is represented by the graph of part **A** [in which the last term (counter-term) is not shown]. In part **B** the irreducible correlations are visualized by vertical segments at the extreme right. The framed graph is cancelled by the counter-term. This is important, because the latter graph is actually a contribution to the reducible correlation $\langle 1\rangle\,\langle 2\,3\rangle$ rather than to the cumulant $\langle\langle 1\,2\,3\rangle\rangle$.

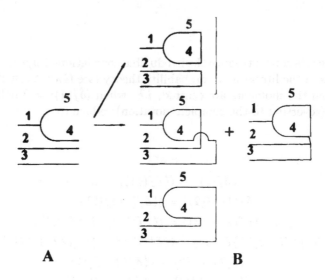

A   **B**

Figure 8.1. Graphs representing: A: the moment equation (8.28); B: the cumulant equation (8.31).

We also note that the remaining two terms in the right hand side involving second moments are equal to each other, because of Eq. (8.16). Hence Eq. (8.31) reduces to:

$$L(1\,4)\ \langle\langle\,\delta f(4)\,\delta f(2)\,\delta f(3)\,\rangle\rangle$$
$$+N(1\,4\,5)\ \{\langle\,\delta f(2)\,\delta f(4)\,\rangle\ \langle\,\delta f(3)\,\delta f(5)\,\rangle$$
$$+\tfrac{1}{2}\,\langle\langle\ \delta f(4)\delta f(5)\delta f(2)\delta f(3)\ \rangle\rangle\} = 0. \tag{8.32}$$

The equations for the higher order cumulants $\langle\langle\,\delta f(1)...\delta f(s)\,\rangle\rangle$, for $s \geq 4$, will not be written down explicitly, because they are not needed in the subsequent developments.

In conclusion, the nonlinear stochastic equation (8.17) is equivalent to an *infinite hierarchy for the cumulants of the random function* $\delta f(1)$.

8.3 The Renormalization Programme

The infinite hierarchy of equations for the moments (or the cumulants) of the random response $\delta f(1)$ cannot, in general, be solved analytically. Therefore, the theory of plasma turbulence consists of finding some approximation schemes for the solution of this problem. These schemes involve, in some form or another, a *closure* of the infinite hierarchy, by which the higher-order cumulants are assumed to be approximately expressed in terms of a finite set of lower-order ones. Such approximations must, however, be done with care in order to avoid difficulties and pitfalls which will be discussed later on.

A general closure scheme can be formally defined as follows. We start with the moment representation (rather than the cumulant representation). We choose a certain positive integer $P \geq 2$, called the *truncation level* and decide, on the basis of a physical argument, that *all moments (or cumulants) of order higher than P can be expressed as functionals (in general, nonlinear) of the moments of order less than or equal to P:*

$$\langle\,[\mathcal{Q}\,\delta f(1)\,\delta f(2)]\,\delta f(3)...\delta f(P+n)\,\rangle$$
$$= \mathcal{F}_{P+n}\,\{\,1,2,..,P+n;\ \langle\,\delta f(a_1)\,\delta f(a_2)\,\rangle,$$
$$\langle\,\delta f(a_1)\,\delta f(a_2)\,\delta f(a_3)\,\rangle,...,\langle\,\delta f(a_1)\,\delta f(a_2)...\delta f(a_P)\,\rangle\,\}\,,$$
$$n \geq 1. \tag{8.33}$$

The symbol on the right hand side denotes a quantity $\mathcal{F}_{P+n}$ that is a function of the variables $1,2,...,P+n$ (*i.e.* of the variables α_1, v_1, $k_1, t_1,...,$ α_{P+n}, v_{P+n}, k_{P+n}, t_{P+n}), which moreover depends functionally on the moments of order $2,3,...$ up to P.[6]

---

[6] $\mathcal{F}$ contains all second-order moments corresponding to couples (a_1, a_2) chosen among the set $(1,2,...,P)$, all third-order moments corresponding to triplets (a_1, a_2, a_3), etc.

Among all the moments, the second order moments play a predominant role. From the practical point of view, they are the only ones that can be measured experimentally. From the theoretical point of view, their knowledge determines the average fluxes of particles and energy, *i.e.* the quantities of major interest in transport theory, our main subject.

Many different truncation schemes have been proposed by the various scientists working in plasma turbulence theory. Although these schemes appear to be *prima facie* different from each other, it is possible to dig out a basic common structure: this will be investigated in the present section.

If the ansatz (8.33) is substituted into the moment hierarchy, (8.27), (8.28),... the latter reduces to a finite set of P equations:

$$L(12)\langle \delta f(2)\delta f(4)\rangle + \tfrac{1}{2} N(1\,2\,3)\langle \mathcal{Q}[\delta f(2)\delta f(3)]\delta f(4)\rangle = 0, \qquad (8.34)$$

$$L(1\,2)\,\langle \delta f(2)\delta f(4)\delta f(5)\rangle$$
$$+\tfrac{1}{2} N(1\,2\,3)\,\langle\,[\mathcal{Q}\,\delta f(2)\delta f(3)]\,\delta f(4)\delta f(5)\,\rangle = 0,\ldots \qquad (8.35)$$

$$L(1\,2)\,\langle \delta f(2)\delta f(4),..,\delta f(P+1)\rangle$$
$$+\tfrac{1}{2} N(1\,2\,3)\langle [\mathcal{Q}\delta f(2)\delta f(3)]\,\delta f(4),...,\delta f(P+1)\rangle = 0, \qquad (8.36)$$

and, finally, the equation for the moment of order P, in which the moment of order $(P+1)$ appearing in the right hand side is expressed by Eq. (8.33):

$$L(1\,2)\,\langle \delta f(2)\delta f(4),...,\delta f(P+2)\rangle$$
$$+\tfrac{1}{2} N(1\,2\,3)\,\mathcal{F}_{P+1}\{2,3,4,..,P+2;\langle \delta f(a_1)\delta f(a_2)\rangle,$$
$$...,\langle \delta f(a_1)\delta f(a_2)...\delta f(a_P)\rangle\} = 0. \qquad (8.37)$$

Given the importance of the second moment, this set of equations is treated according to the following strategy. We start with the last equation of the chain, Eq. (8.37) [which is a *closed* equation for the moment of order P]; its formal solution yields an expression of the moment of order P in terms of the moments of order less than or equal to $P-1$:

$$\langle \delta f(1)\delta f(2)...\delta f(P)\rangle$$
$$= \mathcal{G}_P\{1,2,...,P;\langle \delta f(a_1)\delta f(a_2)\rangle,...,\langle \delta f(a_1)\delta f(a_2)...\delta f(a_{P-1})\rangle\}$$

This expression is then substituted into Eq. (8.36), which can now be solved in order to yield the moment of order $P-1$ as a functional of the moments up to $P - 2$. The process is repeated up to, and including Eq. (8.35), whose solution yields an expression of the third order moment as a functional of the second-order ones alone:

$$\langle \delta f(1)\delta f(2)\delta f(3) \rangle = \mathcal{G}_3 \{1, 2, 3; \langle \delta f(a_1)\delta f(a_2)\rangle \} . \qquad (8.38)$$

This quantity, which results from the elimination of all the higher moments is, in general, very complicated, but it is a functional of the second-order moment *alone*. When this expression is substituted into the first equation (8.34) of the truncated hierarchy [in which the symbol $\mathcal{Q}$ can be removed, see Eq. (8.26), and the dummy variables are renamed], we obtain a *closed nonlinear equation for the second moment:*

$$L(13) \langle \delta f(3)\delta f(2) \rangle + \tfrac{1}{2} N(1\,3\,4) \mathcal{G}_3 \{2, 3, 4; \langle \delta f(a_1)\delta f(a_2)\rangle \} = 0, \qquad (8.39)$$

[where the symbols $a_1, a_2 (\neq a_1)$ may have any value in the set $(1, 2, 3, 4)$]. The explicit form of the functional $\mathcal{G}_3$ depends on the level of truncation P and on the specific closure assumption (8.33) adopted. We could start our study with Eq. (8.39); however, it is highly profitable both for the mathematical treatment and especially for the physical insight, to first investigate more closely the structure of this equation.

The second term in Eq. (8.39) is a nonlinear combination of second moments; it depends on the two "fixed" variables 1 and 2 (3 and 4 are dummy summation variables). It usually happens that among the terms making up this expression there is a group of terms containing the correlation of the response $\delta f(2)$ with some other (dummy) response, say $\delta f(3)$. This group of terms entering $\tfrac{1}{2} N(1\ 3\ 4)\mathcal{G}_3\{2\ 3\ 4\}$ can always be written in the form:

$$\Delta L(1\,3) \langle \delta f(3)\delta f(2) \rangle. \qquad (8.40)$$

These (and only these) terms have the *same form as the linear term* of Eq. (8.39) [see Eq. (8.26)]. There is, however, a major difference: because of the nonlinearity of the functional $\mathcal{G}_3$, *the operator* $\Delta L(1\,3)$ *is necessarily a functional of the second moments* (*other than* $\langle \delta f(3)\delta f(2)\rangle$). Indeed, as the nonlinear functional $\mathcal{G}_3$ is at least quadratic in the correlations, after the explicit extraction of a factor $\langle \delta f(3)\delta f(2)\rangle$ we are left with a functional that depends at least linearly on the other second moments $\langle \delta f(a_1)\delta f(a_2)\rangle$ $(\neq \langle \delta f(3)\delta f(2)\rangle)$. Thus,

$$\Delta L(1\,3) \equiv \Delta L \{1, 3; \langle \delta f(a_1)\delta f(a_2)\rangle \} . \qquad (8.41)$$

For simplicity, in the forthcoming formulae we shall usually not write down explicitly the functional dependence on the correlations, unless we want to stress this feature.

Of course, not all the contributions to the second term of Eq. (8.39) are of the form (8.40). It is therefore always possible to write this term in the form:

$$\tfrac{1}{2} N(1\,3\,4)\, \mathcal{G}_3 \left\{2, 3, 4;\ \langle \delta f(a_1) \delta f(a_2) \rangle \right\}$$
$$= \Delta L\left\{1, 3;\ \langle \delta f(a_1) \delta f(a_2) \rangle \right\} \langle 3\,2 \rangle - H\left\{1, 2;\ \langle \delta f(a_1) \delta f(a_2) \rangle \right\} \quad (8.42)$$

(the minus sign in the last term is chosen for convenience). $H(1\,2)$ is a functional of the second moments that is *not* of the form (8.40). With this decomposition, the equation for the second moment, (8.39) takes the form:

$$\left\{ L(1\,3) + \Delta L(1\,3) \right\} \langle \delta f(3)\, \delta f(2) \rangle = H(1\,2) . \quad (8.43)$$

As a result of these operations we obtained an equation having the *same form* as the initial *linear* equation:

$$L(1\,3)\, \langle \delta f(3)\, \delta f(2) \rangle = 0, \quad (8.44)$$

but containing a modified evolution operator, and also an additional term in the right hand side, which can be looked at, formally, as a source term. The operator

$$\mathcal{L}(1\,3) = L(1\,3) + \Delta L(1\,3) \quad (8.45)$$

is called the *renormalized evolution operator*. The process leading to the *renormalized equation* (8.43) is called a *renormalization process*. The name "renormalization" originates from quantum field theory. The historical aspects of microscopic turbulence theory will be discussed separately in Sec. 8.4.

Next we note that the renormalization process can be carried out in an even more detailed form. Indeed, remembering the form (8.19) of the linear evolution operator, we may similarly collect in the renormalization operator $\Delta L(1\,3)$ all the terms (if any) that end at right with an operator $\Phi(4\,3)$; we thus write:

$$\Delta L(1\,3) = \Sigma \left\{1, 3;\ \langle \delta f(a_1)\, \delta f(a_2) \rangle \right\}$$
$$+ V \left\{1, 4;\ \langle \delta f(a_1)\, \delta f(a_2) \rangle \right\} \Phi(4\,3). \quad (8.46)$$

Needless to say, the operator Σ (which, by definition, cannot contain a factor Φ at its extreme right) as well as V are functionals of the second moments. The *renormalized equation for the second moment* is now written in the form:

$$g^{-1}(1\,3)\, \langle \delta f(3)\, \delta f(2) \rangle + \Delta(1\,4)\, \langle \delta \phi(4)\, \delta f(2) \rangle = H(1\,2). \quad (8.47)$$

This equation has the same form as the *linear* equation (8.26) (with $N \equiv 0$), plus an additional source term. The coefficient $g^{-1}(1\,2)$ is the *inverse renormalized Propagator* and is defined as follows:

$$g^{-1}(1\,2) = g_0^{-1}(1\,2) + \Sigma(1\,2). \qquad (8.48)$$

The *renormalized Vertex operator* $\Delta(1\,2)$ is given by:

$$\Delta(1\,2) = \Delta_0(1\,2) + V(1\,2). \qquad (8.49)$$

The physical content of the renormalized equation of evolution will appear in the forthcoming sections.

With the *formally* linear equation (8.47) we may associate a *renormalized causal Green's operator* $G(1\,2)$ obeying the equation:

$$\left[g^{-1}(1\,3) + \Delta(1\,4)\,\Phi(4\,3)\right] G(3\,2) = I(1\,2), \qquad (8.50)$$

where $I(1\,2)$ is the identity operator:

$$I(1\,2) \rightarrow I_{\mathbf{k}_1\mathbf{k}_2}^{\alpha_1\alpha_2}(\mathbf{v}_1, \mathbf{v}_2, t_1, t_2) = \delta^{\alpha_1\alpha_2}\,\delta(\mathbf{v}_1 - \mathbf{v}_2)\,\delta(\mathbf{k}_1 - \mathbf{k}_2)\,\delta(t_1 - t_2). \qquad (8.51)$$

Eq. (8.50) is sometimes called the *Dyson equation* in the literature (MARTIN et al. 1973, KROMMES 1984). It is actually the renormalized form of Eq. (7.89) of the quasilinear theory. The operator $\Delta L(1\,3) \equiv \Sigma(1\,3) + V(1\,4)\,\Phi(4\,3)$, renormalizing the evolution operator, is called the *mass operator*. We also recall that $G(1\,2)$ is often called by the (inappropriate) name of "response function" in the literature (see Sec. 7.6).

In terms of the Green's operator, the *formal* solution of Eq. (8.47) is:[7]

$$\langle \delta f(1)\,\delta f(2) \rangle = G(1\,3)\,H(3\,2). \qquad (8.52)$$

It can be seen that this equation does not appear to have the correct symmetry. The left hand side is invariant under a permutation of the variables 1 and 2, whereas the right hand side does not, at first sight, have this symmetry. The formula can be written in a symmetric form if we introduce instead of the operator $H(3\,2)$ the operator $F(1\,2)$, defined as follows:

$$H(3\,2) = G(2\,4)\,F(3\,4). \qquad (8.53)$$

The operator $F(3\,4)$ can always be constructed, because the causal Green's operator G is invertible, by definition [see Eq. (8.50)]. In some approximation schemes, and notably in the DIA, as will be seen below, the

---

[7] This form neglects the possible contribution of the initial value of the correlation, which is set equal to zero.

operator H automatically appears in the form (8.53). The solution (8.52) of Eq. (8.47) is now written in the form:

$$\langle \delta f(1)\, \delta f(2)\rangle = G(1\,3)\, G(2\,4)\, F(3\,4). \tag{8.54}$$

The right hand side now has the correct symmetry, provided $F(3\,4)$ is invariant under permutation. The operator $F(3\,4)$ is called the *random noise source*. The justification of this name will appear in the forthcoming developments.

The "solution" (8.54) is however only a formality. Indeed, as Eq. (8.47) is nonlinear, we know that both $G(1\,3)$ and $F(3\,4)$ are still functionals of the unknown second moment $\langle \delta f(a_1)\, \delta f(a_2)\rangle$. Eq. (8.54) is thus merely another guise of eq. (8.47).

We chose here to introduce the idea of the renormalization in a very general way and to derive on this basis (in Secs. 8.5 - 8.6 those relations which appear as identities valid independently of the specific truncation hypothesis used in the renormalization process. The particular form taken by the various renormalized quantities in a special approximation scheme, namely the Direct Interaction Approximation (DIA), are developed in Sec. 8.7.

8.4   Short History of Microscopic Plasma Turbulence Theory

"The entire theory of probability is nothing but tranforming variables. Some probability distribution must be given a priori on a set of elementary events; the problem is to transform it into a probability distribution for the possible outcomes... Mathematics can only derive probabilities of outcomes from a given a priori distribution". These statements are quoted from VAN KAMPEN's (1981) remarkable book on Stochastic Processes in Physics and Chemistry (Chap. I, Excursus, p. 21). They express in a nutshell the essence and the limitations of every probabilistic theory.

When applying this idea to the theory of plasma turbulence, we first recall that the use of the stochastic GKE (or of any other appropriate stochastic kinetic equation) as a basis of description results from our wish of *modelling* a complex process. The latter is, in principle, perfectly determined by the laws of dynamics. However, we very well know that the "elementary" dynamical laws governing the motion of interacting systems give rise to exceedingly tangled trajectories because of the intrinsic instabilities of such systems. The stochastic modeling leads to a description where many (irrelevant?) details are smoothed out, but which has a remarkable predictive power (for probabilities!). According to the opening statement, such a modelling can only work if some probabilistic assumption is made *a priori*. The theory then develops the consequences of this *a priori* assumption. This procedure will be illustrated in all the forthcoming chapters, dealing with the study of the nonlinear equations of plasma dynamics.

The first important work opening the activities in the domain of plasma turbulence is the paper of PINES & BOHM 1952, devoted to the microscopic study of the collective excitations in a plasma. In this work, a concept promised to a flourishing future was introduced: the *Random Phase Approximation (RPA)*. It consists (roughly speaking) of the *a priori* assumption that the potential fluctuations $\delta\phi_{k,\omega}$ for different values of k, ω have phases distributed at random. This amounts to assuming a *Gaussian distribution of the fluctuations:* all odd moments are zero and all even moments reduce to sums of products of second moments.

The RPA and the additional assumption that the electrostatic energy stored in the collective modes is small compared to the thermal energy, $\lambda \equiv (e/T) \left\langle |\delta\phi|^2 \right\rangle^{1/2} \ll 1$, formed the basis of the fundamental papers written independently by VEDENOV, VELIKHOV & SAGDEEV 1961, 1962 and by DRUMMOND & PINES 1962, 1964. It was shown there that an internal instability, described by the stochastic Vlasov equation, leads to a feedback effect on the evolution of the average distribution function. The latter obeys an equation of diffusive type, called the *Quasilinear equation*. The effective diffusion coefficient is a result of wave-particle interactions and is a functional of the spectrum.

This work and its immediate extension including weak wave-wave coupling, became the starting point of an enormous body of literature in the 60's and 70's. Within this framework an impressive number of specific instabilities under various physical conditions and in presence of magnetic fields of various geometries were exhaustively studied. It ended up in a very well organized theoretical body of knowledge, known under the name of *Weak Turbulence Theory*. It is a theory of essentially perturbative type, based on the assumption $\lambda \ll 1$. It has been extensively reviewed in several excellent monographs: KADOMTSEV 1965, KLIMONTOVICH 1967, 1982, SAGDEEV & GALEEV 1969, DAVIDSON 1972, TSYTOVICH 1977, SITENKO 1982.

It soon appeared, however, that the weak turbulence theory could not adequately explain many experimental data. Moreover, the theory is plagued by divergence difficulties. Among the most celebrated objections to the quasilinear theory we quote the work of ADAM et al., 1979 and LAVAL & PESME 1980, 1983 (ALP). They found that when the turbulent trapping time $\tau_{tr} \approx \left(k^2 D_{ql}\right)^{-1/3}$ is smaller than the inverse growth rate of Langmuir waves, the nonlinear growth rate is significantly larger than the linear one. This work was done for a one-dimensional, one-species Vlasov plasma (Langmuir turbulence). On the other hand, numerical simulations (the most recent one by TSUNODA et al., 1987) did not find such a departure. It was only in 1993 that LIANG & DIAMOND 1993 found the erroneous point in the ALP argument and showed that the correct nonlinear growth rate differs only very little from the quasilinear result. Although this specific point is now clarified, all the other difficulties of the quasilinear theory subsist.

In order to cure the latter and to push the formalism towards stronger turbulence conditions, the idea of *renormalization* was introduced in a pioneering paper by DUPREE 1966. It was soon followed by a very clear work of WEINSTOCK 1969, in which the statistical-mechanical foundation of the theory was firmly laid. A long series of works can be attached to this starting point, among which we quote a few: ALTSHUL' & KARPMAN 1966, DUPREE 1967, WEINSTOCK 1970, 1972, RUDAKOV & TSYTOVICH 1971, TAYLOR & MCNAMARA 1971, KONO & ICHIKAWA 1973, MISGUICH & BALESCU 1975 a,b, MONTGOMERY 1975, VACLAVIK 1975, DUPREE & TETREAULT 1978, HORTON & CHOI 1979.

In all the previously quoted papers, the plasma turbulence was treated in the so-called *coherent approximation,* (to be defined in Sec. 8.6). It was again DUPREE 1970, together with KADOMTSEV & POGUTSE 1970, who pointed out the importance of including the *incoherent fluctuations* in a realistic theory of plasma turbulence. They showed that the latter describe the fine structure of the turbulent flow, leading to a *granular structure* in phase space. DUPREE coined the name of *clumps* for this phenomenon. In spite of some controversies in the beginning, the idea of the importance of clumps (in particular for anomalous transport) is now generally accepted. A rather large number of papers has developed this initial idea, among which we quote:[8] DUPREE 1972, 1978, BOUTROS-GHALI & DUPREE 1981, BALESCU & MISGUICH 1984, SUZUKI 1984, TERRY & DIAMOND 1984, ZHANG & BALESCU 1987.

In parallel with these developments, another line of ideas was being intensely developed in the second half of the 70's and the 80's. Its origin actually goes back much earlier, to the work of KRAICHNAN 1959. In this basic paper, he introduced a new type of truncation scheme for hydrodynamic turbulence (*i.e.* for the Navier-Stokes equation), which he called the *direct interaction approximation* (DIA). It was based on a very penetrating statistical analysis of the turbulent flow problem. Although its success in Navier-Stokes turbulence was limited (and required an extension, the so-called Lagrangian DIA), it was shown to be an entirely self-consistent scheme (KRAICHNAN 1961, 1970 a, b, PHYTHIAN 1969). ORSZAG & KRAICHNAN 1967 applied the DIA to the Vlasov equation. It was, however, only later, when it began to be clearly realized that the weak turbulence theory (even in its renormalized version) based on the RPA presented some specific difficulties (see above) that the DIA idea acted like

---

[8] Many more references are given in Chaps. 10 and 13.

a spark in a part of the plasma turbulence community. It started an explosion of papers, among which we quote only a few : MARTIN et al. 1973, the basic review paper of DuBOIS & ESPEDAL 1978, DuBOIS & ROSE 1981, KROMMES & KLEVA 1979, SIMILON 1981, DuBOIS & PESME 1984, OTTAVIANI et al. 1991. An excellent recent review of the DIA is the chapter entitled "Statistical Descriptions and Plasma Physics" by KROMMES 1984, in the Handbook of Plasma Physics. In these works, the general formalism of renormalized plasma turbulence theory was developed as a framework in which the DIA could be naturally inscribed (although the generality of this formalism was not sufficiently stressed).

The DIA incorporates all the results reviewed in the present short outline. Moreover, it does so in a very elegant and systematic way. One may regret, however, as I personally do, that most expositions of this theory are unnecessarily burdened by a very heavy formalism. This tendency started by the work of MARTIN et al. 1973, who constructed a very general theory of the turbulent response, using the full artillery of mathematical methods of quantum field theory: this presentation was then adopted by later authors (DuBois, Krommes,...). I do not deny the elegance and generality of this approach. But it surrounds the DIA by an aura of mathematical sophistication that leads to a double practical result. On one hand, the newcomer has to spend a long time in trying to understand these difficult papers; on the other hand, he is left with the general impression that the DIA is a very complicated physical idea.

In reality, the DIA is a quite simple and transparent approximation scheme. I attempt to prove this statement in this and the next chapters and show that the understanding of the DIA only requires the most elementary concepts of probability theory. A very clear presentation of the DIA (for the Navier Stokes equation) very similar in spirit to the one given here can be found in ORSZAG's 1970 paper.

Before going over to the application of the DIA to the gyrokinetic equation, we briefly discuss an alternative approach to the problem: the *Renormalization Group (RNG)* formalism. Originally introduced in quantum field theory (STUECKELBERG & PETERMANN 1953, BOGOLIUBOV & SHIRKOV 1980 (for a clear recent review, see COLLINS 1984); it obtained a striking success when adapted by WILSON 1971, 1972, WILSON & FISHER 1972 to equilibrium statistical mechanics. It led to the solution of a long-standing problem: the theory of critical phenomena associated with the phase transitions. Reviews of the latter problems can be found in BALESCU 1975, and in the monographs of MA 1976, and of BINNEY et al. 1992. It was realized later that the methods of the RNG could be extended and be useful in the theory of hydrodynamic turbulence (FORSTER et al. 1976, 1977, FOURNIER & FRISCH 1983, YAKHOT & ORSZAG 1986). It should be noted, however, that KRAICHNAN 1982 expresses a very critical point of view towards the RNG applications to hydrodynamics.

In plasma physics, an early group of papers: POUQUET et al. 1978, PELLETIER 1980, FOURNIER et al. 1982, though very interesting, were not immediately followed by other developments. It was only ten years later that the subject was picked up again by three different groups. In these recent works the RNG methods were applied both to the macroscopic MHD description (LONGCOPE & SUDAN 1991, CAMARGO & TASSO 1992) and to the microscopic GKE description (CARATI et al. 1994).

The basic idea of the RNG rests on the study of the behaviour of the evolution equations under the group of dilatations. One first proposes as an "axiom" that in turbulence theory one is not interested in phenomena occurring on very short length scales (of molecular orders of magnitude). One then separates the fluctuations as: $\delta f_{\mathbf{k}} = \delta f_{\mathbf{k}}^{<} + \delta f_{\mathbf{k}}^{>}$ into large scale (= small $\mathbf{k}$) and small scale (= large $\mathbf{k}$) fluctuations; one then tries to eliminate $\delta f_{\mathbf{k}}^{>}$ from the equations of evolution.[9] Very schematically, after elimination of a "shell" of fluctuations (*i.e.* $\delta f_{\mathbf{k}}$ with $k_c > k > k_c - \Delta$), one tries to rescale the coefficients of the dynamical equation in such a way as to ensure its form-invariance. In this way, the renormalization concept, expressed as the functional relation between Eqs. (8.47) and (8.26) enters the RNG theory. Although very elegant and general, the RNG approach cannot be developed *in practice* beyond a level essentially equivalent to the DIA. It will not be further discussed here.

8.5 General Relations between Renormalized Quantities

After this historical interlude, we return to the main line of our presentation. The basic idea of the renormalization programme is to write the equation for the second moment in a *form* as close as possible to its linearized approximation. It is therefore not surprising that we will find among the various quantities entering this equation a number of relations which are straightforward generalizations of the corresponding linear relations.

Consider first the operator $g(1\,2)$. Matrix-multiplying throughout Eq. (8.47) by this operator we find:

$$\langle \delta f(1)\,\delta f(2)\rangle = g(1\,3)\,\{-\Delta(3\,4)\,\langle \delta\phi(4)\delta f(2)\rangle + H(3\,2)\}. \qquad (8.55)$$

This relation expresses a "partial" solution for the correlation function, which has the same status as Eq. (7.81); this justifies the name of *renormalized propagator* given to $g(1\,2)$. From Eq. (8.48) we may also write the latter in the form:

$$g(1\,2) = \left[g_0^{-1} + \Sigma\right]^{-1}(1\,2). \qquad (8.56)$$

---

[9] A rather similar procedure, although in a different context, will be found in Chap. 16.

From Eq. (7.32) we know that the unperturbed propagator describes the very simple free motion of the particles in the direction parallel to the magnetic field. Its main feature is the existence, in its Laplace transform, of a *singularity* which occurs when the parallel particle velocity $v_\parallel$ equals the phase velocity $\omega/k_\parallel$;[10] the unperturbed propagator g_0 has a δ-function peak [see Appendix B, Eq. (B.11)] for this value of $v_\parallel$, expressing the *resonance* between particles and waves. The renormalized propagator g no longer has this property. The operator Σ represents the action of the fluctuating field (*i.e.* of the collective interactions) on the motion of the particles. As a result of this effect, the trajectories of the (previously free) particles become "blurred". Instead of a *ballistic* motion, the particles undergo a *diffusive* motion due to the random deviations produced by the potential $\delta\phi$. [The (widely used) name "ballistic" justly suggests the perfectly deterministic orbit of the bullet shot from a gun].

As a mathematical result, the singularity of the propagator at $v_\parallel = \omega/k_\parallel$ is suppressed and replaced by a hump with finite maximum and finite width on the real axis. This is the important phenomenon of *resonance broadening*, first discovered by DUPREE 1966. It will be studied more explicitly in the Chap. 9.

Coming back to Eq. (8.56), we may transform it into a useful form by using the following identity, valid for any two operators A and B (which, in general, do not commute with each other):

$$\frac{1}{A+B} = \frac{1}{A} - \frac{1}{A}\,B\,\frac{1}{A+B}.$$ (8.57)

The validity of this identity is easily checked by multiplying both sides to the right by $(A+B)$. Taking $A = g_0^{-1}$ and $B = \Sigma$, we find:

$$g(1\,2) = g_0(1\,2) - g_0(1\,3)\,\Sigma(3\,4)\,g(4\,2).$$ (8.58)

This is an elegant integral equation relating the renormalized propagator to the bare propagator. It is often used for producing successive approximations, obtained by iterations of this equation. It should not be forgotten, when using this "linear-looking" equation that the renormalization operator Σ implicitly involves the propagator g, hence the equation is actually highly nonlinear.

As discussed in Sec. 7.6, the propagator g_0 does not provide a complete solution of the linear GKE, because the potential $\delta\phi$ must be expressed self-consistently in terms of the response δf. The complete solution is provided by the unperturbed Green's operator G_0 (or by its Laplace transform R_0) in the form (7.85). G_0 obeys Eq. (7.89) which can be written, in the present system of notations:

---

[10] The complex frequency w is treated as an "almost" real quantity, i.e., $w = \omega + i\epsilon$.

$$[g_0^{-1}(1\,3) + \Delta_0(1\,4)\,\Phi(4\,3)]\,G_0(3\,2) = I(1\,2). \qquad (8.59)$$

It is clear that the *renormalized Green's operator* $G(1\,2)$ defined by Eq. (8.50) formally plays the same role in the nonlinear case: it provides the "solution" of Eq. (8.47) in the form (8.52) or (8.54). The analogy between G and G_0 was pointed out only rather recently (to our knowledge) in the work of KROMMES & KIM 1988. We stress again the fundamental difference between the expressions (8.54) or (8.52) and the analogous relation (7.85) of the linear theory. The latter provides a complete and explicit solution of the linear GKE. On the contrary, in the present case both G and F are functionals of the unknown correlations, hence (8.54) is a nonlinear equation satisfied by $\langle \delta f(1)\,\delta f(2) \rangle$.

Equation (8.50) for the Green's operator can be cast into a slightly different form by left-multiplication with g:

$$G(1\,2) = g(1\,2) - g(1\,3)\,\Delta(3\,4)\,\Phi(4\,5)\,G(5\,2). \qquad (8.60)$$

This form is analogous to Eq. (8.58). The latter provided a link between the renormalized propagator g and the bare one g_0. Here we go one step higher: the renormalized propagator g appears as the "zeroth order approximation" [11] of the renormalized Green's operator G.

Left-multiplying (8.60) by the operator Φ, we note that the combined operator $\Phi(1\,3)G(3\,2)$ obeys, by itself, an integral equation that can be written in the form :

$$\{I(1\,3) + \Phi(1\,5)\,g(5\,6)\,\Delta(6\,3)\}\,\Phi(3\,4)\,G(4\,2) = \Phi(1\,3)\,g(3\,2). \qquad (8.61)$$

In this form we see the natural appearance of the *renormalized dielectric operator* $\varepsilon(1\,3)$:

$$\varepsilon(1\,3) = I(1\,3) + \Phi(1\,5)\,g(5\,6)\,\Delta(6\,3). \qquad (8.62)$$

The analogy with the linear dielectric function (7.75) is striking: with the latter we may associate a linear dielectric operator:[12]

$$\varepsilon_0(1\,3) = I(1\,3) + \Phi(1\,5)\,g_0(5\,6)\,\Delta_0(6\,3) \qquad (8.63)$$

The relation (8.61) can therefore be written in the following form, first obtained by DuBois & ESPEDAL 1978:

$$\Phi(1\,3)\,G(3\,2) = \varepsilon^{-1}(1\,3)\,\Phi(3\,4)\,g(4\,2). \qquad (8.64)$$

---

[11] The idea of "order" is, however, purely formal and is not necessarily related to a specification of order of magnitude of any small parameter.

[12] The relation between dielectric *operator* and dielectric *function* will be made explicit below.

It shows that between the operators ΦG and Φg there exists a very simple (formally linear) relation: the latter can be related to the so-called *principle of superposition of dressed particles* (ROSTOKER 1961).

Substituting Eq (8.64) into (8.60) we obtain an *"explicit" expression of the Green's operator G in terms of the propagator g:*

$$G(1\,2) = g(1\,2) - g(1\,3)\,\Delta(3\,4)\,\varepsilon^{-1}(4\,5)\,\Phi(5\,6)\,g(6\,2). \qquad (8.65)$$

The structural analogy with Landau's form (7.87) of the resolvent of the linear GKE is striking.

Next, it will be useful to obtain a formal solution for the correlation of potential fluctuations, briefly called the *spectrum:* $\langle \delta\phi(1)\,\delta\phi(2)\rangle$. This quantity is important, because it will later appear that all the operators (such as Σ, Δ,...) of the renormalized theory are naturally expressed in terms of the spectrum. In order to obtain an equation for the latter, we invoke in succession Eqs. (8.24) and (8.54) to obtain:

$$\langle \delta\phi(1)\,\delta\phi(2)\rangle = \Phi(1\,3)\,\Phi(2\,4)\,\langle \delta f(3)\,\delta f(4)\rangle.$$
$$= \Phi(1\,3)\,G(3\,5)\,\Phi(2\,4)\,G(4\,6)\,F(5\,6).$$

Using finally Eq. (8.64) we obtain the form:

$$\langle \delta\phi(1)\,\delta\phi(2)\rangle = \varepsilon^{-1}(1\,3)\,\varepsilon^{-1}(2\,4)\,K(3\,4), \qquad (8.66)$$

with:

$$K(3\,4) = \Phi(3\,5)\,g(5\,6)\,\Phi(4\,7)\,g(7\,8)\,F(6\,8). \qquad (8.67)$$

The operator $K(1\,2)$ is sometimes called the *source of incoherent potential fluctuations* for reasons to be discussed in Sec. 8.6.

We end this section with a few general and very useful properties of the dielectric operator. Using our correspondence rules (8.6) and (8.7) as well as (8.9), we find the explicit form of this operator:

$$\varepsilon(1\,2) \rightarrow \varepsilon_{\mathbf{k}_1\mathbf{k}_2}^{\alpha_1\alpha_2}(\mathbf{v}_1, \mathbf{v}_2; t_1, t_2) = \delta(\mathbf{k}_1 - \mathbf{k}_2)\,\varepsilon_{\mathbf{k}_1}^{\alpha_1\alpha_2}(\mathbf{v}_1, \mathbf{v}_2; t_1, t_2), \qquad (8.68)$$

with:

$$\varepsilon_{\mathbf{k}}^{\alpha_1\alpha_2}(\mathbf{v}_1, \mathbf{v}_2; t_1, t_2) = \delta^{\alpha_1\alpha_2}\,\delta(\mathbf{v}_1 - \mathbf{v}_2)\delta(t_1 - t_2) + \chi_{\mathbf{k}}^{\alpha_2}(\mathbf{v}_2; t_1, t_2). \qquad (8.69)$$

The peculiar feature here is that the operator χ is independent of α_1 and of $\mathbf{v}_1$, a property shared with the operator Φ, (8.9):

$$\chi_{\mathbf{k}}^{\alpha_2}(\mathbf{v}_2, t_1, t_2) = \frac{4\pi}{k^2} \sum_{\alpha_3 \alpha_4} \int dv_3 dv_4 \int dt_3 dt_4$$

$$\times e_{\alpha_3} J_{\mathbf{k}_3}^{\alpha_3}(\mathbf{v}_3)\, \delta(t_1 - t_3)\, g_{\mathbf{k}}^{\alpha_3 \alpha_4}(\mathbf{v}_3, \mathbf{v}_4, t_3, t_4)\, \Delta_{\mathbf{k}}^{\alpha_4 \alpha_2}(\mathbf{v}_4, \mathbf{v}_2, t_4, t_2). \tag{8.70}$$

It is now easily found that the **renormalized dielectric function** $\varepsilon_{\mathbf{k}}$ generalizing the linear dielectric function $\varepsilon_{0\mathbf{k}}$ (7.75), is defined as follows:

$$\varepsilon_{\mathbf{k}}(t_1, t_2) = \sum_{\alpha_2} \int dv_2\, \varepsilon_{\mathbf{k}}^{\alpha_1 \alpha_2}(\mathbf{v}_1, \mathbf{v}_2, t_1, t_2) \tag{8.71}$$

If the turbulence is *stationary*, all two-time quantities depend only on the time differences. A Laplace transformation then changes Eqs. (8.69), (8.5) into:

$$\widetilde{\varepsilon}_{\mathbf{k}}(w) = 1 + \sum_{\alpha} \widetilde{\chi}_{\mathbf{k}}^{\alpha}(w), \tag{8.72}$$

with the **polarizability** of species α, $\widetilde{\chi}_{\mathbf{k}}^{\alpha}(w)$ defined as:

$$\widetilde{\chi}_{\mathbf{k}}^{\alpha}(w) = \int dv\, \widetilde{\chi}_{\mathbf{k}}^{\alpha}(\mathbf{v}, w)$$

$$= \frac{4\pi}{k^2} \sum_{\alpha_3 \alpha_4} \int dv\, dv_3 dv_4\, e_{\alpha_3} J_{\mathbf{k}_3}^{\alpha_3}(\mathbf{v}_3)\, \widetilde{g}_{\mathbf{k}}^{\alpha_3 \alpha_4}(\mathbf{v}_3, \mathbf{v}_4, w)\, \widetilde{\Delta}_{\mathbf{k}}^{\alpha_4 \alpha}(\mathbf{v}_4, \mathbf{v}, w). \tag{8.73}$$

The *inverse dielectric operator* $\varepsilon^{-1}(1\,2)$ often appears in the general expressions. It will be shown here that, because of the peculiar structure (8.69), its expression can easily be found. First, because of homogeneity, we have:

$$\varepsilon^{-1}(1\,2) \rightarrow \delta(\mathbf{k}_1 - \mathbf{k}_2)\delta(w_1 - w_2) \left(\widetilde{\varepsilon}^{-1}\right)_{\mathbf{k}_1}^{\alpha_1 \alpha_2} (\mathbf{v}_1, \mathbf{v}_2, w_1). \tag{8.74}$$

This operator must satisfy the condition:

$$\sum_{\alpha_3} \int dv_3\, \widetilde{\varepsilon}_{\mathbf{k}_1}^{\alpha_1 \alpha_3}(\mathbf{v}_1, \mathbf{v}_3, w_1) \left(\widetilde{\varepsilon}^{-1}\right)_{\mathbf{k}_1}^{\alpha_3 \alpha_2} (\mathbf{v}_3, \mathbf{v}_2, w_2)$$

$$= \delta^{\alpha_1 \alpha_2} \delta(\mathbf{k}_1 - \mathbf{k}_2)\, \delta(w_1 - w_2). \tag{8.75}$$

Eq (8.69) suggests the following form:

$$\left(\widetilde{\varepsilon}^{-1}\right)_{\mathbf{k}}^{\alpha_3 \alpha_2}(\mathbf{v}_3, \mathbf{v}_2, w) = \delta^{\alpha_3 \alpha_2}\delta(\mathbf{v}_3 - \mathbf{v}_2) + \lambda_{\mathbf{k}}^{\alpha_2}(\mathbf{v}_2, w).$$

The condition (8.75) then leads to the following equation:

$$\left\{1 + \sum_{\alpha_3} \int d\mathbf{v}_3\, \chi_{\mathbf{k}}^{\alpha_3}(\mathbf{v}, w)\right\} \lambda_{\mathbf{k}}^{\alpha_2}(\mathbf{v}_2, w) + \chi_{\mathbf{k}}^{\alpha_2}(\mathbf{v}_2, w) = 0.$$

The bracketed coefficient is nothing other than the dielectric function (8.71); hence, collecting all these results, we find:

$$\left(\widetilde{\varepsilon}^{-1}\right)_{\mathbf{k}}^{\alpha_1 \alpha_2}(\mathbf{v}_1, \mathbf{v}_2, w) = \delta^{\alpha_1 \alpha_2}\delta(\mathbf{v}_1 - \mathbf{v}_2) - \frac{1}{\widetilde{\varepsilon}_{\mathbf{k}}(w)} \chi_{\mathbf{k}}^{\alpha_2}(\mathbf{v}_2, w). \qquad (8.76)$$

Using eqs (8.71), (8.73) we find the following remarkable relation, which simplifies many forthcoming calculations:

$$\sum_{\alpha_2} \int d\mathbf{v}_2 \left(\widetilde{\varepsilon}^{-1}\right)_{\mathbf{k}}^{\alpha_1 \alpha_2}(\mathbf{v}_1, \mathbf{v}_2, w) = \frac{1}{\widetilde{\varepsilon}_{\mathbf{k}}(w)}. \qquad (8.77)$$

The renormalized dielectric operator was derived and studied by DuBois & Espedal 1978, Krommes & Kleva 1979, Similon 1981, Krommes 1984, Krommes & Kim 1988. A very detailed and careful recent review of the dielectric properties of a plasma near thermodynamic equilibrium in the gyrokinetic description was given by Krommes 1993 [see also Krommes et al. 1986].

8.6   Coherent and Incoherent Responses

We now introduce a classification of the turbulent responses which turns out to be very useful in the theory. This classification can be defined in a quite general way, independently of any truncation approximation.

In order to introduce the idea, we consider the nonlinear stochastic GKE (8.17) as a starting point, together with its linearized approximation:

$$g_0^{-1}(1\,2)\,\delta f(2) + \Delta_0(1\,2)\,\delta\phi(2) = 0.$$

From the latter we immediately find the relation:

$$\delta f^{(0)}(1) = -g_0(1\,2)\,\Delta_0(2\,3)\,\delta\phi(3). \qquad (8.78)$$

Going back to the nonlinear GKE (8.17), the idea of the renormalization programme consists of rewriting it *(if possible)* in the form:

$$g^{-1}(1\,2)\,\delta f(2) + \Delta(1\,2)\,\delta\phi(2) = \delta k(1), \qquad (8.79)$$

where g and Δ are, respectively, the renormalized propagator and the renormalized vertex, and $\delta k(1)$ represents whatever is left, which is not in the form of the left hand side. The right hand side can be (loosely) interpreted as a "stochastic force" driving the fluctuations [by analogy with the Langevin equation (6.8)]. It should, however, be stressed that the existence of the "stochastic force" is not granted *a priori*. This important point will be discussed in detail in Sec. 8.9, in connection with the DIA.

Meanwhile, if $\delta k(1)$ can be neglected for some reason, the solution of the renormalized equation (8.79) is of the same form as the solution (8.78) of the linearized GKE:

$$\delta f^{(c)}(1) = -g(1\,2)\,\Delta(2\,3)\,\delta\phi(3). \tag{8.80}$$

This stochastic function, which has the same form as the corresponding linear paradigm (8.78), will be called the *coherent response*. Using the Poisson equation (8.8) we find

$$\delta\phi(1) = -\,\Phi(1\,2)\,g(2\,3)\,\Delta(3\,4)\,\delta\phi(4).$$

Recalling the definition (8.62) of the renormalized dielectric operator, this can be written as:

$$\varepsilon(1\,2)\,\delta\phi(2) = 0. \tag{8.81}$$

Written explicitly in the Laplace representation (for stationary turbulence), this equation is:

$$\sum_{\alpha_2} \int d\mathbf{v}_2\,\widetilde{\varepsilon}_{\mathbf{k}}^{\alpha_1\alpha_2}(\mathbf{v}_1,\mathbf{v}_2,w)\,\delta\widetilde{\phi}_{\mathbf{k}}(w) = 0,$$

or using (8.71), simply

$$\widetilde{\varepsilon}_{\mathbf{k}}(w)\,\delta\widetilde{\phi}_{\mathbf{k}}(w) = 0. \tag{8.82}$$

This homogeneous equation for $\delta\widetilde{\phi}_{\mathbf{k}}(w)$ has the same form as the corresponding quasilinear equation (7.74). In both cases there exists a non-trivial solution when the dielectric function vanishes:

$$\widetilde{\varepsilon}_{\mathbf{k}}(w) = 0. \tag{8.83}$$

But there appears now an important difference between Eq. (8.83) and the linear dispersion equation (7.76). In the linear case the dispersion equation tells us that there exist non-trivial solutions for the spectrum only for certain discrete values of the frequency (depending on $\mathbf{k}$): $\widetilde{S}_{\mathbf{k}}(w)$ is a line spectrum, with poles at the eigenvalues [or delta-function singularities if the eigenvalues are real), see Eq. (7.78)]. In the nonlinear case, however, the dielectric function depends (at least linearly) on the spectrum. Hence Eq. (8.83) *is a nonlinear equation for the spectrum*, which may have nontrivial solutions different from the quasilinear line spectrum, as discussed at the end of Sec. 7.5.

Besides the spectrum, the various other types of two-point correlations can easily be calculated in the coherent approximation, using (8.80):

$$\left\langle \delta f^{(c)}(1)\, \delta\phi(2) \right\rangle = -\, g(1\,3)\, \Delta(3\,5)\, \langle \delta\phi(5)\, \delta\phi(2) \rangle ,$$

$$\left\langle \delta f^{(c)}(1)\, \delta f^{(c)}(2) \right\rangle = g(1\,3)\, \Delta(3\,5)\, g(2\,4)\, \Delta(4\,6)\, \langle \delta\phi(5)\, \delta\phi(6) \rangle . \tag{8.84}$$

We thus conclude that, *in the coherent approximation, the spectrum* $\langle \delta\phi(1)\, \delta\phi(2) \rangle$ *is the basic second moment:* all other types of correlations are simply expressed in terms of it.

The coherent approximation has been developed in detail by many authors working in plasma turbulence theory (DUPREE 1966, 1967, DUBOIS & ESPEDAL 1978, KROMMES & KLEVA 1979, KROMMES & SIMILON 1980, SIMILON 1981, KROMMES 1984), because it is the simplest approximation of a renormalized nonlinear theory.

Having defined the coherent response by Eq. (8.80), it is always possible to write the exact solution of the nonlinear GKE in the form :

$$\delta f(1) = \delta f^{(c)}(1) + \delta f^{(i)}(1). \tag{8.85}$$

The term $\delta f^{(i)}(1)$ is called the **incoherent response**: it is defined by this equation simply as the difference between the exact and the coherent response. The decomposition (8.85) was first introduced by DUPREE 1970 and by KADOMTSEV & POGUTSE 1970: these were the first authors who stressed the importance of the incoherent fluctuations.

Substituting Eq. (8.85) into the Poisson equation, we find:

$$\delta\phi(1) = -\Phi(1\,2)\, g(2\,3)\, \Delta(3\,4)\, \delta\phi(4) + \Phi(1\,2)\, \delta f^{(i)}(2).$$

Defining the incoherent potential fluctuation $\delta\phi^{(i)}(1)$ as:

$$\delta\phi^{(i)}(1) = \Phi(1\,2)\, \delta f^{(i)}(2), \tag{8.86}$$

the Poisson equation is written in the form:

$$\varepsilon(1\,2)\, \delta\phi(2) = \delta\phi^{(i)}(1). \tag{8.87}$$

The situation is very different from the coherent case (8.81): the Poisson equation is (mathematically) inhomogeneous. The incoherent fluctuations act like a source driving the total potential fluctuations. We easily find a corresponding equation for the spectrum:

$$\varepsilon(1\,3)\, \varepsilon(2\,4)\, \langle \delta\phi(3)\, \delta\phi(4) \rangle = \left\langle \delta\phi^{(i)}(1)\, \delta\phi^{(i)}(2) \right\rangle . \tag{8.88}$$

This equation can be compared to Eq. (8.66): it then appears that the correlation of incoherent fluctuations is identical with the renormalized operator K:

$$\left\langle \delta\phi^{(i)}(1)\,\delta\phi^{(i)}(2) \right\rangle = K(1\,2). \tag{8.89}$$

This relation justifies the name of $K(1\,2)$ as the *source of incoherent potential fluctuations*.

We now derive another interesting expression for the stochastic response $\delta f(1)$. Using Eqs. (8.80) and (8.86), (8.87) we find:

$$\begin{aligned}
\delta f(1) &= \delta f^{(c)}(1) + \delta f^{(i)}(1) \\
&= -g(1\,2)\,\Delta(2\,3)\,\Phi(3\,4)\,\delta f(4) + \delta f^{(i)}(1) \\
&= -g(1\,2)\,\Delta(2\,3)\,\varepsilon^{-1}(3\,4)\,\Phi(4\,5)\,\delta f^{(i)}(5) + \delta f^{(i)}(1) \\
&= \left[-g(1\,4)\,\Delta(4\,5)\,\varepsilon^{-1}(5\,6)\,\Phi(6\,7)\,g(7\,2) + g(1\,2) \right]\,g^{-1}(2\,3)\,\delta f^{(i)}(3).
\end{aligned}$$

Recalling the relation (8.65) we find the remarkable form:

$$\delta f(1) = G(1\,2)\,g^{-1}(2\,3)\,\delta f^{(i)}(3). \tag{8.90}$$

This is a very compact expression showing clearly that the incoherent response is a source driving the total fluctuating response. Calculating now the correlation function we obtain:[13]

$$\langle \delta f(1)\,\delta f(2) \rangle = G(1\,3)\,g^{-1}(3\,5)\,G(2\,4)\,g^{-1}(4\,6)\left\langle \delta f^{(i)}(5)\,\delta f^{(i)}(6) \right\rangle. \tag{8.91}$$

The correlation function is thus entirely determined by the correlation of the incoherent responses. Comparing this expression with Eq. (8.54) we find:

$$F(1\,2) = g^{-1}(1\,3)\,g^{-1}(2\,4)\left\langle \delta f^{(i)}(3)\,\delta f^{(i)}(4) \right\rangle, \tag{8.92}$$

or:

$$\left\langle \delta f^{(i)}(1)\,\delta f^{(i)}(2) \right\rangle = g(1\,3)\,g(2\,4)\,F(3\,4). \tag{8.93}$$

Thus, the correlation of incoherent responses is simply related to the *random noise source* appearing in the renormalized equation for the correlation (8.47). But the latter is provided explicitly by the truncation scheme

[13] It may be interesting to note that this relation "short-circuits" a more complicated relation obtained by a direct use of (8.85) :

$$\langle \delta f\,\delta f \rangle = \left\langle \delta f^{(c)}\,\delta f^{(c)} \right\rangle + \left\langle \delta f^{(c)}\,\delta \tilde{f} \right\rangle + \left\langle \delta \tilde{f}\,\delta f^{(c)} \right\rangle + \left\langle \delta \tilde{f}\,\delta \tilde{f} \right\rangle.$$

This decomposition has led to some contradictory discussions in the early literature about the role of the cross-terms (DUPREE 1972, HUI & DUPREE 1975, DUPREE et al. 1975). These discussions now appear to be irrelevant, in the light of Eq. (8.91).

adopted. Hence Eq. (8.93) yields an "explicit" expression of the incoherent correlation.[14]

8.7 The Direct Interaction Approximation (DIA)

We consider a *nonlinear stochastic equation* of the form (8.20) (which we rewrite here, for convenience):

$$L(1\,2)\,\delta f(2) + \tfrac{1}{2}\,N(1\,2\,3)\,Q\,\delta f(2)\,\delta f(3) = 0, \qquad (8.94)$$

where $L(1\,2)$ is a linear operator, and $N(1\,2\,3)$ a quadratically nonlinear operator, having the symmetry property:

$$N(1\,2\,3) = N(1\,3\,2). \qquad (8.95)$$

The operator Q was defined in Eq. (4.45). Both L and N are given, nonrandom operators.

In the present section we consider a quite general problem. The operators L and N are not necessarily those related to the gyrokinetic equation, as defined in Sec.8.1. The general results obtained here can therefore be applied to different representations of the gyrokinetic equation (*e.g.*, the Fourier representation rather than the time-dependent representation), or even to different problems (*e.g.*, macroscopic models of plasma turbulence, or Navier-Stokes turbulence of neutral fluids).

We now consider the equations for the *moments* of the random variable $\delta f(1)$. By definition, we have $\langle \delta f(1)\rangle = 0$ [Eq. (1.33)]. (In the probabilistic jargon it is said that the moments of δf are *centred*). We then obtain from (8.94) an infinite hierarchy of *moment equations*, of which we rewrite down here explicitly the first two [(8.27), (8.28)]:

$$L(1\,4)\,\langle\,\delta f(4)\,\delta f(2)\,\rangle + \tfrac{1}{2}\,N(1\,3\,4)\,\langle\,\delta f(3)\,\delta f(4)\,\delta f(2)\,\rangle = 0, \qquad (8.96)$$

$$L(1\,4)\,\langle\,\delta f(4)\,\delta f(2)\,\delta f(3)\,\rangle$$
$$+\tfrac{1}{2}\,N(1\,4\,5)\,\langle\,Q\,[\delta f(4)\,\delta f(5)]\,\delta f(2)\,\delta f(3)\,\rangle = 0. \qquad (8.97)$$

At this point KRAICHNAN 1959 (see also KROMMES 1984) introduces the idea of "*maximal randomness*" by stating that "the statistical dependence among the $[\delta f]$ is induced wholly by the nonlinear terms in Eq. (8.94)

---

[14] The term "explicit" must, as usual, be understood between quotation marks, because g and F are both functionals of the correlation function, hence of the incoherent correlation function.

and not at all by the initial conditions or by the external forces that may be acting".

In absence of any information, the most natural assumption is the complete randomness of the stochastic variable $\delta f(1)$. Invoking well-known arguments based on the law of large numbers, one makes this statement more precise by saying that the distribution of $\delta f(1)$ is *Gaussian*. Such a distribution is completely determined by its second moment $\langle \delta f(1)\, \delta f(2) \rangle$: *all odd moments are zero and all even moments are sums of products of second moments*. More simply, the Gaussian distribution is defined by the fact *all cumulants of order three and higher are zero:* $\langle\langle \delta f(1)...\delta f(p) \rangle\rangle = 0, \ p \geq 3$.

It is, however, immediately seen that such an assumption leads to triviality. Indeed, it amounts to a truncation of the chain to its first equation (8.96). Moreover, annulling the third moment reduces this equation to:

$$L(1\,4)\,\langle \delta f(4)\, \delta f(2) \rangle = 0,$$

i.e. to its *linear* approximation. The conclusion is immediate: *if we wish to describe the effects of the nonlinearity, the assumption of a Gaussian a priori distribution is insufficient: we must do better!*

Coming back now to our previous discussion, we wish to reconcile our fundamental "ignorance" with the requirements of the dynamical law. In other words, we wish to prudently assume an *a priori* distribution "as close as possible" to a Gaussian, but on the other hand we want to retain a truly nonlinear dynamics. Kraichnan's assumption of maximal randomness then requires that we retain at least a non-vanishing third-order moment (or cumulant, which amounts to the same). Therefore, the "minimalist" assumption for a non-trivial truncation is to assume:

$$\langle\langle \delta f(1)\, \delta f(2)...\delta f(p) \rangle\rangle = 0, \quad for\ p \geq 4. \tag{8.98}$$

Let us state already that this simple and quasi-evident relation is the essence of the physical content of the DIA. This assumption is related to the so-called *Quasinormal approximation* introduced by MILLIONCHTCHIKOV 1941, PROUDMAN & REED 1954, and TATSUMI 1957 in the context of the Navier-Stokes turbulence. (The latter was, however, limited to one-time correlations.) It was soon shown, however, that the quasi-normal approximation does not lead to a satisfactory theory of turbulence [see the clear and detailed discussion in ORSZAG 1973]. In the DIA, the truncation assumption (8.98) deals with multiple-time correlations. But this assumption by itself is not yet sufficient for a reasonable theory: an additional element necessary for defining the DIA was introduced by KRAICHNAN 1959, and will appear in Eq. (8.101) below.

Even so, there is a potential danger associated with the truncation hypothesis (8.98). There exists a strange theorem which, apparently, is not very well known in the plasma turbulence community. This theorem, due to MARCINKIEWICZ 1939, states that: *The probability distribution function will violate its positive definiteness if its cumulant generating function is a polynomial of degree greater than 2.* In other words (GARDINER 1985, Sec. 2.7), for any physical, definite positive distribution function, *either all but the first two cumulants vanish (= Gaussian process) or there is an infinite number of non-vanishing cumulants.* This theorem was generalized by RAJAGOPAL & SUDARSHAN 1974 to multivariate distribution functions and to probability functionals; these authors also attracted attention on the significance of this theorem for turbulence theory (as well as for a large class of truncation approximations for hierarchical equations). Thus, the truncation (8.98) cannot define a physically acceptable distribution functional of the fluctuations δf: it is *bound to* introduce a positivity violation problem *somewhere* in the theory. This poses the famous problem of the *realizability of a given approximation scheme*: it will be discussed in detail in Sec. 8.9. Let us just state here that the realizability of the DIA will be explicitly demonstrated: it is due to the additional element mentioned above.

We now express the moments in terms of irreducible *cumulants* defined by Eqs. (8.30). The equation for the third moment (or cumulant, which amounts to the same) was derived in Eq. (8.32). But now, in the DIA, Eq. (8.98) implies that the fourth order cumulant is zero; hence Eq. (8.32) reduces to:

$$L(1\,4)\,\langle\delta f(4)\,\delta f(2)\,\delta f(3)\rangle + N(1\,4\,5)\,\langle\delta f(2)\,\delta f(4)\rangle\,\langle\delta f(3)\,\delta f(5)\rangle = 0.$$
$$(8.99)$$

This equation can be formally solved in terms of the causal Green's operator $G_0(1\,2)$ associated with the linear operator $L(1\,2)$:

$$\langle\delta f(1)\,\delta f(2)\,\delta f(3)\rangle = -G_0(1\,6)\,N(6\,4\,5)\,\langle\delta f(2)\,\delta f(4)\rangle\,\langle\delta f(3)\,\delta f(5)\rangle\,.$$
$$(8.100)$$

We now make two important remarks. First, we note that, in the next step of the argument of Sec. 8.3, Eq. (8.100) must be used for the construction of a *renormalized* operator $L(12)+\Delta L(12)$. It is therefore quite natural to assume that the Green's operator entering the renormalized expressions should be taken as the *renormalized Green's operator* $G(1\,2)$ associated with $L(1\,2)+\Delta L(1\,2)$, rather than the linear one, $G_0(1\,2)$. We thus formulate the *second assumption defining the DIA* as follows:

$$G_0(1\,2) \rightarrow G(1\,2). \qquad\qquad (8.101)$$

The exact definition of the renormalized Green's operator must be done self-consistently, *a posteriori* [see (8.107), below].

Next, we note that (8.100) is not the complete expression of the third moment: it only describes the coupling of particle 1 to two others. We must clearly add to this the contributions of the coupling of particles 2 and 3 (Fig. 8.2). We then obtain an expression of the third moment having the correct symmetry under permutation of the three particles. As a result of this discussion, we obtain :

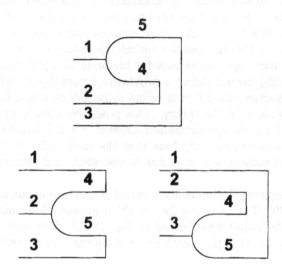

Figure 8.2. Graphs representing the complete solution for the third moment, Eq. (8.102).

$$\langle \delta f(1)\, \delta f(2)\, \delta f(3) \rangle = -\, G(1\,6)\, N(6\,4\,5)\, \langle \delta f(2)\, \delta f(4) \rangle\, \langle \delta f(3)\, \delta f(5) \rangle$$
$$-G(2\,6)\, N(6\,4\,5)\, \langle \delta f(3)\, \delta f(5) \rangle\, \langle \delta f(1)\, \delta f(4) \rangle$$
$$-G(3\,6)\, N(6\,4\,5)\, \langle \delta f(1)\, \delta f(5) \rangle\, \langle \delta f(2)\, \delta f(4) \rangle\,. \tag{8.102}$$

This equation [corresponding to Eq. (8.38)] defines the third moment as a functional of the second moment and of the renormalized Green's operator in the framework of the DIA (SIMILON 1981, OTTAVIANI et al. 1991, HU et al., 1997).

In order to derive the renormalized second moment equation, we substitute Eq. (8.102) into (8.27); using the symmetry property (8.16) and doing some trivial renaming of dummy variables, the equation for the second moment becomes:

$$\{ L(1\,3) - N(1\,4\,5)\, N(6\,7\,3)\, G(5\,6)\, \langle \delta f(4)\delta f(7) \rangle \}\, \langle \delta f(3)\delta f(2) \rangle$$
$$-\tfrac{1}{2}\, N(1\,4\,5)\, N(3\,7\,6)\, G(2\,3)\, \langle \delta f(4)\delta f(6) \rangle\, \langle \delta f(5)\delta f(7) \rangle = 0. \tag{8.103}$$

This equation can be rewritten in the form (8.43) combined with (8.53) as:

$$[L(1\,3) + \Delta L(1\,3)]\,C(3\,2) = G(2\,3)\,F(1\,3), \qquad (8.104)$$

where the renormalization of the linear evolution operator is:

$$\Delta L(1\,3) = -\,N(1\,4\,5)\,N(6\,7\,3)G(5\,6)\,C(4\,7), \qquad (8.105)$$

and the random noise source appears naturally in the form (8.53):

$$F(1\,3) = \tfrac{1}{2}\,N(1\,4\,5)\,N(3\,7\,6)C(4\,6)C(5\,7). \qquad (8.106)$$

The symmetry $F(1\,3) = F(3\,1)$ is manifest.

Eq. (8.104) has exactly the form of Eq. (8.43), combined with (8.53). The DIA formalism is completed with an equation for the Green's operator G, corresponding to Eq. (8.50):

$$[L(1\,3) + \Delta L(1\,3)]\,G(3\,2) = I(1\,2). \qquad (8.107)$$

Eqs. (8.104) - (8.107) completely define the *DIA approximation for the general stochastic nonlinear equation* (8.94). They are made explicit by introducing specific definitions for the operators L and N. Let us stress at this point the importance of the introduction of the *renormalized Green's operator G* into the theory. This step makes the whole difference between an approximation merely based on a low-order perturbative truncation of the hierarchy, such as the quasi-normal approximation (8.98), and the DIA. It amounts to taking account of the nonlinear modification of the dynamics, a feature absent from the quasilinear Green's operator G_0.

8.8   The Renormalized Gyrokinetic Equation

We now apply the Direct Interaction Approximation to the specific case of the Gyrokinetic equation (GKE). We do this in two steps.

In a first step we use the particular definition of the linear operator $L(1\,2)$ in the form (8.19) and of the nonlinear operator $N(1\,2\,3)$ in the form (8.15). This introduces the distinction between particle response $\delta f(1)$ and potential fluctuation $\delta \phi(a) = \Phi(a\,2)\delta f(2)$. This process eventually leads to a separate renormalization of the propagator and of the vertex operator, as was done at the end of Sec. 8.3, Eqs. (8.46) - (8.50).

In order to increase transparency, we adopt from here on the following convention: *a numeral (1,2,..) will be associated with the argument of a particle response, e.g. $\delta f(1)$; a latin letter (a,b,..) will be associated with the argument of a potential fluctuation, e.g. $\delta \phi(a)$.* We also frequently use in forthcoming equations the compact notations $C(1\,2)$, $T(1\,a)$, $T^{\dagger}(a1)$, $S(a\,b)$ defined in Eqs. (8.21) - (8.23) for the four types of second moments. As a typical example of application of this convention, we quote the following relation [see Eq. (8.24)]: $\Phi(1\,3)\,C(3\,2) \rightarrow \Phi(a\,3)\,C(3\,2) = T^\dagger(a\,2)$.

The first term in the left hand side of Eq. (8.104) has the form (8.19): $L(1\,3) = g_0^{-1}(1\,3) + \Delta_0(1\,a)\,\Phi(a\,3)$. In the renormalization $\Delta L(1\,3)$, Eq. (8.105), we split the second nonlinear coupling operator according to Eq. (8.15), with the result:

$$N(6\,7\,3) = M(6\,a\,3)\,\Phi(a\,7) + M(6\,a\,7)\,\Phi(a\,3).$$

The second term has the form of the vertex operator, *i.e.*, it ends with a factor $\Phi(a\,3)$ which, contracted with $C(3\,2)$ produces a mixed correlation $T^\dagger(a\,2)$. The first term is not of this form: it is therefore associated with the inverse propagator. We rewrite thus Eq. (8.104) in the form:

$$g^{-1}(1\,3)\,C(3\,2) + \Delta(1\,a)\,T^\dagger(a\,2) = G(2\,3)\,F(1\,3). \qquad (8.108)$$

The *renormalized inverse propagator* is:

$$g^{-1}(1\,3) = g_0^{-1}(1\,3) + \Sigma(1\,3), \qquad (8.109)$$

with:

$$\Sigma(1\,3) = -\,M(1\,a\,4)\,M(5\,b\,3)$$
$$\times \left\{ G(4\,5)\,S(b\,a) + \Phi(a\,6)\,G(6\,5)\,T^\dagger(b\,4) \right\}. \qquad (8.110)$$

The *renormalized vertex operator* is:

$$\Delta(1\,a) = \Delta_0(1\,a) + V(1\,a), \qquad (8.111)$$

with:

$$V(1\,\mathbf{a}) = -M(1\,b\,3)\,M(4\,a\,5)$$
$$\times \left\{ G(3\,4)\,T(5\,b) + \Phi(b\,6)\,G(6\,4)\,C(3\,5) \right\}. \qquad (8.112)$$

The *random noise source* is:

$$F(1\,3) = M(1\,a\,4)\,M(3\,b\,5)$$
$$\times \left\{ T^\dagger(b\,4)\,T(5\,a) + S(b\,a)\,C(5\,4) \right\}. \qquad (8.113)$$

Finally, the *Green's operator* obeys Eq. (8.50), which we rewrite here for convenience:

$$\left\{ g^{-1}(1\,3) + \Delta(1\,a)\,\Phi(a\,3) \right\}\,G(3\,2) = I(1\,2). \qquad (8.114)$$

In a second step of our development, we may *eliminate the Green's operator* $G(1\,2)$ from the expressions of the renormalized quantities. To this purpose, we express the resolvent G in terms of g, Δ and ε by using Eq. (8.65); whenever the Green's operator appears in the combination ΦG, we use Eq. (8.64). We then group together all the terms containing an explicit factor ε^{-1}. As a result of these transformations, the coefficients in Eqs. (8.110) and (8.112) are now written in the following final form (KROMMES & KLEVA 1979, SIMILON 1981).

The renormalization of the *inverse propagator* is split into two terms, respectively called the *diffusive* term and the *polarization* term:

$$\Sigma(1\,2) = \Sigma^D(1\,2) + \Sigma^P(1\,2). \tag{8.115}$$

The *diffusive* term is:

$$\Sigma^D(1\,2) = -\,M(1\,a\,3)\,g(3\,4)\,M(4\,b\,2)\,S(b\,a), \tag{8.116}$$

and the *polarization* term is:

$$\Sigma^P(1\,2) = -\,M(1\,a\,3)\,\varepsilon^{-1}(a\,b)\,\Phi(b\,4)\,g(4\,5)\,M(5\,c\,2)\,T^\dagger(c\,3)$$
$$+M(1\,a\,3)\,g(3\,4)\,\Delta(4\,c)\,\varepsilon^{-1}(c\,d)\,\Phi(d\,5)\,g(5\,7)\,M(7\,b\,2)\,S(b\,a). \tag{8.117}$$

The *vertex renormalization* is similarly decomposed:

$$V(1\,a) = V^D(1\,a) + V^P(1\,a). \tag{8.118}$$

The *diffusive* term is:

$$V^D(1\,\mathbf{a}) = -\,M(1\,b\,3)\,g(3\,4)\,M(4\,a\,5)\,T(5\,b), \tag{8.119}$$

and the *polarization* term:

$$V^P(1\,\mathbf{a}) = -\,M(1\,b\,3)\,\varepsilon^{-1}(b\,c)\,\Phi(c\,4)\,g(4\,5)\,M(5\,a\,6)\,C(6\,3)$$
$$+M(1\,b\,3)\,g(3\,4)\,\Delta(4\,c)\,\varepsilon^{-1}(c\,d)\,\Phi(d\,5)\,g(5\,6)\,M(6\,a\,7)\,T(7\,b). \tag{8.120}$$

It may appear amazing that such a trivially simple statistical assumption as (8.98) [together with the dynamical assumption (8.101)] ends up in an extremely complex set of equations for the second moment. The reason therefor is that the *a priori statistical assumption* has been made fully consistent with the *nonlinear dynamical equations of evolution*.

It is hoped that the derivation presented here will help the reader to realize that, in spite of its horrendous *prima facie* appearance, the DIA basically contains very simple physical ideas.

8.9   Realizability and Langevin Representation

Before being acceptable as a physically sound approximation, any closure procedure must undergo a certain number of tests. One of the most important of these is the *constraint of realizability*. It consists of proving that *the moments calculated from the approximate theory are indeed moments of an underlying probability distribution function (PDF)* of the fluctuations δf. The latter, an approximation of the true PDF, must possess the properties of a PDF, in particular, being a *positive definite* function. This induces an infinite number of constraints in the form of inequalities to be satisfied by the moments defined from the truncated theory. It is practically impossible to verify all these constraints.

There is, however, a more direct way to show that a given theory (in particular, the DIA) is realizable. It consists of proving the possibility of constructing a *model system* whose fluctuating dynamics is defined by a Langevin equation and a prescribed statistics.[15] If the *exact* equations for the second moment and of the Green's operator coincide with those of the DIA, the realizability is demonstrated. The proof of the realizability of the DIA along these lines was obtained by PHYTHIAN 1969, KRAICHNAN 1970 a, among others. A very clear exposition of the subject is given by ORSZAG 1973; see also KROMMES 1984.

We consider a fluctuating quantity $\delta \overline{f}(1)$ of the same nature as $\delta f(1)$; in other words, this symbol represents a set of variables of the same form as Eq. (8.4) in which $\delta f \to \delta \overline{f}$. It should be clear, however, that $\delta \overline{f}(1)$ is *not* the fluctuating response obeying Eq. (8.20). It is assumed that the stochastic variable $\delta \overline{f}(1)$ has the same second moment as $\delta f(1)$, thus:

$$\langle \delta \overline{f}(1)\, \delta \overline{f}(2) \rangle = \langle \delta f(1)\, \delta f(2) \rangle = C(1\,2). \qquad (8.121)$$

It is supposed to obey the following Langevin equation:

$$L(1\,3)\, \delta \overline{f}(3) + \Delta L(1\,3)\, \delta \overline{f}(3) = \delta k(1), \qquad (8.122)$$

where $L(1\,2)$ is the linear operator defined by Eq. (8.19) and $\Delta L(1\,3)$ is the DIA-renormalization of the latter, defined by Eq. (8.105). $\delta k(1)$ is a "random force", to be defined below. Multiplying both sides by $\delta \overline{f}(2)$, averaging and using Eq. (8.121), we find the following equation for the second moment of $\delta \overline{f}$:

$$[L(1\,3) + \Delta L(1\,3)]\, C(3\,2) = \langle \delta k(1)\, \delta \overline{f}(2) \rangle. \qquad (8.123)$$

The left hand side of this equation coincides with the left hand side of the DIA-renormalized equation (8.104). It follows that the causal Green's function associated with the "model equation" (8.123) is precisely the same

[15] The existence of such a Langevin equation was postulated in Eq. (8.79).

as the operator $G(1\,2)$, obeying the DIA equation (8.107). Hence, the proof of the realizability amounts to showing that one can find a "random force" $\delta k(1)$ such that the right hand side of (8.123) equals the right hand side of the corresponding DIA equation (8.104), with (8.106), *i.e.*, $\langle \delta k(1)\,\delta \overline{f}(2) \rangle = G(2\,3)\,F(1\,3)$.

We now introduce two additional random functions $\delta \xi(1)$, $\delta \xi'(1)$, having the following properties:

- they are mutually independent:

$$\langle \delta \xi(1)\,\delta \xi'(2) \rangle = 0, \tag{8.124}$$

and are statistically independent of $\delta \overline{f}(2)$:

$$\langle \delta \xi(1)\,\delta \overline{f}(2) \rangle = \langle \delta \xi'(1)\,\delta \overline{f}(2) \rangle = 0; \tag{8.125}$$

- their second moment equals $C(1\,2)$:

$$\langle \delta \xi(1)\,\delta \xi(2) \rangle = \langle \delta \xi'(1)\,\delta \xi'(2) \rangle$$
$$= \langle \delta \overline{f}(1)\,\delta \overline{f}(2) \rangle = \langle \delta f(1)\,\delta f(2) \rangle. \tag{8.126}$$

We now define the random force as the following functional of $\delta \xi$, $\delta \xi'$:

$$\delta k(1) = \frac{1}{\sqrt{2}}\,N(1\,3\,4)\,\delta \xi(3)\,\delta \xi'(4). \tag{8.127}$$

It then follows from Eqs. (8.124) - (8.126):

$$\langle \delta k(1)\,\delta k(2) \rangle$$
$$= \frac{1}{2} N(1\,3\,4)\,N(2\,5\,6)\,\langle \delta \xi(3)\,\delta \xi'(4)\,\delta \xi(5)\,\delta \xi'(6) \rangle$$
$$= \frac{1}{2} N(1\,3\,4)\,N(2\,5\,6)\,\langle \delta \xi(3)\,\delta \xi(5) \rangle\,\langle \delta \xi'(4)\,\delta \xi'(6) \rangle$$
$$= \frac{1}{2}\,N(1\,3\,4)\,N(2\,5\,6)\,C(3\,5)\,C(4\,6) = F(1\,2), \tag{8.128}$$

where $F(1\,2)$ is precisely the DIA random noise source of Eq. (8.106). The last step of the proof is immediate. Given that the Green's function of the model equation is the DIA operator $G(1\,2)$, the solution of Eq. (8.122) is:

$$\delta \overline{f}(1) = G(1\,2)\,\delta k(2), \tag{8.129}$$

hence, the right hand side of Eq. (8.123) is:

$$\langle \delta k(1)\,\delta \overline{f}(2) \rangle = G(2\,3)\,\langle \delta k(1)\,\delta k(3) \rangle = G(2\,3)\,F(1\,3). \tag{8.130}$$

Thus, Eq. (8.123) is identical to the DIA equation (8.104). We have thus shown that the approximate DIA correlation function $\mathcal{C}(1\,2)$ obeys the same equation as the exact solution of Eq. (8.123), derived from the model Langevin equation (8.122). This achieves the proof of the realizability of the DIA.

8.10   Energy Conservation

A second stringent test to be passed by the DIA is to ensure energy conservation. It was shown in Sec. 5.4 that the nonlinear GKE conserves the total energy (kinetic + electrostatic). This property should be inherited by the DIA equation in order to make physical sense.

In the original equation, the starting point for the proof was the equation for the mixed one-time moment $T(1\,a) = \langle \delta f(1)\, \delta\phi(a)\rangle$, Eq. (5.29). It was shown that, after multiplication by $(2\pi)^{-3} e_\alpha J_k^\alpha(\mathbf{v})$, integration over $\mathbf{v}, \mathbf{k}$ and summation over α, the linear term produces by itself the correct electrostatic energy balance; the nonlinear term contributes zero to the latter. These results are contained in Eqs. (5.32) - (5.34). These equations can be expressed in our compact notation by introducing the following operator:

$$\mathcal{P}(1\,a) \to \mathcal{P}^{\alpha_1\alpha_a}_{\mathbf{k}_1,\mathbf{k}_a}(\mathbf{v}_1,\mathbf{v}_a; t_1, t_a)$$
$$= e_{\alpha_1}\, \delta^{\alpha_1\alpha_a}\, J^{\alpha_1}_{\mathbf{k}_1}(\mathbf{v}_1)\, \delta(\mathbf{v}_1 - \mathbf{v}_a)\, \delta(t_1 - t_a) \qquad (8.131)$$

We also need a slightly modified definition of the matrix product involving this operator: the time integration relative to the first argument (t_1) is skipped; this "partial" contraction is denoted by the symbol $*$:

$$\mathcal{P}(1\,a) * f(1)$$
$$\to \sum_{\alpha_1} \int d\mathbf{v}_1 d\mathbf{k}_1\, \mathcal{P}^{\alpha_1\alpha_a}_{\mathbf{k}_1,\mathbf{k}_a}(\mathbf{v}_1,\mathbf{v}_a; t_1, t_a)\, f^{\alpha_1}_{\mathbf{k}_1}(\mathbf{v}_1, \mathbf{k}_1, t_1). \qquad (8.132)$$

It is then easily checked that with these rules, Eqs. (5.32), (5.33) and (5.34) are compactly expressed as:

$$\mathcal{P}(1\,a) * \Delta_0(1\,b)\, S(b\,a) = 0 \qquad (8.133)$$

$$\mathcal{P}(1\,a) * M(1\,b\,2)\, \langle \delta\phi(b)\, \delta f(2)\, \delta\phi(a)\rangle = 0 \qquad (8.134)$$

$$\mathcal{P}(1\,a) * g_0^{-1}(1\,2)\, T(2\,a) = 0. \qquad (8.135)$$

The last equation by itself produces the electrostatic energy balance equation, as shown in Sec. 5.4.

We now consider the DIA equation (8.108), together with (8.109) and (8.111): it is transformed into an equation for the mixed correlation $T(3\,a)$ by the action of the operator $\Phi(a\,2)$. By "partial" contraction with $\mathcal{P}$, we obtain:

$$\mathcal{P}(1\,a) * \{ \left[g_0^{-1}(1\,3) + \Sigma(1\,3) \right] T(3\,a)$$
$$+ \left[\Delta_0(1\,b) + V(1\,b) \right] S(b\,a) - \Phi(a\,2)\,G(2\,3)\,F(1\,3) \}$$
$$= 0. \tag{8.136}$$

Given that the electrostatic energy balance is entirely contained in Eq. (8.135), and considering Eq. (8.133), we see that the energy conservation test amounts to showing that:

$$\mathcal{P}(1\,a)*\{\Sigma(1\,3)\,T(3\,a) + V(1\,b)\,S(b\,a) - \Phi(a\,2)\,G(2\,3)\,F(1\,3)\} = 0. \tag{8.137}$$

This is the *fundamental constraint that must be obeyed by the renormalization operators* Σ, V, H in order to ensure the compatibility of the renormalized GKE (8.108) with the energy balance equation (8.135). It was first obtained in the present clear form by SIMILON 1981 (though he did not use the present compact notations). The same author provided the first complete and systematic proof of the following fundamental fact:

The DIA expressions of the renormalized operators Σ, V, H *fulfill exactly the energy conservation constraint.*[16]

Similon's theorem goes actually farther in detail, and this latter aspect significantly increases its importance as a test of consistency. We first note that, using (8.113) and (8.64), we may write:

$$H'(1\,a) \equiv \Phi(a\,2)\,G(2\,3)\,F(1\,3)$$
$$= M(1\,c\,4)\,\varepsilon^{-1}(a\,b)\,\Phi(b\,2)\,g(2\,3)\,M(3\,d\,6)\,\left[T^{\dagger}(d\,4)T(6\,c) + C(6\,4)S(d\,c) \right]$$
$$\equiv H'^{\Sigma}(1\,a) + H'^{V}(1\,a). \tag{8.138}$$

We similarly write down separately the two terms of the polarization parts (8.117), (8.120) of the operators Σ and V. The substitution of all these operators into Eq. (8.137) thus leads to the following form of the constraint:

---

[16] The energy conservation property for the DIA applied to the Navier-Stokes equation was known earlier (see the discussion in ORSZAG 1973).

$$\mathcal{P}(1\,a) * \left\{ \left[\Sigma^D(1\,3) + \Sigma^P_{(1)}(1\,3) + \Sigma^P_{(2)}(1\,3) \right] T(3\,a) \right.$$

$$+ \left. \left[V^D(1\,b) + V^P_{(1)}(1\,b) + V^P_{(2)}(1\,b) \right] S(b\,a) - H'^\Sigma(1\,a) - H'^V(1\,a) \right\}$$

$$= 0. \tag{8.139}$$

Similon's result is that *the cancellations occur in pairs.* Specifically, if we denote by $[A]$ the contribution of the operator A to the left hand side of Eq. (8.139), we obtain the following pairwise cancellations:

$$\left[\Sigma^D \right] + \left[V^D \right] = 0, \tag{8.140}$$

$$\left[\Sigma^P_{(1)} \right] + \left[H'^\Sigma \right] = 0, \tag{8.141}$$

$$\left[V^P_{(1)} \right] + \left[H'^V \right] = 0, \tag{8.142}$$

$$\left[\Sigma^P_{(2)} \right] + \left[V^P_{(2)} \right] = 0. \tag{8.143}$$

The proof consists simply of substituting the expressions (8.113) - (8.120) into (8.139), writing them out explicitly by using (8.7) - (8.11), (8.132) and checking the result. As a sample calculation, we consider Eq. (8.140). Expanding the calculations, we obtain:

$$\left[\Sigma^D \right] + \left[V^D \right]$$

$$= \sum_{\alpha_1,\alpha_2} \int dv_1 dv_2 dk_2 dk_3 \int dt_2\, H(t_1 - t_2)\, H(t_2) J^{\alpha_1}_{k_2}(v_1) J^{\alpha_2}_{k_3 - k_2}(v_2)$$

$$\times \frac{c^2}{B^2} e_{\alpha_1} [k_2 \cdot (k_3 \times b)]^2\, g^{\alpha_1 \alpha_2}_{k_3}(v_1, v_2; t_1, t_2)$$

$$\times \left\{ J^{\alpha_2}_{k_2 - k_3}(v_2) S_{k_2 - k_3}(t_1, t_2) T^{\alpha_2}_{k_2}(v_2; t_2, t_1) \right.$$

$$\left. - J^{\alpha_2}_{k_3 - k_2}(v_2) S_{k_3 - k_2}(t_2, t_1) T^{\alpha_2}_{k_2}(v_2; t_2, t_1) \right\} .$$

The vanishing of this expression results from the symmetry:

$$S_{k_3 - k_2}(t_2, t_1) = \left\langle \delta\phi_{-k_3 + k_2}(t_2) \delta\phi_{k_3 - k_2}(t_1) \right\rangle = S_{k_2 - k_3}(t_1, t_2).$$

The remaining relations (8.141) - (8.143) are proved in a similar way, although the calculations are more tedious.

These relations are of great practical importance. Indeed, as mentioned above, the full DIA is still a very complicated approximation scheme, and many authors try to introduce additional simplifications by neglecting some of the terms. Thus, in his first paper on renormalization, DUPREE 1966 introduced only the diffusive part of the propagator renormalization, Σ^D. In our present language we would say that he assumed : $\Sigma^P = V = H = O$. We now see that this is an inconsistent approximation: because of (8.140), it does not fulfil the energy conservation constraint. If the propagator is renormalized by Σ^D, the vertex must also be renormalized by V^D. This fact was only discovered 12 years later by DUPREE & TETREAULT 1978, who stressed the importance of the energy conservation condition and derived the form of the diffusive vertex renormalization, V^D [they called it "the β-term"].

The approximation consisting of the neglect of the random noise source F, together with the neglect of the polarization parts of Σ and V:

$$F = 0, \quad \Sigma^P = V^P = 0 \tag{8.144}$$

was developed by several authors, *e.g.* DUPREE & TETREAULT 1978, KROMMES & KLEVA 1979, SIMILON 1981. It will be treated in Chap. 9 (see also Sec. 8.6). This scheme leads to the simplest approximation that is fully consistent with the energy conservation constraint. It will be called: the *Renormalized Quasilinear (RQL)* approximation. We note that it is a typically *Coherent Approximation*. On the other hand, the conditions (8.141), (8.142) imply that if $F \neq 0$ is retained in an approximation scheme, the polarization parts $\Sigma_{(1)}^P$ and $V_{(1)}^P$ must also be kept, for consistency with the energy conservation.

8.11 Markovian Closures

In recent years there has been a rather large amount of effort devoted towards further developments of the DIA. We will only mention the study of higher approximations of the renormalization process beyond the DIA, based on the field-theory inspired MARTIN-SIGGIA-ROSE 1973 formalism: they are extremely complicated and do not lead to any practically exploitable result (see *e.g.*, KROMMES 1996).

A quite different approach gave rise to an interesting series of papers: it is the quest of a *Markovian approximation to the DIA* (BOWMAN 1992, OTTAVIANI et al. 1991, HU et al. 1997, KROMMES 1999, among others). The motivation of this work is the following. The DIA ends up in a set of equations for the *two-point, two-time correlation function* $C(1\,2) = \langle \delta f^{\alpha_1}(\mathbf{Y}_1, \mathbf{v}_1; t_1)\, \delta f^{\alpha_2}(\mathbf{Y}_2, \mathbf{v}_2; t_2) \rangle$. The solution of these equations is practically impossible, not only analytically, but also numerically. It would be desirable to introduce an additional approximation in order to deal only with *one-time correlation functions,* and which involves only a *memoryless propagator*. For this correlation function we introduce the following notation: $\widehat{C}(\underline{1}\,\underline{2};t_1) = \langle \delta f^{\alpha_1}(\mathbf{Y}_1, \mathbf{v}_1; t_1)\,\delta f^{\alpha_2}(\mathbf{Y}_2, \mathbf{v}_2; t_1)\rangle$ (the underlined symbol $\underline{1}$ denotes all variables relative to particle 1, except for the time). The numerical solution of such equations would be much simpler. We do not enter the details of these theories here, as they do not introduce any essentially new concept. Let us just quote the ansatz of OTTAVIANI et al. 1991. Instead of Eq. (8.105), these authors propose the following Markovian version:

$$\Delta L^M(1\,3)$$

$$= -N(\underline{1}\,\underline{4}\,\underline{5}; t_1)N(\underline{6}\,\underline{7}\,\underline{3}; t_3)\,\theta(\underline{5}\,\underline{6}; t_1)\,\widehat{C}(\underline{4}\,\underline{7}; t_1)\,\delta(t_1 - t_3), \qquad (8.145)$$

and a similar expression for the random noise source $F(1\,3)$. These expressions introduce the "triad interaction time" $\theta(\underline{5}\,\underline{6}; t_1)$, an object that contains the trace of the memory effects deleted explicitly in the Markovianization process. This quantity introduces difficulties, because it is in general complex, thus preventing a physical interpretation of the results in terms of a real correlation decay time. Several solutions have been proposed for curing these difficulties, by keeping at the same time the constraints of realizability and energy conservation described above. We shall not discuss these questions here, referring the interested reader to the original publications cited above.

Chapter 9

The Coherent Direct Interaction Approximation

9.1 The Renormalized Quasilinear Approximation

The Direct Interaction Approximation (DIA), as presented in Chap. 8, is a quite elegant and internally consistent approximation scheme. Although representing an intrinsically very simple situation [the minimal deviation from Gaussianity], its complete set of equations is of such a degree of complexity that it is beyond any possibility of solution, be it analytical or numerical. It is therefore necessary to make many additional approximations, *within the DIA framework,* in order to obtain explicit results. These additional simplifications cannot always be easily justified, because of the lack of a small parameter. One therefore advances by means of semi-intuitive postulates, trying to keep as much as possible an internal consistency. The results are then tested *a posteriori,* by comparison either with real experimental data or with results of numerical simulations. This way of proceeding is (unfortunately, but inevitably) typical of strong turbulence theory, be it in neutral fluids or in plasmas.

The strongest simplification is obtained by introducing the so-called *coherent approximation,* which was described in general terms in Sec. 8.6. It is obtained by neglecting the random noise source term $F(1\,2)$ in the renormalized DIA equation (8.108) for the two-point, two-time correlation $C(1\,2)$:

$$F(1\,2) = 0; \quad (coherent\,appr.); \tag{9.1}$$

the latter equation thus reduces to:[1]

$$g^{-1}(1\,3)\ C(3\,2) + \Delta(1\,a)\,T(a\,2) = 0. \tag{9.2}$$

[1] We use here systematically the conventions introduced in Sec. 8.7: numerals are associated with arguments of δf, latin letters with arguments of $\delta \phi$.

181

Note that Eq. (8.114) for the Green's operator does not involve the random noise source, hence it remains unchanged: we rewrite it here for convenience:

$$\{g^{-1}(1\,3) + \Delta(1\,a)\,\Phi(a\,3)\}\,G(3\,2) = I(1\,2). \tag{9.3}$$

In this coherent approximation, the moments $C(1\,2)$, $T(a\,1)$, $T^\dagger(1\,a)$, $S(a\,b)$ [defined in Eqs. (8.21) - (8.23)][2] are interrelated, as shown in Eq. (8.84):

$$C(1\,2) = -g(1\,3)\,\Delta(3\,b)T(b\,2), \tag{9.4}$$

$$T(a\,1) = -g(1\,3)\,\Delta(3\,b)\,S(a\,b), \tag{9.5}$$

$$T^\dagger(1\,a) = -g(1\,3)\,\Delta(3\,b)\,S(b\,a). \tag{9.6}$$

The spectrum obeys in this approximation an equation derived from (8.88):

$$\varepsilon(a\,c)\,\varepsilon(b\,d)\,S(c\,d) = 0. \tag{9.7}$$

Finally, the energy conservation constraint (8.140) implies that if F (or H) is set equal to zero, the propagator and the vertex must be consistently approximated by using only the diffusive parts Σ^D, V^D in their renormalization [see Eqs. (8.115) - (8.119)]:

$$g^{-1}(1\,2) = g_0^{-1}(1\,2) + \Sigma^D(1\,2), \tag{9.8}$$

$$\Delta(1\,a) = \Delta_0(1\,a) + V^D(1\,a). \tag{9.9}$$

The neglect of the polarization parts leads to a formidable simplification of the explicit expressions.

Eqs. (9.4) - (9.9) constitute a complete definition of the *coherent approximation* within the DIA. In this scheme, all the relations between turbulent quantities (9.2) - (9.7) have the *same form as in the linear (or quasi-linear) approximation*, developed in Chap. 7. The only role of the nonlinearity is in the replacement of the unperturbed propagator g_0 and vertex Δ_0 by the corresponding renormalized quantities g and Δ. Thus, in spite of their deceivingly simple form, Eqs. (9.4) - (9.7) are *nonlinear* equations for the correlation functions. In order to stress this feature, we shall more appropriately call the coherent approximation the *renormalized quasilinear theory (RQL)*. As mentioned in Sec. 8.10, it is the

---

[2] Remember the conventions defined after Eq. (8.23) about the order of the variables in the mixed correlations T and $T^\dagger$. With the present distinction between letters and numerals, these definitions become more transparent.

simplest, internally consistent approximation. It exhibits some important features of plasma turbulence, but not all. Our purpose here is therefore not the derivation - in a single stroke - of a realistic form of the transport equations, but rather a gradual introduction of the turbulent effects into the anomalous transport theory.

The RQL theory was introduced in the pioneering papers of DUPREE 1966 (for the Vlasov equation) and DUPREE 1967 (for the drift wave turbulence, which is our purpose here). It was further developed in a large number of papers, among which we mention: ALTSHUL' & KARPMAN 1966, WEINSTOCK 1969, RUDAKOV & TSYTOVICH 1971, MISGUICH & BALESCU 1975 a,b, DUPREE & TETREAULT 1978, KROMMES & KLEVA 1979, SIMILON 1981, SALAT 1988.

9.2 Turbulent Resonance Broadening

The forthcoming calculations are conveniently performed in the complete (space and time) Fourier representation. The corresponding expressions are easily obtained from the equations developed in Chap. 8. In order to have a more compact notation, we group together the wave-vector $\mathbf{k}$ and the frequency ω into a "four-vector":

$$\kappa \equiv (\mathbf{k}, \omega). \tag{9.10}$$

We thus use the following change of notation: $\delta \tilde{f}_{\mathbf{k}}(\omega) \to \delta f_\kappa$; we also use the compact notation $\delta(\kappa_1 - \kappa_2) \equiv \delta^3(\mathbf{k}_1 - \mathbf{k}_2)\,\delta(\omega_1 - \omega_2)$.

In developing the RQL theory we first proceed to a closer analysis of the *(causal) propagator* $g(1\,2)$. In the unperturbed case, this operator has a very simple form (8.10), (7.32):[3]

$$(g_0)^{\alpha_1\alpha_2}_{\kappa_1\kappa_2}(\mathbf{v}_1, \mathbf{v}_2) = \mathcal{G}_{0\kappa}(\mathbf{v})\,\delta^{\alpha_1\alpha_2}\delta(\mathbf{v}_1 - \mathbf{v}_2)\,\delta(\kappa_1 - \kappa_2)$$

$$= [-i\,(\omega_1 - k_{1\|}v_{1\|})]^{-1}\delta^{\alpha_1\alpha_2}\delta(\mathbf{v}_1 - \mathbf{v}_2)\,\delta(\kappa_1 - \kappa_2). \tag{9.11}$$

The operator is *diagonal in all variables*: α, $\mathbf{v}$, κ; moreover, the coefficient of the δ's is a function of κ_1, $\mathbf{v}_1$, but is *independent of the species index* α_1.[4] One may wonder how much of this simplicity is retained in the renormalized DIA propagator. The diagonal character in the wave-vector is a result of the quasi-homogeneous and quasi-stationary nature of the reference turbulent state and is therefore retained for all the renormalized DIA

---

[3] In order to be consistent with the *causal* character, it is necessary to consider implicitly that the frequency ω has a small positive imaginary part ($i\epsilon$) which tends to zero. In other words, the time-Fourier transform is the limit of a Laplace transform.

[4] The notation $\widehat{\mathcal{G}}_{0\mathbf{k}}(\mathbf{v}, \omega)$ of Eq. (7.32) is replaced by the simpler one: $\mathcal{G}_{0\kappa}(\mathbf{v})$.

operators. On the other hand, the diagonality in α and $\mathbf{v}$ is, in general, lost.

Consider, however, the special case of the *coherent approximation*, defined by Eq. (9.8). The propagator is a solution of Eq. (8.58):

$$g(1\,2) = g_0(1\,2) - g_0(1\,3)\,\Sigma^D(3\,4)\,g(4\,2). \tag{9.12}$$

We write the explicit form of this equation by using Eqs. (7.32), (9.11), (8.14) and (8.116):

$$g^{\alpha_1\alpha_2}_{\kappa_1\kappa_2}(\mathbf{v}_1,\mathbf{v}_2) = \mathcal{G}_{0\kappa_1}(\mathbf{v}_1)\,\delta^{\alpha_1\alpha_2}\delta(\mathbf{v}_1 - \mathbf{v}_2)\,\delta(\kappa_1 - \kappa_2)$$

$$-\mathcal{G}_{0\kappa_1}(\mathbf{v}_1)\sum_{\alpha_3}\int d\mathbf{v}_3\int d\kappa_3\int d\kappa'\,\mathbf{k}_{\perp 1}\cdot\mathbf{m}_{\mathbf{k}'}\cdot\mathbf{k}_{\perp 3}\,S_{\kappa'}J^{\alpha_1}_{\mathbf{k}'}(\mathbf{v}_1)J^{\alpha_3}_{\mathbf{k}'}(\mathbf{v}_3)$$

$$\times g^{\alpha_1\alpha_3}_{\kappa_1-\kappa',\kappa_3-\kappa'}(\mathbf{v}_1,\mathbf{v}_3)\,g^{\alpha_3\alpha_2}_{\kappa_3,\kappa_2}(\mathbf{v}_3,\mathbf{v}_2). \tag{9.13}$$

We introduced here the tensor $\mathbf{m_k}$ defined as:

$$\mathbf{m_k} = \frac{1}{2\pi}\frac{c^2}{B^2}\,(\mathbf{k}\times\widehat{\mathbf{b}})\,(\mathbf{k}\times\widehat{\mathbf{b}}). \tag{9.14}$$

If Eq. (9.13) is solved by successive iterations, it is immediately seen that, to all orders, the right hand side remains diagonal in α and in $\mathbf{v}$. We thus find the following result, valid in the RQL approximation:

$$g^{\alpha_1\alpha_2}_{\kappa_1,\kappa_2}(\mathbf{v}_1,\mathbf{v}_2) = \mathcal{G}^{\alpha_1}_{\kappa_1}(\mathbf{v}_1)\,\delta^{\alpha_1\alpha_2}\,\delta(\mathbf{v}_1 - \mathbf{v}_2)\,\delta(\kappa_1 - \kappa_2). \tag{9.15}$$

Eq. (9.13) then reduces to:[5]

$$\mathcal{G}^{\alpha}_{\kappa}(\mathbf{v}) = \mathcal{G}_{0\kappa}(\mathbf{v}) - \mathcal{G}_{0\kappa}(\mathbf{v})\,\Sigma^{\alpha}_{\kappa}(\mathbf{v})\,\mathcal{G}^{\alpha}_{\kappa}(\mathbf{v}), \tag{9.16}$$

with:

$$\Sigma^{\alpha}_{\kappa}(\mathbf{v}) \equiv \mathbf{k}_{\perp}\cdot\mathbf{D}^{\alpha}_{\kappa}(\mathbf{v})\cdot\mathbf{k}_{\perp}$$

$$= \int d\kappa'\,\mathbf{k}_{\perp}\cdot\mathbf{m}_{\mathbf{k}'}\cdot\mathbf{k}_{\perp}\,S_{\kappa'}\,[J^{\alpha}_{\mathbf{k}'}(\mathbf{v})]^2\,\mathcal{G}^{\alpha}_{\kappa-\kappa'}(\mathbf{v}). \tag{9.17}$$

Eq. (9.16), combined with (9.17) easily yields a formal solution for the renormalized propagator:

$$\mathcal{G}^{\alpha}_{\kappa}(\mathbf{v}) = \mathcal{G}_{0\kappa}(\mathbf{v})\cdot\frac{1}{1 + [\mathbf{k}_{\perp}\cdot\mathbf{D}^{\alpha}_{\kappa}(\mathbf{v})\cdot\mathbf{k}_{\perp}]\,\mathcal{G}_{0\kappa}(\mathbf{v})}, \tag{9.18}$$

---

[5] From here on, in forthcoming equations of the present chapter we delete the superscript "D" on all diffusive operators.

or, using (9.11):

$$\mathcal{G}_\kappa^\alpha(\mathbf{v}) = \frac{1}{-i\left(\omega - k_\parallel v_\parallel\right) + \mathbf{k}_\perp \cdot \mathsf{D}_\kappa^\alpha(\mathbf{v}) \cdot \mathbf{k}_\perp}. \tag{9.19}$$

Substituting this result into Eq. (9.17), the latter is converted into an integral equation for Σ:

$$\Sigma_\kappa^\alpha(\mathbf{v}) = \int d\kappa' \, \mathbf{k}_\perp \cdot \mathsf{m}_{\mathbf{k}'} \cdot \mathbf{k}_\perp \, S_{\kappa'} \frac{[J_{\mathbf{k}'}^\alpha(\mathbf{v})]^2}{-i\left[\omega - \omega' - (k_\parallel - k_\parallel')v_\parallel\right] + \Sigma_{\kappa-\kappa'}^\alpha(\mathbf{v})}. \tag{9.20}$$

At this point we make an approximation that is not easily justified in full rigour, though it is intuitively clear. The possibility of advancing the analytical treatment relies very strongly on this *"Markovian approximation"*.[6] We thus decide that we are only interested in sufficiently long time scales and length scales; we therefore neglect the small given wave vector **k** and the small given frequency ω (*i.e.* the 4-vector κ) as compared to the integration variable κ' (which takes all possible values in the integral). Thus, in Eq. (9.20) we replace $\Sigma_{\kappa-\kappa'}^\alpha(\mathbf{v}) \to \Sigma_{-\kappa'}^\alpha(\mathbf{v})$ and $g_{0\,\kappa-\kappa'}(\mathbf{v}) \to g_{0\,-\kappa'}(\mathbf{v})$, *i.e.*, $\omega - \omega' \to -\omega'$, $k_\parallel - k_\parallel' \to -k_\parallel'$. [An analysis of the time scales involved here can be found in SALAT 1988]. Using this "Markovian" approximation, Eq. (9.20) reduces to:

$$\mathbf{k}_\perp \cdot \mathsf{D}_\kappa^\alpha(\mathbf{v}) \cdot \mathbf{k}_\perp$$

$$= \int d\kappa' \, \mathbf{k}_\perp \cdot \mathsf{m}_{\mathbf{k}'} \cdot \mathbf{k}_\perp \, S_{\kappa'} \frac{[J_{\mathbf{k}'}^\alpha(\mathbf{v})]^2}{-i\left[-\omega' + k_\parallel' v_\parallel\right] + \mathbf{k}_\perp' \cdot \mathsf{D}_{-\kappa'}^\alpha(\mathbf{v}) \cdot \mathbf{k}_\perp'}. \tag{9.21}$$

Some symmetry considerations allow us to further simplify this result. The tensor character of $\mathsf{D}_\kappa^\alpha(\mathbf{v})$ is determined by the tensor:

$$\mathsf{m}_{\mathbf{k}'} \, S_{\kappa'} = \frac{1}{2\pi}\left(\frac{c}{B}\right)^2 (\mathbf{k}' \times \hat{\mathbf{b}})\,(\mathbf{k}' \times \hat{\mathbf{b}})\, S_{\kappa'} = \frac{1}{2\pi}\left(\frac{c}{B}\right)^2 \mathsf{E}_{\perp\kappa'},$$

where $\mathsf{E}_{\perp\kappa'}$ is the Eulerian correlation function of the perpendicular electric field components $(ik\,\delta\phi_\kappa) \times \hat{\mathbf{b}}$. Clearly, the quantity $(c/B)^2 \mathsf{E}_{\perp\kappa'}$ represents the *autocorrelation of the electric drift velocity* (4.34). It is natural to assume that the (complete) tensor of the electric field correlations is

---

[6] The Markovian approximation is not always justified. Specific examples have been treated by KROMMES & OBERMAN 1976 and by KROMMES & SIMILON 1980, where a problem with k-dependent diffusion coefficient appears, with a logarithmic divergence as $k \to 0$.

isotropic and diagonal in the plane perpendicular to the magnetic field; in other words, this tensor has *gyrotropic symmetry*:

$$\mathbf{E}_\kappa \equiv \mathbf{E}_{\parallel\kappa} + \mathbf{E}_{\perp\kappa} = \mathcal{E}_{\parallel\kappa}\,\widehat{\mathbf{b}}\,\widehat{\mathbf{b}} + \mathcal{E}_{\perp\kappa}\,(\mathbf{e}_x\,\mathbf{e}_x + \mathbf{e}_y\,\mathbf{e}_y)\,, \qquad (9.22)$$

where $\mathcal{E}_{\parallel\kappa}$, $\mathcal{E}_{\perp\kappa}$ are two scalar functions of κ. A simple argument then leads to the conclusion that:

$$\mathcal{E}_{\perp\kappa} = \tfrac{1}{2}\,k_\perp^2\,S_\kappa. \qquad (9.23)$$

Combining Eqs. (9.22), (9.23) we come to the conclusion that the quantity $\mathbf{k}_\perp \cdot \mathbf{D}_\kappa^\alpha(\mathbf{v}) \cdot \mathbf{k}_\perp$ can be expressed in terms of a single *scalar* function $D_D^\alpha(\mathbf{v})$, called the *"Dupree diffusion coefficient"*:[7]

$$\mathbf{k}_\perp \cdot \mathbf{D}_\kappa^\alpha(\mathbf{v}) \cdot \mathbf{k}_\perp = k_\perp^2\,D_D^\alpha(\mathbf{v}). \qquad (9.24)$$

It follows from Eq. (9.21) that the scalar function $D_D^\alpha(\mathbf{v})$ is *independent of κ* (a consequence of the "Markovian" approximation!) and obeys the following integral equation (in which we rename the integration variable $\kappa' \to \kappa$):

$$D_D^\alpha(\mathbf{v}) = \frac{c^2}{B^2}\frac{1}{2\pi}\int d\mathbf{k}d\omega\,\mathcal{E}_{\perp\kappa}\frac{[J_\mathbf{k}^\alpha(\mathbf{v})]^2}{i\left(\omega - k_\parallel v_\parallel\right) + k_\perp^2\,D_D^\alpha(\mathbf{v})}, \qquad (9.25)$$

or, noting that the imaginary part of the integrand is an odd function of κ,

$$D_D^\alpha(\mathbf{v}) = \frac{c^2}{B^2}\frac{1}{2\pi}\int d\mathbf{k}d\omega\,\mathcal{E}_{\perp\kappa}\,[J_\mathbf{k}^\alpha(\mathbf{v})]^2\,\frac{k_\perp^2\,D_D^\alpha(\mathbf{v})}{\left(\omega - k_\parallel v_\parallel\right)^2 + [k_\perp^2\,D_D^\alpha(\mathbf{v})]^2}. \qquad (9.26)$$

This is *Dupree's Integral Equation*, first derived in DUPREE 1967 by a different method. The present derivation shows that Dupree's theory is included as a special case in the Direct Interaction Approximation.

It may be noted that, if the FLR effects are ignored, the only possible dependence of $D_D^\alpha(\mathbf{v})$ on $v_\perp$ would be purely parametric, because the propagator in (9.26) only depends on $v_\parallel$; one can then consistently assume that $D_D^\alpha(\mathbf{v})$ only depends on $v_\parallel$. This was, indeed, the choice of DUPREE 1967, and is certainly valid for the electron diffusion coefficient. For the ions, however, it must be admitted that $D_D^\alpha(\mathbf{v}) = D_D^\alpha(v_\parallel, v_\perp)$. Next, we note that Eq. (9.26) can be written as:

$$D_D^\alpha(\mathbf{v})\left\{1 - \frac{c^2}{B^2}\frac{1}{2\pi}\int d\mathbf{k}d\omega\,k_\perp^2\,\mathcal{E}_{\perp\kappa}\frac{[J_\mathbf{k}^\alpha(\mathbf{v})]^2}{\left(\omega - k_\parallel v_\parallel\right)^2 + [k_\perp^2\,D_D^\alpha(\mathbf{v})]^2}\right\} = 0.$$

---

[7] We introduce a subscript "D" in order to avoid confusion between the Dupree diffusion coefficient and the "ordinary" diffusion coefficient of transport theory. The relation between these quantities will be derived later on.

Clearly, this nonlinear integral equation always admits the trivial solution:

$$D_D^\alpha(\mathbf{v}) = 0. \tag{9.27}$$

On the other hand, Eq. (9.26) *may admit a non-trivial solution*, satisfying the following equation:

$$1 = \frac{c^2}{B^2} \frac{1}{2\pi} \int dk d\omega \, k_\perp^2 \, \mathcal{E}_{\perp\kappa} \frac{[J_k^\alpha(\mathbf{v})]^2}{\left(\omega - k_\parallel v_\parallel\right)^2 + \left[k_\perp^2 \, D_D^\alpha(v_\parallel)\right]^2} \tag{9.28}$$

Before continuing the formal development, let us dwell for a moment on the physical meaning of Dupree's theory. The renormalized propagator (9.19) has the following form in the gyrotropic case:

$$\mathcal{G}_\kappa^\alpha(\mathbf{v}) = \frac{1}{-i\left(\omega - k_\parallel v_\parallel\right) + k_\perp^2 \, D_D^\alpha(\mathbf{v})}. \tag{9.29}$$

It may be said that this function proceeds from the unperturbed propagator $\mathcal{G}_{0\kappa}(\mathbf{v})$, Eq. (7.32) by changing the frequency w as follows:

$$w \to \omega + ik_\perp^2 D_D^\alpha(\mathbf{v}). \tag{9.30}$$

The frequency is thus displaced by a finite amount into the upper *complex* plane [if $D_D^\alpha(\mathbf{v})$ is positive!]. This is the phenomenon of **turbulent resonance broadening**. In order to understand this terminology, we recall that the unperturbed propagator possesses a singularity (a pole) on the real axis; it must therefore be interpreted as a distribution [see Appendix B, Eq. (B.11)]:

$$\mathcal{G}_{0\kappa}(\mathbf{v}) = \frac{1}{-i\left(\omega - k_\parallel v_\parallel + i\epsilon\right)}$$
$$= \pi\delta\left(\omega - k_\parallel v_\parallel\right) + i\mathcal{P}\frac{1}{\omega - k_\parallel v_\parallel}, \quad \epsilon \to 0. \tag{9.31}$$

The real part of this "function" is an infinite delta-peak, representing a sharp *resonance* between the parallel particle velocity $v_\parallel$ and the phase velocity of a wave of frequency ω and wave vector $k_\parallel$: $v_\parallel = \omega/k_\parallel \equiv v_{ph}$.

On the other hand, the real part of the renormalized propagator is:

$$\mathcal{R}e \, \mathcal{G}_\kappa^\alpha(\mathbf{v}) = \frac{k_\perp^2 \, D_D^\alpha(\mathbf{v})}{\left(\omega - k_\parallel v_\parallel\right)^2 + [k_\perp^2 \, D_D^\alpha(\mathbf{v})]^2}. \tag{9.32}$$

The delta-peak has been transformed into a Lorentzian function which has a finite amplitude and a finite width. This function is represented in

Fig. 9.1:[8] the origin of the name: *"resonance broadening"* is now clearly exhibited.

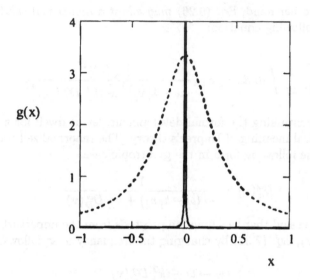

Figure 9.1. The resonance broadening. $x \equiv \omega - k_\parallel v_\parallel$; $d \equiv k_\perp^2 D_D$; $g(x) \equiv \mathcal{R}e\, g_\kappa(\mathbf{v})$.   Solid line: $d = 0.0001$, Dotted line: $d = 0.3$.

The propagator determines the type of motion of the particles. Consider the unperturbed GKE (7.30) in which the (formal) source term, *i.e.*, the vertex, is disregarded:

$$-i\left(\omega - k_\parallel v_\parallel\right) \delta f_\kappa^\alpha(\mathbf{v}) = 0. \tag{9.33}$$

The corresponding equation of evolution in physical space $(\mathbf{Y}, t)$ is simply the Liouville equation for a free particle:

$$\left(\frac{\partial}{\partial t} + v_\parallel \frac{\partial}{\partial Z}\right) \delta f^\alpha(\mathbf{Y}, \mathbf{v}; t) = 0. \tag{9.34}$$

Its general solution, for a given initial condition, is:

$$\delta f^\alpha(X, Y, Z, \mathbf{v}; t) = \delta f^\alpha(X, Y, Z - v_\parallel t, \mathbf{v}; 0), \tag{9.35}$$

which describes the motion of the particles in the Z-direction, with fixed velocity $v_\parallel$, along straight line orbits. This is adequately called a *ballistic motion*.

[8] In this figure the Dupree diffusion coefficient $D_D^\alpha(\mathbf{v}) \approx D_D^\alpha(V_{T\alpha})$ is treated as approximately constant.

If the unperturbed propagator is replaced by the renormalized one (9.29), Eq. (9.33) becomes:

$$\left[-i\omega + ik_\parallel v_\parallel + D_D^\alpha(\mathbf{v})\, k_\perp^2\right]\, \delta f_\kappa^\alpha(\mathbf{v}) = 0, \qquad (9.36)$$

which, by inverse Fourier transformation, leads to:

$$\left(\frac{\partial}{\partial t} + v_\parallel \frac{\partial}{\partial Z} - D_D^\alpha(\mathbf{v})\, \nabla_\perp^2\right)\, \delta f^\alpha(\mathbf{Y}, \mathbf{v}; t) = 0. \qquad (9.37)$$

This is no longer a Hamiltonian (Liouvillian) equation. A *diffusive motion* in the perpendicular direction is now superposed on the ballistic one. The orbits are no longer well defined. Hence, the collective interactions, expressed by the nonlinear coupling of the turbulent fluctuations, have produced a dissipative behaviour which, in turn, has an important influence on the transport. It must be stressed again, however, that Dupree's "diffusion coefficient" $D_D^\alpha(\mathbf{v})$ (a velocity-dependent coefficient in the *microscopic* GKE) should not be confused with the *macroscopic* diffusion coefficient (the coefficient of the density gradient in the transport equation for the particle flux). The connection between these two quantities will be clarified in Sec. 9.3

The renormalization of the vertex operator (8.111), (8.119) is done along the same lines as above. We do not exhibit the detail of this calculation, but quote directly the result (VANDEN EIJNDEN 1992):

$$V_\kappa^\alpha(\mathbf{v}) = \frac{e_\alpha}{T_\alpha}\, F_M^\alpha(\mathbf{v})\, J_\kappa^\alpha(\mathbf{v})\, k_\perp^2\, D_D^\alpha(\mathbf{v})\, V_\kappa^\alpha(\mathbf{v}). \qquad (9.38)$$

This equation shows that whenever $D_D^\alpha(\mathbf{v}) = 0$, we automatically also have $V_\kappa^\alpha(\mathbf{v}) = 0$. This implies that the energy conservation constraint (8.140) is *automatically* satisfied.

Some rather lengthy, but elementary calculations lead to the following form for the function $V_\kappa^\alpha(\mathbf{v}) \approx V_\kappa^\alpha(V_{T\alpha})$

$$V_\kappa^\alpha(V_{T\alpha}) = -\,(1 + h_\kappa^\alpha) \qquad (9.39)$$

The rather complicated function h_κ^α can be approximated in terms of an average wave vector $\overline{\mathbf{k}}$ and an average frequency $\overline{\omega}$:

$$h_\kappa^\alpha \approx \frac{|\overline{k}_\parallel|\, V_{T\alpha}}{\overline{\omega} - |\overline{k}_\parallel|\, V_{T\alpha}}. \qquad (9.40)$$

Recalling Eq. (7.96), we find the following approximate values:

$$\begin{aligned} h_\kappa^\alpha \ll 1, &\implies V_\kappa^i(\mathbf{v}) \approx -1 \quad \textit{ions} \\ h_\kappa^\alpha \approx -1, &\implies V_\kappa^e(\mathbf{v}) \approx 0 \quad \textit{electrons} \end{aligned} \qquad (9.41)$$

Most remarkably, it appears that the electron vertex renormalization is negligibly small, $V_\kappa^e(\mathbf{v}) \approx 0$; but then it follows from the energy conservation constraint (8.140) that the propagator renormalization must also vanish: $\Sigma_\kappa^e(\mathbf{v}) \approx 0$, hence $D_D^e(\mathbf{v}) \approx 0$. This result will be confirmed by another argument in Sec. 9.3.

To summarize, we write the complete renormalized inverse propagator and the renormalized vertex as follows:

$$g_\kappa^\alpha(\mathbf{v}) = \left[-i\left(\omega - k_\parallel v_\parallel\right) + k_\perp^2 D_D(\mathbf{v})\,\varepsilon_\alpha\right]^{-1}$$
$$\Delta_\kappa^\alpha(\mathbf{v}) = \frac{e_\alpha}{T_\alpha}\, F_M^\alpha(\mathbf{v})\, \left\{i\left[\omega - \omega_{*\mathbf{k}}^{T\alpha}(\mathbf{v})\right] - k_\perp^2\, D_D(\mathbf{v})\,\varepsilon_\alpha\right\}. \qquad (9.42)$$

with:

$$\varepsilon_i = 1, \quad \varepsilon_e = 0; \quad D_D(\mathbf{v}) = D_D^i(\mathbf{v}). \qquad (9.43)$$

The function $\omega_{*\mathbf{k}}^{T\alpha}(\mathbf{v})$ was defined in Eq. (8.12).

9.3   The Threshold of Diffusivity

The phenomenon of resonance broadening appears only when the Dupree integral equation (9.26) possesses a nontrivial solution. In order to determine when this condition is satisfied, we must study Eq. (9.28), which is a complicated integral equation for $D_D^\alpha(\mathbf{v})$. A rough estimate of the solution can be obtained by the following approximation (introduced by DUPREE 1967). We may consider the integral in Eq. (9.28) as the average of a function of $\mathbf{k}$ and ω, weighted by the spectrum of electric field fluctuations $\mathcal{E}_{\perp\kappa}$, Eq. (9.23). This expression in turn is approximated by the function evaluated at some average values $\overline{\mathbf{k}}, \overline{\omega}$, times the integral of $\mathcal{E}_{\perp\kappa}$, which represents the square of the total amplitude of the fluctuating electric field, $\overline{E}_\perp$. Thus, for an arbitrary function $F(\mathbf{k}, \omega)$:

$$\frac{1}{2\pi} \int d\mathbf{k}d\omega \tfrac{1}{2}k_\perp^2 S(\mathbf{k}, \omega)\, F(\mathbf{k}, \omega)$$
$$\approx F(\overline{\mathbf{k}}, \overline{\omega})\, \frac{1}{2\pi} \int d\mathbf{k}d\omega \tfrac{1}{2}k_\perp^2 S(\mathbf{k}, \omega) \approx F(\overline{\mathbf{k}}, \overline{\omega})\, \overline{E}_\perp^2. \qquad (9.44)$$

When applied to Eq. (9.28), in which the running velocity is further approximated by the respective thermal velocity, this approximation yields:

$$1 = \frac{c^2}{B^2} \overline{E}_\perp^2 \, \overline{k}_\perp^2 \, \frac{\left[J_0(\overline{k}_\perp \rho_{L\alpha})\right]^2}{\left(\overline{\omega} - \overline{k}_\parallel V_{T\alpha}\right)^2 + \left(\overline{k}_\perp^2 D_D^\alpha\right)^2}, \qquad (9.45)$$

from which we obtain:

$$D_D^\alpha = \frac{1}{\overline{k}_\perp} \frac{c\overline{E}_\perp}{B} J_0(\overline{k}_\perp \rho_{L\alpha}) \left\{ 1 - \frac{\frac{1}{\overline{k}_\perp^2}(\overline{\omega} - \overline{k}_\parallel V_{T\alpha})^2}{\left[\frac{c\overline{E}_\perp}{B} J_0(\overline{k}_\perp \rho_{L\alpha})\right]^2} \right\}^{1/2} . \qquad (9.46)$$

Here $D_D^\alpha \equiv D_D^\alpha(V_{T\alpha})$, and $\rho_{L\alpha} = V_{T\alpha}/|\Omega_\alpha|$ is the thermal Larmor radius.

It now clearly appears that there exists a *threshold for the amplitude of electric field fluctuations*, $E_{\perp c}^\alpha$, below which there is no real solution for the Dupree diffusion coefficient, hence there is no resonance broadening and no associated anomalous transport. The value of the threshold is determined by annulling the argument of the square root in Eq. (9.46):

$$J_0(\overline{k}_\perp \rho_{L\alpha}) \frac{c\overline{E}_{\perp c}^\alpha}{B} = \frac{1}{\overline{k}_\perp} |\overline{\omega} - \overline{k}_\parallel V_{T\alpha}| \qquad (9.47)$$

The left hand side is the average amplitude of the electric drift velocity, modified by the finite Larmor radius effect for the ions.

The emergence of a non-zero diffusion coefficient is a typical *bifurcation phenomenon*. Whenever the "control parameter" $(c\overline{E}_\perp/B)$ reaches the threshold value, the former solution $D_D = 0$ becomes unstable and the system follows a new path, defined by $D_D \neq 0$. It appears that the threshold velocity is different for the ions and for the electrons. The frequency $\overline{\omega}$ is of the order of the drift wave frequency $\omega_{*\overline{k}}^e$; hence, by Eq. (7.96), we find the following thresholds:

$$|J_0(\overline{k}_\perp \rho_{Li})| \frac{c\overline{E}_{\perp c}^i}{B} \approx \frac{\overline{\omega}}{\overline{k}_\perp}, \quad \textit{ions}$$

$$\frac{c\overline{E}_{\perp c}^e}{B} \approx \frac{|\overline{k}_\parallel| V_{Te}}{\overline{k}_\perp}, \quad \textit{electrons} \qquad (9.48)$$

As a result of (7.96) the threshold field for the electrons is higher than for the ions (unless the Bessel function is very small, *i.e.*, the ion Larmor radius is very large): the ions are more sensitive to the resonance broadening mechanism. It follows that for most "reasonable" values of the average drift velocity amplitude $(c\overline{E}_\perp/B)$ above the ion threshold value (but not too large), the electron Dupree diffusion coefficient is zero, in agreement with the result of Sec. 9.2.

When the amplitude of the electric field fluctuations is greater than the threshold value, the Dupree diffusion coefficient is a growing function of $\overline{E}_\perp$, according to Eq. (9.46), as shown in Fig. 9.2.

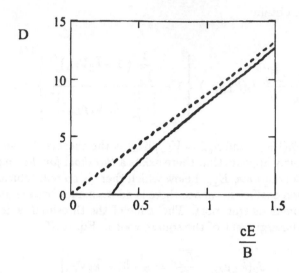

Figure 9.2. Dupree diffusion coefficient. (Numerical data from Table 2.1.) Solid line: Dupree diffusion;   dashed line: Asymptotic Bohm diffusion.

Asymptotically, for large $(c\overline{E}_\perp/B)$, the growth is linear:

$$D^\alpha_{D\,as} \sim D^\alpha_B = J_0(\overline{k}_\perp \rho_{L\alpha}) \frac{c\overline{E}_\perp}{\overline{k}_\perp B}. \tag{9.49}$$

This scaling of the diffusion coefficient, proportional to $(c\overline{E}_\perp/B)$, is typical of the so-called *Bohm diffusion coefficient*. It is to be contrasted with the quasilinear diffusion coefficient, as given by Eq. (7.126) which scales like $(c\overline{E}_\perp/B)^2$. The Bohm scaling is closer to the experimental values observed in fusion devices. It is unfortunately less favourable for fusion, because the diffusion coefficient decreases less strongly with increasing magnetic field, thus allowing more important leaks. It will be shown later, however, that even the Bohm diffusion coefficient is not correct for very large values of $(c\overline{E}_\perp/B)$.

9.4   The Renormalized Dispersion Equation

We now make the theory self-consistent, by using the Poisson equation. In a coherent approximation, such as the RQL, it was shown in Sec. 8.6, Eq. (8.83) that the Poisson equation reduces to a renormalized dispersion equation, $\widetilde{\varepsilon}_k(w_k) = 0$. In order to solve this equation, we may closely follow the treatment of Sec. 7.7. The dispersion equation is written in the form (7.94), in which the ionic function $\widehat{M}^{i(n)}$ is renormalized, but the electronic function $M^{e(n)}$ is not, because, as shown above, $D_D^e = 0$. Thus, the dispersion equation is:

$$1 + M_{\mathbf{k}}^{e(n)}(w_{\mathbf{k}}) + \theta_i^{-1}\left[\Gamma_0 + \widehat{M}_{\mathbf{k}}^{i(n)}(w_{\mathbf{k}})\right] = 0. \tag{9.50}$$

Recalling Eqs. (7.75) and (8.73), we note that the renormalized ionic function $\widehat{M}_{\mathbf{k}}^{i(n)}(w_{\mathbf{k}})$ is obtained by replacing in the quasilinear expression the propagator $\mathcal{G}_{0\kappa}^\alpha(\mathbf{v})$ and the vertex $\Delta_{0\mathbf{k}}^\alpha(\mathbf{v})$ by the corresponding renormalized quantities. As follows from Eq. (9.42) this amounts simply to replacing the complex frequency $w_{\mathbf{k}}$ by $w_{\mathbf{k}} + ik_\perp^2 D_D(\mathbf{v}) \approx w_{\mathbf{k}} + ik_\perp^2 D_D$, where $D_D = D_D^i(V_{Ti})$. We introduce the following dimensionless coefficient, which we assume to be small, of order λ:

$$\lambda d = \frac{k_\perp^2 D_D}{\omega_{*\mathbf{k}}^e}. \tag{9.51}$$

Eq. (7.100) is then renormalized as follows:

$$\widehat{M}_{\mathbf{k}}^{i(n)}(\overline{\omega}) = -\Gamma_0 - \theta_i \frac{\Gamma_0(1 - M\eta_i)}{(\overline{\omega} + i\lambda d)} + O(\lambda^2). \tag{9.52}$$

Expanding the root of the dispersion equation as in (7.101), we find, to dominant order:

$$\overline{\omega}^{NL} = a_0 + \lambda i a_1^{NL} + O(\lambda^2). \tag{9.53}$$

The dispersion equation is now easily solved. The zeroth order, real frequency a_0 appears to be unchanged by the renormalization, and is still given by Eq. (7.102); only the first order part, which is purely imaginary, is changed:

$$a_1^{NL} = a_1 - d, \tag{9.54}$$

where a_1 is the quasilinear value of Eq. (7.102).

From this simple equation follow the important consequences which make the RQL approximation reasonable, in contrast to the simple QL approximation. We consider a situation which is *linearly unstable*, i.e., $a_1 > 0$.[9] The amplitude of the fluctuations is then growing in time. After a certain time, the amplitude of the fluctuating electric field exceeds the threshold defined in the previous section: a positive Dupree diffusion coefficient then develops. As appears from Eq. (9.54) this coefficient will produce a damping that counteracts the instability. Thus, after a certain time, the plasma reaches a *marginally stable* state in which the nonlinear

---

[9] We recall from the discussion of Sec. 7.7 that in the (academic) case $\eta_e = \eta_i = 0$, the drift waves are always unstable. When the temperature gradients are non-zero, this property is no longer always satisfied. At this point we assume that the parameters are such that $a_1 > 0$. The linearly stable case is discussed below.

growth rate a_1^{NL} vanishes. Combining Eqs. (7.102), (9.54) and (9.51), we find the following relation, valid in the marginally stable state:

$$k_\perp^2 D_D = \sqrt{\pi} f \Gamma_0 \left[1 - \Gamma_0(1 - M\eta_i) - \tfrac{1}{2}\eta_e \right] (1 - M\eta_i) \omega_{*\mathbf{k}}^e. \qquad (9.55)$$

If we consider a plasma that is *linearly stable*, any initial fluctuation will decay in time; hence the amplitude of the fluctuations will not reach the threshold value for the emergence of a Dupree diffusion coefficient. The fluctuations will simply decay according to the linear law, the nonlinear resonance broadening does not occur.

The result obtained here can also be considered from another point of view. As mentioned in Sec. 7.5, the quasilinear dispersion equation is a *homogeneous linear* equation for the potential fluctuation, or for the fluctuation spectrum. As a result, the linear dispersion equation can only determine an eigenvalue, *i.e.*, a relation between frequency and wave vector, but leaves the amplitude of the electric field fluctuations undetermined. In the present case, the nonlinear dielectric function depends on the electric field fluctuation spectrum $\mathcal{E}_{\perp\kappa}$ through the Dupree diffusion coefficient. As a result, the dispersion equation can determine (at least partially) the amplitude of the fluctuations $\overline{E}_\perp$ [related to the Dupree diffusion coefficient by Eq. (9.46)] by calling upon the nonlinear saturation mechanism. This information is quite important for the transport theory, as will be shown in Sec. 9.5

9.5   Anomalous Fluxes in the RQL Approximation

We now derive expressions of the anomalous fluxes in the present RQL approximation. We start from a plasma that is linearly unstable; it will evolve by the mechanism described above toward a saturated marginally stable state, in which the growth rate is zero. We determine the anomalous fluxes in this state. These quantities are defined by Eqs. (7.50) and (7.53), in which the functions $M_\kappa^{\alpha(B)}$ are replaced by their renormalized expressions, as was done also in Sec. 9.4.

We discuss first the particle fluxes, defined as in Eq. (7.122), with $\gamma_{\mathbf{k}} = 0$:

$$\Gamma_x^\alpha = -\frac{e_\alpha}{T_\alpha} \frac{c}{B} \frac{1}{2\pi} \int d\mathbf{k} \int d\omega \; ik_y \, S_{\mathbf{k}}(\omega) \, \widehat{M}_{\mathbf{k}}^{\alpha(n)}(\omega)$$

$$= -\frac{e_\alpha}{T_\alpha} \frac{c}{B} \int d\mathbf{k} \; ik_y \, S_{\mathbf{k}} \, \widehat{M}_{\mathbf{k}}^{\alpha(n)}(\omega_{\mathbf{k}}). \qquad (9.56)$$

The second form follows from the fact that in the coherent approximation, $S_{\mathbf{k}}(\omega) \approx 2\pi \, \delta(\omega - \omega_{\mathbf{k}}) S_{\mathbf{k}}$. Before deriving the explicit expression of the fluxes, we note that the argument of Eq. (7.127) is immediately extended to the renormalized case, as a result of the dispersion equation (9.50). Thus, the anomalous particle fluxes are again automatically ambipolar in the RQL approximation:

$$\Gamma_x^i = \Gamma_x^e \qquad (9.57)$$

Consider now the ion particle flux. A simple parity argument shows that only the imaginary part of $\widehat{M}_{\mathbf{k}}^{i(n)}(\omega)$ contributes to the flux:

$$Im \, \widehat{M}_{\mathbf{k}}^{i(n)}(\omega) = \int d\mathbf{v} \, F_M^\alpha(\mathbf{v}) \, J_{\mathbf{k}}^i(\mathbf{v}) \, k_\perp^2 \, D_D(\mathbf{v})$$
$$\times \, \frac{k_\parallel v_\parallel - \omega_{*\mathbf{k}}^{Ti}(\mathbf{v}) + (\omega - k_\parallel v_\parallel) \, \mathcal{V}_\kappa^i(\mathbf{v})}{\left[\omega - k_\parallel v_\parallel\right]^2 + \left[k_\perp^2 D_D(\mathbf{v})\right]^2}. \qquad (9.58)$$

When this form is substituted into Eq. (9.56), the only term which survives the k_y- integration is the one involving $\omega_{*\mathbf{k}}^{iT}(\mathbf{v})$; this quantity, defined in Eqs. (7.17), (7.18), is proportional to k_y, hence yields a non-zero contribution to Eq. (9.56). Thus, the vertex renormalization $\mathcal{V}_\kappa^\alpha$ does not contribute to the fluxes. Noting that, under the integral, k_y^2 can be replaced by $\frac{1}{2}k_\perp^2$, we obtain:

$$\frac{ec}{T_i B} \frac{1}{2\pi} \int d\kappa \, k_y \, S_\kappa \, Im \, \widehat{M}_\kappa^{i(n)}$$
$$= \int d\mathbf{v} \, F_M^i(\mathbf{v}) \left[1 + \left(\frac{v^2}{V_{Ti}^2} - \frac{3}{2}\right) \eta_i\right] D_D(\mathbf{v}) \, L_n^{-1}$$
$$\times \left\{\frac{c^2}{2B^2} \frac{1}{2\pi} \int d\kappa \, k_\perp^4 \, S_\kappa \, \frac{\left[J_{\mathbf{k}}^i(\mathbf{v})\right]^2}{\left[\omega - k_\parallel v_\parallel\right]^2 + \left[k_\perp^2 D_D(\mathbf{v})\right]^2}\right\}. \qquad (9.59)$$

We now note that the last factor is just the right hand side of Dupree's integral equation (9.26). *Whenever this equation possesses a non-trivial solution,* $D_D(\mathbf{v}) \neq 0$, the bracketed factor equals 1, and Eq. (9.56), combined with (9.59), reduces to the following very simple form for the anomalous ion particle flux:

$$\Gamma_x^i = \int d\mathbf{v} \, F_M^i(\mathbf{v}) \, D_D(\mathbf{v}) \left[1 + \left(\frac{v^2}{V_{Ti}^2} - \frac{3}{2}\right) \eta_i\right] \frac{1}{L_n}. \qquad (9.60)$$

Approximating $D_D(\mathbf{v}) \approx D_D(V_{Ti}) \equiv D_D$ as in Eq. (9.46), we find:

$$\Gamma_x^i = D_D X_1 - \frac{1}{2} \frac{n}{T_i} D_D X_3 = \Gamma_x^e. \qquad (9.61)$$

This has clearly the form of a linear transport equation relating the anomalous ion flux to the thermodynamic forces, defined as:

$$X_1 = \frac{n}{L_n} = -\frac{dn}{dx}$$

$$X_2 = \eta_e \frac{T_e}{n} X_1 = -\frac{dT_e}{dx},$$

$$X_3 = \eta_i \frac{T_i}{n} X_1 = -\frac{dT_i}{dx}. \tag{9.62}$$

We may note, in particular, the approximate identity between the Dupree diffusion coefficient (evaluated at $\mathbf{v} = V_{Ti}$) and the macroscopic diffusion coefficient D:

$$D \approx D_D(V_{Ti}). \tag{9.63}$$

The energy fluxes are treated in a similar way. There is, however, no relation analogous to the ambipolarity condition (9.57) relating the two energy fluxes: they are therefore evaluated separately. One finds for the electron flux an expression similar to Eq. (9.59) with $\widehat{M}_{\mathbf{k}}^{i(n)} \to M_{\mathbf{k}}^{e(P)}$. But we found that the electron propagator is not renormalized in the present case, hence $D_D^e \approx 0$. Thus, *there is no anomalous electron energy flux associated with the "universal" drift instability in the RQL approximation.*

The anomalous ion energy flux is evaluated in terms of the renormalized quantity $\widehat{M}_{\mathbf{k}}^{i(P)}$, which differs from $\widehat{M}_{\mathbf{k}}^{i(n)}$ only by an additional factor v^2/V_{Ti}^2 [see (7.54), (7.55)]. Hence the whole argument leading to Eq. (9.60) is immediately extended, with the result:

$$\frac{Q_x^i}{T_i} = \int d\mathbf{v}\, F_M^i(\mathbf{v})\, D_D^i(\mathbf{v}) \frac{v^2}{V_{Ti}^2} \left[1 + \left(\frac{v^2}{V_{Ti}^2} - \frac{3}{2}\right)\eta_i\right]\frac{1}{L_n}, \tag{9.64}$$

thus, with the same approximations as for the particle fluxes:

$$\frac{Q_x^i}{T_i} = D_D X_1 + \frac{1}{2}\frac{n}{T_i} D_D X_3. \tag{9.65}$$

We will see, however, that the simple linear forms (9.61), (9.65) are misleading. A closer analysis reveals an unexpectedly different situation.

9.6   The Marginal Stability Constraint

As it stands, the transport equation (9.61) is still incomplete. Indeed, the transport coefficients are expressed in terms of the Dupree diffusion coefficient which, in turn, depends on the amplitude of the electric field fluctuations $\overline{E}_\perp$ through Eq. (9.46). In order to complete the theory, this quantity must be determined from the Poisson equation, *i.e.*, the dispersion equation.

On the other hand, it was stated from the beginning of the previous section that the fluxes were determined in a situation where the plasma is marginally stable. To the best of our knowledge, the use of the marginal stability constraint in transport theory only appears in the Master thesis of VANDEN EIJNDEN 1992. As shown in Sec. 9.4, this puts a non-trivial constraint on the Dupree diffusion coefficient, which must take a value that compensates exactly the linear growth rate, Eq. (9.55). As a result, *the anomalous transport coefficients depend on the thermodynamic forces.* Indeed, from the definitions of the drift wave frequency ω^e_{*k}, Eqs. (7.18), (2.27) and of the dimensionless quantities f and $\eta_\alpha f$, Eqs. (7.97), (2.19), combined with the definitions (9.62), we find:[10]

$$\omega^e_{*k} = \frac{1}{2} k_y V_{Te} \rho_{Le} \frac{1}{n} X_1, \tag{9.66}$$

and:[11]

$$\lambda f = \frac{k_y \rho_{Le}}{2|k_\parallel|} \frac{1}{n} X_1,$$

$$\lambda \eta_e f = \frac{k_y \rho_{Le}}{2|k_\parallel|} \frac{1}{T_e} X_2,$$

$$\lambda \eta_i f = \frac{k_y \rho_{Le}}{2|k_\parallel|} \frac{1}{T_i} X_3. \tag{9.67}$$

Substituting these values into Eq. (9.55), (using also $k_y^2 \approx \frac{1}{2} k_\perp^2$) we find:

$$D = \frac{\sqrt{\pi}}{8} \frac{\rho_{Le}^2 V_{Te}}{|k_\parallel|} \Gamma_0 \left(\frac{1}{n} X_1 - M \frac{1}{T_i} X_3 \right)$$

$$\times \left[(1 - \Gamma_0) \frac{1}{n} X_1 - \frac{1}{2T_e} X_2 + \Gamma_0 M \frac{1}{T_i} X_3 \right]. \tag{9.68}$$

---

[10] It should be remembered that in the present approximation all quantities are evaluated for an average value of the wave vector, $\overline{k}$, as explained in Sec. 9.3. For simplicity we suppress the overbar in the forthcoming formulae.

[11] One might be surprised to see that all quantities in Eq. (9.66) are proportional to the electron Larmor radius ρ_{Le}, a quantity that has been systematically neglected in most previous expressions. But the smallness of ρ_{Le} is compensated here by the smallness of the ratio $|k_\parallel|/k_y$ in the gyrokinetic ordering. From the numerical data of Table 2.1, one finds a value for $k_y \rho_{Le}/|k_\parallel|$ of a few centimetres. Its ratio to the density gradient scale length is thus of a few percent: small (as it should be, of order λ) but not negligible.

Thus, after substitution into Eqs. (9.61) and (9.65), both *the particle fluxes and the ion heat flux appear to be homogeneous cubic forms in the thermodynamic forces.*

One should however be cautious: the result (9.68) is not general. The treatment described in this chapter should be viewed as a typical example of the methodology of anomalous transport theory based on the gyrokinetic equation. The relation (9.68) clearly results from the saturation of the linear instability by the resonance broadening mechanism. The nonlinearity in the thermodynamic forces X_n results from the fact that the growth rate of the "universal" instability γ_k depends on these forces (because the drift wave instability only exists in spatially inhomogeneous systems). There are, however, many other possible instabilities in a plasma, which have quite different growth rates. In all cases, the system reaches a marginally stable state in which $k_\perp^2 D = \gamma_k$; but, clearly, the expression of the Dupree diffusion coefficient will depend on the characteristics of the particular instability that dominates the evolution. Moreover, the resonance broadening is not the only possible mechanism that can saturate the linear instability. Hence, a direct comparison of Eq. (9.68) with experimental data would not be very significant. It may, however, be concluded from the results of the present chapter that the anomalous transport equations do not, in general, have the simple linear form (9.61), (9.65) with constant transport coefficients, as in the classical or neoclassical transport theory. The nonlinearity of the turbulent process may induce a complicated, generally nonlinear relation between fluxes and thermodynamic forces.

Chapter 10

Mode Coupling and Trajectory Correlation

10.1 Nonlinear Mode Coupling. The Spectrum Balance Equation

The RQL approximation introduced some important features of plasma turbulence, in particular the resonance broadening and the replacement of the ballistic unperturbed propagator by one which describes diffusive motion. The latter can counteract the growth rate of the fluctuations due to linear instabilities, When completed with a marginal stability condition, it leads to a closed set of transport equations. The RQL is, however, insufficient for a complete theory of turbulence. We must therefore investigate more fully the content of the DIA. Unfortunately, we are immediately faced with great difficulties. We must take account of the random noise source which, as will be seen, plays a major role in determining the properties of a turbulent plasma. On the other hand, we must also introduce in the propagator and the vertex renormalizations the polarization contributions, which are quite complicated. In particular, the simple diagonal property (9.15) is lost in the general case, because the polarization part of Σ, (8.117) contains a kernel that introduces a dependence on the species index through $T_\kappa^\alpha(\mathbf{v})$ and $\Delta_\kappa^\alpha(\mathbf{v})$.

One of the difficulties that appears upon trying to handle these expressions (within the DIA) is the fact that all three types of correlation functions, *i.e.*, $C(1\,2)$, $T(a\,1)$ and $S(a\,b)$, appear in the expressions of the renormalized operators (8.115) - (8.120), (8.113). The complete determination of these functions thus requires the simultaneous solution of Eq. (8.108), combined with two other equations for $T(a\,1)$ and for the spectrum $S(a\,b)$.

This difficulty can be short-circuited if one accepts the following approximation scheme. *The renormalized operators are evaluated by using*

*within their expressions (8.115) - (8.120), (8.113) the coherent approxima-
tions (9.4) - (9.6) for the correlations $C(1\,2)$ and $T(a\ 1)$, as well as the
coherent approximation (9.15) for the propagator.* In this way the renormalized operators are entirely expressed in terms of the spectrum $S(a\,b)$
of potential fluctuations.[1] As a bonus, remarkably symmetric forms are
obtained for these operators.

Consider first Eq. (8.117) for the polarization part of the propagator
renormalization, which is written as follows in the Fourier representation:

$$P\Sigma_\kappa^{\alpha_1,\alpha_2}(\mathbf{v}_1,\mathbf{v}_2)$$

$$= -\sum_\beta \int d\mathbf{v}' \int d\kappa' \frac{4\pi e_\beta}{|\mathbf{k}-\mathbf{k}'|} m_{\mathbf{k}\mathbf{k}'} \frac{1}{\varepsilon_{\kappa-\kappa'}} J_{\mathbf{k}'}^{\alpha_2}(\mathbf{v}_2) J_{\mathbf{k}-\mathbf{k}'}^{\beta}(\mathbf{v}') g_{\kappa-\kappa'}^{\beta\alpha_2}(\mathbf{v}',\mathbf{v}_2)$$

$$\times \left[J_{\mathbf{k}-\mathbf{k}'}^{\alpha_1}(\mathbf{v}_1) T_{\kappa'}^{\alpha_1}(\mathbf{v}_1) + \sum_\gamma \int d\mathbf{v}'' J_{\mathbf{k}'}^{\alpha_1}(\mathbf{v}_1) g_{\kappa-\kappa'}^{\alpha_1\gamma}(\mathbf{v}_1,\mathbf{v}'') \Delta_{\kappa-\kappa'}^{\gamma}(\mathbf{v}'') S_{\kappa'} \right]. \tag{10.1}$$

We introduced here the following scalar coupling coefficient related to
the tensor (9.14):

$$m_{\mathbf{k}\mathbf{k}'} = \mathbf{k}_\perp \cdot \mathbf{m}_{\mathbf{k}'} \cdot \mathbf{k}_\perp = \frac{1}{2\pi} \left(\frac{c}{B}\right)^2 \left[\mathbf{k}_\perp \cdot \left(\mathbf{k}' \wedge \hat{\mathbf{b}}\right) \right]^2. \tag{10.2}$$

Upon substitution of the coherent approximation (9.5) for the mixed
correlation $T(a\ 1)$ we find:

$$P\Sigma_\kappa^{\alpha_1\alpha_2}(\mathbf{v}_1,\mathbf{v}_2) = \int d\kappa' \frac{4\pi e_{\alpha_2}}{|\mathbf{k}-\mathbf{k}'|} m_{\mathbf{k}\mathbf{k}'} \frac{1}{\varepsilon_{\kappa-\kappa'}} J_{\kappa'}^{\alpha_2}(\mathbf{v}_2) J_{\kappa-\kappa'}^{\alpha_2}(\mathbf{v}_2) \mathcal{G}_{\kappa-\kappa'}^{\alpha_2}(\mathbf{v}_2)$$

$$\times \left\{ J_{\kappa-\kappa'}^{\alpha_1}(\mathbf{v}_1) \mathcal{G}_{\kappa'}^{\alpha_1}(\mathbf{v}_1) \Delta_{\kappa'}^{\alpha_1}(\mathbf{v}_1) - J_{\kappa'}^{\alpha_1}(\mathbf{v}_1) \mathcal{G}_{\kappa-\kappa'}^{\alpha_1}(\mathbf{v}_1) \Delta_{\kappa-\kappa'}^{\alpha_1}(\mathbf{v}_1) \right\} S_{\kappa'}. \tag{10.3}$$

Combining this result with Eq. (9.17), we write the *total* renormalization of the propagator (8.115) in the following form (in the present approximation):

$$\Sigma_\kappa^{\alpha_1\alpha_2}(\mathbf{v}_1,\mathbf{v}_2) = \int d\kappa' \, m_{\mathbf{k}\mathbf{k}'} J_{\mathbf{k}'}^{\alpha_2}(\mathbf{v}_2) \mathcal{G}_{\kappa-\kappa'}^{\alpha_2}(\mathbf{v}_2) S_{\kappa'} \Xi_{\kappa\kappa'}^{\alpha_1\alpha_2}(\mathbf{v}_1,\mathbf{v}_2) \tag{10.4}$$

where we defined:

---

[1] One could have thought of an opposite strategy. By using Eq. (8.24) one could have
expressed $S(a\,b)$ and $T(a\ 1)$ in terms of $C(1\,2)$, without any approximation. But the
resulting equations would be much more complicated!

$$\Xi_{\kappa\,\kappa'}^{\alpha_1\alpha_2}(\mathbf{v}_1,\mathbf{v}_2) = \delta^{\alpha_1\alpha_2}\delta(\mathbf{v}_1-\mathbf{v}_2)\,J_{\mathbf{k}'}^{\alpha_1}(\mathbf{v}_1)$$

$$+\frac{1}{\varepsilon_{\kappa-\kappa'}}\frac{4\pi e_{\alpha_2}}{|\mathbf{k}-\mathbf{k}'|}\,J_{\mathbf{k}-\mathbf{k}'}^{\alpha_2}(\mathbf{v}_2)\left\{J_{\mathbf{k}-\mathbf{k}'}^{\alpha_1}(\mathbf{v}_1)\,\mathcal{G}_{\kappa'}^{\alpha_1}(\mathbf{v}_1)\,\Delta_{\kappa'}^{\alpha_1}(\mathbf{v}_1)\right.$$

$$\left.-\,J_{\mathbf{k}'}^{\alpha_1}(\mathbf{v}_1)\,\mathcal{G}_{\kappa-\kappa'}^{\alpha_1}(\mathbf{v}_1)\,\Delta_{\kappa-\kappa'}^{\alpha_1}(\mathbf{v}_1)\right\}. \tag{10.5}$$

The first term corresponds to the diffusive contribution and the remaining ones to the polarization part.

A similar calculation performed on the vertex operator leads to a remarkably analogous form, expressed in terms of the *same* operator Ξ:

$$V_{\kappa}^{\alpha_1}(\mathbf{v}_1) = \sum_{\alpha_2}\int d\mathbf{v}_2\int d\kappa'\,m_{\mathbf{k}\mathbf{k}'}\,J_{\mathbf{k}}^{\alpha_2}(\mathbf{v}_2)\,\mathcal{G}_{\kappa-\kappa'}^{\alpha_2}(\mathbf{v}_2)\,S_{\kappa'}$$

$$\times\;\mathcal{G}_{-\kappa'}^{\alpha_2}(\mathbf{v}_2)\,\Delta_{-\kappa'}^{\alpha_2}(\mathbf{v}_2)\,\Xi_{\kappa\,\kappa'}^{\alpha_1\alpha_2}(\mathbf{v}_1,\mathbf{v}_2). \tag{10.6}$$

The analogy in structure between the two renormalized operators in the present approximation is striking.

Finally, the random noise source, defined by Eqs. (8.113), reduces to the following form:

$$F_{\kappa}^{\alpha_1\alpha_2}(\mathbf{v}_1,\mathbf{v}_2) = -\int d\kappa'\,m_{\mathbf{k}\mathbf{k}'}\,J_{\mathbf{k}'}^{\alpha_1}(\mathbf{v}_1)\,\mathcal{G}_{\kappa-\kappa'}^{\alpha_1}(\mathbf{v}_1)\,\Delta_{\kappa-\kappa'}^{\alpha_1}(\mathbf{v}_1)\,S_{\kappa-\kappa'}S_{\kappa'}$$

$$\times\left[J_{\mathbf{k}-\mathbf{k}'}^{\alpha_2}(\mathbf{v}_2)\,\mathcal{G}_{-\kappa'}^{\alpha_2}(\mathbf{v}_2)\,\Delta_{-\kappa'}^{\alpha_2}(\mathbf{v}_2)-J_{-\mathbf{k}'}^{\alpha_2}(\mathbf{v}_2)\,\mathcal{G}_{\kappa-\kappa'}^{\alpha_2}(\mathbf{v}_2)\,\Delta_{\kappa-\kappa'}^{\alpha_2}(\mathbf{v}_2)\right]. \tag{10.7}$$

The first bracketed term is the part associated with the propagator, the second one is the part associated with the vertex in Eqs. (8.141), (8.142).

In conclusion, we reached our goal of expressing all renormalized operators in terms of the spectrum alone. These equations must now be completed by an equation for the spectrum.

A convenient starting point is the set of equations (8.66), (8.67) relating quite generally the spectrum to the random noise source. Using Eqs. (8.10), (8.76), (9.15) as well as the important identity (8.77), Eqs. (8.66), (8.67) reduce to:

$$S_{\kappa} = \sum_{\alpha_1,\alpha_2}\left(\frac{4\pi}{k^2}\right)^2\frac{e_{\alpha_1}e_{\alpha_2}}{\varepsilon_{\kappa}\,\varepsilon_{-\kappa}}\int d\mathbf{v}_1 d\mathbf{v}_2$$

$$\times\,J_{\mathbf{k}}^{\alpha_1}(\mathbf{v}_1)\,\mathcal{G}_{\kappa}^{\alpha_1}(\mathbf{v}_1)\,J_{-\mathbf{k}}^{\alpha_2}(\mathbf{v}_2)\,\mathcal{G}_{-\kappa}^{\alpha_2}(\mathbf{v}_2)\,F_{\kappa}^{\alpha_1\alpha_2}(\mathbf{v}_1,\mathbf{v}_2). \tag{10.8}$$

We now note that the random noise source can be written in the following elegant form:

$$F_\kappa^{\alpha_1 \alpha_2}(\mathbf{v}_1, \mathbf{v}_2) = 2\frac{1}{2\pi} \int d\kappa' d\kappa'' \, \delta(\kappa - \kappa' - \kappa'')$$

$$\times \nu^{\alpha_1}(\kappa|\kappa', \kappa''; \mathbf{v}_1) \, [\nu^{\alpha_2}(\kappa|\kappa', \kappa''; \mathbf{v}_2)]^* \, S_{\kappa'} S_{\kappa''}, \qquad (10.9)$$

where the star denotes complex conjugation, and:

$$\nu^\alpha(\kappa|\kappa', \kappa''; \mathbf{v}) = \frac{c}{2B} \left[\mathbf{k} \cdot (\mathbf{k}' \times \mathbf{b}) \, J_{\mathbf{k}''}^\alpha(\mathbf{v}) \, \mathcal{G}_{\kappa'}^\alpha(\mathbf{v}) \, \Delta_{\kappa'}^\alpha(\mathbf{v}) \right.$$

$$\left. + \mathbf{k} \cdot (\mathbf{k}'' \times \mathbf{b}) \, J_{\mathbf{k}'}^\alpha(\mathbf{v}) \, \mathcal{G}_{\kappa''}^\alpha(\mathbf{v}) \, \Delta_{\kappa''}^\alpha(\mathbf{v}) \right]. \quad (10.10)$$

Note the symmetry property:

$$\nu^\alpha(\kappa|\kappa', \kappa''; \mathbf{v}) = \nu^\alpha(\kappa|\kappa'', \kappa'; \mathbf{v}). \qquad (10.11)$$

The equivalence of (10.9), (10.10) with (10.1) is easily checked, by noting the following symmetries:

$$J_{-\mathbf{k}}^\alpha(\mathbf{v}) = J_{\mathbf{k}}^\alpha(\mathbf{v}), \quad \mathcal{G}_{-\kappa}^\alpha(\mathbf{v}) = [\mathcal{G}_\kappa^\alpha(\mathbf{v})]^*,$$

$$\Delta_{-\kappa}^\alpha(\mathbf{v}) = [\Delta_\kappa^\alpha(\mathbf{v})]^*, \quad \mathbf{k} \cdot (\mathbf{k}' \times \mathbf{b}) = -\mathbf{k} \cdot (\mathbf{k}'' \times \mathbf{b}). \quad (10.12)$$

We now combine Eqs. (10.8) and (10.9) and obtain:

$$S_\kappa = \frac{1}{|\varepsilon_\kappa|^2} \frac{2}{2\pi} \int d\kappa' d\kappa'' \, \delta(\kappa - \kappa' - \kappa'') \, \left| \mu^{(2)}(\kappa|\kappa' \, \kappa'') \right|^2 S_{\kappa'} S_{\kappa''}, \quad (10.13)$$

where:

$$\mu^{(2)}(\kappa|\kappa' \, \kappa'') = \sum_\alpha \frac{4\pi}{k^2} \int d\mathbf{v} \, J_{\mathbf{k}}^\alpha(\mathbf{v}) \, \mathcal{G}_\kappa^\alpha(\mathbf{v}) \, \nu^\alpha(\kappa|\kappa' \, \kappa''; \mathbf{v}) \qquad (10.14)$$

Eq. (10.11) implies the following symmetries:

$$\mu^{(2)}(\kappa|\kappa' \, \kappa'') = \mu^{(2)}(\kappa|\kappa'' \, \kappa'), \qquad (10.15)$$

$$\mu^{(2)}(-\kappa| - \kappa', -\kappa'') = \left[\mu^{(2)}(\kappa|\kappa' \, \kappa'') \right]^*. \qquad (10.16)$$

In the present approximation, we have obtained a *closed equation* (10.13) for the spectrum of potential fluctuations: it is called the *spectrum balance equation*. Eq. (10.13) is a generalization of an equation first derived for special situations, in the limit of weak coupling and vanishing Larmor radius, by many authors, among whom we quote: DRUMMOND

& PINES 1962, GALEEV & KARPMAN 1963, KADOMTSEV & PETVIASHVILI 1963, KADOMTSEV 1965, HORTON & CHOI 1979. In the present, renormalized form it appears in DuBOIS & ESPEDAL 1978, KROMMES & KLEVA 1979 and SITENKO & SOSENKO 1987.

The relatively simple, quadratic nonlinearity of Eq. (10.13) is, as usual, illusory. The coupling coefficient as well as the renormalized dielectric function are strongly nonlinear functionals of the spectrum. Thus, the simple-looking equation (10.13) is actually a formidable nonlinear equation. Its main interest is in clearly exhibiting the structure of the evolution process, and therefore to provide a starting point for further approximations.

A widely used additional simplification scheme is the so-called *weak turbulence approximation (WTA)*. It could be regarded as a first step in a "kind of" perturbation theory applied to the DIA. In the WTA, the propagators and vertex operators *occurring in the renormalization corrections* Σ, V, F in Eqs. (8.113) - (8.120) are replaced by their unperturbed values; a similar procedure is used in the nonlinear wave-coupling coefficients ν^α and $m^{(2)}$ of Eqs. (10.10), (10.16). It is not surprising to find that the resulting expressions have been well known for a long time. The WTA has been, historically, the first form of plasma turbulence theory, developed in the 1960's [First appearance: DRUMMOND & PINES 1962]. There exists a number of excellent text books and review papers (*e.g.*, KADOMTSEV 1965, SAGDEEV & GALEEV 1969, DAVIDSON 1972, HORTON & CHOI 1979) and an enormous number of research papers devoted to its development, which cannot all be cited here.

The weak turbulence form of the spectrum balance equation (10.13) is obtained by replacing in Eqs. (10.10), (10.14): $g \to g_0$, $\Delta \to \Delta_0$. If, moreover, the FLR effects are neglected, one sets $J_k^\alpha(\mathbf{v}) \to 1$, and obtains:

$$\mu_0^{(2)}(\kappa|\kappa'\,\kappa'') = \frac{c}{2B} \sum_\alpha \frac{4\pi e_\alpha}{k^2} \int d\mathbf{v}\, \mathcal{G}_{0\,\kappa}^\alpha(\mathbf{v})\, \mathbf{k} \cdot (\mathbf{k}' \wedge \mathbf{b})$$
$$\times \left[\mathcal{G}_{0\,\kappa'}^\alpha(\mathbf{v}) \Delta_{0\,\kappa'}^\alpha(\mathbf{v}) - \mathcal{G}_{0\,\kappa''}^\alpha(\mathbf{v}) \Delta_{0\,\kappa''}^\alpha(\mathbf{v})\right]. \quad (10.17)$$

Eq. (10.13) reduces to the simpler, truly quadratic form:

$$\varepsilon_{0\,\kappa} S_\kappa = \frac{2}{2\pi\varepsilon_{0\,-\kappa}} \int d\kappa' d\kappa''\, \delta(\kappa - \kappa' - \kappa'') \left|\mu_0^{(2)}(\kappa|\kappa'\,\kappa'')\right|^2 S_{\kappa'} S_{\kappa''}.$$
$$(10.18)$$

The physical process represented by Eq. (10.18) is easily visualized (in first approximation!): it describes the nonlinear coupling of the two "modes" $S_{\kappa'}$, $S_{\kappa''}$ to produce the "mode" S_κ. This is adequately called a *three-wave coupling process*. The strength of the coupling is measured by the coefficient $\left|\mu_0^{(2)}(\kappa|\kappa'\,\kappa'')\right|^2$ and is governed by a "selection rule" expressing the conservation of wave vectors $\mathbf{k}$ and frequencies ω. The process is strongly reminiscent of the phonon dynamics in the theory of lattice vibrations in a crystal (see, *e.g.*, PEIERLS 1965).

We now summarize the results obtained in the present section. In the approximation scheme devised here, the spectrum S_κ is obtained as a solution of the closed spectrum balance equation (10.13). This reduction has been achieved by using the coherent approximation (9.4) - (9.6) for all the correlation functions involved in the expression of the renormalized evolution operators. The present scheme can therefore be called the *quasi-coherent approximation (QCA)*. Let it be stressed, however, that the present scheme goes *beyond the pure coherent approximation*, as discussed in Sec. 8.6: it retains, indeed, the *incoherent fluctuations*. Comparing Eq. (10.13) to (8.88), we see that we have derived an approximate expression for the spectrum of incoherent fluctuations:

$$\tilde{S}_\kappa = 2 \int d\kappa' d\kappa'' \, \delta \left(\kappa - \kappa' - \kappa'' \right) \left| \mu^{(2)}(\kappa | \kappa' \, \kappa'') \right|^2 S_{\kappa'} \, S_{\kappa''}. \qquad (10.19)$$

10.2   Trajectory Correlation

It was shown in Sec. 10.1 that, at the price of some additional approximations, the DIA leads to a closed equation for the spectrum. We now investigate in some detail the evolution of the two-particle correlation function (in detail: *two-point, two-time correlation*) $C(1\,2) \equiv \langle \delta f(1) \, \delta f(2) \rangle$. This study will uncover several important features of the turbulent plasmas. It is indispensable for this purpose to retain the effect of the *incoherent fluctuations*, which play an essential role. We have already seen in Sec. 10.1 that the latter lead to a nonlinear mode coupling in the spectrum balance equation. In the forthcoming treatment we proceed by first writing some very general equations, which are then progressively simplified by introducing additional approximations, making the equations more and more operational.

We start with the complete DIA-renormalized equation (8.108) for the correlation function $C(1\,2)$:

$$g^{-1}(1\,3) \, C(3\,2) + \Delta(1\,a) \, T(a\,2) = G(2\,3) \, F(1\,3). \qquad (10.20)$$

We now write out explicitly the random noise source term as given, in the DIA, by Eq. (8.113):

$$g^{-1}(1\,3) C(3\,2) + \Delta(1\,a) \, T(a\,2)$$
$$= G(2\,3) \, M(1\,a\,4) \, M(3\,b\,5) \left[T(b\,4) \, T^\dagger(5\,a) + S(b\,a) C(5\,4) \right]. \qquad (10.21)$$

Clearly, the first term of the random noise source combines naturally with the vertex, and the second term with the propagator. After left-multiplication by g, we write:

$$C(1\,2) - g(1\,3)\,G(2\,4)\,M(3\,a\,5)\,M(4\,b\,6)\,S(b\,a)\,C(6\,5)$$
$$= -g(1\,3)\,\Delta(3\,a)\,T^{\dagger}(a\,2) + g(1\,3)\,G(2\,4)\,M(3\,a\,5)\,M(4\,b\,6)\,T^{\dagger}(b\,5)\,T(6\,a).$$
$$(10.22)$$

We now approximate the Green's operators by the corresponding propagators: $G(2\,4) \approx g(2\,4)$ [see Eq. (8.60)]. Next, we use the quasi-coherent approximation as in Sec. 10.1, approximating the mixed particle-potential correlation by Eqs. (9.5), (9.6). We also define the quantity $Q(1\,2)$:

$$Q(1\,2) = \Delta(1\,a)\,\Delta(2\,b)\,S(a\,b)$$
$$+M(1\,a\,3)\,M(2\,b\,4)\,g(3\,5)\,g(4\,6)\,\Delta(5\,c)\,\Delta(6\,d)\,S(b\,c)\,S(d\,a).\qquad(10.23)$$

With all these substitutions, Eq. (10.2) is rewritten in the following form:

$$C(1\,2) - g(1\,3)\,g(2\,4)\,M(3\,a\,5)\,M(4\,b\,6)\,S(b\,a)\,C(6\,5)$$
$$= g(1\,3)\,g(2\,4)\,Q(3\,4).\qquad(10.24)$$

This is the basic equation in the approximation adopted here. It is a formally linear equation for the correlation function $C(12)$, with coefficients depending on the spectrum $S(a\,b)$. The latter is determined by the (closed!) spectrum balance equation (10.13) which was derived under the same set of approximations. At the present stage, the main simplification of the original equation (10.21) is the elimination of the mixed correlation $T(1\,a)$ introduced by the quasi-coherent approximation expression (9.5).

We define the following operator, which acts on C in the left hand side of Eq. (10.24):

$$\widetilde{G}^{-1}(1\,2|3\,4) = I(1\,3)\,I(2\,4)$$
$$-g(1\,5)\,g(2\,6)\,M(5\,a\,4)\,M(6\,b\,3)\,S(b\,a).\qquad(10.25)$$

We multiply both sides of (10.24) by the inverse of this operator and define a new operator:

$$G^{(2)}(1\,2|3\,4) = \widetilde{G}(1\,2|5\,6)\,g(5\,3)\,g(6\,4).\qquad(10.26)$$

Eq. (10.24) then possesses the following formal solution:

$$C(1\,2) = G^{(2)}(1\,2|3\,4)\,Q(3\,4).\qquad(10.27)$$

The operator $G^{(2)}(1\,2|3\,4)$ is simply the *propagator* associated with Eq. (10.24): acting on the source term $Q(3\,4)$ it produces the correlation $C(1\,2)$. It must, of course, be distinguished from the propagator $g(1\,2)$ associated with the renormalized stochastic equation (8.79) for the one-particle function $\delta f(1)$. Here we deal with the two-particle function $C(1\,2)$; we therefore call $G^{(2)}(1\,2|3\,4)$ the *two-particle propagator*. The analysis of its structure is very rewarding. We first note that the operator $G^{(2)}(1\,2|3\,4)$ obeys an integral equation that is easily obtained from (10.25) by using the identity (8.57). Combining the result with (10.26) we find:

$$G^{(2)}(1\,2|3\,4) = g(1\,3)\,g(2\,4)$$
$$+\,g(1\,5)\,g(2\,6)\,M(5\,a\,8)\,M(6\,b\,7)\,S(b\,a)\,G^{(2)}(7\,8|3\,4).\qquad(10.28)$$

This equation clearly exhibits the role of the "incoherent fluctuations". If the latter are neglected, *i.e.*, if we put the random noise source equal to zero: $F(1\,2) = 0$, the second term in the right hand side vanishes and we find:

$$G^{(2,c)}(1\,2|3\,4) = g(1\,3)\,g(2\,4).\qquad(10.29)$$

Thus, in the coherent approximation, the two-particle propagator is simply the product of two (renormalized) one-particle propagators. This means that the two particles move, ignoring each other, as if there was no direct interaction between them. It should, however, be clear that the interactions are not absent from Eq. (10.29): they produce the renormalization of the one-particle propagators $g(1\,2)$. The nonlinear interactions of particle 3 (or 4) with the surrounding medium transform the ballistic free motion into a motion of diffusive type, as shown in Sec. 9.2. Thus, Eq. (10.29) describes the *independent diffusion of the two particles*.

The incoherent fluctuations destroy this unrealistically simple picture. The two-particle propagator (10.28) is no longer factorized: the two particles no longer diffuse independently. We shall say that the propagator $G^{(2)}(1\,2|3\,4)$ of Eq. (10.28) describes *correlated diffusive trajectories*. It can also be said that we are here in presence of a *correlated relative diffusion process* that is non-trivially different from independent diffusion.

The structure of the correlated propagator can be vividly illustrated by means of a graph symbolizing the integral equation (10.28) [Fig. 10.1]. Here a reinforced line represents a one-particle propagator, a rectangle with two incoming lines (to the right) and two outgoing lines represents a two-particle propagator. A three-pronged vertex is associated with a coupling factor M and a circle with a spectrum factor S.

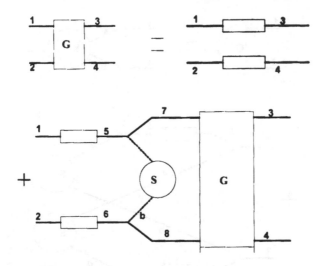

Figure 10.1. Graph representing the integral equation (10.28) for the correlated two-particle propagator $G^{(2)}$.

The incoherent fluctuations modify not only the propagator, but also the source term Q in the equation for the correlation function C. The structure of the source term, as defined by Eq. (10.23), is represented graphically in Fig. 10.2.

If the second term in Eq. (10.23) is neglected, Eq. (10.27) combined with (10.29) yields:

$$C^{(c)}(1\,2) = g(1\,3)\,g(2\,4)\,\Delta(3\,a)\,\Delta(4\,b)\,S(b\,a). \tag{10.30}$$

This is easily recognized as the solution for the correlation function in the completely coherent approximation. Indeed, in the latter case, Eq. (9.2):

$$C^{(c)}(1\,2) = -g(1\,3)\,\Delta(3\,a)\,T(a\,2), \tag{10.31}$$

which, combined with the coherent relation (9.5), is identical with (10.30).

We now note that a modification of the source term in (10.27) produces, of course, a change in the value of the correlation function, but does not influence the nature of the motion, which is solely determined by the propagator $G^{(2)}$. As our interest lies mainly in the latter aspect, we decide to neglect henceforth the complicated second term in Eq. (10.23). This neglect is actually consistent with the quasi-coherent approximation used for the elimination of the mixed correlation $T^{\perp}(1\,a)$ in Eq. (10.2). Indeed,

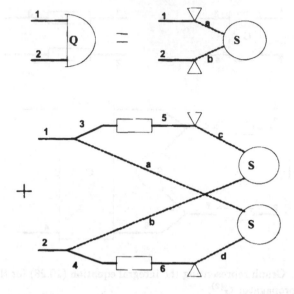

Figure 10.2. Graph representing coherent and incoherent contributions to the source term Q, Eq. (10.23).

in the quasi-coherent approximation the propagator $g(1\,2)$ appearing inside any complex expression is diagonal in $\kappa, \alpha, \mathbf{v}$, because the polarization term Σ^P (8.117) is ignored in the renormalization process. The neglect of the latter contribution requires by Eq. (8.141) the simultaneous neglect of the term associated with the propagator in the random noise source, *i.e.*, the first term in Eq. (8.113): this is precisely the second term of Eq. (10.23). Thus, in the present approximation, the basic equation (10.24), after left-multiplication by an operator g^{-1}, reduces to:

$$g^{-1}(1\,3)\,C(3\,2) - g(2\,4)\,M(1\,a\,5)\,M(4\,b\,6)\,S(b\,a)\,C(6\,5)$$
$$= g(2\,4)\,\Delta(1\,a)\cdot(4\,b)\,S(a\,b). \qquad (10.32)$$

10.3   Relative Diffusion

We now make Eq. (10.32) explicit, using Eqs. (9.15), (8.11), (9.14), (10.2), (8.21), (8.22), (8.25); we also use the form (9.19), (9.20) of the renormalized propagator and obtain:

$$(-i\omega + ik_{\|}v_{1\|})\,C_{\kappa}^{\alpha_1\alpha_2}(\mathbf{v}_1,\mathbf{v}_2)$$

$$+ \int d\kappa'\, m_{\mathbf{kk'}}\, \mathcal{G}_{\kappa-\kappa'}^{\alpha_1}(\mathbf{v}_1)\,[J_{\mathbf{k'}}^{\alpha_1}(\mathbf{v}_1)]^2\, S_{\kappa'}\, C_{\kappa}^{\alpha_1\alpha_2}(\mathbf{v}_1,\mathbf{v}_2)$$

$$- \int d\kappa'\, m_{\mathbf{kk'}}\, \mathcal{G}_{-\kappa}^{\alpha_2}(\mathbf{v}_2)\,[J_{\mathbf{k'}}^{\alpha_1}(\mathbf{v}_1)J_{\mathbf{k'}}^{\alpha_2}(\mathbf{v}_2)]\, S_{\kappa'}\, C_{\kappa-\kappa'}^{\alpha_1\alpha_2}(\mathbf{v}_1,\mathbf{v}_2)$$

$$= \mathcal{G}_{-\kappa}^{\alpha_2}(\mathbf{v}_2)\,\Delta_{\kappa}^{\alpha_1}(\mathbf{v}_1)\,\Delta_{-\kappa}^{\alpha_2}(\mathbf{v}_2)\,S_{\kappa}. \tag{10.33}$$

[The second term comes from $g^{-1}(1\,3)$; in the third term we made use of the symmetry: $C_{\kappa}^{\alpha_1\alpha_2}(\mathbf{v}_1,\mathbf{v}_2) = C_{-\kappa}^{\alpha_2\alpha_1}(\mathbf{v}_2,\mathbf{v}_1)$, which is obvious from Eq. (8.25)]. We now introduce a *Markovian approximation*, as in Sec. 9.1. The second term is changed by the substitution $\mathcal{G}_{\kappa-\kappa'}^{\alpha_1}(\mathbf{v}_1) \to \mathcal{G}_{\kappa'}^{\alpha_1}(\mathbf{v}_1)$ as in (9.21); in the third term we substitute $\mathcal{G}_{-\kappa}^{\alpha_2}(\mathbf{v}_2) \to \mathcal{G}_{-\kappa'}^{\alpha_2}(\mathbf{v}_2)$.[2] We define:

$$\mathsf{D}_{\kappa}^{\alpha_1\alpha_2}(\mathbf{v}_1,\mathbf{v}_2) = m_{\mathbf{k}}\, \mathcal{G}_{-\kappa}^{\alpha_2}(\mathbf{v}_2)\, J_{\mathbf{k}}^{\alpha_1}(\mathbf{v}_1)J_{\mathbf{k}}^{\alpha_2}(\mathbf{v}_2)\, S_{\kappa}. \tag{10.34}$$

From this tensor we derive the following other ones:

$$\mathsf{D}_{\kappa}^{\alpha_1}(\mathbf{v}_1) = \mathsf{D}_{\kappa}^{\alpha_1\alpha_1}(\mathbf{v}_1,\mathbf{v}_1), \tag{10.35}$$

$$\mathsf{D}^{\alpha_1}(\mathbf{v}_1) = \int d\kappa'\, \mathsf{D}_{\kappa'}^{\alpha_1}(\mathbf{v}_1). \tag{10.36}$$

The coefficient defined by Eq. (10.36) is the same as the Dupree coefficient of Eq. (9.21). We also introduce the abbreviation:

$$P_{\kappa}^{\alpha_1\alpha_2}(\mathbf{v}_1,\mathbf{v}_2) = \mathcal{G}_{-\kappa}^{\alpha_2}(\mathbf{v}_2)\,\Delta_{\kappa}^{\alpha_1}(\mathbf{v}_1)\,\Delta_{-\kappa}^{\alpha_2}(\mathbf{v}_2)\,S_{\kappa}. \tag{10.37}$$

Eq. (10.33), Markovianized as explained above, thus becomes:

[2] This operation is not as easily justified [BOUTROS-GHALI & DUPREE 1981]. One uses the fact that

$$C_{\kappa-\kappa'}^{\alpha_1\alpha_2}(\mathbf{v}_1,\mathbf{v}_2) \approx \mathcal{G}_{\kappa-\kappa'}^{\alpha_1}(\mathbf{v}_1)\,\mathcal{G}_{-\kappa+\kappa'}^{\alpha_2}(\mathbf{v}_2)\dots$$

$$\approx \left[\omega - \omega' - (k_{\|} - k_{\|}')v_{1\|}\right]^{-1} \left[-\omega + \omega' + (k_{\|} - k_{\|}')v_{2\|}\right]^{-1}\dots$$

which has a pole at $\omega' + (k_{\|} - k_{\|}')v_{2\|}$. It is then accepted that the propagator $\mathcal{G}_{-\kappa}^{\alpha_2}(\mathbf{v}_2)$ be approximated by its residue at that pole, which yields:

$$\mathcal{G}_{-\kappa}^{\alpha_2}(\mathbf{v}_2) \approx (-\omega + k_{\|}v_{2\|})^{-1} \approx \left[-\omega' - (k_{\|} - k_{\|}')v_{2\|} + k_{\|}v_{2\|}\right]^{-1}$$

$$= \left[-\omega' + k_{\|}'v_{2\|}\right]^{-1} \approx \mathcal{G}_{-\kappa'}^{\alpha_2}(\mathbf{v}_2).$$

$$(-i\omega + ik_\parallel v_{1\parallel}) C_\kappa^{\alpha_1 \alpha_2}(\mathbf{v}_1, \mathbf{v}_2)$$
$$+\mathbf{k}_\perp \cdot \mathbf{D}^{\alpha_1}(\mathbf{v}_1) \cdot \mathbf{k}_\perp \, C_\kappa^{\alpha_1 \alpha_2}(\mathbf{v}_1, \mathbf{v}_2)$$
$$-\int d\kappa' \, (\mathbf{k} - \mathbf{k}')_\perp \cdot \mathbf{D}_{\kappa'}^{\alpha_1 \alpha_2}(\mathbf{v}_1, \mathbf{v}_2) \cdot (\mathbf{k} - \mathbf{k}')_\perp \, C_{\kappa - \kappa'}^{\alpha_1 \alpha_2}(\mathbf{v}_1, \mathbf{v}_2)$$
$$= P_\kappa^{\alpha_1 \alpha_2}(\mathbf{v}_1, \mathbf{v}_2). \tag{10.38}$$

In order to understand the new features introduced by the incoherent fluctuations, we consider, for comparison, the corresponding equation in the coherent RQL approximation. From Eq. (9.36), written for a homogeneous, but not necessarily isotropic state, we obtain, after multiplication by $\delta f_{-\kappa}^{\alpha_2}(\mathbf{v}_2)$ and ensemble-averaging, the following equation:

$$(-i\omega + ik_\parallel v_{1\parallel}) C_\kappa^{\alpha_1 \alpha_2}(\mathbf{v}_1, \mathbf{v}_2) + \mathbf{k}_\perp \cdot \mathbf{D}^{\alpha_1}(\mathbf{v}_1) \cdot \mathbf{k}_\perp \, C_\kappa^{\alpha_1 \alpha_2}(\mathbf{v}_1, \mathbf{v}_2)$$
$$= p_\kappa^{\alpha_1 \alpha_2}(\mathbf{v}_1, \mathbf{v}_2), \quad (RQL). \tag{10.39}$$

where $p_\kappa^{\alpha_1 \alpha_2}(\mathbf{v}_1, \mathbf{v}_2)$ denotes the adequate source term (renormalized vertex), whose form does not matter here. The feature which we wish to emphasize is that the coefficients on the left hand side do not depend on particle 2: *in the RQL approximation the particles 1 and 2 diffuse independently of each other.*

Going over to physical space, we define the *relative space and time correlation function* (for a quasi-homogeneous, quasi-stationary state) [see Eqs. (6.45), (6.46)]:

$$C^{\alpha_1 \alpha_2}(\mathbf{R}, \tau; \mathbf{v}_1, \mathbf{v}_2) = \langle \delta f^{\alpha_1}(\mathbf{Y}, \mathbf{v}_1; t) \, \delta f^{\alpha_2}(\mathbf{Y} + \mathbf{R}, \mathbf{v}_2; t + \tau) \rangle. \tag{10.40}$$

The Fourier transform of this function of $\mathbf{R}$ and τ is:

$$C^{\alpha_1 \alpha_2}(\mathbf{R}, \tau; \mathbf{v}_1, \mathbf{v}_2) = \frac{1}{2\pi} \int d\omega dk \, e^{i\mathbf{k} \cdot \mathbf{R} - i\omega \tau} C_\kappa^{\alpha_1 \alpha_2}(\mathbf{v}_1, \mathbf{v}_2). \tag{10.41}$$

Hence, Eq. (10.39) is transformed into:

$$\{\partial_\tau + v_{1\parallel} \nabla_\parallel - \mathbf{D}^{\alpha_1}(\mathbf{v}_1) : \nabla_\perp \nabla_\perp\} C^{\alpha_1 \alpha_2}(\mathbf{R}, \tau; \mathbf{v}_1, \mathbf{v}_2)$$
$$= P^{\alpha_1 \alpha_2}(\mathbf{R}, \tau; \mathbf{v}_1, \mathbf{v}_2). \quad (RQL) \tag{10.42}$$

(The definition of the right hand side is obvious). Let us stress, for clarity, that here the variables are the *relative* time $\tau \equiv t_2 - t_1$ and the *relative* distance $\mathbf{R} = \mathbf{Y}_2 - \mathbf{Y}_1$ associated with the motion of two particles, 1 and 2. We have also used the gradient with respect to the relative distance:

$$\frac{\partial}{\partial \mathbf{R}} \equiv \nabla = \mathbf{b}\nabla_{\parallel} + \nabla_{\perp}. \qquad (10.43)$$

The relative motion is a combination of convection $(v_{\parallel}\nabla_{\parallel})$ in the parallel direction and diffusion in the perpendicular direction. We stress again the mutual independence of particles 1 and 2: the variables α_2 and $\mathbf{v}_2$ are mere parameters in this equation.

Comparing Eq. (10.38) to the RQL equation (10.39), we note that the incoherent fluctuations add the last term in the left hand side of the former.[3] This term introduces the *trajectory correlations* in three ways:

- The coefficient $D_{\kappa'}^{\alpha_1\alpha_2}(\mathbf{v}_1, \mathbf{v}_2)$ depends on α_2,
- This coefficient also depends on $\mathbf{v}_2$,
- Eq. (10.38) is no longer a purely algebraic equation for $C_{\kappa}^{\alpha_1\alpha_2}(\mathbf{v}_1, \mathbf{v}_2)$, but rather an *integral equation* connecting this Fourier component (for the four-vector κ) to all the other components $C_{\kappa-\kappa'}^{\alpha_1\alpha_2}(\mathbf{v}_1, \mathbf{v}_2)$.

The latter feature is the most important one; it also makes a simple direct solution of the equation impossible. Eq. (10.38) can now be transformed to physical space, by noting that the last term in the left hand side is in the form of a convolution. We define the following tensorial coefficient:

$$D^{\alpha_1\alpha_2}(\mathbf{R}, \tau; \mathbf{v}_1, \mathbf{v}_2) = \frac{1}{2\pi}\int d\kappa \, e^{-i\omega\tau + i\mathbf{k}\cdot\mathbf{R}} \, D_{\kappa}^{\alpha_1\alpha_2}(\mathbf{v}_1, \mathbf{v}_2). \qquad (10.44)$$

Using also the definition (10.41) of the relative space-time correlation function, we find from (10.38):

$$\begin{aligned}
\left(\partial_{\tau} + v_{1\parallel}\nabla_{\parallel}\right) C^{\alpha_1\alpha_2}(\mathbf{R}, \tau; \mathbf{v}_1, \mathbf{v}_2) & \\
-D_{R1}^{\alpha_1\alpha_2}(\mathbf{R}, \tau; \mathbf{v}_1, \mathbf{v}_2) : \nabla_{\perp}\nabla_{\perp}C^{\alpha_1\alpha_2}(\mathbf{R}, \tau; \mathbf{v}_1, \mathbf{v}_2) & \\
= P^{\alpha_1\alpha_2}(\mathbf{R}, \tau; \mathbf{v}_1, \mathbf{v}_2) &
\end{aligned} \qquad (10.45)$$

The right hand side is the inverse Fourier transform of the source term (10.37). We introduced here the *partial relative tensorial diffusion coefficient*:

$$D_{R1}^{\alpha_1\alpha_2}(\mathbf{R}, \tau; \mathbf{v}_1, \mathbf{v}_2) = D^{\alpha_1}(\mathbf{v}_1) - D^{\alpha_1\alpha_2}(\mathbf{R}, \tau; \mathbf{v}_1, \mathbf{v}_2). \qquad (10.46)$$

The *relative diffusion equation* for the relative space-time correlation function $C^{\alpha_1\alpha_2}(\mathbf{R}, \tau; \mathbf{v}_1, \mathbf{v}_2)$ (10.40) is apparently very similar to Eq. (10.42) derived in the RQL approximation: it is a pure partial differential

---

[3] They also modify the source term, in a way that is irrelevant for the present argument.

equation describing an advection-diffusion process in the relative coordinates $\mathbf{R}, \tau$. There is, however, an essential difference: the (partial) relative diffusion coefficient depends on both velocities and both species indices and, most importantly, on the relative distance $\mathbf{R}$ and relative time τ. This consequence of the trajectory correlation introduces a radical change in the physics: it will be analyzed below.

Before continuing the systematic study of Eq. (10.45) we make the following qualitative remarks. We integrate out the parallel relative distance and look for a stationary solution. The relative diffusion equation for particles then reduces to:

$$-D_{R1}^{\alpha_1\alpha_2}(\mathbf{R}_\perp, \tau; \mathbf{v}_1, \mathbf{v}_2) : \nabla_\perp \nabla_\perp C^{\alpha_1\alpha_2}(\mathbf{R}_\perp, \tau; \mathbf{v}_1, \mathbf{v}_2)$$
$$= P^{\alpha_1\alpha_2}(\mathbf{R}_\perp, \tau; \mathbf{v}_1, \mathbf{v}_2). \tag{10.47}$$

We now note the following important property. For particles of the same species $(\alpha_1 = \alpha_2 = \alpha)$ that are initially $(\tau = 0)$ very close together $(\mathbf{r}_\perp \approx 0)$ and have very close values of the velocity $(\mathbf{v}_1 \approx \mathbf{v}_2)$ we find from Eqs. (10.35), (10.36), (10.44):

$$D^{\alpha\alpha}(\mathbf{R}_\perp, \tau; \mathbf{v}_1, \mathbf{v}_1) = D^\alpha(\mathbf{v}_1), \quad \mathbf{R}_\perp \to 0, \ \tau \to 0, \tag{10.48}$$

or, equivalently,

$$D_{R1}^{\alpha_1\alpha_1}(\mathbf{R}_\perp, \tau; \mathbf{v}_1, \mathbf{v}_1) \to 0, \quad \mathbf{R}_\perp \to 0, \ \tau \to 0. \tag{10.49}$$

On the other hand, the source term $P^{\alpha\alpha}(\mathbf{R}_\perp, \tau; \mathbf{v}_1, \mathbf{v}_1)$ does *not* tend to zero in this limit. Hence, the Laplacian of the *like-particle correlation function* and therefore the function itself *is very much enhanced at small distance and small relative velocity by the presence of trajectory correlations.* We thus arrive at a picture of the turbulent plasma as a granular medium rather than a homogeneous one. This is the *clump effect*, first discovered simultaneously by DUPREE 1970, 1972, 1978 and by KADOMTSEV & POGUTSE 1970, 1971. [The description of clumps as enhanced correlations was especially emphasized in MISGUICH & BALESCU 1978]. We postpone the detailed discussion of the clump phenomenon to Chap. 13.

We note that the two particles 1 and 2 play a quite dissymmetric role in Eq. (10.45). The dynamics is essentially dictated by particle 1, whereas particle 2 is "almost" passive: not as much, though, as in the RQL equation (10.39), because the trajectory correlation introduces an influence of 2 on 1. In order to study the physics of the clump phenomenon it is useful to consider a different type of correlation function, which obeys a more symmetric equation. This is the "two-point, one-time", or "relative distance, equal time" correlation function:

$$C^{\alpha_1 \alpha_2}(\mathbf{R}, \mathbf{v}_1, \mathbf{v}_2; t) = \lim_{\tau \to 0} \langle \delta f^{\alpha_1}(\mathbf{Y}, \mathbf{v}_1; t) \, \delta f^{\alpha_2}(\mathbf{Y} + \mathbf{R}, \mathbf{v}_2; t + \tau) \rangle.$$

(10.50)

We skip here the detailed derivation of the equation of evolution of this function: it involves a careful limiting process. Its result is easily understood:

$$\left[\partial_t + (v_{1\parallel} - v_{2\parallel}) \nabla_\parallel \right] C^{\alpha_1 \alpha_2}(\mathbf{R}, \mathbf{v}_1, \mathbf{v}_2; t)$$
$$- D_R^{\alpha_1 \alpha_2}(\mathbf{R}, \mathbf{v}_1, \mathbf{v}_2) : \nabla_\perp \nabla_\perp C^{\alpha_1 \alpha_2}(\mathbf{R}, \mathbf{v}_1, \mathbf{v}_2; t)$$
$$= P_S^{\alpha_1 \alpha_2}(\mathbf{R}, \mathbf{v}_1, \mathbf{v}_2; t).$$

(10.51)

The *relative diffusion coefficient* appearing here is related to the partial diffusion coefficient of Eq. (10.46) as follows:

$$D_R^{\alpha_1 \alpha_2}(\mathbf{R}, \mathbf{v}_1, \mathbf{v}_2) = D_{R1}^{\alpha_1 \alpha_2}(\mathbf{R}, 0; \mathbf{v}_1, \mathbf{v}_2) + D_{R2}^{\alpha_2 \alpha_1}(\mathbf{R}, 0; \mathbf{v}_2, \mathbf{v}_1).$$

(10.52)

$[D_{R2}^{\alpha_2 \alpha_1}(\mathbf{R}, 0; \mathbf{v}_2, \mathbf{v}_1)$ is obtained from $D_{R1}^{\alpha_1 \alpha_2}(\mathbf{R}, 0; \mathbf{v}_1, \mathbf{v}_2)$ by permutation of the indices 1 and 2.] We also introduced the symmetric source term:

$$P_S^{\alpha_1 \alpha_2}(\mathbf{R}, \mathbf{v}_1, \mathbf{v}_2; t) = P^{\alpha_1 \alpha_2}(\mathbf{R}, 0; \mathbf{v}_1, \mathbf{v}_2; t) + P^{\alpha_2 \alpha_1}(\mathbf{R}, 0; \mathbf{v}_2, \mathbf{v}_1; t).$$

(10.53)

Eq. (10.51) was derived (in a slightly less general case) by DUPREE 1970, 1972, 1978. As explained above, the one-time correlation function obeys a more symmetric equation than the relative-time equation (because both $\delta f(1; t)$ and $\delta f(2; t)$ depend on the same time variable). The same qualitative features are present here as in Eq. (10.45), in particular, the *clump effect*. The solution of the relative diffusion equation (10.51) requires a different type of methodology, which will be developed in the forthcoming chapters. We will return to this problem in Chap. 13.

10.4   Partial Conclusions

We arrived here at a kind of limit point. The general purpose in the first part of this book was the derivation of transport theory from first principles. More precisely, we used as a starting point the gyrokinetic equation - which is itself a reasonable approximation of the Liouville equation describing exactly the motion of a system of charged particles in presence of an electromagnetic field. The latter is produced by external sources but also, most significantly, by the charged particles themselves. Hence the gyrokinetic equation is coupled to the Maxwell equations, which make the whole theory self-consistent.

This is, indeed, a very ambitious program: it should yield, *in principle*, the complete explanation of the transport phenomena in a plasma. Unfortunately, it is impossible to implement this program *in practice*, because of its extreme mathematical complexity. Although it is capable of exhibiting the emergence of a turbulent state as a result of the well-known dynamical instability of many-body Hamiltonian systems, it is at present impossible to study this problem (analytically or numerically) beyond a small number of ultra-simplified mathematical models (the most advanced of these is the work of GASPARD 1998 on the Lorentz model of hard spheres). The shortcut used in plasma physics is to consider only *fully developed turbulence*. This implies the assumption of a fluctuating electromagnetic field, which introduces a *superimposed stochasticity* which allowed us to develop a framework for a self-consistent theory based on the microscopic kinetic theory. It resulted in the idea of *renormalization*, a very attractive and fertile one, in principle. But we were blocked again by the great complication of the equations, even in the simplest approximations, such as the DIA. We were able to exhibit, however, a number of important aspects of turbulent plasmas, such as the resonance broadening, the nonlinear wave-coupling, the clumping effect; the latter is the precursor of the more general problem of the formation of coherent structures in a turbulent plasma.

In order to make progress in practice, we will have to limit our ambitions even more. This is a current attitude of the theoretical physicist: at a certain point of complexity in the development of a theory, he has to make a short-circuit by introducing an external hypothesis that allows the continuation of an argument which has become impossibly intricate. This is always a somewhat painful moment, as it implies giving up the exact explanation of some physical processes. The appropriateness of the simplifying modelling can only be tested *a posteriori* by its experimentally verifiable consequences. The approach described above will be at the basis of the developments of the forthcoming chapters.

Chapter 11

Langevin Equations

11.1 A-Langevin Equations

It appeared in the previous chapters that in all (approximate) treatments of a turbulent plasma the most difficult step in the calculation of the transport coefficients was the determination of the spectrum of the electric field fluctuations (or of the potential fluctuations, S_κ). Even in the most advanced versions of the DIA there remained difficulties. In the renormalized quasilinear approximation S_κ could only be determined by imposing an extraneous condition of marginal stability or any other saturation mechanism external to the RQL. In the more complete version, including the incoherent fluctuations, it required the solution of a nonlinear spectrum balance equation which could not be treated analytically, even in the weak coupling approximation. On the other hand, the knowledge of the spectrum is necessary for making the theory self-consistent, *i.e.*, relating the field fluctuations (through Maxwell's or Poisson's equations) to the fluctuations of the particle distribution functions.

Given this situation, we decide from here on to give up the complete self-consistency of the theory. We shall rather *introduce, at an appropriate place, an assumption about the form of the fluctuation spectrum of the turbulent electric and/or magnetic fields driving the plasma and producing the overall turbulence*.[1] This assumption may be introduced through "reasonable" theoretical arguments, or else it may be based on experimental data, thus making the theory semi-phenomenological. We shall also consider occasionally the role of ordinary particle collisions, which are the motor of the dissipation in the classical transport theory of non-turbulent plasmas. They will be modelled here by a simple stochastic process, which

---

[1] Remember the quotation of VAN KAMPEN opening Sec. 8.4: the *a priori* statistical assumption is placed here at a "higher" level. Instead of the sole definition of the ensemble of realizations of δf, we define *independently* a distribution of the potential $\delta \phi$ (or of the magnetic field fluctuations, see below).

can be superposed to the purely anomalous dissipation considered previously. We call this type of theories by the general name of *stochastic dynamics*. Their final result will be called *stochastic transport theory*. These theories take different forms, according to the methods used in their development.

The first tool considered here is based on the concept of *stochastic differential equations*, or *Langevin equations* (LANGEVIN 1908). These were already considered in Chap. 6, Eqs. (6.8) - (6.12), where they appeared as the characteristic equations of the first-order partial differential gyrokinetic equation. In the forthcoming treatment they will rather be taken as a starting point. In order to prepare the way, we begin by recalling some very classical notions of statistical physics. We follow here closely the presentation of our book on Statistical Dynamics, BALESCU 1997.

The problem considered here (as an introduction) concerns a charged test particle moving in presence of a constant magnetic field $\mathbf{B}_0 = B_0\mathbf{e}_z$ and undergoing collisions with the particles in the surrounding medium. The magnetic field acts on the particle via the Lorentz force, oriented in a direction perpendicular to $\mathbf{B}_0$. The collisions are modelled as a force which produces, on the average, a systematic friction proportional to the velocity, but whose instantaneous intensity and direction are random. Thus, contrary to the problems considered earlier, the stochastic element is here provided by the collisions, rather than by the turbulent fluctuations. The Newton equations for the particle positions (*not* the guiding centres!) are obtained by a generalization of Eq. (4.12). We denote the position vector by $\mathbf{q} = (q_x, q_y, q_z)$ and the velocity by $\mathbf{v} = (v_x, v_y, v_z)$:

$$\frac{d\mathbf{q}(t)}{dt} = \mathbf{v}(t), \tag{11.1}$$

$$\frac{d\mathbf{v}_\perp(t)}{dt} = \Omega \left[\mathbf{v}_\perp(t) \times \mathbf{e}_z\right] - \nu\mathbf{v}_\perp(t) + \mathbf{a}(t), \tag{11.2}$$

$$\frac{dv_z(t)}{dt} = -\nu v_z(t) + a_z(t). \tag{11.3}$$

Here and in forthcoming equations, for any 3D-vector $\mathbf{A}$, we denote by $\mathbf{A}_\perp$ the two-component vector (A_x, A_y) perpendicular to the magnetic field. Ω denotes the *Larmor frequency* of a particle of mass m and charge e in the magnetic field of strength B_0, Eq. (4.18). As explained above, the collisional force has an average component $-m\nu\mathbf{v}(t)$ and a random, fluctuating component $m\mathbf{a}(t)$. The constant ν (which has the dimension of an inverse time) is identified with the *collision frequency*. The equations for the ions and for the electrons have the same form: they differ only by the values of the mass m, the charge e and the collision frequency ν. We may therefore simplify the notations by suppressing the species index α whenever it is not indispensable.

In the present philosophy the collisions are modelled by a stochastic process, hence the function $\mathbf{a}(t)$ is not determined: it is only defined by its statistical properties. This implies that the object $\mathbf{a}(t)$ cannot assume any numerical values. Only its average, or the average of functions of $\mathbf{a}$, have definite values. It follows that the same property is induced upon the unknown quantities $\mathbf{q}(t)$ and $\mathbf{v}(t)$. Thus, Eqs. (11.1) - (11.3) are *stochastic differential equations*, that must be treated by methods different from ordinary differential equations, as will be shown below.

Particularly interesting are averages of products of random quantities (*moments*), which were defined in Sec. 6.3. Among these, a special importance is attached to the second moments, or autocorrelation functions $\alpha(t, t') = \langle A(t) A(t') \rangle$. In order to characterize completely the stochastic process $A(t)$, we need to know the n-time correlation functions, *i.e.*, $\langle A(t_1) A(t_2)...A(t_n) \rangle$, for all integer values of n. The simplest situation arises when the probability distribution of $A(t)$ is Gaussian with vanishing average, $\langle A(t) \rangle = 0$. In that case, it is well known that all odd order moments of $A(t)$ vanish, and all even order moments are expressed as sums of products of the two-time correlations $\alpha(t, t')$ (see Sec. 8.2). In other words, a Gaussian process is completely defined by its first two moments. We recall that a *stationary stochastic process* is one for which the autocorrelation function depends only on the difference between the two times: $\alpha(t, t') = \alpha(t - t')$: in this case the relation between two events depends only on the time interval between them, whatever the initial time considered.

We now return to the set of stochastic differential equations (11.2), (11.3). In order to make it a definite mathematical object, the "primary" random function appearing in it [in our case, the acceleration $\mathbf{a}(t)$] must be defined statistically *a priori*: this requires the specification of its probability distribution, considered as a *given property*. The strategy of solution is then as follows (a very clear exposition of the solution of Langevin equations, for neutral molecules, is found in RESIBOIS & DE LEENER 1977). The equation is solved *formally* as an ordinary one: we thus obtain an expression of the unknown $\mathbf{v}(t)$ as a functional of $\mathbf{a}(t)$ (which is a source term in the equation). From this expression, the average of any function of $\mathbf{v}(t)$ can be calculated in terms of averages of functions of $\mathbf{a}(t)$, whose statistical properties are known.

In our case, we assume that $\mathbf{a}(t)$ is a *stationary, delta-correlated Gaussian process (white noise)*. This means that the first and second moments of the acceleration are:

$$\langle a_r(t) \rangle = 0, \quad \langle a_r(t_0) a_s(t_0 + t) \rangle = A \delta_{rs} \delta(t), \quad r, s = x, y, z; \quad (11.4)$$

the constant A will be determined below. The second moment appearing here is the *acceleration autocorrelation function*. Eqs. (11.2), (11.3), completed with (11.4) are characteristic examples of LANGEVIN EQUATIONS,

or more specifically: *A-Langevin equations* (*A* stands for: "acceleration").[2]
A thorough treatment is given in COFFEY et al. 1996.

The following initial condition is assumed: *the velocity of the test particle has, at time zero, the value* $\mathbf{v}(0) = \mathbf{v}_0$, *with probability 1.* Eqs. (11.2)
and (11.3) are solved *for a given realization of the fluctuating acceleration* $\mathbf{a}(t)$: the equations are then treated as ordinary differential equations.
Their solution is straightforward (details are omitted here) and can easily
be checked: it is best expressed in terms of a propagator $\mathbf{G}(t)$ (a tensor!):

$$\mathbf{v}(t) = \mathbf{G}(t) \cdot \mathbf{v}_0 + \int_0^t d\theta\, \mathbf{G}(\theta) \cdot \mathbf{a}(t - \theta). \qquad (11.5)$$

The propagator is defined by the following matrix:

$$\mathbf{G}(\theta) = \begin{pmatrix} e^{-\nu\theta}\cos(\Omega\theta) & e^{-\nu\theta}\sin(\Omega\theta) & 0 \\ e^{-\nu\theta}\sin(\Omega\theta) & e^{-\nu\theta}\cos(\Omega\theta) & 0 \\ 0 & 0 & e^{-\nu\theta} \end{pmatrix}, \quad \theta > 0. \qquad (11.6)$$

This solution represents the well-known motion of the test particle describing a circular helix around a magnetic field line in absence of collisions.
But, unlike the Hamiltonian systems considered in Chap. 4, the dissipative
term introduced by the collisions produces (on the average) a damping of
this motion. As explained above, the velocity $\mathbf{v}(t)$ is a random quantity,
just like $\mathbf{a}(t)$: its statistical properties are determined by those of the acceleration. Thus, averaging both sides of Eq. (11.5) over the ensemble of
realizations of the fluctuating force and using (11.4), we find:

$$\langle \mathbf{v}(t) \rangle^{(0)} = \mathbf{G}(t) \cdot \mathbf{v}_0. \qquad (11.7)$$

(The superscript 0 means that the average is calculated for a given, deterministic initial condition $\mathbf{v}_0$). The perpendicular *velocity autocorrelation
function* $R_{xx}^{(0)}(t_1, t_2)$, for $t_1 > t_2$, is obtained as:[3]

$$R_{xx}^{(0)}(t_1, t_2) \equiv \langle v_x(t_1) v_x(t_2) \rangle^{(0)}$$
$$= \exp[-\nu(t_1 + t_2)] \left[\cos(\Omega t_1)\cos(\Omega t_2)v_{0x}^2 + \sin(\Omega t_1)\sin(\Omega t_2)v_{0y}^2\right]$$
$$+ \frac{A}{2\nu} \exp[-\nu(t_1 - t_2)] \left[1 - \exp(-2\nu t_2)\right] \cos[\Omega(t_1 - t_2)], \qquad (11.8)$$

where the constant A (whose value is still at our disposal!) was introduced
in Eq. (11.4). The parallel autocorrelation function $R_{zz}^{(0)}$ is obtained by

---

[2] The Langevin equation was originally introduced for the study of the Brownian motion
of a heavy particle undergoing collisions with the light particles of a surrounding fluid
(LANGEVIN 1908).
[3] Because of the gyrotropy of our problem, the x- and y-coordinates are equivalent. It
is therefore sufficient to consider a single perpendicular direction, say x.

setting $\Omega = 0$ in this formula. We note that the solution obtained here for the velocity - contrary to the acceleration - is no longer a stationary process: its autocorrelation function depends on both $(t_1 - t_2)$ and $(t_1 + t_2)$: this is physically unreasonable because nothing in the physical situation allows us to distinguish a privileged time. Also, the intrinsic physical quantities (such as the transport coefficients) that will be related to the autocorrelation would depend on the initial value of the velocity, which is physically absurd. This question was clearly discussed by RESIBOIS & DE LEENER 1977. Up to this point we considered the initial velocity as a given, deterministic quantity. In other words, $R_{xx}^{(0)}(t_1, t_2)$ is a *conditional average*, subject to the constraint that $\mathbf{v}(t = 0) = \mathbf{v}_0$. The test particle in our problem represents, however, an undistinguished particle, chosen at random among the 10^{23} particles of a plasma, which we may assume to be (at least locally) in thermal equilibrium. We are thus naturally led to complete our problem by introducing ASSUMPTION A: we consider *the initial velocity* $\mathbf{v}_0$ *as a random variable, whose probability distribution is Gaussian:*

$$\varphi(\mathbf{v}_0) = \pi^{-3/2} V_T^{-3} \exp\left(-\frac{v_0^2}{V_T^2}\right), \qquad (11.9)$$

where V_T is the *thermal velocity* (2.24):

We now take the point of view that *the physically observable quantities are defined as double averages over both* $\mathbf{a}(t)$ *and* $\mathbf{v}_0$. The doubly averaged velocity, obtained from (11.7) [this quantity is denoted by an overbar], then becomes simply:

$$\left\langle \overline{\mathbf{v}(t)} \right\rangle = 0. \qquad (11.10)$$

The perpendicular velocity autocorrelation function, averaged over $\mathbf{v}_0$, is:

$$R_{xx}(t_1, t_2) \equiv \left\langle \overline{v_x(t_1)\, v_x(t_2)} \right\rangle = \int d\mathbf{v}_0\, \varphi(\mathbf{v}_0)\, R_{xx}^{(0)}(t_1, t_2)$$

$$= \frac{1}{2}\left\{\frac{A}{\nu} e^{-\nu|t_1 - t_2|} + \left(V_T^2 - \frac{A}{\nu}\right) e^{-\nu(t_1 + t_2)}\right\} \cos[\Omega(t_1 - t_2)],$$

$$t_1 > 0, \quad t_2 > 0. \qquad (11.11)$$

Next, we introduce ASSUMPTION B: *We require the velocity* $\mathbf{v}(t)$ *to be a stationary random process.* This is realized by using the freedom that is still at our disposal: we choose the value of the constant A in such a way as to eliminate the terms depending on $t_1 + t_2$ in Eq. (11.11):

$$A = \nu V_T^2. \qquad (11.12)$$

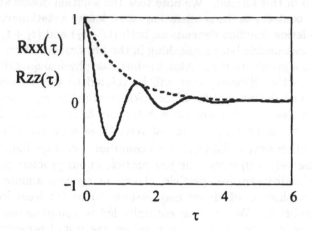

Figure 11.1. Velocity autocorrelation vs. reduced time: $\tau = \nu t$. Solid: $R_{xx}(\tau)$ [for $\Omega/\nu = 4$]; dashed: $R_{zz}(\tau)$.

The velocity autocorrelation functions are then reduced to the following expressions, which only depend on the difference of the times: $t = t_1 - t_2$, and thus describe a stationary process:

$$R_{xx}(t) = \tfrac{1}{2} V_T^2 \exp[-\nu |t|] \cos(\Omega t) , \qquad (11.13)$$

$$R_{zz}(t) = \tfrac{1}{2} V_T^2 \exp[-\nu |t|] . \qquad (11.14)$$

These functions (scaled with $\tfrac{1}{2} V_T^2$) are plotted in Fig. 11.1 against the reduced time $\tau = \nu t$ [the reduced Larmor frequency is $\omega = \Omega/\nu$]. It will be seen below that their exponential decay is an important feature, ensuring the diffusive character of the process.

We now determine the instantaneous position $\mathbf{q}(t)$ of the test particle: this requires the solution of Eq. (11.1), or simply:

$$\frac{dq_x(t)}{dt} = v_x(t), \quad \frac{dq_z(t)}{dt} = v_z(t), \qquad (11.15)$$

to be solved with the initial condition: $\mathbf{q}(0) = \mathbf{q}_0$. For the perpendicular coordinate we thus find:

$$q_x(t) = q_{x0} + \int_0^t dt'\, v_x(t') \equiv q_{x0} + \delta q_x(t). \qquad (11.16)$$

From this result we find, upon ensemble-averaging, the first moments of the *displacement* $\delta \mathbf{q}(t) = \mathbf{q}(t) - \mathbf{q}(0)$:

$$\langle \delta q_x(t) \rangle = 0, \qquad (11.17)$$

$$\langle \delta q_x^2(t) \rangle = \int_0^t dt_1 \int_0^t dt_2 \left\langle \overline{v_x(t_1) v_x(t_2)} \right\rangle$$

$$= 2 \int_0^t dt' \, (t - t') \, R_{xx}(t'), \quad t > 0. \qquad (11.18)$$

We thus find that, *on average, the particle remains in its initial position (because the average velocity is zero), but it undergoes random motions in space, resulting in a mean square displacement that is non-zero, and is a function of time.* This situation is typical of a *diffusive process.* We find here the simplest case of the relation between diffusion (a typical transport process) and random motions. Such a relation appeared already, in a much more sophisticated way, in the previous chapters (Chaps. 6, 9, 10). The connection in the present case is obtained by the argument developed in the next section.

11.2 Mean Square Displacement and Diffusion Coefficient

Consider now the macroscopic description of the plasma. The *density profile* $n(\mathbf{x}, t)$, *i.e.,* the average particle density of the plasma at point $\mathbf{x} = (x, y, z)$ (in physical space) and time t obeys the continuity equation (2.2):

$$\partial_t n(\mathbf{x}, t) = -\boldsymbol{\nabla} \cdot \boldsymbol{\Gamma}(\mathbf{x}, t). \qquad (11.19)$$

In a rather general case, this equation is closed by assuming that the particle flux consists of two contributions: a *convective* part, characterized by a velocity $\mathbf{V}(t)$, and a *diffusive* part that is linearly related to the density gradient, which is its associated thermodynamic force (*Fick's law*):

$$\boldsymbol{\Gamma}(\mathbf{x}, t) = \mathbf{V}(t) \, n(\mathbf{x}, t) - \mathsf{D}(t) \cdot \boldsymbol{\nabla} n(\mathbf{x}, t). \qquad (11.20)$$

$\mathsf{D}(t)$ is the *running diffusion tensor.* In most cases, it tends asymptotically, for long times, toward a constant tensor, whose diagonal elements are positive. In the case under consideration, *i.e.,* a plasma in presence of a constant magnetic field, the running diffusion tensor is gyrotropic, having the following form:

$$\mathsf{D}(t) = D_\perp(t) \, (\mathbf{e}_x \, \mathbf{e}_x + \mathbf{e}_y \, \mathbf{e}_y) + D_\parallel(t) \, \mathbf{e}_z \, \mathbf{e}_z, \qquad (11.21)$$

where the quantities $D_\perp(t)$, $D_\parallel(t)$ are, respectively, the perpendicular and the parallel running diffusion coefficients. If we choose to orient the x-axis along the density gradient, the *convection-diffusion equation*, also called *advection-diffusion equation* is derived by combining (11.19) with (11.20):

$$\partial_t n(\mathbf{x},t) = -\mathbf{V}(t) \cdot \boldsymbol{\nabla} n(\mathbf{x},t)$$
$$+ \left[D_\perp(t) \left(\nabla_x^2 + \nabla_y^2 \right) + D_\parallel(t) \, \nabla_z^2 \right] n(\mathbf{x};t). \tag{11.22}$$

We consider here the simple boundary condition: $n(\mathbf{x};t) \to 0$ for $|\mathbf{x}| \to \infty$. We now note that $n(\mathbf{x},t)$ can also be interpreted as the probability density of finding a particle at point $\mathbf{q} = \mathbf{x}$, at time t (normalized to the total number of particles, N). We may then calculate the average displacement of a particle ($d_t \equiv d/dt$) in the x-direction:

$$d_t \langle x(t) \rangle = d_t \, N^{-1} \int d\mathbf{x} \, x \, n(\mathbf{x},t)$$

$$= N^{-1} \int d\mathbf{x} \, x \left[-V_x(t) \, \nabla_x + D_\perp(t) \, \nabla_x^2 \right] n(\mathbf{x},t) = V_x(t). \tag{11.23}$$

(The last equality results from an integration by parts). In the same way, we obtain from Eq. (11.22):[4]

$$d_t \langle x^2(t) \rangle = N^{-1} \int d\mathbf{x} \, x^2 \left[-V_x(t) \, \nabla_x + D_\perp(t) \, \nabla_x^2 \right] n(\mathbf{x},t)$$

$$= 2V_x(t) \int d\mathbf{x} \, x \, n(\mathbf{x},t) + 2D_\perp(t) \, N^{-1} \int d\mathbf{x} \, n(\mathbf{x},t)$$

$$= 2 \left[d_t \langle x(t) \rangle \right] \langle x(t) \rangle + 2 \, D_\perp(t). \tag{11.24}$$

We define now the *mean square displacement (MSD)* in the x-direction as:

$$\langle \delta x^2(t) \rangle = \langle x^2(t) \rangle - \langle x(t) \rangle^2 = \left\langle \left[x(t) - \langle x(t) \rangle \right]^2 \right\rangle. \tag{11.25}$$

We then obtain from (11.24):

$$D_\perp(t) = \frac{1}{2} \frac{d}{dt} \langle \delta x^2(t) \rangle \tag{11.26}$$

This beautiful *relation between the mean square displacement and the running diffusion coefficient* was derived by EINSTEIN 1905, 1906 for

---

[4] In all forthcoming formulae we use the abbreviated notation: $\delta x^2(t) \equiv [\delta x(t)]^2$.

a somewhat simpler situation. It appears that in most (*but not all!*) situations, the MSD behaves linearly in the limit of long time: $\langle \delta x^2(t) \rangle \sim 2D_\perp t$, $t \to \infty$, with a constant positive $D_\perp$. The latter is then identified with the ordinary diffusion coefficient of macroscopic theory. Eq. (11.26) becomes in this case:

$$D_\perp = \frac{1}{2t} \langle \delta x^2(t) \rangle, \quad t \to \infty \qquad (11.27)$$

The Einstein relation is most often quoted in this form; it should, however, be clear that this is only valid asymptotically. More generally, the diffusion coefficient is defined as:

$$D_\perp = \frac{1}{2} \frac{d}{dt} \langle \delta x^2(t) \rangle, \quad t \to \infty \qquad (11.28)$$

The argument developed above referred to diffusion in a single (x) dimension. It is easily generalized for diffusion in d dimensions with the MSD defined as: $\langle \delta |\mathbf{x}(t)|^2 \rangle = \langle \delta x_1^2(t) + ... + \delta x_d^2(t) \rangle$:

$$D(t) = \frac{1}{2d} \frac{d}{dt} \langle \delta |\mathbf{x}(t)|^2 \rangle. \quad (d \text{ dimensions}) \qquad (11.29)$$

We now return to our original problem of Sec. 11.1. Having calculated the MSD from the Langevin equation in Eq. (11.18), we obtain the running collisional perpendicular diffusion coefficient $\chi_\perp(t)$ as:[5]

$$\chi_\perp(t) = \int_0^t dt' R_{xx}(t'). \qquad (11.30)$$

The asymptotic diffusion coefficient is thus obtained in the limit $t \to \infty$:

$$\chi_\perp = \int_0^\infty dt' R_{xx}(t') = \int_0^\infty dt' \langle \overline{v_x(t_0) v_x(t_0 + t')} \rangle \qquad (11.31)$$

This equation is a special case of the celebrated *Green-Kubo formulae* expressing the transport coefficients as time-integrals of autocorrelation functions of microscopic fluxes, averaged in equilibrium (GREEN 1951, KUBO 1957, RESIBOIS & DE LEENER 1977, BALESCU 1997).

Let us show that Eq. (11.31) is equivalent to the general formulae (6.74), (6.75), as applied to the present case. The latter (disregarding the irrelevant term $\mathcal{L}^{Va}$) involves the Lagrangian autocorrelation of the drift velocity field $\delta \mathbf{V}^E[\mathbf{x}(t), t]$. In the present case we deal with the velocity of a point particle; thus, instead of the Lagrangian autocorrelation $\mathcal{L}_{xx}^{VV}(t)$, we

---

[5] We use the notation χ for the purely collisional (classical) diffusion coefficient, in order to distinguish it from the anomalous diffusion coefficient denoted later by D.

have to deal with the much simpler velocity autocorrelation $\langle v_x(0)\, v_x(t) \rangle$. Assuming that the reference distribution function $F^\alpha(\mathbf{x}, \mathbf{v}, t)$ is a local Maxwellian with constant temperature, we have $F^\alpha(\mathbf{x}, \mathbf{v}, t) = n_0(\mathbf{x})\varphi_M^\alpha(\mathbf{v})$ [Eq. (6.79) with $V_{T\alpha} = const$], Eq. (6.74) reduces to:

$$\Gamma_x(t) = -\int_0^t dt'\, \langle v_x(0)\, v_x(t') \rangle\, \nabla_x n_0(X)\, \varphi_M(\mathbf{v})$$

$$= -\left\{ \int_0^t dt'\, \langle v_x(0)\, v_x(t') \rangle\, \varphi_M(\mathbf{v}) \right\}\, \nabla_x n_0(\mathbf{x}),$$

where we recognize the velocity autocorrelation $R_{xx}(t)$ defined by (11.31). Comparing this expression with the macroscopic relation (11.20) for the particle flux [with $\mathbf{V}(t) = 0$]:

$$\Gamma_x(t) = -\chi_{xx}(t)\, \nabla_x n_0(\mathbf{x}),$$

we find that the running diffusion coefficient $\chi_\perp(t) \equiv \chi_{xx}(t)$ is precisely given by Eqs. (11.30), (11.31).

The properties of the velocity autocorrelation function determine the type of transport process going on in the system. If the autocorrelation function ensures the convergence of the time integral, the process is *diffusive*: this is the most frequent case met in practice. If the integral in (11.31) diverges, we deal with a *superdiffusive* process. It may also happen that the integral vanishes: we then have a *subdiffusive* process. Note that the latter case requires the velocity autocorrelation to be positive in some domain and negative in the complementary region, and that the areas under these two regions be equal: this is a very delicate condition indeed.

Using Eq. (11.13) we obtain explicitly:

$$\chi_\perp(t) = \tfrac{1}{2} V_T^2\, \frac{\nu + [\Omega \sin(\Omega t) - \nu \cos(\Omega t)] \exp(-\nu t)}{\nu^2 + \Omega^2}. \tag{11.32}$$

The asymptotic perpendicular diffusion coefficient is:

$$\chi_\perp = \tfrac{1}{2} V_T^2\, \frac{\nu}{\nu^2 + \Omega^2}. \tag{11.33}$$

The parallel running diffusion coefficient is obtained by a similar calculation from Eq. (11.14):

$$\chi_\parallel(t) = \tfrac{1}{2} V_T^2\, \frac{1 - \exp(-\nu t)}{\nu}, \tag{11.34}$$

and the asymptotic parallel diffusion coefficient:

$$\chi_\parallel = \frac{V_T^2}{2\nu}. \tag{11.35}$$

Eqs. (11.33), (11.35) are precisely identical to the expressions obtained as a first approximation from classical kinetic theory (BALESCU 1988 a). This result proves the following remarkable fact: *The collision model based on a Gaussian white noise for the fluctuating friction force yields a result that is equivalent, in first approximation, to the solution of the kinetic equation in the hydrodynamic regime.* The stochastic treatment based on the Langevin equations is, indeed, completely independent of any kinetic equation; but it cannot provide us with the dependence of the collision frequency on the physical parameters: ν is here a free parameter. The previous statement justifies the use of stochastic methods in problems where the kinetic equation becomes prohibitively complicated.

We conclude this section with a discussion of the dependence of the diffusion coefficients on the magnetic field (BALESCU 1988). Clearly, the parallel coefficient is independent of Ω (hence of B): the Lorentz force is zero in the direction of the field. The perpendicular diffusion coefficient is a strongly decreasing function of Ω. In the limit of a very strong magnetic field (*i.e.*: $\Omega/\nu \gg 1$) it tends to zero like:

$$\chi_\perp \simeq \frac{V_T^2 \nu}{2\Omega^2}, \quad \Omega/\nu \gg 1. \tag{11.36}$$

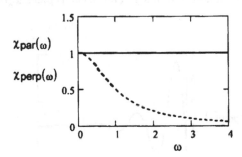

Figure 11.2. Collisional diffusion coefficients (scaled with $\chi_\parallel$) vs. reduced Larmor frequency $\omega = \Omega/\nu$.

This behaviour reflects the fact that the particles spiral around the magnetic field lines: for a very strong field they tend to stick to the field lines. This also explains the different dependence of the diffusion coefficients on the collision frequency. In the parallel direction the collisions hinder the motion of the particles (through friction), hence the diffusion coefficient is inversely proportional to the collision frequency, $\chi_\parallel \propto \nu^{-1}$. In the perpendicular direction, the particles can move across the (strong) magnetic field only if the collisions decorrelate them from the field lines,

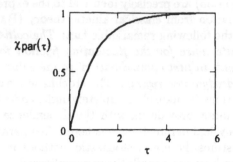

Figure 11.3. Parallel running diffusion coefficient $\chi_\parallel(\tau)$ (scaled with $\chi_\parallel$) vs. $\tau = \nu t$.

hence $\chi_\perp \propto \nu$. This behaviour, illustrated in Fig. 11.2, is very important for understanding the behaviour of a plasma in a magnetic field.

We show in Fig. 11.3 the behaviour of the parallel running diffusion coefficient: starting linearly from zero, it tends exponentially towards its asymptotic constant, positive value. The characteristic time involved is, indeed, the inverse collision frequency. This is the typical *diffusive behaviour* described above.

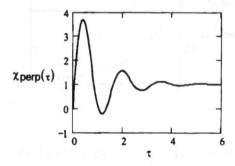

Figure 11.4. Perpendicular running diffusion coefficient $\chi_\perp(\tau)$ (scaled with $\chi_\perp$) vs. $\tau = \nu t$. $\Omega/\nu = 4$.

The perpendicular running diffusion coefficient shown in Fig. 11.4 also tends to an asymptotic positive value; but the behaviour is here oscillatory. In particular, this example shows that nothing prevents the *running* diffusion coefficient from being negative for some limited time.

The A-Langevin equations provide the most exhaustive treatment of the diffusion coefficients based on stochastic equations. The limitations of the method are also apparent. The A-Langevin equations cannot yield the complete set of transport coefficients (including thermal conductivity, thermodiffusion, etc.). Nevertheless, in more complicated problems they provide valuable informations about the transport mechanisms.

11.3   V-Langevin Equations

When considering more complicated problems than the purely collisional (classical) diffusion, the A-Langevin equations can no longer be integrated analytically. It is therefore very useful to note that one can "short-circuit" the first steps of the treatment of the collisional noise by introducing an alternative type of Langevin equations. The price to be paid is that some information has to be injected into the theory from outside, rather than being derived from more fundamental principles.

In order to illustrate the idea, we consider here a simpler limiting situation of a *very strong constant magnetic field*. In that case, the equations of motion can be written in the *drift approximation* described in detail in Chap. 4. We even consider here the extreme case in which the Larmor radius is vanishingly small and the position of the particles can be identified with the position of the guiding centres: $\mathbf{q}(t) \approx \mathbf{Y}(t)$. In the case of a constant magnetic field there is no motion at all in the perpendicular direction, but only a uniform motion in the parallel direction. To this deterministic component is, however, superposed a random velocity due to the collisions, which produce diffusion both in the parallel and in the perpendicular directions. But, as seen above, the coefficient $\chi_\perp$ is very much smaller than $\chi_\parallel$ in the strong field limit (see Fig. 11.2). We therefore set $\chi_\perp \approx 0$, thus excluding completely any motion in the perpendicular direction. The motion of a test particle (*i.e.*, test guiding centre) is now described by the trivially simple equation:

$$\frac{dZ(t)}{dt} = v_\parallel(t). \tag{11.37}$$

The parallel velocity $v_\parallel(t)$ appearing here is a *random function of time*, that must be defined statistically. We assume that it is a *Gaussian process*, which is therefore completely defined by its first two moments. For simplicity, we assume that the average parallel velocity is zero. As for the second moment, *i.e.*, the velocity autocorrelation, we take its value from the result of the A-Langevin equation (11.14):

$$\langle v_\parallel(t) \rangle = 0, \quad \langle v_\parallel(t_0)\, v_\parallel(t_0 + t) \rangle = R_{zz}(t) = \tfrac{1}{2} V_T^2\, e^{-\nu|t|}. \tag{11.38}$$

Thus, in this "V-Langevin equation", the form of the velocity autocorrelation function is not *derived*, but rather is *injected* from outside. We note that the velocity is not delta-correlated like the acceleration in the A-Langevin equation. Rather, its autocorrelation has a finite range in time, of order ν^{-1}, *i.e.*, the inverse collision frequency, or the average time between two collisions, called the relaxation time in kinetic theory.

The solution of this V-Langevin equation (for a given realization) is trivially simple:

$$Z(t) = Z(0) + \int_0^t dt_1 \, v_{\|}(t_1). \tag{11.39}$$

The average position is thus:

$$\langle Z(t) \rangle = Z(0). \tag{11.40}$$

The mean square displacement is:

$$\langle \delta Z^2(t) \rangle = \int_0^t dt_1 \int_0^t dt_2 \, \langle v_{\|}(t_1) \, v_{\|}(t_2) \rangle$$
$$= V_T^2 \int_0^t dt_1 \, (t - t_1) \, e^{-\nu t_1}. \tag{11.41}$$

We thus recover the result (11.34):

$$\chi_{\|}(t) = \tfrac{1}{2} \, \partial_\tau \, \langle \delta Z^2(t) \rangle = \frac{V_T^2}{2\nu} \, (1 - e^{-\nu t}). \tag{11.42}$$

The diffusion coefficient was obtained in a much simpler way, because two steps in the derivation from the A-Langevin equation were avoided: the average over the stochastic acceleration and the average over the initial velocities. In the present V-Langevin equation, the treatment of the collisional noise involves a single averaging process over the fluctuating velocity. This simplification will be precious in more complicated problems.

A natural generalization of the V-Langevin equation will appear as a paradigm for a large number of problems of turbulent plasma physics and also of neutral fluid dynamics. The generalization concerns in the first place a passage to two or three dimensions, and especially the consideration of a *velocity field* depending not only on time, but also on the instantaneous position of the test particle.

We thus define the prototype of a **V-Langevin equation** as a *set of first-order stochastic differential equations for the position of a test particle in d dimensions, moving in a fluctuating velocity field.*[6]

$$\frac{d\mathbf{Y}(t)}{dt} = \mathbf{v}[\,\mathbf{Y}(t), t], \tag{11.43}$$

---

[6] In comparison with Chap. 6.1, we use here and in forthcoming developments the more compact notation: $\mathbf{Y}(t) \equiv \mathbf{Y}(t|\mathbf{Y}_0, 0)$.

to be solved with the deterministic initial condition:[7]

$$\mathbf{Y}(0) = \mathbf{Y}_0. \tag{11.44}$$

The velocity field is defined statistically as a coloured homogeneous and stationary Gaussian random process, with the following two Eulerian moments:

$$\langle \mathbf{v}(\mathbf{Y}, t) \rangle = \mathbf{V}, \tag{11.45}$$

$$\langle \delta \mathbf{v}(\mathbf{Y}, t_0) \, \delta \mathbf{v}(\mathbf{Y} + \mathbf{R}, t_0 + t) \rangle = \mathbf{C}(R, |t|). \tag{11.46}$$

To this fluctuating field may be added a deterministic component of the velocity.

The V-Langevin equations appear in many problems of plasma physics, each one defined by a given choice of the statistical quantities $\mathbf{V}$ and $\mathbf{C}$. We have already met with a first example: the *drift motion of guiding centres in a fluctuating electric field* (with the parallel acceleration δa^α neglected) introduced the equations of motion (6.8) - (6.12) as characteristic equations of the gyrokinetic equation. In Chap. 6, however, the statistical definition of the velocity field was left open at this stage: this point will be further discussed in the next section. The V-Langevin equations for the electrostatic drift motion will be further studied in Chaps. 14 and 16. A physically different problem that also leads to V-Langevin equations is the *motion of charged particles in a fluctuating magnetic field*: this will be studied in the next sections of the present chapter (and further in Chap. 15).

11.4 Statistical Definition of Magnetic Field Fluctuations

We now introduce a problem that is in the mainline of the present book: *anomalous transport produced by magnetic field fluctuations*. Instead of the constant magnetic field considered in the previous chapters, we assume that this field has a small fluctuating component. The origin of these fluctuations may be either external (irregularities in the coils producing the field) or, more interestingly, internal, due to instabilities that destroy the regular magnetic surfaces in a confined plasma. Typically, a *tearing instability* leads to the formation of magnetic islands which may overlap, thus creating regions where the magnetic field is stochastic, and must be described statistically.

---

[7] A generalization for the case of a random initial condition can also be considered (see *e.g.* the treatment of the A-Langevin equation) but will not be used here.

More specifically, we consider a strong "reference" magnetic field $\mathbf{B}_0$, constant in space and time, which undergoes small fluctuations in a perpendicular direction (the fluctuations in a direction parallel to $\mathbf{B}_0$ are masked by the strong reference field and may be neglected). The total magnetic field is thus expressed, in a shearless slab geometry with the z-axis directed along the reference field, as: $\mathbf{B} = B_0(\mathbf{e}_z + \overline{\beta}\mathbf{b})$. Here $\overline{\beta}$ is a dimensionless number measuring the characteristic amplitude of the fluctuations, relative to the main magnetic field B_0. Their space- and time-dependence is described by the function $\mathbf{b}(\mathbf{X}, Z, t)$. As will be seen below, it is convenient to consider separately the "parallel" coordinate Z, directed along the reference field $\mathbf{B}_0$, and the "perpendicular" coordinates forming a two-component vector $\mathbf{X} = (X, Y)$.[8] In general, the fluctuation $\mathbf{b}$ may depend in an arbitrary way on space and time; here, however, as in most existing works, this quantity is assumed to be time-independent ("*frozen turbulence*"). Clearly, the quantity $\mathbf{b}$ is dimensionless; hence it may only depend on the coordinates scaled with appropriate characteristic lengths λ_r ($r = x, y, z$). We impose some reasonable symmetry requirements; one of them is *gyrotropy*, *i.e.*, isotropy in the plane perpendicular to the unperturbed magnetic field (this is equivalent to assuming the system to possess cylindrical symmetry). It then follows that the two characteristic lengths in the perpendicular direction must be equal to each other: $\lambda_x = \lambda_y = \lambda_\perp$; as for the third length, we denote it by $\lambda_z = \lambda_\parallel$. These two characteristic lengths will be called the *perpendicular (resp. parallel) correlation lengths* (for a reason to be clarified below). The scaled coordinates are now defined as follows in terms of the dimensional coordinates X, Y, Z:

$$x = \frac{X}{\lambda_\perp}, \quad y = \frac{Y}{\lambda_\perp}, \quad z = \frac{Z}{\lambda_\parallel}. \tag{11.47}$$

Summarizing, we assume the following form of the magnetic field:

$$\mathbf{B}(X, Y, Z) = B_0[\mathbf{e}_z + \overline{\beta}b_x(x, y, z)\,\mathbf{e}_x + \overline{\beta}b_y(x, y, z)\,\mathbf{e}_y] \tag{11.48}$$

The magnetic field must satisfy the constraint (imposed by Maxwell's equations):

$$\nabla \cdot \mathbf{B} = B_0\overline{\beta}\,(\nabla \cdot \mathbf{b}) = 0. \tag{11.49}$$

In order to satisfy automatically this constraint, we represent the fluctuating field in terms of a vector potential $\mathbf{A}(\mathbf{X}, Z) = \overline{\beta}\lambda_\perp\widetilde{\mathbf{a}}(\mathbf{x}, z)$ having only a z-component:

$$\widetilde{\mathbf{a}}(\mathbf{X}, Z) = \psi(x, z)\,\mathbf{e}_z. \tag{11.50}$$

---

[8] For brevity, we prefer here the shorter notations $\mathbf{X}$, $\mathbf{x}$ to the usual, heavier notations $\mathbf{X}_\perp$, $\mathbf{x}_\perp$.

From the well-known relation: $\mathbf{B}(\mathbf{X}, Z) = \nabla \times \mathbf{A}(\mathbf{X}, Z)$ follows:

$$b_x(\mathbf{x}, z) = \frac{\partial \psi(\mathbf{x}, z)}{\partial y}, \quad b_y(\mathbf{x}, z) = -\frac{\partial \psi(\mathbf{x}, z)}{\partial x}. \quad (11.51)$$

It is interesting for the forthcoming problems to write down the equations for the magnetic field lines. These are derived from the well-known geometric equations: $dX/\bar{\beta} B_0 b_x = dY/\bar{\beta} B_0 b_y = dZ/B_0$, which are rewritten by taking Z as an independent parameter:

$$\frac{dX(Z)}{dZ} = \bar{\beta} b_x[\mathbf{x}(z), z],$$

$$\frac{dY(Z)}{dZ} = \bar{\beta} b_y[\mathbf{x}(z), z]. \quad (11.52)$$

The form (11.48) holds in a given realization. But the quantity $b(x, y, z)$ [hence, also the magnetic field $\mathbf{B}(X, Y, Z)$] is a stochastic quantity; it must therefore be defined statistically. In order to cope with the zero-divergence constraint, we introduce the primary definition at the level of the potential $\psi(\mathbf{x}, z)$. Our main *a priori* assumption will be the following: the fluctuating potential $\psi(x, y, z)$ is a *Gaussian stochastic process*, supposed to be *spatially homogeneous* and *gyrotropic*. This process is thus defined by its first two moments:

$$\langle \psi(\mathbf{x}, z) \rangle = 0,$$
$$\langle \psi(\mathbf{x}_0, z_0)\, \psi(\mathbf{x}_0 + \mathbf{x}, z_0 + z) \rangle = \langle \psi(\mathbf{0}, 0)\, \psi(\mathbf{x}, z) \rangle = \overline{A}(\mathbf{x}, z), \quad (11.53)$$

where the scalar function $\overline{A}(\mathbf{x}, z)$ depends only on the **length** of the perpendicular component of the relative distance, $r = \sqrt{x^2 + y^2}$, and on its z-component. $\bar{\beta}^2 \lambda_\perp^2 \overline{A}(r, z)$ is thus the *Eulerian autocorrelation function of the potential*. Unless otherwise specified, it will always be assumed that this autocorrelation function is *factorized*, into a factor depending only on $\mathbf{x}$ and a factor depending on z; thus:

$$\langle \psi(\mathbf{0}, 0)\, \psi(\mathbf{x}, z) \rangle = A(\mathbf{x})\, B_\parallel(z). \quad (11.54)$$

It will be seen that this property appears rather "naturally" in applications.

A useful role will be played by the Fourier transform of the of the correlation function, expressed in terms of the dimensionless wave vector $\mathbf{k} = \lambda_\perp \mathbf{K}$, $k_\parallel = \lambda_\parallel K_z$. Defining:

$$\psi(\mathbf{x}, z) = \int d\mathbf{k}\, dk_\parallel\; e^{i\mathbf{k} \cdot \mathbf{x} + ik_\parallel z} \widehat{\psi}(\mathbf{k}, k_\parallel),. \quad (11.55)$$

the Eulerian autocorrelation becomes:

$$\left\langle \widehat{\psi}(\mathbf{k},k_\parallel)\widehat{\psi}(\mathbf{k}',k_\parallel') \right\rangle = \widehat{A}(k)\,\widehat{B}_\parallel(k_\parallel)\,\delta(\mathbf{k}+\mathbf{k}')\delta(k_\parallel+k_\parallel'). \tag{11.56}$$

In the present chapter we use a specific model, in order to obtain explicit results. The two factors of the Eulerian potential autocorrelation function are taken to be:

$$A(\mathbf{x}) = \exp\left[-\tfrac{1}{2}\left(x^2+y^2\right)\right], \quad B_\parallel(z) = \exp\left(-\tfrac{1}{2}z^2\right). \tag{11.57}$$

The corresponding Fourier representation is:

$$\widehat{A}(k) = (2\pi)^{-1}\exp\left(-\tfrac{1}{2}k^2\right), \quad \widehat{B}_\parallel(k_\parallel) = (2\pi)^{-1/2}\exp\left(-\tfrac{1}{2}k_\parallel^2\right). \tag{11.58}$$

It now clearly appears that the two characteristic lengths: the *perpendicular correlation length*, $\lambda_\perp$, and the *parallel correlation length*, $\lambda_\parallel$ measure the spatial range of the correlations in the two directions (*i.e.*, for $R \gg \lambda_\perp$, hence $r \gg 1$, etc., the magnetic potential correlation is negligibly small).

We now consider other types of Eulerian correlations that will be needed below. The mixed correlations of potential and magnetic field are defined as:

$$\langle\psi(\mathbf{0},0)\,b_n(\mathbf{x},z)\rangle = C_{\psi n}(\mathbf{x})B_\parallel(z),$$
$$\langle b_n(\mathbf{0},0)\,\psi(\mathbf{x},z)\rangle = C_{n\psi}(\mathbf{x})B_\parallel(z), \quad n=x,y. \tag{11.59}$$

Using Eqs. (11.51) and (11.54) we relate these quantities to the potential correlation:

$$C_{\psi x} = -C_{x\psi} = \frac{\partial}{\partial y}A(x,y), \quad C_{\psi y} = -C_{y\psi} = -\frac{\partial}{\partial x}A(x,y). \tag{11.60}$$

Next, we consider the four possible Eulerian correlations of the magnetic field components: they form a 2×2 matrix $\mathcal{B}_{mn}$ $(m,n=x,y)$ which is easily derived from Eqs. (11.50) and (11.54). The (Gaussian) statistics of the magnetic field is thus defined by:

$$\langle b_m(\mathbf{x})\rangle = 0, \quad \langle b_m(\mathbf{0},0)\,b_n(\mathbf{x},z)\rangle = \mathcal{B}_{mn}(\mathbf{x})\,B_\parallel(z), \quad m,n=x,y. \tag{11.61}$$

The function $\mathcal{B}_{mn}$ depends only on the relative perpendicular distance $\mathbf{x}$ (homogeneous turbulence) and is related to the potential autocorrelation as follows:

$$B_{xx} = -\frac{\partial^2}{\partial y^2} A(x,y), \quad B_{yy} = -\frac{\partial^2}{\partial x^2} A(x,y),$$

$$B_{xy} = B_{yx} = \frac{\partial^2}{\partial x \partial y} A(x,y). \tag{11.62}$$

In the special case defined by Eq. (11.57), we find (WANG et al. 1995):

$$C_{\psi x} = -C_{x\psi} = -y \exp(-\tfrac{1}{2} r^2),$$

$$C_{\psi y} = -C_{y\psi} = x \exp(-\tfrac{1}{2} r^2), \tag{11.63}$$

and:

$$B_{xx} = (1 - y^2) \exp(-\tfrac{1}{2} r^2),$$

$$B_{yy} = (1 - x^2) \exp(-\tfrac{1}{2} r^2),$$

$$B_{xy} = B_{yx} = xy \exp(-\tfrac{1}{2} r^2). \tag{11.64}$$

The prefactor of the exponential is quite important: it ensures the compatibility of this quantity with the constraint (11.49). Note that, due to this factor (hence, to the divergence-free nature of the magnetic field), the magnetic field autocorrelation is *not* a diagonal tensor. [9]

The Fourier transform of the magnetic field correlation tensor has the following form:

$$\left\langle \widehat{b}_m(\mathbf{k}, k_\parallel) \, \widehat{b}_n(\mathbf{k}', k_\parallel') \right\rangle = \widehat{B}_{mn}(\mathbf{k}) \, \widehat{B}_\parallel(k_\parallel) \, \delta(\mathbf{k} + \mathbf{k}') \delta(k_\parallel + k_\parallel'), \tag{11.65}$$

with:

$$\widehat{B}_{mn}(\mathbf{k}) = (k^2 \delta_{mn} - k_m k_n) \widehat{A}(\mathbf{k}),$$

$$\widehat{B}_\parallel(k_\parallel) = (2\pi)^{-1/2} \exp(-\tfrac{1}{2} k_\parallel^2). \tag{11.66}$$

11.5   Langevin Equation for Test Particle in a Fluctuating Magnetic Field

We now single out a *test particle* in a plasma, moving in the magnetic field (11.48). If the average magnetic field B_0 is sufficiently strong, the position

---

[9] It may be noted that in several comparable works (e.g., ISICHENKO 1991, RAX & WHITE 1992) a diagonal correlation tensor is assumed, thus violating the condition (11.49); the non-diagonal character is correctly used in ZIMBARDO et al. 1984.

of the particle can be identified with the position of its guiding centre $(\mathbf{X}, Z) = (X, Y, Z)$, whose motion is determined by the superposition of two factors:

- a) The *dominant part of the guiding centre motion, parallel to the perturbed field.* The effect of the perpendicular drift motion (whose velocity is small compared to the parallel velocity) will be studied in Chap. 15.
- b) The *collisions with the particles of the background plasma.* The latter are modelled by assuming that they produce a random variation of the velocity, with components $v_{\parallel}$ ($\mathbf{v}_{\perp}$) in the parallel (perpendicular) direction. This process is defined statistically in the same way as in the collisional V-Langevin equation, (11.38).

The equation of motion of the test particle is then obtained from the equations of the magnetic field lines (11.52). The coordinates of the guiding centre moving along this line obey the following equations of motion:

$$\frac{d\mathbf{X}(t)}{dt} = \overline{\beta}\, \mathbf{b}[\mathbf{x}(t), z(t)]\, v_{\parallel}(t) + \mathbf{v}_{\perp}(t), \qquad (11.67)$$

$$\frac{dZ(t)}{dt} = v_{\parallel}(t), \qquad (11.68)$$

where $\mathbf{X}(t) \equiv \mathbf{X}[Z(t)]$. These equations, completed with the statistical definitions of the magnetic field and of the velocity, are a typical case of *V-Langevin equations*. There is, however, an important difference between these equations and Eq. (11.37) of the simple collisional theory. Instead of the single random force of the latter, we deal here with *three random driving factors*: $\mathbf{b}(\mathbf{x}, z), v_{\parallel}(t), \mathbf{v}_{\perp}(t)$; in other words, we have here a *triply stochastic problem.* Moreover, we note that the collisions introduce an *additive noise*, whereas the magnetic fluctuations produce a *multiplicative noise.*

The main difficulty of the present problem lies in Eq. (11.67): in each realization, this is a highly nonlinear equation, because of the dependence of b on $\mathbf{x}(t), z(t)$. For this reason, the analytical treatment of this equation requires some restrictive approximations (KADOMTSEV & POGUTSE 1979, RECHESTER & ROSENBLUTH 1978, KROMMES et al. 1983, ISICHENKO 1991, WANG et al., 1995). This general problem will be discussed in Chap. 15. Meanwhile, we demonstrate the use of the V-Langevin equation by simplifying drastically the problem: we construct a model which, although being academic, has the advantage of being analytically soluble. We introduce the following additional assumptions in Eqs. (11.67) - (11.68).

- c) The perpendicular random collisional velocity $\mathbf{v}_{\perp}(t)$ is set equal to zero. This implies that the collisional diffusion coefficient $\chi_{\perp}$ is assumed negligibly small. (This assumption will be lifted in Chap. 15.)

- d) The *perpendicular correlation length is assumed to be infinite:* $\lambda_\perp = \infty$ (or, at least, very large compared to the parallel one).

Assumption d) is the most crucial one. It implies that the scaled coordinates $\mathbf{x} = \lambda_\perp^{-1} \mathbf{X}$ are set equal to zero [see Eq. (11.47)]. As a result, the fluctuating field only depends on the parallel coordinate: $\mathbf{b}(\mathbf{x},z) \to \mathbf{b}(z)$. In this case it is easily seen that Eq. (11.67) becomes a *linear differential equation* (for a given realization of the magnetic field). As will be seen below, Assumption d) implies that the particles cannot decorrelate from the field lines. Indeed, all decorrelation mechanisms, *i.e.*, perpendicular drift motions and perpendicular collisions, have been neglected in the present model.

The V-Langevin equations (11.67), (11.68) now reduce to:

$$\frac{dX(t)}{dt} = \overline{\beta}\, b_x[z(t)]\, v_\parallel(t), \tag{11.69}$$

$$\frac{dY(t)}{dt} = \overline{\beta}\, b_y[z(t)]\, v_\parallel(t) \tag{11.70}$$

$$\frac{dZ(t)}{dt} = v_\parallel(t). \tag{11.71}$$

The initial condition is supposed to be deterministic; thus, *in all realizations*:

$$X(0) = X_0, \quad Y(0) = Y_0, \quad Z(0) = Z_0. \tag{11.72}$$

Our model thus consists of a set of three *doubly stochastic equations*: these involve the fluctuating magnetic field $\mathbf{b}(z)$ and the fluctuating parallel velocity $v_\parallel(t)$ modelling the collisions. The total magnetic field is represented as:

$$\mathbf{B}(z) = B_0 \left[\mathbf{e}_z + \overline{\beta}\, b_x(z)\, \mathbf{e}_x + \overline{\beta}\, b_y(z)\, \mathbf{e}_y \right]. \tag{11.73}$$

The fluctuating field $\mathbf{b}(z)$ is defined statistically by assuming that it is a homogeneous, gyrotropic and stationary Gaussian process, whose first two moments are derived from Eqs. (11.61), (11.64) by setting $\mathbf{x} = 0$:

$$\langle \mathbf{b}(z) \rangle_b = 0, \quad \langle b_m(0)\, b_n(z) \rangle_b = \delta_{mn}\, \mathcal{B}_\parallel(z) \quad m, n = x, y. \tag{11.74}$$

Whenever necessary for clarity, a subscript on a bracket denotes the quantity that is averaged over. Thus, in the present simple model, the magnetic field Eulerian correlation reduces to a diagonal 2×2 tensor, proportional to the unit tensor. The function $\mathcal{B}_\parallel(z)$ was defined in Eq. (11.57).

For the forthcoming calculations, it is useful to introduce the Fourier representation of the fluctuating field:

$$b_m(z) = \int dk\, e^{ikz} \hat{b}_m(k). \qquad (11.75)$$

The correlation of the Fourier components is:

$$\left\langle \hat{b}_m(k)\hat{b}_n(k') \right\rangle_b = \delta_{mn}\, \hat{B}_\parallel(k)\, \delta(k+k'). \qquad (11.76)$$

The function $\hat{B}_\parallel(k)$ is called the *spectral density of the magnetic fluctuations*, or, briefly, the *spectrum*: it was defined in Eq. (11.66).

The parallel fluctuating velocity $v_\parallel(t)$ is modelled as in Eq. (11.38) in order to represent the effect of the collisions: it is assumed to be a stationary Gaussian process with:

$$\langle v_\parallel(t) \rangle_\parallel = 0 \qquad \langle v_\parallel(t_0)\, v_\parallel(t_0+t) \rangle_\parallel = \tfrac{1}{2} V_T^2 \exp(-\nu|t|), \qquad (11.77)$$

where V_T is the thermal velocity, and ν is the collision frequency; the subscript $\parallel$ denotes an average over $v_\parallel$.

Finally, it is assumed that the magnetic fluctuations and the velocity fluctuations are statistically independent. This implies the following form for the Eulerian correlation of the magnetic field at a *fixed position* and the velocity at an arbitrary time:

$$\langle b_m(z)\, v_\parallel(t) \rangle_{b,\parallel} = 0, \qquad (11.78)$$

where the two subscripts denote an average over both b and $v_\parallel$. Our problem is now completely specified, and we are ready for its solution.

11.6 Magnetic Subdiffusion

We now solve the stochastic initial value problem defined by Eqs. (11.69) - (11.78) (BALESCU et al. 1994). We will be mainly interested in the motion of the particles in the X-direction, due to the combined effect of the fluctuating magnetic field $b(z)$ and of the parallel collisions. We are looking, more specifically, for an expression of the cross-field running diffusion coefficient, to be denoted by $D^\infty(t) \equiv D_{xx}(t)$.[10]

The guiding centre position coordinate is split into an average part $\mathbf{X}(t) = \langle \mathbf{X}(t) \rangle_{b,\parallel} + \delta\mathbf{X}(t)$. With the deterministic initial condition (11.72) we find, in a given realization:

$$\langle Z(t) \rangle_{b,\parallel} = Z_0, \qquad \langle X(t) \rangle_{b,\parallel} = X_0,$$

$$\delta Z(t) = \int_0^t dt_1\, v_\parallel(t_1), \quad \delta X(t) = \overline{\beta} \int_0^t dt_1\, b_x[z(t_1)]\, v_\parallel(t_1). \qquad (11.79)$$

---

[10] The superscript "∞" suggests the infinite perpendicular correlation length $\lambda_\perp = \infty$.

The quantity of interest is the mean square displacement (MSD) in the X-direction. We thus square the last expression and average over both b_x and $v_\parallel$:

$$\langle \delta X^2(t) \rangle = \overline{\beta}^2 \int_0^t dt_1 \int_0^t dt_2 \, \langle b_x[z(t_1)] \, v_\parallel(t_1) \, b_x[z(t_2)] \, v_\parallel(t_2) \rangle_{b,\parallel}. \quad (11.80)$$

We are now confronted with the problem of Lagrangian correlations, discussed in some detail in Sec. 6.4. In the present case [with $b_x = b_x(z)$] the problem can be solved exactly, as will be shown below. Using Eqs. (11.75), (11.76), (11.79), we find:

$$\langle \delta X^2(t) \rangle = \overline{\beta}^2 \int_0^t dt_1 \int_0^t dt_2 \int dk_1 \int dk_2$$
$$\times \left\langle \widehat{b}(k_1) \widehat{b}(k_2) \exp\left\{ \lambda_\parallel^{-1}[ik_1 Z(t_1) + ik_2 Z(t_2)] \right\} v_\parallel(t_1) v_\parallel(t_2) \right\rangle_{b,\parallel}.$$
$$(11.81)$$

We now note that $Z(t)$, as given by Eq. (11.79), does not depend on b; hence, because of Eq. (11.78), the average of the two factors $\widehat{b}$ can be factored out from the rest and, using (11.76), we find:

$$\langle \delta X^2(t) \rangle = \overline{\beta}^2 \int_0^t dt_1 \int_0^t dt_2 \int dk \widehat{B}_\parallel(k) \, Z(k, t_1, t_2)$$
$$= 2\overline{\beta}^2 \int_0^t dt_1 \int_0^{t_1} dt_2 \int dk \widehat{B}_\parallel(k) \, Z(k, t_1 - t_2),$$

where $\widehat{B}_\parallel(k)$ is the spectrum defined in Eq. (11.66). Note that the function Z depends only on the difference $t_1 - t_2$ (as shown below); we thus find:

$$\langle \delta X^2(t) \rangle = 2\overline{\beta}^2 \int_0^t dt' \int dk \, (t - t') \, \widehat{B}_\parallel(k) \, Z(k, t'), \quad (11.82)$$

with:

$$Z(k, t) = \left\langle \exp\left[i\lambda_\parallel^{-1} k \int_{t_0}^{t_0+t} dt_1 \, v_\parallel(t_1) \right] v_\parallel(t_0 + t) \, v_\parallel(t_0) \right\rangle_\parallel. \quad (11.83)$$

Next, we obtain from (11.26) the expression of the perpendicular running diffusion coefficient $D^\infty(t)$ in the limit $\lambda_\perp \to \infty$:

$$D^\infty(t) = \overline{\beta}^2 \int_0^t dt_1 \int dk\, \widetilde{B}_\parallel(k)\, \mathcal{Z}(k, t_1). \qquad (11.84)$$

The integrand in this expression has a simple physical meaning. Comparing this equation with Eq. (11.30), we see that the integrand is the Lagrangian autocorrelation of the x-components of the velocity (in a stationary turbulence):

$$\mathcal{L}_{xx}(t) = \langle v_x[z(0), 0]\, v_x[z(t), t] \rangle = \overline{\beta}^2 \int dk\, \widehat{B}_\parallel(k)\, \mathcal{Z}(k, t). \qquad (11.85)$$

This interpretation is confirmed by starting from (11.80), and recalling that the x-component of the velocity is defined by the V-Langevin equation (11.69) as: $v_x[z(t), t] = \overline{\beta}\, b_x[z(t)]\, v_\parallel(t)$. Eq. (11.84) is also consistent with the more general expression (6.74) of the particle flux in terms of the Lagrangian drift velocity autocorrelation $\mathcal{L}^{VV}(t)$.

The function $\mathcal{Z}(k, t)$ can be calculated exactly, as follows:

- Expand the exponential in Eq. (11.83) in a Taylor series;
- express the resulting high-order moments as sums of products of second-order moments, using the Gaussian character of $v_\parallel(t)$;
- evaluate the integrals, by using the following results, derived from Eq. (11.77):

$$\int_{t_0}^{t_0+t} dt_1\, \langle v_\parallel(t_0 + t)\, v_\parallel(t_1) \rangle_\parallel = \chi_\parallel\, e^{\nu t} \varphi(\nu t);$$

$$\int_{t_0}^{t_0+t} dt_1\, \langle v_\parallel(t_0)\, v_\parallel(t_1) \rangle_\parallel = \chi_\parallel\, \varphi(\nu t);$$

$$\int_{t_0}^{t_0+t} dt_1 \int_{t_0}^{t_0+t} dt_2\, \langle v_\parallel(t_1)\, v_\parallel(t_2) \rangle_\parallel = 2\frac{\chi_\parallel}{\nu}\, \mu(\nu t); \qquad (11.86)$$

where $\chi_\parallel$ is the parallel collisional diffusion coefficient (11.35);

- re-sum the resulting series, which has an obvious exponential form:

$$\mathcal{Z}(k; t)$$

$$= \left\{ \nu \chi_\parallel \exp(-\nu|t|) - \lambda_\parallel^{-2} k^2 \chi_\parallel^2\, \varphi^2(\nu t) \right\} \exp\left[-\frac{k^2 \chi_\parallel}{\lambda_\parallel^2\, \nu}\, \mu(\nu t) \right]. \qquad (11.87)$$

Here the following functions were introduced:

$$\varphi(x) = 1 - e^{-|x|}, \quad \mu(x) = x - \varphi(x). \tag{11.88}$$

After substitution of Eqs. (11.87) and (11.66) into (11.82) the integration over k is elementary; we are left with:

$$\langle \delta X^2(t) \rangle = 2 \int_0^t dt_1 \, (t - t_1) \, \mathcal{L}^\infty(\nu t_1), \tag{11.89}$$

with:

$$\mathcal{L}^\infty(\tau) = \epsilon \frac{1}{\left[1 + 2\overline{\chi}_\parallel \, \mu(\tau)\right]^{1/2}} \left\{ e^{-|\tau|} - \overline{\chi}_\parallel \frac{\varphi^2(\tau)}{1 + 2\overline{\chi}_\parallel \, \mu(\tau)} \right\}. \tag{11.90}$$

We recall that this is the autocorrelation of the perpendicular velocities (for $\lambda_\perp = \infty$). The MSD depends on two parameters. The dimensional parameter ϵ is proportional to the intensity of the magnetic fluctuations:

$$\epsilon = \tfrac{1}{2} V_T^2 \overline{\beta}^2 = \nu \chi_\parallel \overline{\beta}^2. \tag{11.91}$$

The dimensionless parameter $\overline{\chi}_\parallel$ is interpreted as half the square of the ratio of the collisional mean free path $\lambda_{mfp} = V_T/\nu$ to the squared parallel correlation length of the magnetic fluctuations, $\lambda_\parallel$; it is therefore a measure of the collisionality of the plasma (a small $\overline{\chi}_\parallel$ corresponds to a highly collisional plasma):

$$\overline{\chi}_\parallel = \frac{\lambda_{mfp}^2}{2\lambda_\parallel^2} = \frac{V_T^2}{2\nu^2 \lambda_\parallel^2} = \frac{\chi_\parallel}{\nu \lambda_\parallel^2}. \tag{11.92}$$

The simplest way for the evaluation of Eq. (11.89) consists of differentiating it twice with respect to time. Recalling Eq. (11.26), we get:

$$\tfrac{1}{2} \frac{d^2}{dt^2} \langle \delta X^2(t) \rangle \equiv \frac{d}{dt} D^\infty(t) = \mathcal{L}^\infty(\nu t). \tag{11.93}$$

This simple differential equation should be solved with the initial condition: $D^\infty(0) = 0$, $\langle \delta X^2(0) \rangle = 0$. These initial data are required by the original form (11.89). The solution is immediate if one notes that the equation can be written as follows (with $\tau = \nu t$):

$$\frac{d D^\infty(t)}{\nu \, dt} = \frac{1}{\overline{\chi}_\parallel} \frac{d^2 T(\tau)}{d\tau^2}, \tag{11.94}$$

with:

$$T(\tau) = \epsilon \left[1 + 2\overline{\chi}_\parallel \mu(\tau)\right]^{1/2}. \tag{11.95}$$

We then find:

$$D^{\infty}(t) = \chi_{\|} \, \overline{\beta}^2 \, \frac{\varphi(\nu t)}{[1 + 2\overline{\chi}_{\|} \, \mu(\nu t)]^{1/2}} \tag{11.96}$$

and:

$$\langle \delta X^2(t) \rangle = 2\overline{\beta}^2 \lambda_{\|}^2 \left\{ \left[1 + 2\overline{\chi}_{\|} \, \mu(\nu t)\right]^{1/2} - 1 \right\}. \tag{11.97}$$

These are the final results of our study. *The anomalous perpendicular running diffusion coefficient $D^{\infty}(t)$ and the MSD have been obtained from an exact solution of the V-Langevin problem (11.69) - (11.78).*

It is important to study the behaviour of these quantities in two limiting situations. For *short time*, $\tau \equiv \nu t \ll 1$, $\varphi(\tau) = \tau + O(\tau^2)$, $\mu(\tau) = \frac{1}{2}\tau^2 + O(\tau^3)$, hence:

$$D^{\infty}(t) = \epsilon t, \quad \langle \delta X^2(t) \rangle = \epsilon t^2, \quad \nu t \ll 1. \tag{11.98}$$

This is a typically *ballistic, non-diffusive* behaviour.

In the opposite limit of *long time*, $\varphi(\tau) \sim 1$, $\mu(\tau) \sim \tau$, hence:

$$D^{\infty}(t) = \chi_{\|} \frac{\overline{\beta}^2}{(2\overline{\chi}_{\|}\nu)^{1/2}} \frac{1}{\sqrt{t}}, \quad \nu t \gg 1$$

$$\langle \delta X^2(t) \rangle = 4\chi_{\|} \frac{\overline{\beta}^2}{(2\overline{\chi}_{\|}\nu)^{1/2}} \sqrt{t}, \quad \nu t \gg 1. \tag{11.99}$$

This is, for our purpose, the most interesting result. It shows that the MSD does not tend asymptotically to a linear function of time. Our model system thus exhibits *strange diffusion behaviour*. This concept is defined as follows (SHLESINGER et al. 1993, BALESCU 1997). Assume that for a given stochastic system the MSD behaves asymptotically like a power law:

$$\langle \delta X^2(t) \rangle = A t^{\mu}. \tag{11.100}$$

μ is called the *diffusion exponent*. Whenever $\mu = 1$, the system exhibits (normal) *diffusive behaviour*. For every other value of μ (or for a dependence different from a power law, such as, *e.g.*, $\ln t$) the system exhibits *strange diffusive behaviour*. More specifically, when $0 < \mu < 1$, the behaviour is called *subdiffusive*. In this case the asymptotic diffusion coefficient, as defined by Eq. (11.28) is *zero*. When $1 < \mu < \infty$, the behaviour is *superdiffusive;* the asymptotic diffusion coefficient is *infinite*.

The present problem provides us with a remarkably simple example of a *subdiffusive behaviour*, with $\mu = 1/2$. The (normalized) running diffusion

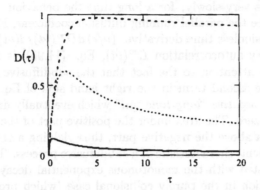

Figure 11.5. Dimensionless running diffusion coefficient $D(\tau) = (\chi_\| \beta^2)^{-1} D^\infty(\tau; \overline{\chi}_\|)$, vs. $\tau = \nu t$. solid: $\overline{\chi}_\| = 100$ (weak collisionality); dotted: $\overline{\chi}_\| = 1$ (intermediate collisionality); dash-dotted: $\overline{\chi}_\| = 0.01$ (strong collisionality).

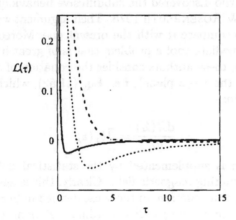

Figure 11.6. Velocity autocorrelation function $\mathcal{L}^\infty(\tau)$ vs. $\tau = \nu t$, for three values of the collisionality. Note: $\mathcal{L}^\infty(0) = 1$ for all $\overline{\chi}_\|$. Solid: $\overline{\chi}_\| = 100$, dotted: $\overline{\chi}_\| = 1$, dash-dotted: $\overline{\chi}_\| = 0.01$.

coefficient $(\nu/\epsilon) D^\infty(\nu t)$ is shown in Fig. 11.5. As can be seen from Eq. (11.96), its shape depends on the collisionality parameter $\overline{\chi}_\|$. For weak collisionality (like in high-temperature plasmas), after passing through a rather low maximum, the running diffusion coefficient decays rather rapidly to zero. For strong collisionality, this quantity approaches unity rather quickly, and then decays very slowly: for a long time the behaviour "simulates" a diffusive one (see the parallel running diffusion coefficient, Fig. 11.3).

The dimensionless time derivative $(\nu/\epsilon)\, dD^{\infty}(\nu t)/d(\nu t)$, *i.e.*, the Lagrangian velocity autocorrelation $\mathcal{L}^{\infty}(\nu t)$, Eq. (11.90)] is shown in Fig. 11.6. We draw attention to the fact that the subdiffusive $\sqrt{t}$-behaviour results from the second term in the right hand side of Eq. (11.90): the latter creates a *negative "long-time tail"* which eventually draws the function $\mathcal{L}^{\infty}(\nu t)$ to zero. The area under the positive part of the curve cancels exactly the area above the negative part, thus yielding a zero asymptotic diffusion coefficient, characteristic of a subdiffusive process. This behaviour is to be contrasted with the monotonous exponential decay of the velocity autocorrelation in the purely collisional case (which produces normal diffusive behaviour). Thus, *the negative long-time algebraic tail of the Lagrangian velocity correlation is a clear signature of the strange, subdiffusive behaviour.*

11.7   Magnetic Field Line Diffusion

The first authors who discovered the subdiffusive behaviour described here were RECHESTER & ROSENBLUTH 1978. Their argument was semi-intuitive: it is interesting to compare it with the present one. Moreover, they introduced as an intermediate tool a problem that is of great interest in itself.

In a first step, these authors consider the equation of a magnetic field line (projected on the $x - z$ plane), *i.e.*, Eq. (11.52), which reduces in the present situation to:

$$\frac{dX(Z)}{dZ} = \overline{\beta}\, b_x(z). \tag{11.101}$$

This equation is supplemented by the statistical definition (11.74), (11.57) of the fluctuating magnetic field. Clearly, this is again a typical example of a *V-Langevin equation*, in the sense defined in Sec. 11.3. The role of time is played here by the parallel coordinate Z along the unperturbed field.

The stochastic differential equation is solved by the same methods as in the previous section, with the result:

$$
\begin{aligned}
\langle \delta X^2(Z) \rangle_b &= \lambda_{\parallel}^2 \overline{\beta}^2 \int_0^z dz_1 \int_0^z dz_2 \, \langle b(z_1)\, b(z_2) \rangle_b \\
&= 2\lambda_{\parallel}^2 \overline{\beta}^2 \int_0^z dz' \, (z - z') \exp\left(-\tfrac{1}{2} z'^{\,2}\right) \\
&= 2\lambda_{\parallel}^2 \overline{\beta}^2 \left\{ \sqrt{\frac{\pi}{2}}\, z \,\mathrm{erf}\left(\frac{z}{\sqrt{2}}\right) + \left[\exp\left(-\frac{z^2}{2}\right) - 1\right] \right\}.
\end{aligned}
$$
$$\tag{11.102}$$

For large $z = Z/\lambda_\parallel$, this function behaves as:

$$\langle \delta X^2(Z) \rangle_b \sim 2\sqrt{\frac{\pi}{2}}\, \lambda_\parallel \overline{\beta}^2\, Z. \qquad (11.103)$$

Thus, the MSD, considered as a function of Z, has a *normal diffusive behaviour*. The magnetic fluctuations produce a spatial diffusion of the magnetic field lines, with a diffusion coefficient:

$$d_m^\infty = \tfrac{1}{2}\, \frac{d}{dZ}\, \langle \delta X^2(Z) \rangle_b \Big|_{z \to \infty} = \sqrt{\frac{\pi}{2}}\, \overline{\beta}^2 \lambda_\parallel, \quad \lambda_\perp \to \infty. \qquad (11.104)$$

The *magnetic diffusion coefficient* d_m^∞ has the dimension of a length.

In a second step, we consider charged particles (guiding centres) moving in this magnetic field. If the perpendicular collisional diffusion is neglected, the particles are tied to the (perturbed) magnetic field lines. This results in a parallel MSD given *asymptotically* by:

$$\langle \delta Z^2(t) \rangle_\parallel = 2\chi_\parallel t \qquad (11.105)$$

The parallel motion of the particles, combined with the perpendicular diffusion of the field lines to which they are attached, results in a perpendicular dispersion of the particles. RECHESTER & ROSENBLUTH 1978 estimated the latter by substituting in the *asymptotic* Eq. (11.103) the length Z defined by the *asymptotic* relation (11.105) as follows:

$$Z \to \overline{Z}(t) = \langle \delta Z^2(t) \rangle_\parallel^{1/2} = \sqrt{2\chi_\parallel t}. \qquad (11.106)$$

As a result, they find the perpendicular MSD as a function of time:

$$\langle \delta X^2(t) \rangle_{RR} \equiv \langle \delta X^2[\overline{Z}(t)] \rangle_{b,\parallel} = 2\sqrt{\pi}\overline{\beta}^2 \lambda_\parallel \sqrt{\chi_\parallel t}. \qquad (11.107)$$

We thus recovered the subdiffusive behaviour obtained in the previous section: $\langle \delta X^2(t) \rangle \propto t^\mu$, with $\mu = 1/2$. The corresponding running diffusion coefficient is [using Eqs. (11.92) and 11.35)]:

$$D_{RR}(t) = \frac{1}{2}\overline{\beta}^2 \lambda_\parallel \sqrt{\frac{\pi \chi_\parallel}{t}} = \sqrt{\frac{\pi}{2}}\, \chi_\parallel \frac{\overline{\beta}^2}{(\gamma\nu)^{1/2}}\, \frac{1}{\sqrt{t}}. \qquad (11.108)$$

The argument gives a physically clear picture of the process; the quantitative estimate is, however, not rigorous. It results from the substitution of an asymptotic result $[\overline{Z}(t)]$ into an asymptotic result $[\langle \delta X^2(Z) \rangle]$. On the contrary, the calculation of the previous section involves a single asymptotic limit (11.99) of the *exact* expression (11.97) of $\langle \delta X^2(t) \rangle_{b,\parallel}$. Comparing Eqs. (11.99) and (11.108) we find:

$$D_{RR}(t) = \sqrt{\frac{\pi}{2}}\, D^\infty(t). \tag{11.109}$$

Thus, the semi-quantitative Rechester-Rosenbluth argument yields the correct scaling of the MSD, but differs by a numerical factor ($\sqrt{\pi/2} \approx 1.25$) from the exact result.

11.8   The Concept of a Hybrid Kinetic Equation

The Langevin equations (in particular, the V-Langevin equations) are a very useful tool in the study of anomalous transport problems, but they are limited in scope. As illustrated in the previous chapter, they can provide very complete information about the mean square displacement, hence about the anomalous diffusion coefficient of a plasma. But they cannot produce *global* quantities describing the plasma as a whole. A very important question that can be asked is the spatial distribution of matter at all times, in other words, the shape of the *spatial density profile*.[11] Such a question can be answered by solving a kinetic equation, *i.e.*, Eq. (4.43), to which a collision term should be added, and which are completed with Maxwell's equations for self-consistency. But, as appears clearly from the first part of the book, this ambitious program very rapidly meets with enormous calculational difficulties whenever not too trivial problems are considered. From here on, as explained at the beginning of Sec. 11.1, we change rather drastically the philosophy of our approach to statistical dynamics. In order to obtain concrete results, we replace the complex processes by simple models. The latter are introduced intuitively by making reasonable assumptions; the results must then be justified *a posteriori* by numerical or real experiments.

Our starting point here is a kinetic equation for the reduced one-particle distribution function. The latter should describe the competition of two driving forces: the fluctuating electric and/or magnetic field and the individual particle collisions. The treatment of the first factor is common to all authors who used a kinetic approach to the present problem (KROMMES 1978, DIAMOND et al. 1980, ROSE 1982, KROMMES et al. 1983, BYKOV & TOPTYGUIN 1992, LAVAL 1993). The evolution is described by a Liouville operator valid for a given realization of the fluctuating field. The treatment of the collisions is more varied. For instance, in KROMMES et al. 1983, a

---

[11] This question could, in principle, be answered by using the Langevin equation: one should then calculate *all* the moments, and then reconstruct from these the instantaneous distribution function. Such a procedure would be very cumbersome and, most often, impracticable.

Fokker-Planck term is used; in LAVAL 1993, a simple relaxation term of Krook type is used. It appears that the type of modelization of the collisions may introduce serious differences in the results. This question was analyzed in detail in BALESCU et al. 1994 and will not be further discussed here. We shall therefore start in the most cautious way.

We illustrate the general method by considering the same problem as in the previous sections, *i.e.*, the motion of charged particles (more precisely: guiding centres) in a fluctuating magnetic field. The kinetic equation is constructed by using the same idea as in classical statistical mechanics (BALESCU 1997) and developed for turbulent plasmas in Chap. 4. The Liouville equation is a first order partial differential equation whose characteristic equations are Hamilton's equations of motion. Here we reverse the argument: *starting from the equations of motion, we construct a partial differential equation of which the former are the characteristic equations*. Thus, Eqs. (11.69) - (11.71) are the characteristic equations of the following partial differential equation:

$$\partial_t f(X, Y, Z, t) + v_\parallel(t) \, \nabla_Z f(X, Y, Z, t)$$
$$+ \overline{\beta} \, v_\parallel(t) \, \mathbf{b}(z) \cdot \nabla_\perp f(X, Y, Z, t) = 0. \tag{11.110}$$

This equation looks rather different from the usual kinetic equations of statistical mechanics, which describe the evolution of a distribution function in the complete phase space $(\mathbf{X}, \mathbf{v})$: the function f appearing here does not depend on the velocity. On the other hand, although Eq. (11.110) is supposed to describe the effect of collisions, it does not contain any collision term. This is in agreement with the philosophy of the Langevin equations. Thus, Eq. (11.110) is declared to be a *stochastic equation*, that must be completed by a statistical definition of its coefficients.

- The *collisions* are modelled by assuming that the velocity of the particle (or guiding centre) is not to be considered as a phase space variable (as in usual statistical mechanics), but rather as a *given function of time, defined statistically as a stationary Gaussian process whose moments are given by Eq. (11.77)*;
- The *magnetic fluctuations* are modelled by defining the functions $b_m(z)$ as a *homogeneous Gaussian process whose moments are given by Eq. (11.74), combined with (11.57)*;
- The two stochastic processes defined above are mutually independent.

Eq. (11.110) together with the statistical definition of the collisions and of the magnetic fluctuations constitutes a closed set of equations. Note that *the physical information in these equations is exactly the same as in the corresponding Langevin equations*. But the partial differential equation gives access to more global information than the latter.

Eq. (11.110) is an example of a class of equations that will be called *hybrid kinetic equations (HKE)* (BALESCU et al. 1994).

A general HKE is defined as a partial differential equation of the form:

$$\partial_t f(\mathbf{Y}, t) + \mathbf{v}(\mathbf{Y}, t) \cdot \nabla f(\mathbf{Y}, t) = 0; \qquad (11.111)$$

in which the vectorial coefficient $\mathbf{v}(\mathbf{Y}, t)$ *is specified statistically as a given stochastic process, e.g., a homogeneous and stationary Gaussian process,*[12] *with*

$$\langle \mathbf{v}(\mathbf{Y}, t) \rangle = \mathbf{V}, \quad \langle \delta \mathbf{v}(\mathbf{Y}, t_0) \, \delta \mathbf{v}(\mathbf{Y} + \mathbf{R}, t_0 + t) \rangle = \mathbf{C}(\mathbf{R}, |t|). \qquad (11.112)$$

The characteristic equations of the first-order partial differential equation (11.111) are the V-Langevin equations (11.43).

The name given to these hybrid kinetic equations suggests a compromise between the rigorous dynamics of the true kinetic equations and the tools of the theory of stochastic processes, in particular, the Langevin equations.

In principle, any equation of the form (11.111) can be transformed into a hybrid kinetic equation by completing it with a statistical prescription of the form (11.112). For instance, the set of equations (6.3) - (6.4) (that can be recombined into a single equation for the distribution function f^α) are turned into a HKE by completing it with an *a priori* statistical definition of the coefficient $\delta \mathcal{K}^\alpha$. Thus, the main difference between the problem of the gyrokinetic equation considered in Chap. 4 and the corresponding HKE is that the latter is no longer closed by the Poisson equation (which ensures self-consistency), but rather by an externally injected semi-intuitive assumption which simplifies, in principle, the mathematical treatment. The problem of drift-wave turbulence treated with such stochastic tools will be reconsidered in Chap. 14. Meanwhile we continue the study of the simpler problem defined by Eq. (11.110).

11.9 The Hybrid Kinetic Equation for Magnetic Fluctuations

We introduce the usual decomposition of the distribution function:

$$f(X, Y, Z, t) = F(X, Y, Z, t) + \delta f(X, Y, Z, t), \qquad (11.113)$$

where $F(X, Y, Z, t) = \langle f(X, Y, Z, t) \rangle_{b, \|}$ is the (doubly) ensemble-averaged distribution function and $\delta f(X, Y, Z, t)$ is the fluctuation, whose average is

[12] Various generalizations could also be considered.

obviously zero: $\langle \delta f \rangle_{b,\parallel} = 0$. The hybrid kinetic equation is split into two coupled equations for F and δf. The equation for the former is obtained by ensemble-averaging Eq. (11.110):

$$\partial_t F(X,Y,Z,t) + \nabla_Z \langle v_\parallel(t)\, \delta f(X,Y,Z,t) \rangle_{b,\parallel}$$
$$+ \overline{\beta}\, \nabla_\perp \cdot \langle v_\parallel(t)\, \mathbf{b}(z)\, \delta f(X,Y,Z,t) \rangle_{b,\parallel} = 0. \qquad (11.114)$$

The equation for the fluctuation is:

$$\partial_t \delta f + v_\parallel(t)\, \nabla_Z \delta f + \overline{\beta}\, v_\parallel(t)\, b_x(z)\, \nabla_X \delta f + \overline{\beta}\, v_\parallel(t)\, b_y(z)\, \nabla_Y \delta f$$
$$- \nabla_Z \langle v_\parallel(t)\, \delta f \rangle_{b,\parallel} - \overline{\beta}\, \nabla_X \langle v_\parallel(t)\, b_x(z)\, \delta f \rangle_{b,\parallel} - \overline{\beta}\, \nabla_Y \langle v_\parallel(t)\, b_y(z)\, \delta f \rangle_{b,\parallel}$$
$$+ v_\parallel(t)\, \nabla_Z F + \overline{\beta}\, v_\parallel(t)\, b_x(z)\, \nabla_X F + \overline{\beta}\, v_\parallel(t)\, b_y(z)\, \nabla_Y F$$
$$= 0. \qquad (11.115)$$

We now assume, for simplicity, that the average distribution function only depends on the coordinate X and on time (*i.e.*, that the density gradient is perpendicular to the magnetic field):

$$F(X,Y,Z,t) = n(X,t). \qquad (11.116)$$

The function $n(X,t)$ is called the *density profile*. The coordinate Y is then clearly irrelevant, and can be integrated out in both equations (11.114), (11.115). The latter equation reduces to:[13]

$$\left[\partial_t + v_\parallel(t)\, \nabla_Z + \overline{\beta}\, v_\parallel(t)\, b_x(z)\, \nabla_X \right] \delta f(X,Z,t)$$
$$= -\overline{\beta}\, v_\parallel(t)\, b_x(z)\, \nabla_X n(X,t). \qquad (11.117)$$

Consider first Eq. (11.114), which reduces to:[14]

$$\partial_t n(X,t) = -\overline{\beta}\, \nabla_X \langle v_\parallel(t)\, b_x(z)\, \delta f(X,Z,t) \rangle_{b,\parallel}. \qquad (11.118)$$

We note that this equation has precisely the form of the macroscopic continuity equation. The average anomalous particle flux in the x-direction is thus identified as:

$$\Gamma_x = \langle v_x(t)\, \delta f(X,Z,t) \rangle_{b,\parallel} \equiv \overline{\beta}\, \left\langle b_x(z)\, \langle v_\parallel(t)\, \delta f(X,Z,t) \rangle_\parallel \right\rangle_b. \qquad (11.119)$$

---

[13] The terms involving $\langle \eta_\parallel(t)\, \delta f \rangle_{b,\parallel}$ and $\langle \eta_\parallel(t)\, b_x(z)\, \delta f \rangle_{b,\parallel}$ are not written, because they do not contribute to Eq. (11.114).

[14] It will be shown *a posteriori* that the average in the second term is independent of z.

The inner average in Eq. (11.119) is nothing other than the particle flux in the parallel, z-direction [see Eq. (5.10)]:

$$\delta\Gamma_z = \langle v_\parallel(t)\, \delta f(X, Z, t)\rangle_\parallel. \tag{11.120}$$

This is still a fluctuating quantity, because of the dependence of δf on b. Eq. (11.119) can thus be rewritten as:

$$\Gamma_x = \overline{\beta}\ \langle b_x(z)\, \delta\Gamma_z\rangle_b. \tag{11.121}$$

This is a well known formula (LIEWER 1985). It is the analog of Eq. (5.13) for the anomalous flux due to magnetic (rather than electrostatic) fluctuations. Its evaluation requires the solution of Eq. (11.117). This process will lead to a closed equation for the density profile. This problem will be deferred to Sec. 12.8, where an efficient method of solution will be derived.

11.10   Hybrid Kinetic Equation for Collisional Diffusion

Before closing this chapter, we revert to the much simpler problem of purely collisional diffusion in a non-fluctuating field. It will illustrate some interesting properties of the HKE, which will be further discussed in the next chapter. The V-Langevin equation used as a starting point here is the trivially simple equation (11.37), completed with the statistical definition (11.38). The HKE's, corresponding to Eqs. (11.117), (11.118) are derived in the same way as in the previous section:

$$\frac{\partial}{\partial t}n(Z, t) + \frac{\partial}{\partial Z}\langle v_\parallel(t)\, \delta f(Z, t)\rangle = 0. \tag{11.122}$$

$$\partial_t \delta f + v_\parallel(t)\frac{\partial}{\partial Z}\delta f = -v_\parallel(t)\frac{\partial}{\partial Z}n. \tag{11.123}$$

Eq. (11.123) is easily solved. The propagator of this equation is:

$$g(Z, t|Z', t') = \delta\left[Z - Z' - \int_{t'}^{t} dt_1\, v_\parallel(t_1)\right]. \tag{11.124}$$

Hence, the solution of Eq. (11.123) is:

$$\delta f(Z, t) = \int dZ'\, g(Z, t|Z', 0)\, \delta f(Z'; 0)$$

$$- \int_0^t dt_1 \int dZ'\, g(Z, t|Z', t_1)\, v_\parallel(t_1)\frac{\partial}{\partial Z'}n(Z', t_1). \tag{11.125}$$

The first term goes rapidly to zero for a well-localized initial condition, and will be neglected (see Sec. 6.2). The second term is easily integrated over Z'; the result is then substituted into (11.122):

$$\partial_t n(Z,t) = \frac{\partial}{\partial Z} \int_0^t dt_1 \left\langle v_\parallel(t)\, v_\parallel(t_1)\, \frac{\partial}{\partial Z}\, n\left[Z - v_\parallel t_1, t_1\right] \right\rangle. \qquad (11.126)$$

This is the exact equation of evolution of the density profile following from the hybrid kinetic equation. It is rather surprising to note that the trivially simple V-Langevin equation (11.37) leads to such a complicated equation. The most serious difficulty comes from the displacement of the Z variable in the right hand side; indeed, the integrand is the average of the unknown function $(n[...])$ of the stochastic variable $v_\parallel$! To make sense of this expression, we Taylor expand it as follows [see Eq. (6.26)]:

$$\langle...\rangle = \left\langle v_\parallel(t)\, v_\parallel(t_1) \right\rangle \frac{\partial}{\partial Z}\, n(Z, t_1)$$

$$+ \int dt_2 \int dt_3 \left\langle v_\parallel(t)\, v_\parallel(t_1) v_\parallel(t_2)\, v_\parallel(t_3) \right\rangle \frac{\partial^3}{\partial Z^3}\, n(Z, t_1) + ... \qquad (11.127)$$

Thus, Eq. (11.126) is equivalent to a partial differential equation of infinite order. The coefficients of the successive derivatives involve velocity moments of increasingly high order. Considering the last term in (11.127), we note that, for a Gaussian process, the fourth order moment is a sum of products of two second order moments. Because of the *exponential form of the velocity autocorrelations* (11.38), the higher moments decrease much faster than the second moment. The fast decay of the memory kernel thus fully justifies the (asymptotic) neglect of all velocity moments of order higher than 2; keeping only the first term in (11.127), we obtain:

$$\partial_t n(Z,t) = \frac{\partial}{\partial Z} \int_0^t dt_1\, R_{zz}(t_1)\, \frac{\partial}{\partial Z}\, n(Z, t - t_1). \qquad (11.128)$$

[Here we reverted to the notation $R_{zz}(t)$ (11.38) for the velocity correlation, and changed the integration variable: $t_1 \rightarrow t - t_1$]. This looks "almost" like a diffusion equation, except for a major difference: it is a *non-Markovian diffusion equation*. Indeed, the rate of change of $n(Z,t)$ at time t depends on the past values of the density profile. We are faced with the same problem as in kinetic theory (Sec. 6.2). The advantage of the HKE model is that here we *know explicitly* (more correctly: we *assumed explicitly*) that the velocity correlation decays *exponentially* with a characteristic time equal to the collisional relaxation time: $\tau_R = \nu^{-1}$. On the other hand, the density profile varies very little over such a time. We are thus justified in introducing *asymptotically*, for $t \gg \nu^{-1}$, the **Markovian approximation**:

i) the retardation is neglected in the right hand side: $n(Z, t - t_1) \approx n(Z, t)$;

ii) the upper limit of integration is extended to infinity.

The resulting equation of evolution for the density profile is then:

$$\partial_t n(Z, t) = \frac{\partial}{\partial Z} \chi_\parallel \frac{\partial}{\partial Z} n(Z, t). \qquad (11.129)$$

This is an *ordinary diffusion equation*, with a constant collisional diffusion coefficient:

$$\chi_\parallel = \int_0^\infty dt_1 \, R_{zz}(t_1). \qquad (11.130)$$

Not surprisingly, we recovered once more the expression (11.31) [with $\perp \to \parallel$, and $x \to z$] which was obtained from the V-Langevin equation.[15]

The matter treated in the present section is interesting for several reasons:

- It shows that the hybrid kinetic equation is a useful concept for the study of stochastic processes. In particular, it allows us to obtain directly the density profile.
- It shows that even the derivation of the simplest diffusion equation poses non-trivial problems which are clearly displayed in the hybrid kinetic equation formalism.
- The hybrid kinetic equation will appear as an ideal tool for the study of situations where the assumptions necessary for the diffusion equation are no longer satisfied. This will open the door for the *strange transport* problems that will be discussed in Chap. 12.

---

[15] Had we not implemented the approximation *ii*) above, we would have obtained a diffusion equation with a time-dependent diffusion coefficient, which is none other than the running diffusion coefficient $\chi_\parallel(t)$. We know, of course, that the latter tends exponentially to the constant $\chi_\parallel$.

Chapter 12

Random Walks and Transport

12.1 Classical Random Walks

We now go one step further in the stochastic modelling of the transport processes. In complex problems, involving motion of particles through strongly inhomogeneous, partially disordered or turbulent media, even the Langevin or the hybrid kinetic equation approaches could become prohibitive. In such cases a more radical modelling could be useful. The description of the motion by "semi-deterministic" laws (*i.e.*, a Newton equation with random forces) is abandoned; rather, the evolution is described by a set of displacements ruled by purely probabilistic prescriptions. This method has been applied to a large variety of problems since the classical works of Einstein, Smoluchowski and Langevin, mentioned in Chap. 11, and has been revived in recent years by very important new developments. A very comprehensive modern presentation will be found in WEISS 1994 and also in the review paper by BOUCHAUD & GEORGES 1990. In the present chapter we follow closely the beautiful review article of MONTROLL & SHLESINGER 1984, and especially Chap. 12 of our book BALESCU 1997, as well as our paper BALESCU 1995. Part of the presentation follows the very recent paper by VAN MILLIGEN et al. 2004 a. Recent detailed reviews on the relation between random walks and strange (anomalous) transport are HAVLIN & BENAVRAHAM 1987 and METZLER & KLAFTER 2000.

We start with the consideration of the very old problem of the *random walk of a particle in a d-dimensional space* (the role of dimensionality in these problems is often non-trivial). The cartesian coordinates of a point in this space are denoted by a d-dimensional vector x. In its classical (*discrete time*) version the random walk is a *discontinuous process*: there exists a "quantum" of time t_0. Imagine a metronome oscillating with a period t_0. At every beat of the metronome (*i.e.*, at each time $t = \nu t_0$, ν: integer), the particle performs a *jump* described by a vector r of arbitrary length and arbitrary direction. It is assumed that these jumps are statistically independent of each other. The *transition probability density, or probability distribution function (PDF) of a jump described by a vector* **r** or, briefly, the *jump PDF* is denoted by $f(\mathbf{r})$. It is a prescribed function, assumed independent of time, characteristic of the problem.

Let $n_\nu(\mathbf{x})$ be the *probability density (PDF) of finding the particle at* **x** *after* ν *steps, knowing that it started certainly in* $\mathbf{x} = 0$ *at time* $t = 0$: $n_0(\mathbf{x}) = \delta(\mathbf{x})$. Because of the mutual independence of the jumps, an equation for $n_\nu(\mathbf{x})$ is easily established. Indeed, the PDF of finding the particle in **x** at time $\nu + 1$ equals the PDF of being in $\mathbf{x}'$ at time ν, times the probability of a jump $(\mathbf{x} - \mathbf{x}')$, summed over all previous positions $\mathbf{x}'$:

$$n_{\nu+1}(\mathbf{x}) = \int d^d\mathbf{x}'\, f(\mathbf{x} - \mathbf{x}')\, n_\nu(\mathbf{x}'); \tag{12.1}$$

($d^d\mathbf{x}$ is the volume element in d dimensions). This recurrence relation is very easily solved. Being in the form of a convolution, it is naturally treated by a Fourier transformation. The Fourier transform $\widehat{n}_\nu(\mathbf{k})$ of a PDF is called its *characteristic function*. The characteristic function of the *jump PDF*, $\widehat{f}(\mathbf{k})$, is called the *structure function* of the random walk. These quantities are defined as follows:[1]

$$\left\{ \begin{array}{c} \widehat{n}_\nu(\mathbf{k}) \\ \widehat{f}(\mathbf{k}) \end{array} \right\} = \int d^d\mathbf{x}\; e^{i\mathbf{k}\cdot\mathbf{x}} \left\{ \begin{array}{c} n_\nu(\mathbf{x}) \\ f(\mathbf{x}) \end{array} \right\}. \tag{12.2}$$

Before continuing, we derive some simple, but very important properties of the characteristic functions. Let $n_\nu(\mathbf{x})$ be any PDF, normalized to one, and denote by $\widehat{n}_\nu(\mathbf{k})$ its characteristic function. Eq. (12.2) can also be interpreted as an average, weighted with $n_\nu(\mathbf{x})$:

$$\widehat{n}_\nu(\mathbf{k}) = \left\langle e^{i\mathbf{k}\cdot\mathbf{x}} \right\rangle. \tag{12.3}$$

From here we obtain the following relations, relating the *moments* of the PDF to the derivatives of its characteristic function, evaluated at $\mathbf{k} = 0$. The normalization condition determines the value of the characteristic function at the origin:

$$1 = \widehat{n}_\nu(\mathbf{k} = 0), \quad \forall \nu. \tag{12.4}$$

The average position at "time" ν is obtained as follows:

---

[1] It may be noted that the definition of the Fourier transform in the context of random walks is slightly different from our usual convention [*e.g.*, (6.51), (6.52)]. The factor $(2\pi)^{-d}$ is attached here to the **k**-integral, rather than to the **x**-integral (in the inversion formula). Moreover, the sign of the argument ($i\mathbf{k}\cdot\mathbf{x}$) is opposite. Both conventions are, indeed, correct and equivalent. The present choice yields more "natural" formulae in this problem, such as the expression (12.3) or the normalization condition (12.4), which would otherwise carry inelegant 2π-factors.

$$\langle \mathbf{x} \rangle_\nu = -i \, \frac{\partial}{\partial \mathbf{k}} \, \widehat{n}_\nu(\mathbf{k}) \bigg|_{\mathbf{k}=0}. \tag{12.5}$$

More generally, for an arbitrary moment, *i.e.* the average of a product of p components of $\mathbf{x}$ we find:

$$\langle x_{a_1} x_{a_2} ... x_{a_p} \rangle_\nu = i^{-p} \, \frac{\partial^p}{\partial k_{a_1} \partial k_{a_2} ... \partial k_{a_p}} \widehat{n}_\nu(\mathbf{k}) \bigg|_{\mathbf{k}=0}. \tag{12.6}$$

Of special importance is the *mean square displacement (MSD)*, *i.e.* the average of the squared distance from the origin (in d dimensions) $r^2 = \sum_{i=1}^d x_i^2$:

$$\langle r^2 \rangle_\nu = - \frac{\partial}{\partial \mathbf{k}} \cdot \frac{\partial}{\partial \mathbf{k}} \widehat{n}_\nu(\mathbf{k}) \bigg|_{\mathbf{k}=0}. \tag{12.7}$$

We now return to Eq. (12.1) which, upon Fourier transformation becomes:

$$\widehat{n}_{\nu+1}(\mathbf{k}) = \widehat{f}(\mathbf{k}) \, \widehat{n}_\nu(\mathbf{k}), \tag{12.8}$$

with the initial condition: $\widehat{n}_0(\mathbf{k}) = 1$. We then find: $\widehat{n}_1(\mathbf{k}) = \widehat{n}(\mathbf{k})$, $\widehat{n}_2(\mathbf{k}) = \widehat{f}(\mathbf{k}) \widehat{f}(\mathbf{k})$, and in general:

$$\widehat{n}_\nu(\mathbf{k}) = \left[\widehat{f}(\mathbf{k}) \right]^\nu, \tag{12.9}$$

and, finally:

$$n_\nu(\mathbf{x}) = (2\pi)^{-d} \int d^d k \; e^{-i\mathbf{k}\cdot\mathbf{x}} \left[\widehat{f}(\mathbf{k}) \right]^\nu. \tag{12.10}$$

This is the complete solution of the classical random walk problem. Clearly, the shape of the density profile will be entirely determined by the structure function $\widehat{f}(\mathbf{k})$.

12.2 Stable Distributions

For simplicity, we first consider only *symmetric random walks*, characterized by even functions of $\mathbf{x}$: $n(\mathbf{x}) = n(x)$, and $\widehat{f}(\mathbf{k}) = \widehat{f}(k)$ [as usual: $|\mathbf{x}| \equiv x$, $|\mathbf{k}| \equiv k$]. An important class of solutions obtains when the structure function corresponds to a so-called *stable distribution*. These distributions are defined by the following form-invariance property:

$$\widehat{f}(a_1 k) \, \widehat{f}(a_2 k) = \widehat{f}(ak), \tag{12.11}$$

where a_1, a_2 are given positive constants and a is a function of the latter. It is shown in Appendix D that the solutions of this functional equation are the characteristic functions of the *Lévy distributions:*

$$\widehat{f}(\mathbf{k}) = \exp\left(-C\,|k|^\alpha\right), \quad 0 < \alpha \leq 2. \tag{12.12}$$

(The limitation is due to the fact that for $\alpha > 2$ the jump PDF $f(\mathbf{x})$ is no longer positive definite). As these concepts are not very familiar in the plasma physics community, we collected some useful properties of the Lévy distributions in Appendix D. Eq. (12.11) is generalized as follows:

$$\prod_{m=1}^{\nu} \widehat{f}(a_m \mathbf{k}) = \widehat{f}(a\mathbf{k}) \tag{12.13}$$

with the composition law (for the Lévy distribution):

$$a = \left(\sum_{m=1}^{\nu} a_m^\alpha\right)^{1/\alpha}. \tag{12.14}$$

We find, in particular from Eq. (12.12):

$$\left[\widehat{f}(\mathbf{k})\right]^\nu = \widehat{f}(\nu^{1/\alpha}\mathbf{k}). \tag{12.15}$$

Consider now a random walk characterized by the structure function (12.12). Combining Eqs. (12.9) and (12.15) we find the Fourier transform of the PDF for finding the particle at $\mathbf{x}$ after ν steps:

$$\widehat{n}_\nu(\mathbf{k}) = \exp(-\nu\,Ck^\alpha). \tag{12.16}$$

The characteristic function keeps at all times the Lévy form; the argument of the exponential in its characteristic function is simply multiplied by ν, *i.e.*, it is proportional to time. Thus the random walk has in this case an important *self-similarity property.*

The Lévy characteristic function for the value $\alpha = 2$ is simply the Gaussian PDF. In this case the inverse Fourier transform of Eq. (12.16) can be obtained analytically, and is again a Gaussian (in $\mathbf{x}$):

$$n_\nu(\mathbf{x}) = \frac{1}{(4\pi\,\nu C)^{d/2}} \exp\left(-\frac{x^2}{4\,C\nu}\right), \quad \alpha = 2. \tag{12.17}$$

The corresponding MSD is easily found from Eq. (12.7):

$$\langle x^2\rangle = 2dC\,\nu, \quad \alpha = 2. \tag{12.18}$$

This MSD grows proportionally to time. We thus found the following important property [see Eqs. (11.27), (11.100) with $\mu = 1$]: *A random walk defined by a Gaussian structure function exhibits **diffusive behaviour**.*

Consider now the case when the structure function corresponds to a Lévy distribution with $0 < \alpha < 2$. In this case the PDF, *i.e.*, the inverse transform of (12.16), has in general no simple analytic form (except for $\alpha = 1$, which corresponds to a Lorentz, or Cauchy distribution, see Appendix D). The main qualitative feature of these Lévy distributions is the presence of a *long tail* in x. As a result, *the MSD is infinite for all values of $0 < \alpha < 2$*. Indeed, applying Eq. (12.7) to (12.16) we find that $\langle x^2 \rangle \sim k^{\alpha-2}|_{k=0} = \infty$. For this reason, these stable Lévy distributions were for a long time considered as "mathematical monsters" without any physical relevance. It is only recently that the interest of the physicists was revived, as will be seen in forthcoming sections and in Chap. 17..

We now revert to the general case of an asymmetric PDF, *i.e.* one for which the first moment differs from zero. We may ask the question: *Why should the stable distributions play a special role in the theory of random walks (or, more generally, in physics)?* Consider a random walk defined by an arbitrary structure function, of which we only require that it be an *analytic function near* $\mathbf{k} = 0$:

$$\widehat{f}(\mathbf{k}) = \int d^d x \, e^{i\mathbf{k}\cdot\mathbf{x}} f(\mathbf{x}) = \langle e^{i\mathbf{k}\cdot\mathbf{x}} \rangle, \tag{12.19}$$

where the function $f(\mathbf{x})$ is *not* specified. Expanding the exponential, we obtain:

$$\widehat{f}(\mathbf{k}) = 1 + i\mathbf{k}\cdot\langle\mathbf{x}\rangle - \tfrac{1}{2}\mathbf{k}\mathbf{k}:\langle\mathbf{x}\mathbf{x}\rangle + O(k^3). \tag{12.20}$$

For simplicity, we consider here the case of an *isotropic* step size PDF, for which:

$$\langle x_i x_j \rangle = \delta_{ij}\frac{1}{d}\langle x_n x_n \rangle = \delta_{ij}\frac{1}{d}\langle x^2 \rangle;$$

(12.20) then reduces to:

$$\widehat{f}(\mathbf{k}) = 1 + i\mathbf{k}\cdot\langle\mathbf{x}\rangle - \frac{1}{2d}k^2\langle x^2 \rangle + O(k^3). \tag{12.21}$$

Our assumption of analyticity of $\widehat{f}(\mathbf{k})$ implies that the coefficients of this series, hence the *moments* of $f(\mathbf{x})$ are finite. It is well known, on the other hand, that an arbitrary structure function can be represented, for small k, in terms of *cumulants*, as follows [see Eq. (8.2)]:

$$\widehat{f}(\mathbf{k}) = \exp\left[i\mathbf{k}\cdot\langle\mathbf{x}\rangle - \frac{1}{2d}k^2\langle\langle x^2 \rangle\rangle + ...\right], \tag{12.22}$$

where the second cumulant is defined as follows:

$$\langle\langle x^2 \rangle\rangle = \langle x^2 \rangle - |\langle\mathbf{x}\rangle|^2 = \langle |\mathbf{x}-\langle\mathbf{x}\rangle|^2 \rangle. \tag{12.23}$$

Hence, *in the limit of small* **k**, any analytic structure function tends to the form (12.22).

We now introduce the "true" time variable, $t = \nu t_0$, and calculate the density profile, which is renamed as $n(\mathbf{x}, t) = n_\nu(\mathbf{x})$. Using (12.10) and (12.22) we find:

$$n(\mathbf{x}, t)$$

$$= (2\pi)^{-2d} \int d^d \mathbf{k}\, e^{-i\mathbf{k}\cdot\mathbf{x}} \exp\left[i\,\nu t_0\,\mathbf{k}\cdot\frac{\langle\mathbf{x}\rangle}{t_0} - \frac{1}{2d}k^2\,\nu t_0\,\frac{\langle\langle x^2\rangle\rangle}{t_0} + ...\right].$$

$$(12.24)$$

Define:[2]

$$\mathbf{V} = \frac{\langle\mathbf{x}\rangle}{t_0}, \quad D = \frac{\langle\langle x^2\rangle\rangle}{2d\,t_0}. \qquad (12.25)$$

The second equation is nothing other than the celebrated *Einstein relation*, which we already met in Eq. (11.27). This result shows that the *classical random walk model reproduces exactly the classical diffusion coefficient obtained by the classical A-Langevin or V-Langevin equations, hence by kinetic theory.*

Before continuing the mainstream of our presentation, we note that the Einstein relation provides us with the basis for a very widely used phenomenological formula for the diffusion coefficient. The so-called *mixing length approximation* consists of estimating the diffusion coefficient associated with a given plasma instability by using in Eq. (12.25) a characteristic "step length" and a characteristic time related to that instability. The former could be, for instance, the Larmor radius, or the width of a banana orbit (in toroidal geometry), or a correlation length; the latter could be taken as the linear growth rate, or some correlation time (see *e.g.* NISHIKAWA & WAKATANI 1990, Chap. 13). This procedure is a very crude way of associating a random walk picture to an instability. The choice of these quantities is largely based on intuitive arguments. Needless to say that such estimates can at best yield an order of magnitude, which should be confirmed by experiment or by a deeper analysis.

The random walk theory provides us directly with the global quantity, *i.e.*, the *density profile*, from which *all* moments can be simply calculated by using Eq. (12.6). We thus find:

$$n(\mathbf{x}, t) = (2\pi)^{-2d} \int d^d \mathbf{k}\, e^{-i\mathbf{k}\cdot(\mathbf{x} - \mathbf{V}t)} e^{-k^2 Dt}, \qquad (12.26)$$

---

[2] Keep in mind that $\langle\mathbf{x}\rangle$ and $\langle\langle r^2\rangle\rangle$ are cumulants of the *transition* PDF $f(\mathbf{x})$, not of $n(\mathbf{x}, t)$!

which yields:

$$n(\mathbf{x}, t) = \frac{1}{(4\pi D t)^{d/2}} \exp\left[-\frac{|\mathbf{x} - \mathbf{V}t|^2}{4Dt}\right].$$ (12.27)

This function is called a *Gaussian packet*. It represents a density profile which, at time zero, is concentrated at the origin; for $t > 0$ the profile spreads out (*diffusion*) and at the same time its maximum moves with a constant velocity $\mathbf{V}$ (*advection*). The shape of the profile at various times is determined by the *Péclet number, Pe*, which measures the relative importance of the advection, compared to the diffusion:

$$Pe = \frac{VL}{D},$$ (12.28)

where L is a characteristic length. In Fig. 12.1 the evolution of the density profile in time (for $d = 1$) is shown for two values of the Péclet number. For small values of Pe, the maximum of the profile moves very little, the main effect is the spreading of matter in space. For large Pe, the maximum moves quickly, while the particles diffuse.

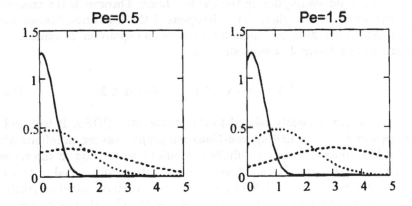

Figure 12.1. Density profiles for two values of the Péclet number. Solid: $t = 0.1$, Dotted: $t = 0.7$, Dashed: $t = 2$.

Eq. (12.27) was derived in the limit $k \to 0$, which corresponds to $|\mathbf{x} - \mathbf{V}t| \to \infty$; thus, for every fixed $\mathbf{x}$, (12.27) represents an asymptotic limit for long times.

The calculation performed above is simply a derivation of the celebrated *central limit theorem* of probability theory. It is formulated as follows in the present context: *Any PDF of a random walk having an analytic structure function tends asymptotically to a Gaussian packet.* This is an answer to our question about the special importance of the Gaussian distribution. It should be noted that the Gaussian packet is the fundamental solution of the *advection-diffusion equation:*

$$\partial_t n(\mathbf{x}, t) = D \nabla^2 n(\mathbf{x}, t) - \mathbf{V} \cdot \nabla n(\mathbf{x}, t), \qquad (12.29)$$

i.e., a solution corresponding to the initial condition $n(\mathbf{x}, 0) = \delta(\mathbf{x})$. For $\mathbf{V} = 0$ (*i.e.,* for a symmetric PDF), this equation reduces to the ordinary diffusion equation.

We found that any *regular* random walk behaves asymptotically as a diffusive process. Its MSD, measured with respect to the maximum of the PDF is:

$$\left\langle |\mathbf{x} - \mathbf{V}t|^2 \right\rangle = 2d\,D\,t \qquad (12.30)$$

Hence D, as given by (12.25) is identified with the diffusion coefficient (11.27).[3]

The main assumption in the Central Limit Theorem is the analyticity of the structure function. What happens if this condition breaks down? Typically a non-analytic function may have an expansion for small k which contains non-integral powers, such as:

$$\tilde{f}(\mathbf{k}) = 1 - Ck^\alpha + ..., \quad 0 < \alpha < 2. \qquad (12.31)$$

This case is clearly related to the stable Lévy PDF's. It will lead asymptotically to a definitely *non-Gaussian profile,* having an infinite MSD. This PDF does not obey a diffusion equation: we are now in the realm of *strange diffusion.* In recent years many authors developed in this context a generalized class of "strange diffusion" equations based on *fractional differential equations* (SHLESINGER et al. 1993). This alternative approach to strange transport will not be developed here; the interested reader is directed to the excellent review paper of METZLER & KLAFTER 2000, which also contains a very extensive bibliography. We already met with a typical example of strange, subdiffusive process in Sec. 11.6. That problem will be revisited below, as it is an analytically soluble paradigm. The theory of the strange diffusion processes will be discussed in the framework of a generalization of the usual random walks in the next sections.

---

[3] The presence of the factor $2d$ in Eqs. (12.25), (12.30) is due to our present consideration of the square of the d-dimensional vector $(\mathbf{x} - \mathbf{V}t)$, which is appropriate for an isotropic problem. In Eq. (11.27) we considered diffusion in a single direction x, thus $d = 1$.

12.3 Continuous Time Random Walks (CTRW)

The idealization of the diffusion process by a random walk in which the jumps take place at regular, well defined times $t = \nu t_0$ is not very realistic. We now introduce an extremely fruitful generalization of the random walk concept in which time appears as a continuous variable.

In the *Continuous Time Random Walk (CTRW)* model, a particle makes a jump r of arbitrary length and arbitrary direction in a d-dimensional space, then remains at its new position for a finite time t of random duration, after which it makes a new jump, etc. It is again assumed that the jumps are statistically independent. This concept was introduced by MONTROLL & WEISS 1965; there exist excellent bibliographical sources for the study of CTRW, among which we cite: MONTROLL & SHLESINGER 1984, BOUCHAUD & GEORGES 1990, SHLESINGER et al. 1993, WEISS 1994, SHLESINGER 1996. In the present section we start by deriving a very general form of the theory (which will be used in Chap. 17). We then simplify the final equations by making appropriate restrictions. The theory presented here was developed by VAN MILLIGEN et al. 2004 a, whose treatment we follow here very closely.

Like for any random walk problem, the first thing to be specified is the basic statistical rule defining the way the random walker goes from one site to the next. (In the classical problem this implies the definition of the jump PDF $f(r)$ and of the requirement that a step be taken at every time unit). In the CTRW model as defined above, we need a more complete specification by means of a joint PDF. We consider the following basic event: a particle arrives at point x' at exactly time t', then remains immobile there during a time $t - t'$; at time t it makes an instantaneous jump $x - x'$ to point x. The PDF of this composite process will be denoted by $\xi(x - x', x'; t - t', t')$. This function is assumed to be factorized as follows:

$$\xi(x - x', x'; t - t', t') = f(x - x', x'; t)\ \psi(t - t'; x'). \qquad (12.32)$$

$f(x - x'; x', t)$ is the *PDF of a jump* $x - x'$: it generalizes the jump PDF $f(x - x')$ introduced in Sec. 12.1; it is supposed to depend on the starting point x' and on time. It has the obvious property:

$$\int d^d x\ f(x - x', x'; t) = 1, \quad \forall x',\ \forall t. \qquad (12.33)$$

$\psi(t - t'; x')$ is the *PDF of a pause of duration* $t - t'$ *at point* x' *between two successive jumps*: it is the characteristic signature of a CTRW: it will be briefly called the *waiting time PDF*.

Our purpose is the derivation of a *Master Equation* describing the evolution of the density profile $n(x, t)$, defined as *the PDF that a particle starting in* $\mathbf{x} = 0$ *at time* $t = 0$, *be in* $\mathbf{x}$ *at time* t. This equation must be of the general form of a gain-loss balance equation:

$$\partial_t n(\mathbf{x}, t) = \int_0^t dt' \int d^d\mathbf{x}' \ [-K(\mathbf{x}, \mathbf{x}'-\mathbf{x}; t', t - t') \, n(\mathbf{x}, t') $$
$$+ \ K(\mathbf{x}', \mathbf{x} - \mathbf{x}' ; t', t - t') \, n(\mathbf{x}', t')]. \qquad (12.34)$$

$K(\mathbf{x}', \mathbf{x} - \mathbf{x}'; t', t - t')$ represents the transition PDF of finding at $\mathbf{x}$ at time t a particle that was in $\mathbf{x}'$ at time t'. The first bracketed term represents the contribution of particles leaving $\mathbf{x}$ between time 0 and t, the second term is the contribution of particles arriving at $\mathbf{x}$ during that time. In order to ensure particle conservation we must have:

$$\int_0^\infty d\tau \int d^d\mathbf{x} \ K(\mathbf{x}', \mathbf{x} - \mathbf{x}'; t', \tau) = 1, \qquad \forall \mathbf{x}', \ \forall t'. \qquad (12.35)$$

We must now relate K to the given functions f and ψ. Let $q(\mathbf{x}; t')$ be the PDF that the particle arrives in $\mathbf{x}$ at time t' as a result of a last jump in a series, after which it remains in $\mathbf{x}$; let also $\zeta(t - t'; \mathbf{x})$ be the PDF that the particle remains immobile in $\mathbf{x}$ between times t' and t after the jump. Then, clearly:

$$n(\mathbf{x}, t) = \int_0^t dt' \, \zeta(t - t'; \mathbf{x}) \, q(\mathbf{x}; t'). \qquad (12.36)$$

We now calculate the two factors entering the right hand side. We first note that the PDF of at least one step being taken sometime in the interval $(0, t)$ is:

$$\int_0^t dt' \, \psi(t'; \mathbf{x});$$

hence, the PDF of the particle remaining immobile at $\mathbf{x}$ during the time t after a jump is:

$$\zeta(t; \mathbf{x}) = 1 - \int_0^t dt' \, \psi(t'; \mathbf{x}). \qquad (12.37)$$

In order to calculate $q(\mathbf{x}, t)$ we must take into account all independent ways by which the particle can arrive in $\mathbf{x}$ after a jump: it may reach $\mathbf{x}$ after one step, two steps, etc. For a j-step process, $q_{(j)}(\mathbf{x}; t)$ is given by the PDF $q_{(j-1)}(\mathbf{x}'; t')$ of reaching $\mathbf{x}'$ at time t' after $j - 1$ steps, times the joint transition PDF $\xi(\mathbf{x} - \mathbf{x}', \mathbf{x}'; t - t', t')$ that the j-th step was taken at time t. Thus,

$$q(\mathbf{x};t) = \sum_{j=0}^{\infty} q_{(j)}(\mathbf{x};t),$$

with:

$$q_{(j)}(\mathbf{x};t) = \int_0^t dt' \int d^d\mathbf{x}'\ \xi(\mathbf{x}-\mathbf{x}',\mathbf{x}';t-t',t')\ q_{(j-1)}(\mathbf{x}';t'), \qquad (12.38)$$

and the condition:

$$q_{(0)}(\mathbf{x};t) = \delta(\mathbf{x})\,\delta(t). \qquad (12.39)$$

Summing both sides of Eq. (12.38) over j, we obtain the integral equation:

$$q(\mathbf{x},t) - \delta(\mathbf{x})\delta(t) = \int_0^t dt' \int d^d\mathbf{x}'\ \xi(\mathbf{x}-\mathbf{x}',\mathbf{x}';t-t',t')\ q(\mathbf{x}';t').$$

Introducing also the form (12.32) we obtain:

$$q(\mathbf{x};t) - \delta(\mathbf{x})\delta(t)$$
$$= \int d\mathbf{x}'\ f(\mathbf{x}-\mathbf{x}';\mathbf{x}',t) \int_0^t dt'\psi(t-t';\mathbf{x}')\ q(\mathbf{x}';t'). \qquad (12.40)$$

Let us call provisionally $J(t;\mathbf{x}')$ the integral over t' in the right hand side: it is a convolution. It is therefore natural to use Laplace transforms in order to simplify the equations. We introduce the direct and the inverse Laplace transforms for any function of time $F(t)$ ($t \geqslant 0$) as follows:

$$\widetilde{\psi}(s) = \int_0^{\infty} dt\, e^{-st}\, \psi(t), \qquad (12.41)$$

$$\psi(t) = \frac{1}{2\pi i} \int_{\Gamma} ds\, e^{st}\, \widetilde{\psi}(s), \qquad (12.42)$$

where $\overline{\Gamma}$ is the usual Bromwich contour in the complex s-plane for the inversion of the Laplace transformation, *i.e.*, a parallel to the imaginary axis lying to the right of all singularities of the integrand.[4]

The integral $J(t;\mathbf{x}')$ is thus transformed as follows

---

[4] Here we use the familiar notation s for the (complex) Laplace variable, rather than the notation $-iw$ used in Chap. 6 and later in the first part of the book. The contour of integration Γ introduced in the latter case [see (6.60) and Fig. 6.1] is replaced here by the more usual form $\overline{\Gamma}$ of the Bromwich contour.

$$\widetilde{J}(s;\mathbf{x}') = \widetilde{\psi}(s;\mathbf{x}')\,\widetilde{q}(\mathbf{x}';s). \tag{12.43}$$

Eq. (12.37) yields:

$$\widetilde{\zeta}(s,\mathbf{x}) = \frac{1}{s}\,[1 - \widetilde{\psi}(s,\mathbf{x})]. \tag{12.44}$$

On the other hand, from the Laplace transform of Eq. (12.36) we obtain:

$$\widetilde{n}(\mathbf{x};s) = \widetilde{\zeta}(s,\mathbf{x})\,\widetilde{q}(\mathbf{x};s) = \frac{1}{s}\,[1 - \widetilde{\psi}(s,\mathbf{x})]\,\widetilde{q}(\mathbf{x};s). \tag{12.45}$$

We may thus eliminate $\widetilde{q}(\mathbf{x}';s)$ from Eq. (12.43):

$$\widetilde{J}(s;\mathbf{x}') = \widetilde{\phi}(s;\mathbf{x}')\,\widetilde{n}(\mathbf{x}';s),$$

with:

$$\widetilde{\phi}(s;\mathbf{x}') = \frac{s\,\widetilde{\psi}(s;\mathbf{x}')}{1 - \widetilde{\psi}(s,\mathbf{x}')}. \tag{12.46}$$

After inverse Laplace transformation and substitution into Eq. (12.40), the latter becomes:

$$q(\mathbf{x};t) - \delta(\mathbf{x})\delta(t)$$
$$= \int d^d\mathbf{x}'\; f(\mathbf{x} - \mathbf{x}';\mathbf{x}',t) \int_0^t dt'\phi(t - t';\mathbf{x}')\,n(\mathbf{x}';t'). \tag{12.47}$$

A few manipulations allow us to transform this into a closed equation for $n(\mathbf{x},t)$. A Laplace transformation of this equation yields:

$$\widetilde{q}(\mathbf{x};s) - \delta(\mathbf{x}) = \widetilde{g}(\mathbf{x};s), \tag{12.48}$$

where $\widetilde{g}(\mathbf{x};s)$ is the Laplace transform of the right hand side. Multiplying both sides by $s\widetilde{\zeta}(s;\mathbf{x})$, and expressing $\widetilde{q}$ in terms of $\widetilde{n}$ by Eq. (12.45), we may write:

$$[s\widetilde{n}(\mathbf{x},s) - \delta(\mathbf{x})] - \delta(\mathbf{x})\left[s\widetilde{\zeta}(s;\mathbf{x})-1\right] = \left[s\widetilde{\zeta}(s;\mathbf{x}) - 1 + 1\right]\widetilde{g}(\mathbf{x};s). \tag{12.49}$$

Combining Eqs. (12.45) and (12.48) we have:

$$\widetilde{g}(\mathbf{x};s) = \frac{\widetilde{n}(\mathbf{x},s)}{\widetilde{\zeta}(s;\mathbf{x})} - \delta(\mathbf{x}). \tag{12.50}$$

Hence, using (12.44) we obtain::

$$[s\widetilde{\zeta}(s;\mathbf{x})-1]\ \widetilde{g}(\mathbf{x};s) = -\widetilde{\psi}(s;\mathbf{x})\left[\frac{s\ \widetilde{n}(\mathbf{x},s)}{1-\widetilde{\psi}(x;\mathbf{x})} - \delta(\mathbf{x})\right]$$

$$= -\widetilde{\phi}(s;\mathbf{x})\ \widetilde{n}(\mathbf{x};s) + \widetilde{\psi}(s;\mathbf{x})\ \delta(\mathbf{x}).$$

Using this result in the inverse Laplace transform of Eq. (12.49), we obtain:

$$\partial_t n(\mathbf{x},t) = \delta(\mathbf{x})\ \partial_t\zeta(t;\mathbf{x})$$

$$+ \int d^d\mathbf{x}'\ f(\mathbf{x}-\mathbf{x}';\mathbf{x}',t)\int_0^t dt'\phi(t-t';\mathbf{x}')\ n(\mathbf{x}',t')$$

$$- \int_0^t dt'\phi(t-t';\mathbf{x})\ n(\mathbf{x},t') + \psi(t;\mathbf{x})\ \delta(\mathbf{x}).$$

Using the inverse Laplace transform of Eq. (12.44), we see that the first and the last term in the right hand side cancel each other, and we are left with the final form:

$$\partial_t n(\mathbf{x},t) = -\int_0^t dt'\phi(t-t';\mathbf{x})\ n(\mathbf{x},t')$$

$$+ \int d\mathbf{x}'\ f(\mathbf{x}-\mathbf{x}';\mathbf{x}',t)\int_0^t dt'\phi(t-t';\mathbf{x}')\ n(\mathbf{x}',t'). \tag{12.51}$$

This equation has precisely the form of a Master Equation (12.34), with the transition probability:

$$K(\mathbf{x}',\mathbf{x}-\ \mathbf{x}'\ ;t',t-t') = f(\mathbf{x}-\mathbf{x}';\mathbf{x}',t)\ \phi(t-t';\mathbf{x}'). \tag{12.52}$$

This general form of the *vMSC Master Equation* will be further used in Chap. 17. At present we introduce two additional assumptions. The PDF of a jump $\mathbf{x}-\mathbf{x}'$ is supposed to be independent of the starting point $\mathbf{x}'$ and of time: it is thus the same function $f(\mathbf{x}-\mathbf{x}')$ as defined in Sec. 12.1 for the "classical" random walk. On the other hand, the PDF of a pause of duration $t-t'$ is supposed to be independent of the position of the "waiting" particle: $\psi(t-t')$. A CTRW defined by these forms of the functions f and ψ will be called "standard". The Master Equation (12.51) then reduces to:

$$\partial_t n(\mathbf{x},t) = \int_0^t dt'\ \phi(t-t')\left[-n(\mathbf{x},t') + \int d^d\mathbf{x}'\ f(\mathbf{x}-\mathbf{x}')\,n(\mathbf{x}',t')\right].$$

$$\tag{12.53}$$

This *MS-master equation* was derived by MONTROLL & SHLESINGER in 1984: it governs the evolution of the density profile in a standard CTRW. Its most characteristic feature is its *non-Markovian character, both in time and in space:* the rate of change of $n(\mathbf{x}, t)$ at point $\mathbf{x}$ and time t is determined by the spatial environment and by the past history. The effective importance of these features is determined by the range of the functions $f(\mathbf{x})$ and $\phi(t)$.

We now note that in this standard case *both* integrals over t and over $\mathbf{x}'$ are convolutions. We may therefore introduce a Laplace transform in time as before, and also a Fourier transform in space: this operation transforms Eq. (12.53) into a purely algebraic equation:

$$s\widehat{\widetilde{n}}(\mathbf{k}, s) - 1 = -\widetilde{\phi}(s) \left[1 - \widehat{f}(\mathbf{k})\right] \widehat{\widetilde{n}}(\mathbf{k}, s). \qquad (12.54)$$

This equation is immediately solved:

$$\widehat{\widetilde{n}}(\mathbf{k}, s) = \frac{1}{s + \widetilde{\phi}(s) \left[1 - \widehat{f}(\mathbf{k})\right]}. \qquad (12.55)$$

An equivalent form is obtained by using Eq. (12.46):

$$\widehat{\widetilde{n}}(\mathbf{k}, s) = \frac{1 - \widetilde{\psi}(s)}{s} \frac{1}{1 - \widetilde{\psi}(s)\,\widehat{f}(\mathbf{k})}. \qquad (12.56)$$

Transforming back to space and time, we find the density profile:

$$n(\mathbf{x}, t) = (2\pi)^{-d} \int d^d k \ e^{-i\mathbf{k}\cdot\mathbf{x}} (2\pi i)^{-1} \int_{\Gamma} ds\, e^{st} \ \frac{1}{s + \widetilde{\phi}(s) \left[1 - \widehat{f}(\mathbf{k})\right]}. \qquad (12.57)$$

Eq. (12.55) or (12.57) is called the *Montroll-Weiss equation:* it yields the complete solution of the Continuous Time Random Walk problem. When the structure function $\widehat{f}(\mathbf{k})$ and the waiting time distribution $\widetilde{\psi}(s)$, hence $\widetilde{\phi}(s)$, are given, the density profile is determined by quadratures. The result of the latter operation is, however, not always expressible in analytic form!

In order to make the connection with the classical random walks, we consider an important special case. Let $\widehat{f}(\mathbf{k})$ be an isotropic structure function of the form given in Eq. (12.21), which we rewrite here for convenience:

$$\widehat{f}(\mathbf{k}) = 1 + i\mathbf{k}\cdot \langle \mathbf{x} \rangle - (2d)^{-1}k^2 \langle x^2 \rangle, \quad k \to 0. \qquad (12.58)$$

Let also $\widetilde{\psi}(s)$ be of the form:

$$\tilde{\psi}(s) = 1 - s \langle t \rangle + O(s^2), \quad s \to 0, \tag{12.59}$$

hence:

$$\tilde{\phi}(s) = \frac{1 - s \langle t \rangle}{\langle t \rangle}, \quad s \to 0. \tag{12.60}$$

Clearly, $\langle t \rangle$ is the first moment of the waiting time distribution, *i.e.*, the average length of the pause between two steps, *assumed to be finite*. We substitute these expressions into (12.56) and evaluate the latter in the limit $k \to 0, s \to 0$:

$$\widehat{\tilde{n}}(\mathbf{k}, s) \approx \frac{1}{s + \dfrac{1 - s \langle t \rangle}{\langle t \rangle} [1 - 1 - i\mathbf{k} \cdot \langle \mathbf{x} \rangle + (2d)^{-1} k^2 \langle x^2 \rangle]}.$$

Expanding this fraction and neglecting in the denominator the terms of order sk, sk^2, as being of subdominant order, we find:

$$\widehat{\tilde{n}}(\mathbf{k}, s) = \frac{1}{s - i\mathbf{k} \cdot \mathbf{V} + D k^2} \tag{12.61}$$

with:

$$\mathbf{V} = \frac{\langle \mathbf{x} \rangle}{\langle t \rangle}, \quad D = \frac{\langle x^2 \rangle}{2d \langle t \rangle}. \tag{12.62}$$

The inverse Laplace transform of (12.61) is obtained from the residue of the right hand side at the pole $s = i\mathbf{k} \cdot \mathbf{V} - D k^2$:

$$\widehat{n}(\mathbf{k}, t) = \exp \left(i\mathbf{k} \cdot \mathbf{V}t - Dk^2 t \right). \tag{12.63}$$

The inverse Fourier transform of this function is precisely the Gaussian packet (12.27). This result proves the generalized *Central Limit Theorem* for the Continuous Time Random Walks:

For any CTRW defined by a jump PDF $f(\mathbf{x})$ having at least finite first and second moments, and a waiting time PDF $\psi(t)$ having at least a finite first moment, the density profile $n(\mathbf{x}, t)$ tends to a Gaussian packet for long times and/or large distances.

Although the asymptotic Gaussian packet is the most frequently encountered case in physics, the non-Gaussian processes have attracted much interest in recent years because of their unusual properties and of their relevance in some important physical problems. These are realized whenever the conditions of the Central Limit Theorem are not satisfied, *i.e.*, when either $f(\mathbf{x})$, or $\psi(t)$, or both, have a *long tail*. We mention briefly the first case, and study the second one in detail in the forthcoming sections.

In the first case we assume that $\tilde{\psi}(s)$ has the regular form (12.59), but the structure function is non-analytic:

$$\hat{f}(\mathbf{k}) = 1 - B\,k^{\alpha} + ..., \quad 0 < \alpha < 2, \quad k \to 0, \tag{12.64}$$

[where, as usual, $k = (\mathbf{k} \cdot \mathbf{k})^{1/2}$]. In this case we find an expression for the Fourier-Laplace transform of the density profile similar to (12.61):

$$\hat{\tilde{n}}(\mathbf{k}; s) = \frac{1}{s + b\,k^{\alpha}}, \tag{12.65}$$

where $b = B/\langle t \rangle$. This function represents a *Lévy packet*, *i.e.*, a long-tailed Lévy-like distribution. This problem will be revisited in Chap. 17.

12.4   The Standard Long-Tail CTRW (SLT-CTRW)

We now introduce a class of CTRW's that is of particular physical importance in transport theory (BALESCU 1995). It is defined by a symmetric and analytic structure function near the origin:

$$\hat{f}(\mathbf{k}) = 1 - \frac{1}{2d}\,\sigma^2\,k^2 + ..., \quad k \to 0. \tag{12.66}$$

On the other hand, the Laplace transform of the waiting time PDF is non-analytic near $s = 0$

$$\tilde{\psi}(s) = 1 - \tau_D^{\beta}\,s^{\beta} + ..., \quad 0 < \beta < 1, \quad s \to 0. \tag{12.67}$$

A CTRW defined by Eqs. (12.66), (12.67) will be called a *Standard Long-Tail CTRW*. It is determined by four constant parameters:

- the *dimensionality* of the process: d,
- the *exponent* β,
- the *characteristic length* σ,
- the *characteristic time* τ_D.

Note that τ_D is *not* to be confused with the average duration of the pauses $\langle t \rangle$, which is infinite:

$$\langle t \rangle = \left. \frac{d\hat{\psi}(s)}{ds} \right|_{s=0} = -\alpha\tau_D^{\beta}\,s^{\beta-1}\big|_{s=0} = \infty. \tag{12.68}$$

The length σ and the time τ_D allow us to introduce dimensionless variables: x/σ, t/τ_D, $k\sigma$, $s\tau_D$. We thus rewrite Eqs. (12.66) and (12.67) as follows, without introducing new notations for the dimensionless variables:

$$\hat{f}(\mathbf{k}) = 1 - (2d)^{-1}k^2 + ..., \quad k \to 0, \tag{12.69}$$

$$\widetilde{\psi}(s) = 1 - s^\beta + ..., \quad 0 < \beta < 1, \quad s \to 0. \tag{12.70}$$

$$\widetilde{\phi}(s) = s\,(s^{-\beta} - 1) \tag{12.71}$$

We now consider the inverse Laplace transform of $\widetilde{\psi}(s)$: this point must be treated rather carefully. The inverse Laplace transform of the constant 1 is the Dirac delta function $\delta(t)$. The transformation of s^β requires the application of a so-called *Tauberian theorem* (DOETSCH 1955, WEISS 1994). We give here the result:

$$\mathcal{L}_t^{-1}\left[s^\beta\right] = \frac{1}{\Gamma(-\beta)}\,t^{-1-\beta}, \quad s > 0. \tag{12.72}$$

Using also the identity $\Gamma(z + 1) = z\Gamma(z)$, we find the waiting time PDF:

$$\psi(t) = \frac{\beta}{\Gamma(1 - \beta)}\,t^{-1-\beta} + ..., \quad t \to \infty. \tag{12.73}$$

The Montroll-Weiss equation (12.55) yields the following asymptotic expression for the Fourier transform of the density profile:

$$\widehat{n}(\mathbf{k}, t) = \mathcal{L}_t^{-1}\left\{\frac{1}{s^{1-\beta}(2d)^{-1}k^2 + s + O(sk^2)}\right\}. \tag{12.74}$$

The mean square displacement of this distribution is obtained by using Eq. (12.7):

$$\langle r^2(t) \rangle = \frac{1}{\Gamma(1 + \beta)}\,t^\beta, \quad 0 < \beta < 1. \tag{12.75}$$

The SLT-CTRW thus describes a typical case of **strange transport**.[5] This name will denote a transport process for which the MSD behaves asymptotically like a power law, with an exponent different from 1:

$$\langle r^2(t) \rangle = Bt^\mu, \quad t \to \infty. \tag{12.76}$$

---

[5] In the literature on dynamical systems theory this concept is most frequently called "anomalous transport". The latter name is, however, also used for a quite different concept in the plasma physics literature, (and also in the present book) to denote a "normal", diffusive transport mechanism ($\langle \delta x^2(t) \rangle \sim t$) whose transport coefficients are determined by turbulent fluctuations (possibly combined with the effect of collisions) rather than by the collisions alone (as in "classical" transport (see the discussion of Sec. 1.1). We thus prefer to use a different name for the processes described here. The name "strange kinetics" first appeared in a paper by SHLESINGER et al. 1993 (see also BALESCU 1997).

The exponent μ will be called the *diffusion exponent*.[6] We thus have the following cases:

$\mu < 1$	SUBDIFFUSIVE	strange
$\mu = 1$	DIFFUSIVE	normal
$\mu > 1$	SUPERDIFFUSIVE	strange
$\mu = 2$	BALLISTIC	strange

The SLT-CTRW thus describes a typical *subdiffusive process* with $\mu = \beta < 1$: it is this feature which makes it particularly interesting for us.

12.5   SLT-CTRW: The Density Profile

We now show that it is possible to derive an analytical asymptotic expression for the density profile $n(\mathbf{x}, t)$ of a SLT-CTRW. Starting from the Montroll-Weiss equation (12.74) we find:

$$\tilde{n}(\mathbf{x}, s) = s^{\beta-1} (2\pi)^{-d} \int d^d k\, e^{-i\mathbf{k}\cdot\mathbf{x}} \frac{1}{s^\beta + \frac{1}{2d}k^2}. \tag{12.77}$$

We now show that the inversion of the F-L integrals of this expression can be done analytically (BALL et al.[7] 1987, BOUCHAUD & GEORGES 1990, KLAFTER & ZUMOFEN 1994). The following simple trick is used first:

$$\frac{1}{g} = \int_0^\infty d\xi\, e^{-g\xi}.$$

We thus find:

$$\tilde{n}(\mathbf{x}, s) = \frac{s^{\beta-1}}{(2\pi)^d} \int d^d k\, e^{-i\mathbf{k}\cdot\mathbf{x}} \int_0^\infty d\xi\, \exp\left\{-\xi\left[s^\beta + \frac{1}{2d}k^2\right]\right\}$$

$$= s^{\beta-1} \int_0^\infty d\xi\, e^{-\xi s^\beta} (2\pi)^{-d} \int d^d k\, \exp\left(-i\mathbf{k}\cdot\mathbf{x} - \frac{\xi}{2d}k^2\right). \tag{12.78}$$

The integration over $\mathbf{k}$ is trivial; the resulting integral over ξ is tabulated in PRUDNIKOV et al. 1988:

$$\tilde{n}(\mathbf{x}, s) = \frac{(2d)^{(2+d)/4}}{(2\pi)^{d/2}} s^{-1+\beta-\beta(2-d)/4}$$

$$\times x^{1-(d/2)} K_{1-(d/2)}\left[\sqrt{2d}\, s^{\beta/2}\, x\right]. \tag{12.79}$$

---

[6] In the literature, the diffusion exponent is frequently denoted by $\nu = \frac{1}{2}\alpha$. With this convention, the diffusive case corresponds to $\nu = \frac{1}{2}$. This is a matter of pure convenience. By reading a paper, one must make sure of the convention adopted by the author.

[7] Some misprints in this paper have been corrected here.

where $K_n(z)$ is the modified Bessel function of order n. We note that the density profile only depends on the length of the vector **x**: $x = (\mathbf{x} \cdot \mathbf{x})^{1/2}$. The inverse Laplace transform of this quantity cannot be calculated exactly, but an important relation can be derived as follows. We write the density profile in the form:

$$n(x,t) = a \int_{\Gamma} ds \, e^{st} s^{\beta-1-\beta(2-d)/2} \left[s^{\beta/2} x \right]^{1-(d/2)}$$
$$\times \, K_{1-(d/2)} \left[\sqrt{2d} \, s^{\beta/2} x \right],$$

where a is a constant, involving factors $(2d)$ and (2π). We change the integration variable from s to $\eta = st$, and introduce a dimensionless variable q as follows:

$$q = \frac{x}{t^{\beta/2}}. \tag{12.80}$$

We then obtain:

$$n(x,t) = a \int d\eta \, \frac{1}{t} e^{\eta} \left(\frac{\eta}{t} \right)^{-1+(\beta d/2)} (q\eta^{\beta/2})^{1-(d/2)}$$
$$\times \, K_{1-(d/2)} \left(\sqrt{2d} \, q \, \eta^{\beta/2} \right).$$

Thus, calling:

$$F(q) = a q^{1-(d/2)} \int d\eta \, \eta^{-1+\beta(2+d)/4} K_{1-(d/2)} \left(\sqrt{2d} \, q \, \eta^{\beta/2} \right), \tag{12.81}$$

we obtain the exact general form of the density profile:

$$n(x,t) = t^{-\beta d/2} F(q). \tag{12.82}$$

Eq. (12.82) is a typical *scaling relation*. It tells us, in particular, that the density profile depends on r only through the combination q defined in Eq. (12.80); q is called the *similarity variable*. The density profile is thus expressed as a product of two factors:

- The first (dimensionless) factor is an explicit function of time alone;
- The second (dimensionless) factor is a function of the similarity variable q alone; its specific form depends on the diffusion exponent β and on the dimensionality d.

Although the function $F(q)$ cannot be calculated analytically in general, its asymptotic behaviour can be obtained as follows. The modified

Bessel function has a well-known asymptotic expansion (ABRAMOWITZ & STEGUN 1972):

$$K_\nu(z) \sim \sqrt{\frac{\pi}{2z}} e^{-z} [1 + O(z^{-1})], \quad \forall \nu. \tag{12.83}$$

Substituting this expression into (12.81) we find:

$$F(q) \sim q^{1-(1+d)/2} \int d\eta\, \eta^{-1+\beta(1+d)/4} \exp\left(\eta - \sqrt{2d}\, q\, \eta^{\beta/2}\right). \tag{12.84}$$

This integral is evaluated by the method of steepest descent. Omitting the slowly varying (power law) prefactor, we find the asymptotic form:

$$F(q) \sim \exp\left(-bq^\delta\right), \quad \delta = \frac{2}{2-\beta}, \quad q \to \infty. \tag{12.85}$$

We thus find a characteristic *stretched exponential* behaviour. We note that in the case $\beta = 1$, *i.e.* in the diffusive case, $\delta = 2$ and $F(q)$ reduces to a Gaussian and (12.82) reduces to the Gaussian packet (12.27). We thus recover the prediction of the Central Limit Theorem. This limiting case allows us to fix the value of the constant $b = 1/4$.

Another interesting consequence of the scaling law (12.82) concerns *the moments of the density profile*. These are obtained as follows:

$$\langle x^{2p}(t)\rangle = \int d^d x\, x^{2p}\, a\, t^{-\beta d/2} F(q).$$

We make the change of integration variable $\mathbf{x} \to t^{\beta/2}\mathbf{y}$ [note that $\mathbf{y} \cdot \mathbf{y} = q^2$]:

$$\langle x^{2p}(t)\rangle = t^{(d+2p)\beta/2} t^{-\beta d/2} a \int d^d y\, y^{2p} F(y).$$

Noting that the last integral over $\mathbf{y}$ is a mere constant, we find:

$$\langle x^{2p}(t)\rangle = M_p\, t^{p\beta}. \tag{12.86}$$

We thus obtained a *scaling law* for all the moments. Note that the scaling exponent $p\beta$ is independent of the dimensionality d of the space. Moreover, the form of the function $F(q)$ does not influence the value of this exponent: it merely determines the value of the constant M_p. The latter depends on the order p, on the diffusion exponent β and on the dimensionality d. The two parameters defining the SLT-CTRW, *i.e.* σ and τ_D, appear in (12.86) as natural units of length and time, respectively. For $p = 1$, Eq. (12.86) reduces, of course, to (12.75). In this particular case the constant M_p is independent of d. The present calculation cannot yield the value of the coefficient M_p; the latter will be obtained explicitly, however, in Sec. 12.6.

We now specialize these results to the *one-dimensional SLT-CTRW* ($d = 1$), which is of interest in many applications; on the other hand, it is possible to derive completely explicit expressions in this case. In particular, it is known (ABRAMOWITZ & STEGUN 1972) that the modified Bessel function $K_{1/2}(z)$ is expressed in terms of elementary functions:

$$K_{1/2}(z) = \sqrt{\frac{\pi}{2}}\, z^{-1/2}\, e^{-z}. \tag{12.87}$$

Note that this expression is exact, whereas the similar Eq. (12.83) is merely an asymptotic approximation for $\nu \neq 1/2$. Using this expression in Eq. (12.79) with $d = 1$, we find:

$$\tilde{n}(x, s) = \frac{1}{\sqrt{2}}\, s^{(\beta-2)/2}\, e^{-\bar{q}}, \quad d = 1, \tag{12.88}$$

where:

$$\bar{q} = \sqrt{2}\, s^{\beta/2}\, x. \tag{12.89}$$

The steepest descent method then leads to:

$$n(x, t) = A\, t^{-\beta/2}\, \exp\left(-\frac{q^{\delta}}{4}\right).$$

In the present case, the constant A can be calculated exactly, using the normalization condition for $n(x; t)$:

$$1 = \int_{-\infty}^{\infty} dx\, A\, t^{-\beta/2}\, \exp\left(-\frac{x^{\delta}}{4t^{\beta\delta/2}}\right).$$

Changing the integration variable to ζ, defined as $\zeta^{\delta} =$ (argument of the exponential), the integral is brought to a form tabulated by PRUDNIKOV et al. 1988:

$$\int_0^{\infty} dx\, x^{\beta-1}\, \exp(-px^{\mu}) = \frac{1}{\mu}\, p^{-(\beta/\mu)}\, \Gamma\left[\frac{\beta}{\mu}\right]. \tag{12.90}$$

A simple calculation then leads to:

$$n(x, t) = \frac{\delta}{4^{(2+\delta)/2\delta}\, \Gamma\left[\frac{1}{\delta}\right]}\, t^{-\beta/2}\, \exp\left(-\frac{q^{\delta}}{4}\right).$$

This expression can be further simplified by using the form of δ, Eq. (12.85) and some identities involving the Γ-function:

$$n(x,t) = \frac{1}{2^{3-\beta}\, \Gamma\left(\frac{4-\beta}{2}\right)}\, t^{-\beta/2} \exp\left(-\frac{q^{2/(2-\beta)}}{4}\right), \quad q \gg 1, \quad d = 1.$$

(12.91)

This function represents a *long-tailed density profile*. It is further studied in the forthcoming sections, where it is compared to a related Gaussian packet.

12.6   SLT-CTRW: The Non-Markovian Diffusion Equation

The Montroll-Shlesinger Master Equation (12.53) was derived for a rather general CTRW. Our purpose here is to derive its specific form for the SLT-CTRW and to study some of its properties.

We consider the SLT-CTRW defined by Eqs. (12.69) and (12.70). Its asymptotic Fourier-Laplace density profile was obtained from the Montroll-Weiss equation (12.55) in the limit $\tau_D s \ll 1$, $\sigma k \ll 1$:

$$\widehat{\widetilde{n}}(\mathbf{k},s) = \frac{s^{\beta-1}}{s^\beta + \frac{1}{2d}k^2}.$$

(12.92)

It is easily checked that this function satisfies the master equation (12.53) with the kernel $\widetilde{\phi}(s)$ defined by (12.70), which is simplified by retaining only the leading term for $s \ll 1$:

$$\widetilde{\phi}(s) = s^{1-\beta}.$$

(12.93)

The asymptotic Fourier-Laplace master equation (12.54) becomes:

$$s\,\widehat{\widetilde{n}}(\mathbf{k},s) - 1 = -\widetilde{\phi}(s)\,\frac{1}{2d}\,k^2\,\widehat{\widetilde{n}}(\mathbf{k},s).$$

(12.94)

The inverse Fourier transform of this equation is [we recall that the initial value of the density profile is: $n(\mathbf{x}, t = 0) = \delta(\mathbf{x})$]:

$$s\widetilde{n}(\mathbf{x}, s) - \delta(\mathbf{x}) = (2d)^{-1}\,\widetilde{\phi}(s)\,\nabla^2\widetilde{n}(\mathbf{x}, s).$$

(12.95)

This looks very much like a Laplace-transformed diffusion equation; the diffusion coefficient in Laplace space is, however, s-dependent. As a result, the inverse Laplace transform of (12.95) is an *integral equation*:

$$\partial_t n(\mathbf{x}, t) = \int_0^t dt_1\, H(t_1)\, \nabla^2 n(\mathbf{x}, t - t_1),$$

(12.96)

with:

$$H(t) = (2d)^{-1} \phi(t). \qquad (12.97)$$

Our main conclusion at this point is: *The density profile of a Standard Long-Tail CTRW obeys a linear, non-Markovian equation.* Given the obvious resemblance of this equation with the diffusion equation, we call it the **non-Markovian diffusion equation**. Note that, in contrast to the general case (12.54) this equation is Markovian with respect to the spatial variable: this is because of the analytic form of the structure function $\hat{f}(\mathbf{k})$: the term $(-k^2)$ yields the Laplace operator in physical space. On the other hand, the rate of change of the density profile is influenced by its past history. Given the *long tail* of the memory kernel $H(t)$, the range of effective influence extends far into the past. It is this feature that is responsible of the non-Gaussian character of the density profile and of the strange transport law.

The non-Markovian feature of this equation should not be surprising to us. We have already seen in Sec. 6.2 that the non-Markovianity is a general property of the kinetic equations of turbulent plasmas: see Eq. (6.2). It is also a general property of the kinetic equations derived from the Liouville equation in non-equilibrium statistical mechanics (BALESCU 1975, 1997). In these cases, the memory kernel is most often a rapidly (exponentially) decaying function of time. This implies the existence of an intrinsic time-scale. It justifies a Markovian approximation which neglects the memory effects after a time much longer than the range of this kernel. In the present case, however, the long tail of this kernel does not allow such an approximation: there exists no intrinsic time-scale. The great interest of the SLT-CTRW lies in the possibility to study in detail the **memory effects** on the evolution of the density profile.

In order to calculate the kernel $H(t)$ we note that $\tilde{\phi}(s)$ can be simply related to the Laplace transform of the waiting time distribution, $\tilde{\psi}(s)$ defined in (12.70):

$$\tilde{\phi}(s) = s^{1-\beta} = \left[1 - \tilde{\psi}_{1-\beta}(s)\right], \qquad (12.98)$$

where $\tilde{\psi}_{1-\beta}(s)$ denotes the function $\tilde{\psi}(s)$ in which the index β is changed to $1 - \beta$. The inverse Laplace transform of 1 is $\delta(t)$; the transform of the second term is given *asymptotically* by the Tauberian theorem (12.72) [we note, indeed that $1 - \beta > 0$ for $0 < \beta < 1$]:

$$H(t) = (2d)^{-1} \left[\delta(t) - \frac{1-\beta}{\Gamma(\beta)} \frac{1}{t^{2-\beta}}\right]. \qquad (12.99)$$

The delta-function is zero for $t > 0$; we keep it here, however, because it will help us in the derivation of a useful form of the equation of evolution. Substituting this form into (12.96) we find:

$$\partial_t n(\mathbf{x}, t) = D_0 \, \nabla^2 n(\mathbf{x}, t) - H_0 \int_0^t dt_1 \, \frac{1}{t_1^{2-\beta}} \, \nabla^2 n(\mathbf{x}, t - t_1), \quad (12.100)$$

where:

$$D_0 = (2d)^{-1}, \quad H_0 = \frac{1 - \beta}{\Gamma(\beta)} \, D_0. \quad (12.101)$$

Eq. (12.100) describes the evolution as a superposition of an ordinary diffusion term, with a diffusion coefficient D_0, and a *non-Markovian process* represented by the second term. The characteristic feature of the kernel in (12.100) is its long tail: it decreases very slowly, as an inverse power law (note that $2 - \beta > 0$). A very interesting feature appears in the definition of the coefficient H_0: this coefficient vanishes for $\beta = 1$. Thus, *in the diffusive case, $\beta = 1$, the non-Markovian term in (12.100) disappears, and we are left with a pure diffusion equation.*

As it stands, however, Eq. (12.100) does not really make sense; indeed, the kernel $t^{-2+\beta}$ diverges at the lower limit $t = 0$. This is not surprising, because the expression of $H(t)$ obtained in (12.99) is an *asymptotic* expression, valid for long times, and certainly inapplicable for $t \to 0$. In order to cure this difficulty, we introduce a *lower cut-off* $t_{\min}$ in the integral of Eq. (12.100), which will be determined below. The equation of evolution is then written in the following, final form:

$$\partial_t n(\mathbf{x}, t) = D_0 \, \nabla^2 n(\mathbf{x}, t) - H_0 \int_{t_{\min}}^t dt_1 \, \frac{1}{t_1^{2-\beta}} \, \nabla^2 n(\mathbf{x}, t - t_1). \quad (12.102)$$

In order to define a criterion for the determination of the cut-off, we consider the moments of this equation, beginning with the second moment, *i.e.*, the MSD. This moment was determined independently in Eq. (12.75) and was shown to scale asymptotically as $\langle x^2(t) \rangle \sim t^\beta$, thus: $\partial_t \langle x^2(t) \rangle \sim t^{\beta-1}$. We now calculate its rate of change by using Eq. (12.102):

$$\partial_t \langle x^2(t) \rangle = \int d^d x \, x^2 \left\{ D_0 \, \nabla^2 n(\mathbf{x}, t) - H_0 \int_{t_{\min}}^t dt_1 \, \frac{1}{t_1^{2-\beta}} \, \nabla^2 n(\mathbf{x}, t - t_1) \right\}.$$

We perform an integration by parts over x and note that:

$$\int d^d x \, x^2 \, \nabla^2 n(\mathbf{x}, t) = 2d, \quad \forall t.$$

We thus obtain:

$$\partial_t \left\langle x^2(t) \right\rangle = 2dD_0 \left\{ 1 + \frac{H_0}{1-\alpha} \frac{t^{\beta-1} - t_{\min}^{\beta-1}}{\tau_D^{\alpha-1}} \right\}. \tag{12.103}$$

This expression contains a term which has the correct scaling, $\sim t^{\beta-1}$, and a term that is constant, and which would dominate for $\beta < 1$ and $t \to \infty$. The latter term is thus spurious. But, we can use our freedom for defining the cut-off $t_{\min}$ in such a way as to annul this term; this leads to the value:

$$t_{\min} = \left[\frac{H_0}{(1-\beta) D_0} \right]^{1/(1-\beta)}. \tag{12.104}$$

When Eq. (12.101) is used, we obtain:

$$t_{\min} = \left[\Gamma(\alpha) \right]^{-1/(1-\beta)}. \tag{12.105}$$

With this choice of the cut-off, we obtain:

$$\partial_t \left\langle x^2(t) \right\rangle = 2dD_0 \frac{1}{\Gamma(\beta)} t^{\beta-1},$$

and consequently, using (12.101):

$$\left\langle x^2(t) \right\rangle = 2dD_0 \frac{1}{\beta\Gamma(\beta)} t^\beta = \frac{1}{\Gamma(1+\beta)} t^\beta.$$

This is identical to Eq. (12.75). As can be seen from Eq. (12.103), the "correct" term originates from the non-Markovian contribution. This shows that the diffusion term in (12.102) is, in a sense, spurious: it compensates the effect originating from the inaccurate short-time domain: the really physical mechanism of evolution is in the long-tail non-Markovian operator.

The validity of Eq. (12.102) is, however, not yet completely settled. Although it yields the correct second moment of the density profile, we should also worry about the higher order moments! In order to determine the latter, we first note the following identity:

$$\nabla^2 x^{2p} = 2p \left(2p + d - 2 \right) x^{2p-2}. \tag{12.106}$$

We then derive from (12.102) the following recursion relation for the moments:

$$\partial_t \left\langle x^{2p}(t) \right\rangle = 2p \left(2p + d - 2 \right) D_0$$

$$\times \left\{ \left\langle x^{2p-2}(t) \right\rangle - \frac{1-\beta}{\Gamma(\beta)} \int_{t_{\min}}^t dt_1 \frac{1}{t_1^{2-\beta}} \left\langle x^{2p-2}(t - t_1) \right\rangle \right\}. \tag{12.107}$$

The general scaling of the moments was derived in (12.86):

$$\langle x^{2p}(t) \rangle = M_p \, t^{p\beta}, \tag{12.108}$$

which, upon substitution into (12.6) yields:

$$p\beta M_p t^{p\beta-1} = 2p\,(2p+d-2)D_0$$
$$\times \left\{ M_{p-1} t^{(p-1)\beta} - \frac{1-\beta}{\Gamma(\beta)} \tau_D^{1-\beta} M_{p-1} \int_{t_{\min}}^{t} dt_1 \, t_1^{\beta-2} (t-t_1)^{(p-1)\beta} \right\}. \tag{12.109}$$

The value of the integral can be obtained (*e.g.*, by the MATHEMATICA program):

$$\int_{t_{\min}}^{t} dt' \, t'^{\beta-2} (t-t')^{(p-1)\beta} = \frac{\Gamma(\beta-1)\,\Gamma\left[1+\beta(p-1)\right]}{\Gamma\left[\beta+\beta(p-1)\right]} \, t^{p\beta-1}$$
$$+ \frac{t_{\min}}{1-\beta} \, t^{\beta(p-1)} \, {}_2F_1\left[-1+\beta, -\beta(p-1), \beta, (t_{\min}/t)\right],$$

where ${}_2F_1[\ldots]$ is a hypergeometric function which, for small values of its last argument, behaves as $1 + O(t_{\min}/t)$. Thus, asymptotically, Eq. (12.6) yields:

$$p\beta C_p t^{p\beta-1} = 2p\,(2p+d-2)D_0\,C_{p-1}$$
$$\times \left\{ t^{(p-1)\beta} - (1-\beta)\frac{\Gamma(\beta-1)\Gamma\left[1+\beta(p-1)\right]}{\Gamma(\beta)\Gamma(p\beta)} t^{p\beta-1} - t_{\min}^{\beta-1} \frac{1}{\Gamma(\beta)} t^{(p-1)\beta} \right\}. \tag{12.110}$$

We see again in the right hand side two terms proportional to $t^{(p-1)\beta}$ which do not have the same scaling as the left hand side. Their coefficient must therefore be put to zero, thus:

$$1 - t_{\min}^{\beta-1} \frac{1}{\Gamma(\beta)} = 0.$$

This yields the *same value for* $t_{\min}$ *as Eq.* (12.105). This result is very important: it guarantees that *the non-Markovian diffusion equation (12.102), combined with the definition (12.105) for the cut-off, yields the correct asymptotic values for all the moments of the density profile, hence for the profile itself.*

It is now easily seen that, using some identities for the Γ-functions, (12.6) reduces to:

$$M_p = 2(p-1)(2p+d-1)\frac{\Gamma[\beta(p-1)]}{\Gamma(\beta p)}D_0\,M_{p-1}.$$

This recurrence relation is easily solved; using the value (12.101) for D_0, we find:

$$M_p = \frac{d(d+2)...(d+2p-2)}{d^p}\frac{(p-1)!}{\beta\,\Gamma(\beta p)}. \qquad (12.111)$$

This coefficient gives the final solution for *all the moments of the SLT-CTRW*, for arbitrary diffusion exponent β $(0 < \beta \leq 1)$, arbitrary order p and arbitrary dimensionality d. This is the result announced after Eq. (12.86).

It is interesting to discuss some special cases. Consider first the MSD, *i.e.*, the case $p = 1$, in arbitrary dimensionality. Recalling that $\beta\Gamma(\beta) = \Gamma(\beta+1)$, we see that (12.111) reduces to:

$$M_1 = \frac{1}{\Gamma(\beta+1)}, \quad p=1, \quad \forall d. \qquad (12.112)$$

This result agrees with (12.75). Thus, the MSD is the only moment whose value is independent of the dimensionality d.

Next, we consider the *diffusive case*, $\beta = 1$, for arbitrary dimensionality. In this case we have $\beta\Gamma(p\beta) = \Gamma(p) = (p-1)!$ and Eq. (12.111) reduces to:

$$M_p = \frac{d(d+2)...(d+2p-2)}{d^p}, \quad \beta=1, \quad \forall p. \qquad (12.113)$$

This is a quite interesting case. In order to derive the moments in the diffusive case, one may start from Eq. (12.100) in which the non-Markovian term is deleted, because $H_0 \sim (1 - \beta)$. We are then left with an ordinary diffusion equation, from which we derive a recursion equation which is simply (12.6) without the non-Markovian term:

$$\partial_t\left\langle x^{2p}(t)\right\rangle = 2p(2p+d-2)\,D_0\left\langle x^{2p-2}(t)\right\rangle \qquad (12.114)$$

For $p = 1$, this equation reduces to $\partial_t\left\langle x^2(t)\right\rangle = 2dD_0$, with the well-known diffusive form for the MSD: $\left\langle x^2(t)\right\rangle = 2dD_0 t$. From this starting point, the recursion equation (12.114) is solved by successive iteration, yielding the same solution as (12.113). The surprising feature is that in the derivation of (12.111), at the step (12.6), the contribution of the diffusive term, *i.e.*, the first term in the right hand side of (12.6), is exactly cancelled by the third term. Hence, the expression (12.111), valid for arbitrary $\beta < 1$, results entirely from the non-Markovian term in the equation of evolution. Nevertheless, when extrapolated to $\beta = 1$, the result connects smoothly to the result obtained from the "pure" diffusion equation.

Another special case that will be of interest in applications is the one-dimensional case $(d = 1)$ for arbitrary β. In this case we note:

$$\frac{d(d+2)...(d+2p-2)\ (p-1)!}{d^p} = 1 \cdot 3 \cdot 5...(2p-1) \cdot 1 \cdot 2 \cdot 3...(p-1)$$

$$= 1 \cdot 3 \cdot 5...(2p-1) \cdot \frac{2 \cdot 4...(2p-2)}{2^{p-1}} = \frac{(2p-1)!}{2^{p-1}}.$$

Thus:

$$M_p = \frac{(2p-1)!}{2^{p-1}} \frac{1}{\beta\Gamma(p\beta)}, \quad d = 1, \quad \forall\beta, \forall p. \tag{12.115}$$

12.7 Magnetic Subdiffusion and SLT-CTRW

- ### Construction of the SLT-CTRW

We now come back to the problem of magnetic turbulence treated in Chap. 11 by the V-Langevin equation and by the hybrid kinetic equation. It describes the motion of guiding centres in the presence of magnetic field fluctuations and of collisional diffusion in the parallel direction. In the limit of an infinite perpendicular correlation length of the stochastic field, it was shown that the dispersion of the particles is subdiffusive, and that the MSD could be calculated analytically.

Quite naturally, we may ask ourselves *whether this subdiffusive behaviour could be described by a continuous time random walk (CTRW) model*, just as the diffusive behaviour produced by collisions can be described by a classical random walk. Given that many features of this problem can be calculated analytically, this limiting model appears as a remarkable "laboratory" for testing the CTRW model.

Our starting point will be Eq. (11.99) for the MSD: it shows that this quantity is asymptotically proportional to $\sqrt{t}$. In view of our study of CTRW's, it appears natural to try to interpret this process in terms of a Standard Long Tail CTRW (SLT-CTRW), which always describes a subdiffusive process. In order to define a SLT-CTRW, we need to specify four parameters: β, d, σ and τ_D [see Eqs. (12.66), (12.67)].[8] These should be related to the parameters at our disposal, *i.e.*, the collisional parallel diffusion coefficient $\chi_{\parallel}$ and the magnetic fluctuation characteristics, *i.e.*, the parallel correlation length, $\lambda_{\parallel}$ and the amplitude of the fluctuations, $\overline{\beta}$.

- The value of the diffusion exponent is obvious: $\beta = 1/2$.
- The main interest in this book is anomalous transport of plasmas across the magnetic surfaces, *i.e.*, in the x-direction of a slab model. Hence we consider the diffusion problem in one dimension: $d = 1$.

---

[8] We now switch back to dimensional variables x and t.

- In order to define a unit of time, we could first think of the inverse collision frequency, ν^{-1}; this, however, does not properly describe the problem, as it takes no account of the turbulence. A natural definition is the time it takes for an initially localized bunch of particles to diffuse collisionally in the z-direction over a distance equal to the parallel correlation length of the fluctuations $\lambda_{\parallel}$:

$$\tau_D = \frac{\lambda_{\parallel}^2}{2\chi_{\parallel}} = \frac{\lambda_{\parallel}^2 \nu}{V_T^2} = \frac{1}{\overline{\chi}_{\parallel} \nu}. \tag{12.116}$$

The second form is obtained by Eq. (11.35), the third one from Eq. (11.92). Thus, the time τ_D differs from ν^{-1} by the factor $\overline{\chi}_{\parallel}^{-1}$, which is large when the plasma is highly collisional.

- The remaining parameter, σ, is obtained by comparing the asymptotic expression (11.99) of the MSD obtained from Langevin with the expression (12.75) obtained from the CTRW [note that for $\beta = 1/2$, $1/\Gamma(1+\beta) = 2/\sqrt{\pi}$]:[9]

$$\langle \delta X^2(t) \rangle = \frac{4\chi_{\parallel} \overline{\beta}^2}{(\overline{\chi}_{\parallel} \nu)^{1/2}} t^{1/2} = \frac{2}{\sqrt{\pi}} \sigma^2 \left(\frac{t}{\tau_D}\right)^{1/2}. \tag{12.117}$$

Using Eq. (12.116) we obtain:

$$\sigma^2 = \sqrt{\pi} \overline{\beta}^2 \lambda_{\parallel}^2. \tag{12.118}$$

The characteristic perpendicular jump length σ of the CTRW is thus determined by the characteristics of the β magnetic fluctuations. This is clear, because the only mechanism producing a perpendicular displacement of the particles (tied to the magnetic field lines) is the radial diffusion of these lines.

We collect these results in the following table defining the SLT-CTRW modelling the magnetic subdiffusion of the guiding centres:

$$
\begin{array}{|c|}
\hline
\beta = 1/2 \\
\hline
d = 1 \\
\hline
\tau_D = \dfrac{\lambda_{\parallel}^2}{2\chi_{\parallel}} = \dfrac{1}{\overline{\chi}_{\parallel} \nu} \\
\hline
\sigma^2 = \sqrt{\pi} \overline{\beta}^2 \lambda_{\parallel}^2 \\
\hline
\end{array}
\tag{12.119}
$$

It is thus easy to adapt a SLT-CTRW to our test problem and note that its characteristic parameters have a reasonable physical meaning. The

---

[9] We use the notations of Chap. 11: $\mathbf{X}$ is the (dimensional) coordinate of a guiding centre, $\mathbf{K}$ is the corresponding (dimensional) wave-vector. In Eq. (12.75) we replace $r \to X/\sigma$, $t \to t/\tau_D$.

MSD is, however, a rather weak criterion, as many different models may lead to the same MSD. A much more sensitive criterion is the density profile, which will presently be considered.

The general asymptotic solution of the SLT-CTRW was derived in the previous section; in one dimension it was given in Eq. (12.91) which, for $\beta = \frac{1}{2}$ reduces to:

$$n_{SLT}(q,t) = \frac{1}{\sigma} \frac{1}{2^{5/2} \, \Gamma(\frac{7}{4})} \left(\frac{\tau_D}{t}\right)^{1/4} \exp\left(-\frac{q^{4/3}}{4}\right). \qquad (12.120)$$

One recognizes the general form of the *scaling law* (12.82). The similarity variable q, defined in (12.80) is here:

$$q = \left(\frac{\tau_D}{t}\right)^{1/4} \frac{X}{\sigma}. \qquad (12.121)$$

Alternatively, the SLT profile can also be characterized as the fundamental solution of the *non-Markovian diffusion equation* (12.102), with the cut-off (12.105); this equation reduces in the present case to:

$$\partial_t n_{SLT}(X,t)$$
$$= D_0 \, \nabla_X^2 \, n_{SLT}(X,t) - H_0 \int_{\tau_D/\pi}^{t} dt_1 \left(\frac{\tau_D}{t_1}\right)^{3/2} \nabla_X^2 \, n_{SLT}(X, t-t_1),$$
$$\qquad (12.122)$$

with:

$$D_0 = \frac{\sigma^2}{2\tau_D}, \qquad H_0 = \frac{1}{2\sqrt{\pi}\tau_D} D_0. \qquad (12.123)$$

We note the slowly decaying, long-tailed memory kernel which decreases like $t^{-3/2}$.

These results will now be compared with the V-Langevin treatment. The latter can be considered at several "levels of sophistication".

- **Rechester - Rosenbluth Theory and SLT-CTRW**

A first result is obtained by translating into an equation of evolution the argument described qualitatively in the pioneering work of RECHESTER & ROSENBLUTH 1978. As shown in Sec. 11.7, the magnetic fluctuations produce a spatial diffusion of the magnetic field lines. This process can be described by a normal diffusion equation in which the role of time is played by the Z-coordinate:

$$\frac{\partial}{\partial Z} \bar{n}(X,Z) = d_m^\infty \, \nabla_X^2 \, \bar{n}(X,Z). \qquad (12.124)$$

This is a purely geometrical problem (WANG et al. 1995): $\bar{n}(X, Z)$ is the PDF for a point, starting at $Z = 0$ on a magnetic field line, to be found at a perpendicular distance X from the latter at "time" Z. The magnetic diffusion coefficient d_m^∞ was derived in Eq. (11.104). In a next step, Rechester and Rosenbluth estimate the perpendicular *particle MSD* by substituting into the magnetic MSD, Eq. (11.103) the asymptotic value (11.106) and obtain the result (11.108) which, expressed in terms of SLT-CTRW parameters [using Eqs. (12.118) and (12.123)] is:

$$D_{RR}(t) = \frac{D_0}{\sqrt{2}} \left(\frac{\tau_D}{t} \right)^{1/2} .$$

We know, however, that this coefficient is wrong by a factor $\sqrt{\pi/2}$. We therefore rather use the correct diffusion coefficient $D^\infty(t)$ of Eq. (11.99):

$$D^\infty(t) = \frac{D_0}{\sqrt{\pi}} \left(\frac{\tau_D}{t} \right)^{1/2} . \tag{12.125}$$

A better treatment in this framework is obtained by making the change of variables (11.106) (corrected by the factor $\sqrt{2/\pi}$) in the geometrical diffusion equation (12.124), which is thus transformed into an equation of evolution in (true) time:

$$\frac{\partial}{\partial t} n_{RR}(X, t) = D^\infty(t) \nabla_X^2 n_{RR}(X, t), \tag{12.126}$$

where $n_{RR}(x, t) = \bar{n}[x, z(t)]$. Thus, the Rechester-Rosenbluth argument leads to an *"ordinary" diffusion equation with a time-dependent diffusion coefficient* $D^\infty(t)$, which coincides with the running diffusion coefficient (11.99) derived from the V-Langevin equation.

In conclusion, the Rechester-Rosenbluth argument yields the same MSD and the same running diffusion coefficient as the exact solution of the V-Langevin equation and as the SLT-CTRW model. Nevertheless, *the equation of evolution (12.126) is completely different from the equation (12.7) resulting from the SLT-CTRW description*. Eq. (12.126) is an ordinary diffusion equation with a time-dependent coefficient, whereas (12.7) is a non-Markovian diffusion equation. The memory effect is thus absent from Eq. (12.126). It may therefore be expected that the density profiles predicted by the two equations be quite different.

Eq. (12.126) is easily solved. The change of variables: $t \to T^2$, $\nu(X, T) = n_{RR}(X, T^2)$ transforms it into:

$$\partial_T \nu(X, T) = 2 \sqrt{\frac{\tau_D}{\pi}} D_0 \nabla_X^2 \nu(X, T). \tag{12.127}$$

The fundamental solution of this *ordinary* diffusion equation is:

$$\nu(X, T) = \frac{1}{\left(4\sqrt{\pi\tau_D}\, D_0\, T\right)^{1/2}} \exp\left(-\frac{\sqrt{\pi}}{8\, D_0\sqrt{\tau_D}\, T}\, X^2\right); \qquad (12.128)$$

hence, transforming back to t and using Eqs. (12.123):

$$n_{RR}(X, t) = \frac{1}{\sigma} \frac{1}{(2\sqrt{\pi})^{1/2}} \left(\frac{\tau_D}{t}\right)^{1/4} \exp\left[-\frac{\sqrt{\pi}}{4} \left(\frac{\tau_D}{t}\right)^{1/2} \frac{X^2}{\sigma^2}\right]. \quad (12.129)$$

In terms of the similarity variable (12.121) this takes the simpler form:

$$n_{RR}(q, t) = \frac{1}{\sigma} \frac{1}{(2\sqrt{\pi})^{1/2}} \left(\frac{\tau_D}{t}\right)^{1/4} \exp\left(-\frac{\sqrt{\pi}}{4}\, q^2\right). \qquad (12.130)$$

Thus, the density profile obtained by the Rechester-Rosenbluth argument is quite different from the one predicted by the SLT-CTRW, Eq. (12.120). The former is simply a Gaussian packet, whose square width increases more slowly than in the "usual" diffusive Gaussian (12.18) ($\sim \sqrt{t}$ instead of $\sim t$). Nevertheless, both profiles have an important feature in common. Indeed, both $n_{SLT}(q, t)$, Eq. (12.120) and $n_{RR}(q, t)$, Eq. (12.130) are of the general scaling form (12.82) (for $d = 1$, $\beta = 1/2$); they differ only by the form of the function $F(q)$.

- **Markovianization of the SLT-CTRW Diffusion Equation**

We may find a deeper relationship between the SLT-CTRW and the RR profiles or, equivalently, between the non-Markovian diffusion equation (12.7) and Eq. (12.126), of which they are fundamental solutions. Non-Markovian equations appeared already in several places in this book. In kinetic theory we derived the kinetic equation (6.2) which contained explicitly memory effects; in the theory of the hybrid kinetic equation we found Eq. (11.128) which is very similar to (12.102). In all these cases, we used a *Markovian approximation* for the simplification of the equations. The argument was based on the existence of two separate time scales: the memory kernel was supposed to decay (exponentially) in a time which is very short compared to the characteristic time of variation of the unknown function. The retardation in the latter can then be neglected, and the equation is Markovianized.

We may try to apply this procedure to Eq. (12.102) (without asking questions about its validity at this point!): we consider here the general case (arbitrary values for $0 < \beta < 1$ and d). We replace $n(\mathbf{X}, t - t_1)$ by $n(\mathbf{X}, t)$ under the integral, and denote by $n_M(\mathbf{X}, t)$ the solution of the resulting Markovianized equation:

$$\partial_t n_M(\mathbf{X}, t) = D_0 \left[1 - A(t) \right] \nabla^2 n_M(\mathbf{X}, t), \qquad (12.131)$$

where:

$$1 - A(t) = 1 - \frac{1 - \beta}{\tau_D \, \Gamma(\beta)} \int_{t_{\min}}^{t} dt_1 \left(\frac{\tau_D}{t_1} \right)^{2 - \beta}$$

$$= 1 - \frac{1}{\Gamma(\beta)} \left(\frac{\tau_D}{t_{\min}} \right)^{1 - \beta} + \frac{1}{\Gamma(\beta)} \left(\frac{\tau_D}{t} \right)^{1 - \beta}. \qquad (12.132)$$

[Note that we did not take the more extreme step of letting the upper limit of the integral go to infinity!] The two first terms on the right hand side cancel each other exactly for the value (12.105) of the cut-off, and we are left with:

$$\partial_t n_M(\mathbf{X}, t) = \frac{D_0}{\Gamma(\beta)} \left(\frac{\tau_D}{t} \right)^{1 - \beta} \nabla^2 n_M(\mathbf{X}, t). \qquad (12.133)$$

Thus, the Markovianized equation of evolution is a diffusion equation with a time-dependent diffusion coefficient. Note that this coefficient tends asymptotically to zero like $t^{-1+\beta}$: Eq. (12.133) describes, indeed, a *subdiffusive process*, as we already know. It is for this reason that the "complete" Markovianization (*i.e.*, letting $t \to \infty$ in the integral) would lead to a trivial result here. *The complete Markovianization is only applicable for a diffusive process.*

Eq. (12.133) can be solved analytically by the same procedure as above. The change of variable $t \to T = t^\beta$, $\nu(\mathbf{X}, T) = n_M(\mathbf{X}, T^{1/\beta})$ transforms it into a diffusion equation with a constant coefficient:

$$\partial_T \nu(\mathbf{X}, T) = \mathcal{D} \nabla^2 \nu(\mathbf{X}, T), \qquad (12.134)$$

with:

$$\mathcal{D} = \frac{D_0}{\Gamma(\beta + 1)} \tau_D^{1 - \beta}. \qquad (12.135)$$

The fundamental solution of this equation is given in Eq. (12.27) [for $\mathbf{V} = 0$]; in the resulting equation we revert to the original time variable and find:

$$n_M(q, t) = \sigma^{-d} \left[\frac{\Gamma(\beta + 1) \, d}{\pi} \right]^{d/2} \left(\frac{\tau_D}{t} \right)^{\beta d/2} \exp \left[-\frac{\Gamma(\beta + 1) \, d}{2} q^2 \right].$$
$$(12.136)$$

This function has the correct scaling form (12.82), with the similarity variable q defined by (12.80). The scaling function $F(q)$ is, however, quite different from (12.85): *the Markovianized solution is a Gaussian profile instead of the stretched exponential* $\exp(-bq^\delta)$ *with* $\delta = 2/(2 - \beta)$. Note, however, that the MSD of this process goes like t^β, which is in agreement with the subdiffusive character. In the specific case studied above, the Markovian density profile reduces to:

$$n_M(q,t) = \frac{1}{\sigma} \frac{1}{(2\sqrt{\pi})^{1/2}} \left(\frac{\tau_D}{t}\right)^{1/4} \exp\left(-\frac{\sqrt{\pi}}{4} q^2\right)$$

$$= n_{RR}(q,t), \quad \beta = \tfrac{1}{2}, d = 1. \tag{12.137}$$

The result is precisely the profile (12.130) obtained from the Rechester-Rosenbluth argument. We have thus identified the nature of the approximation leading to the latter: *the R-R theory results from the Markovianization of the exact non-Markovian diffusion equation (12.102).*

The Markovian approximation is usually performed as an asymptotic approximation for integral equations in which the memory kernel is a rapidly decaying function of time, typically of exponential type. The latter defines an intrinsic characteristic memory time τ_M, whereas the inverse rate of change of the density profile defines a second characteristic time τ_R. Whenever $\tau_M \ll \tau_R$, the Markovianization is fully justified as an asymptotic approximation. But in the present case of the SLT-CTRW, the decay of the memory kernel is very slow: it is characterized by the power-law long tail $t^{\beta-2}$: there exists no intrinsic characteristic time τ_M, hence the Markovianization cannot be justified. It may also be noted that Eq. (12.136) cannot be obtained from the correct solution of the non-Markovian equation by any limiting process.

The exact SLT-CTRW solution $n_{SLT}(q,t)$, Eq. (12.120) and the Markovianized solution $n_M(q;t)$, Eq. (12.137) are shown in Fig. 12.2 as functions of the similarity variable q. In this representation the width of the curves remains unchanged, but their height decreases in time as $t^{-1/4}$. The long tail of the SLT-CTRW curve is clearly apparent.

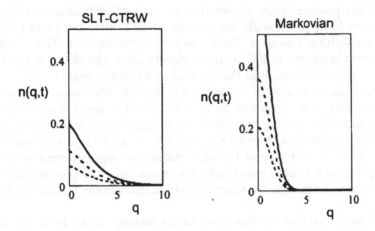

Figure 12.2. Density profiles vs. similarity variable, for the exact SLT-CTRW solution (12.120) and for the Markovian approximation (12.137), at different reduced times $\bar{t} = t/\tau_D$. Solid: $\bar{t} = 1$, dash-dotted: $\bar{t} = 10$, dashed: $\bar{t} = 100$.

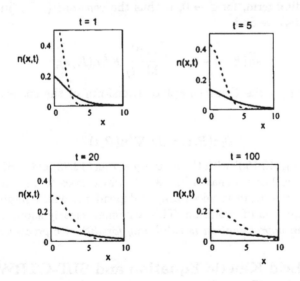

Figure 12.3. Density profiles vs. position, for the exact SLT-CTRW solution and for the Markovian approximation, at different reduced times $\bar{t}$. Solid: exact SLT-CTRW, dash-dotted: Markovian approximation

In Fig. 12.3 the two solutions are shown at various times as functions of the position. Here we see that the SLT-CTRW solution is at all times much lower in the bulk, but extends much farther in x (long tail) as compared with the Gaussian, Markovianized approximation. The width of both curves increases as $t^{1/4}$ in time. Nevertheless, the shape of the two curves is quite different, even though their MSD is the same!

The main conclusion of this section is that *the Markovian approximation is completely inadequate in the SLT-CTRW.* The deep reason thereof is the long tail of the waiting time distribution. Indeed, the resulting memory kernel has a power law dependence on time, which has *no characteristic length* ($\langle t \rangle = \infty$). We cannot therefore define two separate time scales as in the previously treated non-Markovian equations. It is precisely in this feature that the *strange transport* differs fundamentally from the normal one.

We may also look at this question as follows. Consider a *diffusive CTRW* defined by Eqs. (12.58), (12.59). It obeys, in Fourier-Laplace representation, the equation of evolution (12.54), with the memory kernel:

$$\widetilde{\phi}(s) = \frac{s\widetilde{\psi}(s)}{1 - \widetilde{\psi}(s)} = \frac{1}{\langle t \rangle} - s = \frac{1}{\langle t \rangle}\left[1 + O(s)\right]. \qquad (12.138)$$

The leading term, for $s \to 0$, is thus the *constant* $\langle t \rangle^{-1}$. Substituting this into (12.54) we find:

$$s\widehat{\widetilde{n}}(K, s) - 1 = -\frac{\langle R^2 \rangle}{2d\,\langle t \rangle}\,K^2\,\widehat{\widetilde{n}}(K, s),$$

which is precisely the Fourier-Laplace transform of the correct diffusion equation:

$$\partial_t n(R, t) = D\,\nabla^2 n(R, t). \qquad (12.139)$$

Comparing (12.138) with the memory kernel (12.93) of the SLT-CTRW we see that the latter behaves like $s^{1-\beta}$, which goes to zero for $s \to 0$. Thus, the leading term is necessarily s-dependent, which implies a non-Markovian equation of evolution. This argument clearly shows again that *the Markovian approximation is valid only for diffusive processes.*

12.8   Hybrid Kinetic Equation and SLT-CTRW Subdiffusion Equation

In the previous section we investigated the possibility of adapting a SLT-CTRW model to the magnetic turbulent subdiffusion problem, by choosing appropriate parameters (12.119) in order to match the MSD calculated (exactly!) from the V-Langevin equation. We saw that this choice alone does not uniquely define the density profile. We now consider the direct solution of the Hybrid Kinetic Equation (HKE) (11.118) (in which it is assumed that the density profile only depends on X, *i.e.*, $d = 1$), which is rewritten here, for convenience:

$$\partial_t n(X,t) = -\bar{\beta} \, \nabla_X \, \langle v_\parallel(t) \, b_x(z) \, \delta f(X,Z,t) \rangle_{b,\parallel} \, . \tag{12.140}$$

The corresponding equation for the fluctuation is obtained from Eq. (11.117):

$$\begin{aligned}
[\partial_t &+ v_\parallel(t) \, \nabla_Z + \bar{\beta} \, v_\parallel(t) \, b_x(z) \, \nabla_X] \, \delta f(X,Z,t) \\
&= -\bar{\beta} \, v_\parallel(t) \, b_x(z) \, \nabla_X \, n(X,t).
\end{aligned} \tag{12.141}$$

This linear, inhomogeneous partial differential equation is solved by the usual method of characteristics [see also Sec. 6.2]. The propagator is given by:

$$\mathcal{G}(X,Z,t|X',Z',t') = \delta \, [X - X' - F_2(t,t')] \, \delta \, [Z - Z' - F_1(t,t')], \tag{12.142}$$

with:

$$F_1(t,t') = \int_{t'}^{t} dt_1 \, v_\parallel(t_1),$$

$$F_2(t,t') = \bar{\beta} \int_{t'}^{t} dt_1 \, v_\parallel(t_1) \, b_x[z - \lambda_\parallel^{-1} F_1(t,t_1)]. \tag{12.143}$$

Assuming $\delta f(X,Z,0) = 0$, we find the solution:

$$\delta f(X,Z,t) = -\bar{\beta} \int_0^t dt_1 \, v_\parallel(t_1) \, b_x[z - \lambda_\parallel^{-1} F_1(t,t_1)] \, \nabla_X n[X - F_2(t,t_1), \, t_1]. \tag{12.144}$$

This result is substituted into Eq. (12.140):

$$\begin{aligned}
\partial_t n(X,t) = \bar{\beta}^2 \nabla_X \int_0^t dt_1 \, &\langle v_\parallel(t) \, b_x(z) \, v_\parallel(t_1) \, b_x[z - \lambda_\parallel^{-1} F_1(t,t_1)] \\
&\times \, \nabla_X \, n[X - F_2(t,t_1), \, t_1] \rangle_{b,\parallel} \, .
\end{aligned} \tag{12.145}$$

This equation can be transformed by using the Fourier representation (11.75) and the explicit expression (12.143):

$$\partial_t n(X,t) = \bar{\beta}^2 \int_0^t dt_1 \int dk_1 dk_2 \left\langle \hat{b}_x(k_1)\hat{b}_x(k_2) v_\|(t)v_\|(t_1) \right.$$

$$\times \exp\left[i(k_1 + k_2)z - ik_2\lambda_\|^{-1} \int_{t_1}^t dt_2\, v_\|(t_2) \right]$$

$$\times \nabla_X^2\, n \left\{ X - \bar{\beta} \int_{t_1}^t dt_3\, v_\|(t_3) \int dk'\, \hat{b}_x(k') \right.$$

$$\left. \left. \times \exp\left[ik'\left(z - \lambda_\|^{-1} \int_{t_3}^t dt'v_\|(t') \right) \right], t_1 \right\} \right\rangle_{b,\|}. \qquad (12.146)$$

This is the *exact* equation of evolution of the density profile resulting from the HKE (12.140), (12.141), or, equivalently, from the V-Langevin equations (11.69), (11.71). As can be seen, it is a quite complicated equation, involving a highly nonlinear averaging over both stochastic functions $v_\|(t)$ and $b_x(z)$. This feature results from the displacement of X in the density profile: as a result, this function cannot be taken out of the average.[10] Clearly, in spite of some similarities, this equation *is NOT of the form of a CTRW equation (12.96)*. Nevertheless, as can easily be checked, upon multiplication by X^2 and integration over X, this equation yields the correct MSD and running diffusion coefficient, just like the SLT-CTRW equation (12.7), or its Markovianized version (12.126). This shows again that the determination of the density profile is a much more complicated problem than that of the MSD.

In order to go further explicitly, we need some additional simplifying assumptions. Whenever the correlation length $\lambda_\|$ is much shorter than the gradient length, the argument of Sec. 6.2 allows us to neglect the displacement of X in the argument of the density profile . The remaining operations are now easy, and one obtains, using the definition (11.83):

$$\partial_t n(X,t) = \bar{\beta}^2 \int_0^t dt_1 \int dk\, \hat{B}_\|(k)\, Z(k,t_1)\, \nabla_X^2\, n(X, t - t_1). \qquad (12.147)$$

This equation has the form of a non-Markovian diffusion equation, just like (12.96) which resulted from the CTRW theory:

$$\partial_t n(X,t) = \nabla_X^2 \int_0^t dt_1\, H(t_1)\, n(X, t - t_1), \qquad (12.148)$$

with:

---

[10] A similar situation, though in a much simpler context, appeared in the HKE for collisional diffusion, Eq. (11.126), and also in Eq. (6.25).

$$H(t) = \beta^2 \int dk \, \hat{B}_{\|}(k) \, Z(k, t). \tag{12.149}$$

It should be recalled that this function is nothing other than the $R_{xx}(\tau)$ component of the Lagrangian velocity autocorrelation tensor, as shown in Eq. (11.85).

We thus reached an important result: *the Hybrid Kinetic Equation describing simple magnetic turbulence is approximately equivalent to a SLT-CTRW process*. This statement will need, however, a refinement, to be discussed below.

We now investigate more closely the relation between the two processes. The Lagrangian velocity autocorrelation $H(t)$ was calculated analytically in Sec. 11.6, Eq. (11.90). The result was denoted there by the function $H(t) = \mathcal{L}^{\infty}(\nu t)$; it is convenient in the present context to use dimensionless variables scaled with the SLT-CTRW parameters (12.119):

$$\xi = \frac{X}{\sigma}, \quad \bar{t} = \frac{t}{\tau_D} = \overline{\chi}_{\|} \nu t. \tag{12.150}$$

The non-Markovian diffusion equation (12.148) is then written in the form:

$$\partial_{\theta} n(\xi, \bar{t}) = \nabla_{\xi}^2 \frac{1}{2\sqrt{\pi}} \int_0^{\bar{t}} d\bar{t}_1 \, G(\bar{t}_1, \overline{\chi}_{\|}) \, n(\xi, \bar{t} - \bar{t}_1), \tag{12.151}$$

with:

$$G(\bar{t}, \overline{\chi}_{\|}) = \frac{1}{\overline{\chi}_{\|} \left[1 + \overline{\chi}_{\|} \, \mu(\bar{t}/\overline{\chi}_{\|}) \right]^{1/2}}$$
$$\times \left[e^{-\bar{t}/\overline{\chi}_{\|}} - \frac{\overline{\chi}_{\|}}{2} \frac{\varphi^2(\bar{t}/\overline{\chi}_{\|})}{1 + \overline{\chi}_{\|} \, \mu(\bar{t}/\overline{\chi}_{\|})} \right]. \tag{12.152}$$

The functions $\varphi(t)$ and $\mu(t)$ were defined in Eq. (11.88). An important difference with the SLT-CTRW Eq. (12.96) is that the memory kernel $G(\bar{t}, \overline{\chi}_{\|})$ depends, besides the time variable $\bar{t}$, also on the collisionality parameter $\overline{\chi}_{\|}$. Its shape thus depends on the value of the latter. This is shown in Fig. 12.4, where the function $G(\bar{t}, \overline{\chi}_{\|})$ is plotted for three values of $\overline{\chi}_{\|}$: $\overline{\chi}_{\|} = 3$ (weak collisionality), $\overline{\chi}_{\|} = 1$ (medium collisionality) and $\overline{\chi}_{\|} = 1/3$ (strong collisionality). It can be checked that this figure is consistent with Fig. 11.6 in spite of its apparent distortion; indeed, the relation between the two functions is: $R(\bar{t}, \overline{\chi}_{\|}) = \overline{\chi}_{\|} \, G(\overline{\chi}_{\|} \bar{t}, \overline{\chi}_{\|})$.

This memory kernel possesses many interesting features. First, we note that $G(\bar{t}, \overline{\chi}_{\|})$ *is a regular function, devoid of any singularity over the whole range* $0 \leq \bar{t} < \infty$, *and for* $0 < \overline{\chi}_{\parallel} < \infty$. This is in contrast with the kernel $H(t)$ appearing in the SLT-CTRW equation (12.96), which diverges for $t \rightarrow 0$. The reason of the difference lies in the fact that Eq. (12.151) results from an exact solution of the V-Langevin (or HKE) equation, which covers the whole range of time, whereas Eq. (12.96) was derived as an *asymptotic approximation*, valid only for times $\bar{t} \gg 1$. Moreover, as the asymptotic decay is very slow (power law), the equation will be accurate only for "very long" times.

The function $G(\bar{t}, \overline{\chi}_{\parallel})$ can be expanded for small values of $\bar{t} \ll \overline{\chi}_{\parallel}$:

$$G(\bar{t}, \overline{\chi}_{\parallel}) = \frac{1}{\overline{\chi}_{\parallel}} - \frac{1}{\overline{\chi}_{\parallel}^2} \bar{t} + \left(\frac{1}{2\overline{\chi}_{\parallel}^3} - \frac{1}{\overline{\chi}_{\parallel}^2} \right) \bar{t}^2 + ..., \quad \bar{t} \ll \overline{\chi}_{\parallel}.$$

(12.153)

The initial value is thus: $G(\bar{t}, \overline{\chi}_{\parallel}) = \overline{\chi}_{\parallel}^{-1}$: it is large (but finite!) for strong collisionality. As can be seen from Fig. 12.5, the three-term approximation is satisfactory only very near to the origin.

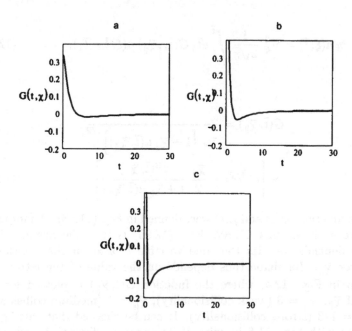

Figure 12.4. The memory kernel $G(\bar{t}, \overline{\chi}_{\parallel})$, Eq. (12.152) for three different values of the collisionality parameter. (a): $\overline{\chi}_{\parallel} = 3$, (b): $\overline{\chi}_{\parallel} = 1$, (c): $\overline{\chi}_{\parallel} = 1/3$.

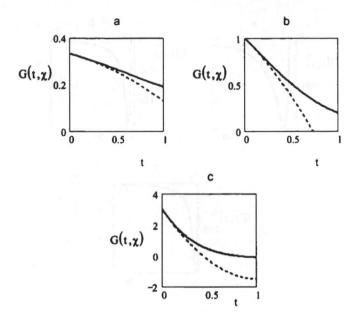

Figure 12.5. The small-$\bar{t}$ approximation of the memory kernel. (a): $\overline{\chi}_{\parallel} = 3$, (b): $\overline{\chi}_{\parallel} = 1$, (c): $\overline{\chi}_{\parallel} = 1/3$.

In the opposite limit, $\bar{t} \gg \overline{\chi}_{\parallel}$, the dominant term is remarkably simple:

$$G(\bar{t}, \overline{\chi}_{\parallel}) \sim -\frac{1}{2\bar{t}^{3/2}}, \quad \bar{t} \gg \overline{\chi}_{\parallel}. \tag{12.154}$$

The asymptotic approach to the exact form is shown in Fig. 12.6. This approximation is excellent over the whole ascending (negative) part of the memory curve.

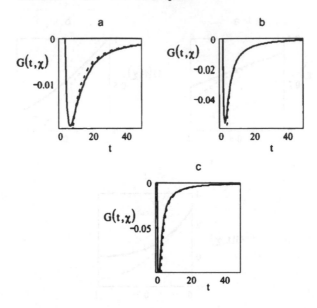

Figure 12.6. The asymptotic approximation of the memory kernel. (a): $\overline{\chi}_\parallel = 3$, (b): $\overline{\chi}_\parallel = 1$, (c): $\overline{\chi}_\parallel = 1/3$.

Another conspicuous property of the memory kernel is its dependence on $\overline{\chi}_\parallel$. Consider, indeed, the SLT-CTRW non-Markovian diffusion equation (12.7). Written in terms of the dimensionless variables $\bar{t}$ and ξ, it takes the form:

$$\partial_{\bar{t}} n(\xi, \bar{t}) = \frac{1}{2} \nabla_\xi^2 n(\xi; \bar{t}) - \frac{1}{4\sqrt{\pi}} \int_{\pi^{-1}}^{\bar{t}} d\bar{t}_1\, \bar{t}_1^{-3/2}\, \nabla_\xi^2 n(\xi, \bar{t} - \bar{t}_1). \qquad (12.155)$$

This equation contains no parameter at all. In contrast, the solution of the HKE depends on the collisionality parameter $\overline{\chi}_\parallel$. This feature, together with the different shape of the memory kernel (in particular the convergence at $\bar{t} = 0$), shows that the quasilinear HKE (12.151) cannot be equivalent to the SLT-CTRW model for all times. It is clear, however, that a comparison of the two models over the whole range of times is unfair: the SLT-CTRW model is, indeed, explicitly an asymptotic approximation, valid for $\bar{t} \gg 1$, and the comparison should only be made in this range. It appears from Eq. (12.154) that, *for $\bar{t} \gg 1$, the memory kernel $G(\bar{t}, \overline{\chi}_\parallel)/2\sqrt{\pi}$ of the HKE becomes independent of $\overline{\chi}_\parallel$ and becomes identical to the SLT-CTRW kernel: $(4\sqrt{\pi})^{-1}\bar{t}^{-3/2}$.*

We now try to construct a convenient asymptotic approximation to the HKE. We pick up a cut-off time $\bar{t}_{\min}$, whose value will be determined *a posteriori*, and write Eq. (12.151) as:

$$\partial_{\bar{t}} n(\xi, \bar{t}) = \frac{1}{2\sqrt{\pi}} \int_0^{\bar{t}_{\min}} d\bar{t}_1 \, G(\bar{t}_1, \overline{X}_{\parallel}) \, \nabla_\xi^2 n(\xi, \bar{t} - \bar{t}_1)$$

$$+ \frac{1}{2\sqrt{\pi}} \int_{\bar{t}_{\min}}^{\bar{t}} d\bar{t}_1 \, G(\bar{t}_1, \overline{X}_{\parallel}) \, \nabla_\xi^2 n(\xi, \bar{t} - \bar{t}_1). \tag{12.156}$$

Assuming $\bar{t} \gg \bar{t}_{\min}$, we may approximate in the first term $n(\xi, \bar{t}-\bar{t}_1) \rightarrow n(\xi, \bar{t})$. In the second term we may use the asymptotic (γ-independent!) approximation: $G(\bar{t}, \overline{X}_{\parallel}) \rightarrow -1/(2\bar{t}^{3/2})$; thus:

$$\partial_{\bar{t}} n(\xi, \bar{t}) = \frac{1}{2\sqrt{\pi}} \int_0^{\bar{t}_{\min}} d\bar{t}_1 \, G(\bar{t}_1, \overline{X}_{\parallel}) \, \nabla_\xi^2 n(\xi, \bar{t})$$

$$- \frac{1}{4\sqrt{\pi}} \int_{\bar{t}_{\min}}^{\bar{t}} d\bar{t}_1 \, \bar{t}_1^{-3/2} \, \nabla_\xi^2 n(\xi, \bar{t}_1 - \bar{t}). \tag{12.157}$$

In the first term, the integral only involves the memory kernel. As the latter converges at $\bar{t} = 0$, the integral is merely a finite constant, that can be calculated analytically as in Eqs. (11.93) - (11.96). The result, expressed in reduced variables, is:

$$K(\bar{t}_{\min}, \overline{X}_{\parallel}) = \frac{1}{2\sqrt{\pi}} \int_0^{\bar{t}_{\min}} d\bar{t}_1 \, G(\bar{t}_1, \overline{X}_{\parallel})$$

$$= \frac{1}{2\sqrt{\pi}} \frac{\varphi(\bar{t}_{\min}/\overline{X}_{\parallel})}{\left[1 + \gamma \mu(\bar{t}_{\min}/\overline{X}_{\parallel})\right]^{1/2}}. \tag{12.158}$$

Eq. (12.157) is now written as:

$$\partial_{\bar{t}} n(\xi, \bar{t}) = K(\bar{t}_{\min}, \overline{X}_{\parallel}) \, \nabla_\xi^2 n(\xi, \bar{t}) - \frac{1}{4\sqrt{\pi}} \int_{\bar{t}_{\min}}^{\bar{t}} d\bar{t}_1 \, \bar{t}_1^{-3/2} \, \nabla_\xi^2 n(\xi, \bar{t} - \bar{t}_1).$$

$$\tag{12.159}$$

This equation is very similar to the SLT-CTRW equation (12.155). We must now determine the cut-off time $\bar{t}_{\min}$: the procedure is similar to the one used for Eq. (12.102). Multiplying both sides by ξ^2 and integrating (12.159) over ξ, we obtain:

$$\partial_{\bar{t}} \langle \xi^2(\bar{t}) \rangle = 2K(\bar{t}_{\min}, \overline{X}_{\parallel}) - \frac{1}{\sqrt{\pi}} \bar{t}_{\min}^{-1/2} + \frac{1}{\sqrt{\pi}} \bar{t}^{-1/2}. \tag{12.160}$$

The last term equals the properly reduced asymptotic value (11.99). We thus fix the value of $\bar{t}_{min}$ by annulling the sum of the first two terms in the right hand side:

$$2\sqrt{\pi}\,K(\bar{t}_{min}, \overline{\chi}_{\parallel}) = \bar{t}_{min}^{-1/2}. \tag{12.161}$$

Substituting Eq. (12.158) into (12.161), squaring the latter and introducing the notation: $m = \bar{t}_{min}/\overline{\chi}_{\parallel}$, we find the following equation for the determination of m:

$$\Phi(m, \overline{\chi}_{\parallel}) \equiv -(1+2m)\,e^{-m} + m\,e^{-2m} + 1 - \overline{\chi}_{\parallel}^{-1} = 0. \tag{12.162}$$

This transcendental equation for m must be solved numerically. Note first that:

$$\Phi(0, \overline{\chi}_{\parallel}) = -\overline{\chi}_{\parallel}^{-1} < 0. \tag{12.163}$$

On the other hand, the function tends asymptotically toward:

$$\Phi(\infty, \gamma) = 1 - \frac{1}{\overline{\chi}_{\parallel}}. \tag{12.164}$$

The graph of the function $\Phi(m, \overline{\chi}_{\parallel})$ shows that, after a shallow minimum, this function tends monotonically towards its asymptotic value. Therefore, if $\overline{\chi}_{\parallel} > 1$, the asymptote is positive and the curve $\Phi(m, \overline{\chi}_{\parallel})$ crosses the m-axis in order to join its negative initial value to the positive final value: there is a real root to Eq. (12.162). For $\overline{\chi}_{\parallel} < 1$, however, the curve remains below the m-axis and there is no real root. For $\overline{\chi}_{\parallel} = 1$, the curve is asymptotically tangent to the m-axis (see Fig. 12.7).

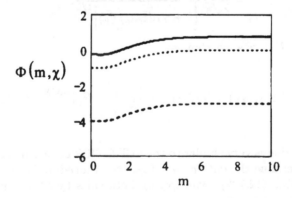

Figure 12.7. The function $\Phi(m, \overline{\chi}_{\parallel})$. Solid: $\overline{\chi}_{\parallel} = 4$,   dotted: $\overline{\chi}_{\parallel} = 1$,   dashed: $\overline{\chi}_{\parallel} = 0.25$.

The numerical solution of Eq. (12.162) is plotted against $\overline{\chi}_{\parallel}$ in Fig. 12.8. As expected, there is no solution for $\gamma < 1$, *i.e.*, for highly collisional plasmas. The solution is infinite for $\overline{\chi}_{\parallel} = 1$. For $\overline{\chi}_{\parallel} > 2$, the solution is surprisingly close to linear: it is well fitted by the following function:

$$\overline{t}_{\min} = 2.848 + 0.976\,\overline{\chi}_{\parallel}, \quad \overline{\chi}_{\parallel} > 2. \tag{12.165}$$

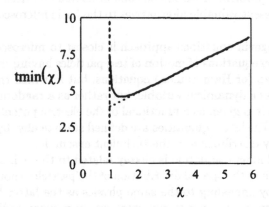

Figure 12.8. The cut-off time $\overline{t}_{\min}$ as a function of the collisionality parameter $\overline{\chi}_{\parallel}$. The dotted line is the linear fit, Eq. (12.165).

In conclusion, *for weakly collisional plasmas, $\overline{\chi}_{\parallel} > 1$, there exists a real cut-off, hence a finite value for $K(\overline{t}_{\min}, \overline{\chi}_{\parallel})$. Eq. (12.159), which has the same form as the SLT-CTRW equation (12.155), is thus validated as an asymptotic approximation.*

For the case of *strongly collisional plasmas*, $\overline{\chi}_{\parallel} \ll 1$, a similar asymptotic approximation can be constructed. As can be seen from Fig. 12.4, the memory function is very strongly peaked at the origin; its value at the origin is $\overline{\chi}_{\parallel}^{-1}$, which tends to infinity for small $\overline{\chi}_{\parallel}$ [see Eq. (12.153)]. On the other hand, its negative tail is quite well represented by Eq. (12.154) (see Fig. 12.6). We may thus use a rough approximation of the form:

$$G(\overline{t}, \overline{\chi}_{\parallel}) \approx \sqrt{\pi}\,\delta(\overline{t}) - \frac{1}{2}\overline{t}^{-3/2}\,\Theta(1 - \overline{t}_{\min}), \quad \overline{\chi}_{\parallel} \ll 1, \tag{12.166}$$

where $\Theta(x)$ is the Heaviside function, and $\overline{t}_{\min}$ is a new cut-off determined by the same method as above. When this form is substituted into Eq. (12.151), the result is exactly the SLT-CTRW equation (12.155). Note that in this limit the parameter $\overline{\chi}_{\parallel}$ has disappeared from the equation.

In conclusion, *both for weakly collisional and for strongly collisional plasmas, the SLT-CTRW provides a very good asymptotic representation of the exact quasilinear equation for the density profile.*

12.9   Conclusions

In Chaps. 11 and 12 we illustrated the application of three different approaches to the problem of turbulence and of anomalous transport in plasmas. They represent valuable alternatives to the basic microscopic, dynamical treatment.

The V-Langevin equations approach is closest to microscopic dynamics. It is based on equations of motion of test particles having the same form as the Newtonian (or Hamiltonian) equations, but in which the velocity is not considered as a dynamical variable, but rather as a random variable. In each realization it is given as a functional of the electric potential, or of the magnetic field. The latter quantities are defined statistically, by prescribing their probability distribution in the turbulent ensemble.

The hybrid kinetic equation is closely related to the V-Langevin equations. It determines the spatial distribution of the particles (not the velocity distribution!), by appealing to the same physics as the latter.[11]

The random walk theories represent an even more radical simplification. The dynamical equations are replaced by a purely probabilistic game based on predetermined transition probabilities. The random walk modelling has long been used in transport theory in its crudest version, illustrated by the "mixing length" concept (Sec. 12.1). The more refined CTRW version has been introduced more recently; it is probably promised to new developments.

All three approaches are examples of what could be called *stochastic transport theory*. The latter is less rigorous than kinetic theory: it is by essence a non-selfconsistent theory. Instead of relating the potential (or the magnetic field) fluctuations to the particles distribution (*i.e.*, the electric charge distribution) through Maxwell's equations, the former quantities are defined *a priori* by an independent statistical assumption. This short-circuit introduces an enormous simplification. This can be appreciated by comparing the extremely complicated self-consistent theories based on renormalization (Chaps. 8 - 10) with the physically transparent pictures of Chaps. 11 and 12. The price to be paid for simplicity is the statistical definition of the potential, or the establishment of rules of the random walk game. These require a strong physical intuition, and an *a posteriori* validation by experiment or by numerical simulation. In the forthcoming

---

[11]Note that the hybrid kinetic equation attached to the A-Langevin equations is the complete Vlasov equation, determining the distribution function in phase space $(\mathbf{x}, \mathbf{v})$. The potential is defined independently by a statistical assumption, instead of being determined self-consistently by the Poisson equation.

chapters we continue to exploit the possibilities of the stochastic transport theory.

Chapter 13

Relative Diffusion and Clumps

13.1 Langevin Equations for Relative Motion

We now come back to the subject treated in Chap. 10, in particular to
the effect of trajectory correlation on transport in a turbulent plasma. The
main result obtained there was Eq. (10.51) for the one-time, two-particle
correlation function. It was shown, without solving the equation, that this
correlation function has a strong peak for small relative distance and small
relative velocity. This *clump effect* implies that the turbulent plasma has,
statistically speaking, a rugged structure. Nearby particles tend to stick
together (for a certain time) before escaping and closely meeting another
particle, etc. This produces a dynamic *granular structure* of the charge
distribution, hence of the potential. This structure has a deep influence
on the transport in the turbulent plasma, as will be shown in the present
and in the forthcoming chapters. It should be said immediately that there
are other causes producing mesoscopic coherent structures in a turbulent
medium (holes, vortices,...): we cannot discuss all of these, for lack of space.

The formation of clumps may be studied in a "fundamental" way by
trying to solve Eq. (10.51). This approach has been initiated by DUPREE
1970, 1972, KADOMTSEV & POGUTSE 1970, 1971. They described the
phenomenon as the formation of a granular structure (or of "macroparti-
cles") [see also ENGELMANN & MORRONE 1972]. An alternative (though
equivalent) point of view stresses the enhancement of correlations at short
distance; this approach, also based on the kinetic theory, is reviewed in
HUI & DUPREE 1975, MISGUICH & BALESCU 1978, 1982, BALESCU &
MISGUICH 1984 (where additional references to previous literature can be
found). The kinetic theoretical treatment proves, however, to be an extremely difficult task. It will now be shown that an application of the
V-Langevin philosophy (implying a single appeal to a phenomenological
assumption) provides us with a physically transparent picture of the clump
phenomenon: this approach was developed by MISGUICH et al 1987.

The starting point is provided by Eqs. (6.8) - (6.11). Assuming that the plasma is collisionless, the motion of a guiding centre along the magnetic field does not influence the perpendicular motion. We thus restrict ourselves to the latter and rewrite Eq. (6.8) (suppressing the subscript "$\perp$") more explicitly as:

$$\frac{d\mathbf{Y}(t)}{dt} = \frac{c}{B} \left\{ \delta\mathbf{E}\left[\mathbf{Y}(t), t\right] \times \mathbf{b} \right\}. \tag{13.1}$$

This is a highly nonlinear stochastic differential equation of V-Langevin type of the form (11.43). The nonlinearity originates from the fact that the field $\delta\mathbf{E}$ must be evaluated at the instantaneous guiding centre position $\mathbf{Y}(t)$.

In the present case we have to treat the *relative motion* of two particles. (From here on we follow closely the work of MISGUICH et al., 1987). For simplicity, we consider that the two particles 1 and 2 are of the same species, and neglect the FLR effects (we thus identify the particle's position with its guiding centre's). It is in this case that the clump effect is most clearly exhibited. We then introduce the centre-of-mass coordinate $\mathbf{S}$, the relative distance $\mathbf{R}$, and the relative velocity $\mathbf{g}$ of the pair of particles:

$$\mathbf{S}(t) = \tfrac{1}{2}\left[\mathbf{Y}_1(t) + \mathbf{Y}_2(t)\right], \quad \mathbf{R}(t) = \mathbf{Y}_1(t) - \mathbf{Y}_2(t),$$

$$\mathbf{g}(t) = \frac{d}{dt}\left[\mathbf{Y}_1(t) - \mathbf{Y}_2(t)\right]. \tag{13.2}$$

We assume that the two particles start with a given value of the relative distance, but that the orientation of the vector $\mathbf{R}$ is random, hence $\langle\mathbf{R}(0)\rangle = 0$. Moreover, the average relative velocity is likewise zero. We also note that the relative velocity is perpendicular to the magnetic field. Hence the equation of motion of the pair is:

$$\frac{d\mathbf{R}(t)}{dt} \equiv \frac{d\,\delta\mathbf{R}(t)}{dt} = \delta\mathbf{g}(t)$$

$$= \frac{c}{B}\left\{\delta\mathbf{E}\left[\mathbf{Y}_1(t), t\right] - \delta\mathbf{E}\left[\mathbf{Y}_2(t), t\right]\right\} \times \mathbf{b}. \tag{13.3}$$

As usual, for any V-Langevin-type theory, *the statistical properties of the fluctuating field must be prescribed a priori:* such theories are never self-consistent. The relevant correlations will be defined below, by making use of experimental data.

We suppose that the two particles are released at time $t = 0$ with a deterministic initial condition, $\delta\mathbf{R}(0) = \mathbf{0}$. We now formally integrate Eq. (13.3):

$$\delta\mathbf{R}(t) = \int_0^t dt_1\,\delta\mathbf{g}(t_1). \tag{13.4}$$

We multiply both sides by $\delta g(t)$ and ensemble-average the result:

$$\frac{d \langle \delta R^2(t) \rangle}{dt} = 2 \langle \delta \mathbf{R}(t) \cdot \delta \mathbf{g}(t) \rangle = 2 \int_0^t dt_1 \, \langle \delta \mathbf{g}(t) \cdot \delta \mathbf{g}(t_1) \rangle . \qquad (13.5)$$

As shown in Sec. 11.2, the mean square displacement of the individual coordinate in a "normal" two-dimensonal ($d = 2$) diffusive process tends asymptotically toward a quantity growing *linearly* in time: its coefficient is related to the (absolute) diffusion coefficient:

$$\langle \delta Y^2(t) \rangle = 2dDt, \ \ t \to \infty \qquad (13.6)$$

In the present problem of relative motion, the mean square of the fluctuating relative distance is:

$$\langle \delta R^2(t) \rangle = \left[\langle \delta Y_1^2(t) \rangle + \langle \delta Y_2^2(t) \rangle - 2 \langle \delta \mathbf{Y}_1(t) \cdot \delta \mathbf{Y}_2(t) \rangle \right] . \qquad (13.7)$$

A *running relative diffusion coefficient* $D_R(t)$ is defined as a natural extension of Eq. (11.29):

$$D_R(t) = \frac{1}{2d} \frac{d}{dt} \langle \delta R^2(t) \rangle . \qquad (13.8)$$

As the distance between the particles grows, which happens as t becomes sufficiently large, one expects the instantaneous positions to become statistically independent, hence $\langle \delta \mathbf{Y}_1(t) \cdot \delta \mathbf{Y}_2(t) \rangle \to 0$ as $t \to \infty$, and therefore:

$$\langle \delta R^2(t) \rangle = \left[\langle \delta Y_1^2(t) \rangle + \langle \delta Y_2^2(t) \rangle \right], \ \ t \to \infty, \qquad (13.9)$$

hence:

$$\frac{1}{4} \frac{d}{dt} \langle \delta R^2(t) \rangle \equiv D_R(t) \to 2D, \ \ t \to \infty. \qquad (13.10)$$

where D is the absolute, independent particle diffusion coefficient, defined in Eq. (13.6).

We now introduce a "trick" which seems trivial, but will prove quite useful, if not for a complete solution of the problem, at least for providing a reasonable model. We define the *normalized relative velocity Lagrangian correlation function* as follows:

$$L(t, t_1) = \frac{\langle \delta \mathbf{g}(t) \cdot \delta \mathbf{g}(t_1) \rangle}{\langle \delta g^2(t) \rangle} . \qquad (13.11)$$

We then obtain from (13.5):

$$\frac{d}{dt}\left\langle \delta R^2(t) \right\rangle = 2\left\langle \delta g^2(t) \right\rangle \int_0^t dt_1\, L(t, t_1). \qquad (13.12)$$

This is a generalization to the relative motion problem of the so-called *Taylor theorem* (TAYLOR 1922, CSANADY 1973, MCCOMB 1992) used in Navier-Stokes turbulence theory. Introducing also a *Lagrangian time-scale* by the equation:

$$t_L(t) = \int_0^t dt_1\, L(t, t_1), \qquad (13.13)$$

we rewrite Eq. (13.12) as follows:

$$\frac{d}{dt}\left\langle \delta R^2(t) \right\rangle = 2t_L(t)\left\langle \delta g^2(t) \right\rangle. \qquad (13.14)$$

This equation is to be solved with the initial condition:

$$\left\langle \delta R^2(0) \right\rangle = R_0^2. \qquad (13.15)$$

We will see below how a strategy of solution can emerge from this nicely factorized form of the equation.

We go back to Eq. (13.3) and use the Fourier representation of the fluctuations and some trigonometric identities to obtain:

$$\delta g(t) = \frac{c}{B} \int dk\, [\delta E_k(t) \times b]\left[e^{ik \cdot Y_1(t)} - e^{ik \cdot Y_2(t)} \right]$$

$$= -2i\frac{c}{B}\int dk\, [k \times b]\, \delta\phi_k(t) \sin\left[\tfrac{1}{2}k \cdot R(t)\right] e^{ik \cdot S(t)}. \qquad (13.16)$$

Next, we calculate the (unnormalized) relative velocity correlation:

$$\langle \delta g(t)\, \delta g(t_1) \rangle = 4\frac{c^2}{B^2}\int dk_1 \int dk_2\, [k_1 \times b]\, [k_2 \times b]$$

$$\times \left\langle \delta\phi_{k_1}(t)\, \delta\phi_{k_2}(t_1)\, e^{ik_1 \cdot S(t) + ik_2 \cdot S(t_1)} \right.$$

$$\times \left. \sin\left[\tfrac{1}{2}k_1 \cdot R(t)\right] \sin\left[\tfrac{1}{2}k_2 \cdot R(t_1)\right] \right\rangle. \qquad (13.17)$$

We now adopt the **Corrsin approximation** introduced in Sec. 6.4. According to this conjecture, the complete Lagrangian correlation can be factored as follows:

$$\langle \delta \mathbf{g}(t)\, \delta \mathbf{g}(t_1) \rangle$$

$$= 4\frac{c^2}{B^2} \int d\mathbf{k}_1 \int d\mathbf{k}_2 \; [\mathbf{k}_1 \times \mathbf{b}] \, [\mathbf{k}_2 \times \mathbf{b}] \, \langle \delta\phi_{\mathbf{k}_1}(t)\, \delta\phi_{\mathbf{k}_2}(t_1) \rangle$$

$$\times \left\langle e^{i\mathbf{k}_1 \cdot \mathbf{S}(t) + i\mathbf{k}_2 \cdot \mathbf{S}(t_1)} \; \sin\left[\tfrac{1}{2}\mathbf{k}_1 \cdot \mathbf{R}(t)\right] \sin\left[\tfrac{1}{2}\mathbf{k}_2 \cdot \mathbf{R}(t_1)\right] \right\rangle. \qquad (13.18)$$

We now use the quasi-homogeneity assumption [Eq. (8.25), without time-ordering], as well as the fact that the drift wave spectrum contains only very small parallel wave vectors:

$$\langle \delta\phi_{\mathbf{k}_1}(t)\, \delta\phi_{\mathbf{k}_2}(t_1) \rangle = \delta(\mathbf{k}_1 + \mathbf{k}_2)\, \delta(k_{1\parallel}) S_{\mathbf{k}_{1\perp}}(t, t_1), \qquad (13.19)$$

and the following trigonometric identities:

$$\sin A \sin B = \tfrac{1}{2}\left[\cos(A - B) - \cos(A + B)\right],$$
$$\cos(A + B) = \cos\left[2A - (A - B)\right]$$
$$= \cos 2A \cos(A - B) + \sin 2A \sin(A - B).$$

We introduce the following abbreviations:

$$\Delta \mathbf{R}(t, t_1) = \mathbf{R}(t) - \mathbf{R}(t_1), \quad \Delta \mathbf{S}(t, t_1) = \mathbf{S}(t) - \mathbf{S}(t_1). \qquad (13.20)$$

Combining all these equations, considering the trace of the tensor in (13.18), and integrating over $k_{\parallel}$, we obtain:

$$\langle \delta\mathbf{g}(t) \cdot \delta\mathbf{g}(t_1) \rangle = 4\pi \frac{c^2}{B^2} \int dk\, k\, k^2\, S_k(t, t_1) \left\langle e^{i\mathbf{k} \cdot \Delta\mathbf{S}(t, t_1)} \cos \frac{\mathbf{k} \cdot \Delta\mathbf{R}(t, t_1)}{2} \right.$$

$$\left. \times \left\{ 1 - \left[\cos \mathbf{k} \cdot \mathbf{R}(t) + \sin \mathbf{k} \cdot \mathbf{R}(t) \tan \frac{\mathbf{k} \cdot \Delta\mathbf{R}(t, t_1)}{2}\right] \right\} \right\rangle.$$

$$(13.21)$$

We now take the limit $t_1 \to t$ which yields the quantity appearing in the right hand side of (13.14):

$$\langle \delta g^2(t) \rangle = 4\pi \frac{c^2}{B^2} \int dk\, k^3\, S_k(t, t_1)\, \langle 1 - \cos \mathbf{k} \cdot \mathbf{R}(t) \rangle. \qquad (13.22)$$

Here one clearly sees the effect of trajectory correlations through the presence of the term $\cos \mathbf{k} \cdot \mathbf{R}(t)$. Further development of this equation requires the evaluation of the average of the cosine: this is done in the *second cumulant approximation* [see Eq. (12.24)]:

$$\langle \cos \mathbf{k} \cdot \mathbf{R}(t) \rangle = \mathcal{R}e \left\langle e^{i\mathbf{k} \cdot \mathbf{R}(t)} \right\rangle$$
$$\cong \cos \mathbf{k} \cdot \langle \mathbf{R}(t) \rangle \exp \left[-\tfrac{1}{2} \mathbf{k}\,\mathbf{k} \, \langle \delta \mathbf{R}(t) \, \delta \mathbf{R}(t) \rangle \right]$$
$$= \exp \left[-\tfrac{1}{4} k^2 \, \langle \delta R^2(t) \rangle \right]. \tag{13.23}$$

In going to the last expression we used the following facts:

- $\langle \mathbf{R}(t) \rangle = 0$,
- the fluctuation spectrum is assumed to be *gyrotropic* (*i.e.*, isotropic in the plane perpendicular to **B**),
- The spectrum is abbreviated as follows: $S_{\mathbf{k}}(t,t) \equiv S_{\mathbf{k}}$; [this function is very slowly dependent on t in a quasi-stationary turbulent state].

Eq. (13.14) is now rewritten as follows, using (13.22), (13.23):

$$\frac{d}{dt} \langle \delta R^2(t) \rangle = 8\pi t_L(t) \frac{c^2}{B^2} \int dk \, k^3 \, S_{\mathbf{k}} \left\{ 1 - \exp \left[-\tfrac{1}{4} k^2 \, \langle \delta R^2(t) \rangle \right] \right\}. \tag{13.24}$$

The use of our additional approximations (mainly the Corrsin conjecture and the second cumulant approximation) led us to a *closed differential equation for the mean square relative distance*. This equation is *nonlinear* because of the *trajectory correlation*; the nonlinearity is, however, very peculiar. As $\langle \delta R^2(t) \rangle$ is, presumably, a monotonously increasing function of time, it appears that, for $t \to \infty$, the exponential in the integrand tends to zero and the equation becomes linear, because the trajectory correlation effect vanishes. This is consistent with Eq. (13.10).

In order to solve Eq. (13.24) we need a knowledge of the spectrum $S_{\mathbf{k}}$ and of the function $t_R(t)$. At this point we inject the phenomenological assumption typical for any Langevin-type theory. We adopt an empirical expression for the spectrum of the potential fluctuations: this is a power-law form based on the experimental data found in the TFR tokamak (TFR GROUP & TRUC 1984).[1] Noting that only the perpendicular wave-vector spectrum really matters (see the discussion above), the TFR spectrum of the potential fluctuations (adapted to our notations) is:

$$S_{\mathbf{k}} = \begin{cases} \sigma \, k_*^{-2} k^{-4}, & k_{\min} \leq k \leq k_* \\ \sigma \, k^{-6}, & k_* \leq k \leq k_{\max} \end{cases} \tag{13.25}$$

[1] Although these data are relatively old, they constitute an exhaustive and consistent set of measurements. The important feature here is the power-law form of the spectrum. Other sets of characteristic exponents, resulting from experiments with different machines could be used and the calculations could readily be adapted.

This function depends on a characteristic amplitude σ, a minimum and a maximum wave-vector and a "transition wave-vector" k_*. It is useful to introduce the following dimensionless quantities:

$$k = \frac{k_\perp}{k_*}, \quad \mu = \frac{k_{min}}{k_*}, \quad \nu = \frac{k_*}{k_{max}}, \quad S(k) = \frac{S_{k_\perp}}{\sigma\, k_*^{-6}}. \qquad (13.26)$$

The reduced spectrum is then:

$$S(k) = \begin{cases} k^{-4}, & \mu \leq k \leq 1 \\ k^{-6}, & 1 \leq k \leq \nu^{-1} \end{cases} \qquad (13.27)$$

The experimental TFR values are: $\mu = 0.4$, $\nu = 0.31$ ($\nu^{-1} = 3.2$). The dimensionless spectrum $S(k)$ is shown in Fig. 13.1.

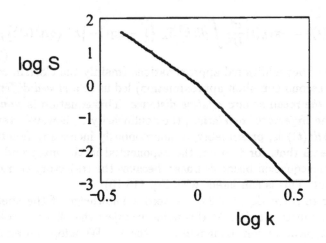

Figure 13.1. The reduced spectrum $S(k)$ as a function of the reduced wave vector k, based on experimental data from the TFR tokamak (Fontenay-aux Roses, France).

If we also introduce the dimensionless mean square relative displacement as:

$$Z(t) = \tfrac{1}{4} k_*^2 \left\langle \delta R^2(t) \right\rangle, \qquad (13.28)$$

Eq. (13.24) reduces to:

$$\frac{d}{dt}Z(t) = 2\pi \frac{c^2}{B^2}\sigma\, t_L(t)$$

$$\times \left\{ \int_\mu^1 dk\, k^{-1}\left[1 - e^{-k^2 Z(t)}\right] + \int_1^{1/\nu} dk\, k^{-3}\left[1 - e^{-k^2 Z(t)}\right] \right\} \qquad (13.29)$$

The k-integrals can be expressed in terms of generalized exponential-integral functions $E_n(x)$ [ABRAMOWITZ & STEGUN 1972]:

$$E_n(x) = \int_1^\infty dt\, t^{-n}\, e^{-xt} = x^{n-1}\int_x^\infty dy\, y^{-n}\, e^{-y}. \qquad (13.30)$$

Among the properties of these functions, the following ones will be used:

$$E_2(x) = e^{-x} - xE_1(x). \qquad (13.31)$$

The small-x expansion of $E_1(x)$ is: [note the logarithmic singularity at the origin!]:

$$E_1(x) = -C - \ln x - \sum_{n=1}^{\infty}(-)^n \frac{x^n}{n\cdot n!}, \qquad (13.32)$$

where C is Euler's constant: $C = 0.577...$ The asymptotic expansion of the exponential integral is:

$$E_1(x) \sim \frac{e^{-x}}{x}\left(1 - \frac{1}{x} + ...\right). \qquad (13.33)$$

Using (13.30), (13.31), we write Eq. (13.29) in the form:

$$\frac{d}{dt}Z(t) = \left(1 - \nu^2 - \ln\mu^2\right) \pi\frac{c^2}{B^2}\sigma\, t_L(t)\left\{1 - G\left[Z(t)\right]\right\}, \qquad (13.34)$$

with:

$$G(Z) = \frac{1}{1 - \nu^2 - \ln\mu^2}$$

$$\times \left\{e^{-Z} - \nu^2 e^{-Z/\nu^2} + ZE_1(Z/\nu^2) - (1+Z)E_1(Z) + E_1(\mu^2 Z)\right\}. \qquad (13.35)$$

The initial condition for Eq. (13.34) is:

$$Z(0) = Z_0 = \tfrac{1}{4}k_*^2 R_0^2. \qquad (13.36)$$

Finally, we need to find a way of dealing with the unknown function $t_L(t)$. We note that this factor simply defines a time-scale (which is, in general, time-dependent) for the evolution. We may therefore, without any approximation, separate the problem into two successive steps. We first define a "dimensionless scaled time" $T(t)$ by the equation:

$$\frac{d}{dt}T(t) = \left(1 - \nu^2 - \ln\mu^2\right)\ \pi\frac{c^2}{B^2}\sigma\,t_L(t), \quad T(0) = 0. \tag{13.37}$$

In terms of this scaled time, Eqs. (13.34), (13.36) are written as:

$$\frac{d}{dT}Z(T) = 1 - G\left[Z\left(T\right)\right], \quad Z(0) = Z_0. \tag{13.38}$$

13.2 The Clump Effect

Our strategy is now defined as follows. We first solve the nonlinear equation (13.38) and determine Z as a function of T. This provides us with the qualitative features of the solution, as will be seen below. Next, we go around the difficult (but not the most important) part of the problem by making reasonable model assumptions about the shape of the function $t_L(t)$. Eq. (13.37) then becomes trivially simple. Substitution of its solution into the function $Z\left[T(t)\right]$ yields the final solution of Eq. (13.34) The graph of the function $Z(t)$ differs from the one of $Z(T)$ merely by a (generally inhomogeneous) dilatation of the time scale.

The function $G(Z)$ is shown in Fig. 13.2 for the default values of the parameters μ, ν characterizing the spectrum. Clearly, the function $G(Z)$, describing the trajectory correlation, tends to zero for large Z, as was already obvious in Eq. (13.24). Hence this result confirms the "normal" character of the diffusion process for large T, when $Z(T) \sim T$.

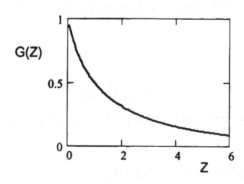

Figure 13.2. The function $G(Z)$, Eq. (13.1); $\mu = 0.4$, $\nu = 0.31$.

A short-time *analytical* solution can be obtained for the case $\nu = 0$ and $Z \ll 1$. The function $G(Z)$ is expanded, using Eq. (13.32), yielding:

$$(1 - \ln \mu^2) \frac{dZ^{(0)}(T)}{dT} = Z^{(0)}(T) \left[(2 - C - \mu^2) - \ln Z^{(0)}(T) \right],$$

$$Z^{(0)} \ll 1. \qquad (13.39)$$

Note that this equation is still nonlinear in $Z^{(0)}$; it is, however, linear in $\ln Z^{(0)}(T)$, hence it is easily solved:

$$Z^{(0)}(T) = \gamma(\mu) \left(\frac{Z_0}{\gamma(\mu)} \right)^{\exp[-T/(1 - \ln \mu^2)]}, \qquad (13.40)$$

where:

$$\gamma(\mu) = 2 - C - \mu^2. \qquad (13.41)$$

The graph of this solution (in log-log representation), with the initial value $Z_0 = 10^{-6}$ is shown in Fig. 13.3.

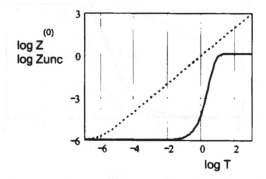

Figure 13.3. Mean square displacement $Z^{(0)}(T)$ (13.28). Short-time analytical solution $Z^{(0)}(T)$, Eq. (13.40) [solid line], compared to the uncorrelated solution, Eq. (13.42) [dotted line]. Parameters: $\mu = 0.4$, $\nu = 0$, $Z_0^{(0)} = 10^{-6}$.

It is important to compare it with the graph corresponding to uncorrelated trajectories, $G(Z) = 0$, which yields:

$$Z_{unc}(T) = Z_0 + T. \qquad (13.42)$$

Note that the characteristics of the spectrum: μ, ν do not appear in $Z_{unc}(T)$. The spectrum determines the value of the (uncorrelated) diffusion coefficient, which is hidden here in the asymptotic relation between $T(t)$ and t.

We note already the large departure of $Z^{(0)}(T)$ from $Z_{unc}(T)$, showing the importance of the trajectory correlation effect. The details of the behaviour will be discussed below. Meanwhile we note that the short-time solution saturates at a constant value $Z^{(0)}_\infty = 2 - C - \mu^2 > 1$, instead of behaving like (13.42). However, as $Z^{(0)}_\infty > 1$, this part of the curve lies outside the domain of validity of the expanded equation (13.40).

In order to obtain a correct picture over the whole domain of T, we solve Eq. (13.38) numerically for the experimental values of μ and ν: the result is shown in Fig. 13.4 in a log-log representation.. Also shown is the (numerical) solution of the equation for $\nu = 0$: this allows us to see that the analytical solution (13.40) is quite good up to $T \approx 10$. Finally, the uncorrelated $Z_{unc}(T)$ is also represented as a straight line.

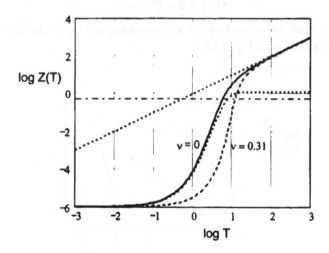

Figure 13.4. Mean square displacement $Z(T)$. Solid line: numerical solution of Eq. (13.34) for $\mu = 0.4$, $\nu = 0.31$; Dashed line: numerical solution for $\nu = 0$; Dotted line: analytical solution $Z^{(0)}(T)$, Eq. (13.40); Dash-dotted straight line: uncorrelated solution, $Z_{unc}(T)$, Eq. (13.42). The horizontal line corresponds to $Z = \frac{1}{2}$; it yields a clump lifetime $T_{cl} = 7.76$.

Next we represent in Fig. 13.5 the *running relative diffusion coefficient:*

$$D_R(T) = \frac{dZ(T)}{dT} = 1 - G[Z(T)]. \qquad (13.43)$$

The *clump effect* is now manifest. It has three aspects.

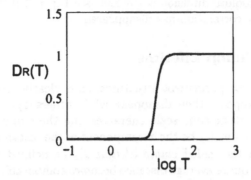

Figure 13.5. Running relative diffusion coefficient $D_R(T)$. Reference values of μ and ν.

- For *short times*, the average relative distance between the two particles increases *much more slowly* than if the particles diffused independently. In other words, as a result of trajectory correlation, the particles tend to "stick" together, forming a *clump*. This results in a grainy microstructure of the turbulent plasma, which was clearly demonstrated in the numerical simulations of DUPREE et al., 1975. This effect also explains the enhancement of the two-point correlation function mentioned after Eq. (10.49) and discussed in detail in MIS-GUICH & BALESCU 1978. The intuitive reason of this behaviour is the fact that particles starting initially very close together (distance shorter than the typical dominant wavelength of the turbulent spectrum) feel for a long time practically the same fluctuating electric field. They will therefore tend to travel together for a relatively long time. In this short-time regime the particles exhibit an almost *subdiffusive behaviour*. The running relative diffusion coefficient is typically: $D_R(T = 1) = 3.7 \ 10^{-6}$. The motion is, indeed, quasi-ballistic.
- At some *intermediate time*, the two particles separate very suddenly and very quickly. At the same time the running relative diffusion coefficient rises sharply from almost zero to its final asymptotic value $D_R = 1$. This approximately exponential separation lasts for a limited, rather brief time (For $T = 2$, $D = 1.4 \ 10^{-3}$, and for $T = 4$, $D_R = 0.97$).
- After this exponential separation, a *long-time diffusive regime* sets in, in which the particles diffuse independently of each other. The mean square displacement grows linearly in time. Alternatively, the relative diffusion coefficient reaches a constant value $D_R(T) = 1$, equal to

twice the absolute diffusion coefficient [see Eq. (13.10)].[2] Every trace of trajectory correlation has disappeared.

13.3   The Clump Lifetime

The clumps are not permanent structures, as is clearly seen in Fig. 13.4 they suddenly appear, then disappear while others appear in different places. An important time scale characterizing the clump effect can be defined as follows. Let R_x be the x-component of the distance; $\langle \delta R_x^2(T) \rangle = \frac{1}{2} \langle \delta R_\perp^2(T) \rangle$. Let the scaled *clump lifetime* T_{cl} be defined as the time at which the mean square average distance becomes comparable with the characteristic wavelength: $\langle \delta R_x^2(T_{cl}) \rangle = k_*^{-2}$, which implies, by Eq. (13.28):

$$Z(T_{cl}) = \frac{1}{2}. \tag{13.44}$$

T_{cl} is a function of the characteristics of the spectrum (here: μ, ν) and also of the initial separation. From our qualitative discussion, one expects T_{cl} to be longer, the smaller the initial separation. From the analytic solution for $\nu = 0$, (13.40) one easily finds:

$$T_{cl} = (1 - \ln \mu^2) \ln \frac{\ln \gamma(\mu) - \ln Z_0}{\ln \gamma(\mu) + \ln 2}. \tag{13.45}$$

As appears from Fig. 13.4, for $Z = \frac{1}{2}$ the analytic solution is still quite good, hence the value (13.45) is quite reliable. For $\nu > 0$ the clump lifetime is somewhat longer, as can also be seen from Fig. 13.4. The clump lifetime (for $\nu = 0$) is shown in Fig. 13.6: it is, indeed, a slowly decreasing function of Z_0 for sufficiently small values of this quantity. It then drops rapidly for $Z_0 > 10^{-1}$: there is no clump effect for large initial separation.

13.4   Real Time Dependence

The last step in our strategy is the passage from the scaled time T to the real time t. This requires the solution of Eq. (13.37) which, in turn, requires the knowledge of the Lagrangian time scale $t_L(t)$, Eq. (13.13), hence of the normalized relative velocity Lagrangian correlation $L(t, t_1)$, Eq. (13.11). Instead of a complete solution of this very difficult problem, we introduce a reasonable model and fit the constants to the experimental data. Moreover, MISGUICH et al. 1987, checked that several different models give very similar results. Given this insensitivity, we choose here only the simplest such model, assuming that the Lagrangian correlation function decays exponentially, with a constant correlation time θ_c:

---

[2] The factor 2 is hidden in the normalization coefficients.

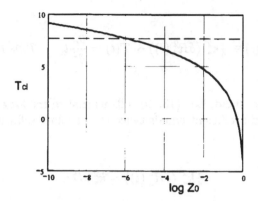

Figure 13.6. The clump lifetime *vs.* the initial separation. For $\log Z_0 = -6$, one finds $T_{cl} = 7.70$, in reasonable agreement with Fig. 13.4.

$$L(t, t_1) = \exp\left(-\frac{t - t_1}{\theta_c}\right). \tag{13.46}$$

One then immediately obtains:

$$t_L(t) = \theta_c \left(1 - e^{-t/\theta_c}\right), \tag{13.47}$$

and:

$$T(t) = \left(\frac{\theta_c}{\vartheta}\right)^2 \left(e^{-t/\theta_c} - 1 + \frac{t}{\theta_c}\right), \tag{13.48}$$

where the characteristic time ϑ is related to the spectrum as follows:

$$\frac{1}{\vartheta^2} = \left(1 - \nu^2 - \ln \mu^2\right) \pi \frac{c^2}{B^2} \sigma. \tag{13.49}$$

We note the limiting values:

$$T(t) \approx \frac{1}{2} \left(\frac{t}{\vartheta}\right)^2, \quad t \ll \theta_c$$

$$T(t) \sim \frac{\theta_c}{\vartheta^2} t, \quad t \gg \theta_c. \tag{13.50}$$

We now express the time parameters in terms of the characteristics of the spectrum, by equating two equivalent expressions of the mean square displacement. We note that the asymptotic limit of $Z(T)$ is the uncorrelated diffusion (13.42) (in which Z_0 can be neglected for long T); this is re-expressed by using Eqs. (13.50) and (13.28) as a function of t:

$$Z(t) \equiv \tfrac{1}{4}k_*^2 \left\langle \delta R_\perp^2(t) \right\rangle \sim T(t) = \frac{\theta_c}{\vartheta^2} t, \quad T \gg 1.$$

On the other hand, Eq. (13.10) tells us that in the long time limit the relative diffusion coefficient equals twice the absolute diffusion coefficient:

$$\tfrac{1}{4}k_*^2 \left\langle \delta R_\perp^2(t) \right\rangle \sim k_*^2 \, 2Dt.$$

Combining these two equations, we find the following relation:

$$\frac{\theta_c}{\vartheta^2} = 2k_*^2 \, D \tag{13.51}$$

At this point we may eliminate one of the two externally introduced parameters, by choosing: $\vartheta = \theta_c$ [this choice ensures that, asymptotically, $T(t) \to (t/\theta_c)$]. Eq. (13.51) then reduces to:

$$\frac{1}{\theta_c} = 2k_*^2 D. \tag{13.52}$$

The single remaining parameter θ_c in the model contains the characteristic properties of the turbulent spectrum, through D and k_*. It is just a time scale factor. In terms of the reduced time $\tilde{t} = t/\theta_c$, the relation between T and t reduces to:

$$T(\tilde{t}) = e^{-\tilde{t}} - 1 + \tilde{t}. \tag{13.53}$$

In Fig. 13.7, $\log Z(\tilde{t})$ is plotted against the "physical" reduced time, $\log \tilde{t}$ and compared to $Z(T)$ plotted against T. As expected, we find that the transformation $T \to t$ produces an inhomogeneous dilatation of the time scale, introducing a slight deformation of the solution graph for short times; the qualitative shape of the curve remains quite similar to Fig. 13.4.

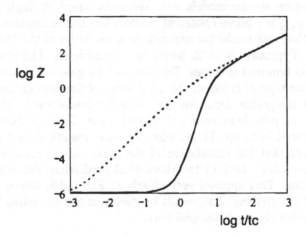

Figure 13.7. Solution $Z(t/t_c)$ of Eq. (13.38) in terms of the real time variable t/t_c, Eq. (13.48), in log-log representation (solid line), compared with the solution $Z(T)$ in terms of the scaled time T (dashed line). Also shown is the uncorrelated solution $Z_{unc}(t/t_c)$ (dotted line).

13.5 Numerical Simulation

We thus achieved a semi-empirical model theory of relative diffusion in which the parameters (to be introduced from outside) are the characteristics of the spectrum of potential fluctuations and the absolute diffusion coefficient. This model theory gives a very pictorial description of the clump effect in plasma turbulence. The skeptical reader is, however, entitled to note that the transparency has been obtained at the price of many successive approximations that were not strongly justified.

Having in mind initially a check of this theory, PETTINI et al., 1988 did a large scale simulation of a Hamiltonian system described by Eq. (13.1) or, explicitly,

$$\frac{d}{dt}\begin{pmatrix} x \\ y \end{pmatrix} = \frac{c}{B}\begin{pmatrix} -\nabla_y \Phi(x,y,t) \\ \nabla_x \Phi(x,y,t) \end{pmatrix}. \tag{13.54}$$

Here (x,y) play the role of (q,p) for a Hamiltonian $H = (c/B)\Phi$. They adopted the following model for the potential:

$$\Phi(x,y,t) = \frac{a}{2\pi}\sum_{m=1}^{N}\sum_{n=1}^{N}\frac{1}{(n^2+m^2)^{3/2}}\sin\left[\frac{2\pi}{L}(nx+my)-\omega t+\varphi_{nm}\right],$$

$$(n^2+m^2 \leq N^2) \tag{13.55}$$

(Alternative, similar models were also considered). A single frequency ω was used; φ_{nm} is a phase chosen at random for each realization. It is easily checked that this model corresponds to a spectrum of the form (13.27) with $\mu = 1$. Typically, $N = 25$ waves were considered. The statistics involved several hundred particles. The simulations give very interesting details about the type of orbits of the $1\frac{1}{2}$ degrees of freedom system (13.54), which is not integrable. One can thus note the transition to chaos when the potential amplitude exceeds a threshold value. These problems will be further discussed in Chap. 14. Among the many results of that paper, we retain the fact that the calculation of the mean square relative distance provides a beautiful check of the theoretical predictions described in the present section. This appears very clearly in Fig. 13.8, whose similarity with Fig. 13.7 is striking. This result by itself justifies the reliability of the relative diffusion theory developed here.

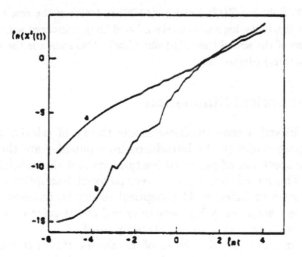

Figure 13.8. Mean square relative displacement as a function of time, obtained by numerical simulation. a: uncorrelated diffusion, b: correlated diffusion. (Reprinted with permission from PETTINI et al., *Phys. Rev.* A **38**, 344 (1988). Copyright 1988 by the American Physical Society.)

Chapter 14

Diffusion in a Stochastic Electrostatic Field

14.1 Statement of the problem

In Chap. 13 we described a mechanism that creates granularity of the turbulent plasma by enhancing the two-particle correlation at small relative distances. As can be clearly seen in Fig. 13.5, the *relative* diffusion of the two particles is almost zero during the clump lifetime. For times exceeding the latter, the particles diffuse independently. There exist other possible mechanisms of creating mesoscopic structures in the turbulent plasma; they will not be further described here. The net result of these processes is a *rugged distribution of charges, hence of the potential* in the medium.

In Chap. 13 the final, absolute diffusion of the particles was taken as a phenomenological input in the theory. A completely self-consistent theory, such as one based on a renormalized kinetic theory sketched in Chap. 10, would determine both the relative and the absolute diffusion coefficients from first principles. We saw, however, that such a theory would be horrendously complicated. One fact is sure. If relative diffusion produces mesoscopic structure, the latter influences the absolute diffusion. In a smooth potential the individual particles have relatively regular trajectories that suffer inflections due to sudden localized potential fluctuations, whose overall effect is a diffusive motion. Whenever mesoscopic structures are present, they form obstacles which perturb the particle trajectories for a finite time. This *trapping effect* slows down the overall motion of the particles and results in a reduction of the diffusion coefficient. The study of this effect in various situations will be the object of Chaps. 14 - 16.

The tool used in this study will be the V-Langevin equation which appeared repeatedly in this book. We found the first occurrence of Langevin equations in Chap. 6, where they described the motion of guiding centres in a fluctuating electric field and a constant magnetic field [Eqs. (6.8)

315

- (6.11)]. They were used for the derivation of the "collisionless" kinetic equation for the average distribution function. It should be stressed that *in a given realization of the electric potential* these Langevin equations for the guiding centre position $Y(t)$ and the corresponding velocity $v(t)$ have a Hamiltonian structure.

The Langevin equations were treated in detail in Chap. 11, together with the associated hybrid kinetic equations. The treatment of clumps in Chap. 13 was also based on Langevin equations. In all those cases the treatment was deliberately "stochastic": the basic fluctuating potential was defined statistically as a *Gaussian process with prescribed Eulerian correlation functions*. In order to obtain macroscopic transport coefficients we had to solve the difficult problem of the *calculation of Lagrangian correlations* of appropriate quantities. More specifically, a *relation between the Eulerian and the Lagrangian correlations* is needed for the determination of the transport coefficients. As explained in Chap. 6, the exact solution of this problem would require the solution of the equations of motion of the particles, an unattainable goal. Except for the oversimplified problem treated in Chap. 11, which can be solved exactly, most previous attempts for solving this problem analytically were based on the celebrated *Corrsin approximation*, presented in Chap. 6. It is, however, well known that this approximation can only be justified in situations where the effect of the nonlinearity is small. In the first sections of the present chapter, we briefly present the application of the latter approximation to a typical anomalous transport problem. When considering strongly nonlinear problems, a radically different approach appears necessary. A method, called the *decorrelation trajectory* method, has the potential of treating strongly nonlinear problems. Although such an approximation cannot be justified exactly, it should be judged by its final results.

We define a reference problem by a set of Langevin equations for the following set of unknowns: a 3-dimensional position vector $Y = (X, Z)$ $[X \equiv Y_\perp = (X, Y)]$:[1]

$$\frac{dX(t)}{dt} = v[X(t), Z(t), t] + \eta_\perp(t), \quad X(0) = 0, \qquad (14.1)$$

$$\frac{dZ(t)}{dt} = v_\parallel[X(t), Z(t), t] + \eta_\parallel(t), \quad Z(0) = 0, \qquad (14.2)$$

$$\frac{dv_\parallel(t)}{dt} = a[X(t), Z(t), t], \quad v_\parallel(0) = 0. \qquad (14.3)$$

Various problems of plasma physics (and also of neutral fluid dynamics) are described by these equations or by special simpler cases. It is

[1] We recall that we are using here, as in Chap. 11 a simpler notation than in Chap. 6: the initial condition is not written explicitly in the arguments; thus: $X(t) \equiv X(t|0; 0)$.

immediately seen that they are of the same form as Eqs. (6.8) - (6.11) which describe the motion of a guiding centre in a fluctuating electric field deriving from a gyro-averaged potential $\overline{\delta\phi}(\mathbf{X}, Z, t)$: the two sets are almost identical; in the present case, however, the effect of the Coulomb collisions is added through the random functions $\eta_\perp(t)$, $\eta_\parallel(t)$.[2] Thus, the following interpretation leads to the problem of collisionless drift wave turbulence:

$$\mathbf{Y}(t) \to \mathbf{Y}(t),$$

$$\mathbf{v}(t) \to \delta\mathbf{V}^E(t) = \frac{c}{B_0}\{\mathbf{b} \times \boldsymbol{\nabla}\overline{\delta\phi}[\mathbf{Y}(t), t)]\}$$

$$a(t) \to \delta a^\alpha(t) = \frac{e_\alpha}{m_\alpha}\{\mathbf{b} \cdot \boldsymbol{\nabla}\overline{\delta\phi}[\mathbf{Y}(t), t)]\}. \qquad (14.4)$$

Simpler cases are obtained by neglecting the parallel acceleration $\delta a^\alpha(t)$, and/or neglecting the effect of collisions (as in Chap. 6).

Another interesting case is obtained by setting $a(t) = 0$ and $\eta(t) = 0$, suppressing the time dependence, and replacing the role of time by the parallel coordinate Z. One then obtains the equations (11.52) describing the spatial fluctuations of the field lines of a magnetic field deriving from a (time-independent) fluctuating vector potential $\mathbf{A}(\mathbf{X}, Z) = \beta\lambda_\perp\psi(\mathbf{X}, Z)\,\mathbf{e}_z$:

$$t \to Z$$

$$\mathbf{X}(t) \to \mathbf{X}(Z),$$

$$\mathbf{v}[\mathbf{X}(t), Z(t), t] \to \beta\,\mathbf{b}[\mathbf{X}(Z), Z] = -\beta\,\mathbf{e}_z \times \boldsymbol{\nabla}\psi[\mathbf{X}(Z), Z]. \qquad (14.5)$$

In both cases the incompressibility constraint holds:

$$\boldsymbol{\nabla} \cdot \mathbf{v}(\mathbf{X}, t) = 0. \qquad (14.6)$$

Note, however, that the equations for the particle motion in a fluctuating magnetic field (11.67), (11.68) are not of the type (14.1), (14.2) because of the presence of the multiplicative noise term in the former. In that case the method explained here needs an extension that will be treated in Chap. 15.

We start our exposition with the simplest problem, considering afterwards various additional features. This simplest and physically most transparent case is the problem of drift motion of a particle (or guiding centre) in a fluctuating electric field, in presence of a very strong magnetic field, which makes the motion quasi-two-dimensional. In a first step, the collisions will be neglected. It is immediately noted that this problem is

---

[2] In the absence of potential fluctuations, Eq. (14.2) reduces to Eq. (11.37), in which $v_\parallel \to \eta_\parallel$, with the associated statistical definition (11.38) describing the effect of the collisions.

equivalent to the diffusion of magnetic field lines, represented by the identification (14.5).

14.2   Collisionless Drift Motion in Two Dimensions

The Langevin equations (14.1) - (14.3) reduce to the first of these, with $\eta(t) = 0$ for the (two-dimensional) coordinates $\mathbf{X}(t) = [X(t), Y(t)]$ of a guiding centre in the plane perpendicular to the strong constant magnetic field $\mathbf{B} = B_0\,\mathbf{e}_z$. The fluctuating (gyrophase-averaged) electrostatic potential, $\overline{\delta\phi}(\mathbf{X}, t)$, will be denoted by the simpler notation:

$$\overline{\delta\phi}(\mathbf{X}, t) \rightarrow \varepsilon\,\phi(\mathbf{X}, t). \tag{14.7}$$

Here ε is a parameter, having the dimension of the electrostatic potential,[3] and measuring its typical amplitude. The V-Langevin equation for the guiding centre position is:

$$\frac{d\mathbf{X}(t)}{dt} = \frac{c\varepsilon}{B_0}\left. \left[\mathbf{e}_z \times \frac{\partial}{\partial \mathbf{X}}\phi(\mathbf{X}, t)\right]\right|_{\mathbf{X}=\mathbf{X}(t)}. \tag{14.8}$$

$\phi(\mathbf{X}; t)$ is a dimensionless function; as such it can only depend on dimensionless coordinates. The Eulerian correlation of the potential introduces both a characteristic (perpendicular) *correlation length* $\lambda_\perp$ and a characteristic *correlation time* τ_c, measuring the extent (in space and time) of the effectively correlated zone [see Eqs. (6.48), (6.49)]. With the three dimensional parameters $\varepsilon, \lambda_\perp, \tau_c$ we may construct a dimensionless quantity called the **Kubo number**. It is defined as the distance covered by a particle with typical velocity V during the correlation time τ_c, divided by the correlation length $\lambda_\perp$. But the drift velocity is related to the electric potential, hence K is proportional to the amplitude of the potential fluctuations, and inversely proportional to the magnetic field:

$$K = \frac{V\tau_c}{\lambda_\perp} = \frac{c}{B_0}\frac{\varepsilon\tau_c}{\lambda_\perp^2}. \tag{14.9}$$

It is easily checked that the V-Langevin equation (14.8) takes a very simple form if the following dimensionless coordinates and time are used:

$$\mathbf{x} = \lambda_\perp^{-1}\mathbf{X},$$

$$\theta = K\frac{t}{\tau_c} = \frac{V}{\lambda_\perp}t \equiv \frac{t}{\tau_{fl}}, \tag{14.10}$$

---

[3] This is in contrast to the corresponding magnetic parameter β of Eq. (11.48) which is dimensionless (because the dimension of the magnetic field is included in B_0).

where the components of the dimensionless vector x are (x, y) and the *flight time* $\tau_{fl} = \lambda_\perp/V$. This scaling is chosen because it is independent of the correlation time τ_c, hence it remains meaningful even in the limit of *"frozen turbulence"*, $\tau_c \to \infty$. It will be seen that this limiting case plays an important role in the forthcoming theory. Eq. (14.8) is then transformed to the following equations for the dimensionless scaled variables

$$\frac{dx(\theta)}{d\theta} = -\left.\frac{\partial\phi(x,y,\theta)}{\partial y}\right|_{x=x(\theta)} \equiv v_x[x(\theta), y(\theta), \theta], \quad x(0) = 0,$$

$$\frac{dy(\theta)}{d\theta} = \left.\frac{\partial\phi(x,y,\theta)}{\partial x}\right|_{x=x(\theta)} \equiv v_y[x(\theta), y(\theta), \theta], \quad y(0) = 0. \quad (14.11)$$

The reduced, dimensionless velocity field is thus defined as:

$$v_x(x,y,\theta) = -\frac{\partial\phi(x,y,\theta)}{\partial y}, \quad v_y(x,y,\theta) = \frac{\partial\phi(x,y,\theta)}{\partial x}. \quad (14.12)$$

It is related to the dimensional, physical velocity $V = dX(t)/dt$ by the relation: $V(X,t) = (\lambda_\perp/\tau_c) K v(X/\lambda_\perp, Kt/\tau_c)$. It follows from Eq. (14.6) that the equations of motion (14.11) have a Hamiltonian structure, (x, y) playing the role of canonical variables (p, q). The scalar potential function is the Hamiltonian:

$$H(x,y,\theta) = \phi(x,y,\theta). \quad (14.13)$$

The Hamiltonian system is, in general (*i.e.*, for finite K), non-autonomous (because of the dependence of ϕ on θ), hence the Hamiltonian is not conserved in time. The situation in which ϕ is independent of time (*"frozen turbulence"*) will, however, be an interesting reference case. In that case ϕ is a constant of the motion:

$$\phi[x(\theta), y(\theta)] = \phi(0,0) = const. \quad (frozen) \quad (14.14)$$

The fluctuating potential is defined statistically by assuming *a priori* that it is a Gaussian, centred, homogeneous, isotropic and stationary stochastic process defined by the factorized Eulerian autocorrelation function:

$$\langle\phi(0,0)\,\phi(x,\theta)\rangle = \mathcal{E}(x)\,T(\theta/K) \equiv \mathcal{E}(x)T(t/\tau_c). \quad (14.15)$$

The two functions $\mathcal{E}$ and T are not specified explicitly at this point. It is only assumed that both have a maximum at the origin: $\mathcal{E}(0) = 1$, $T(0) = 1$, $\nabla_y\mathcal{E}(y)|_0 = 0$, $\partial_\theta T(\theta)|_0 = 0$, and are decreasing monotonously from this maximum to zero at infinity.

The Eulerian correlations of the reduced velocity components, and the mixed potential - velocity correlations are defined as follows (see also Sec. 11.4):

$$\langle \phi(0,0)\, v_j(\mathbf{x},\theta) \rangle = \mathcal{E}_{\phi|j}(\mathbf{x})\, T(\theta/K)$$
$$\langle v_j(0,0)\, \phi(\mathbf{x},\theta) \rangle = \mathcal{E}_{j|\phi}(\mathbf{x})\, T(\theta/K)$$
$$\langle v_j(0,0)\, v_n(\mathbf{x},\theta) \rangle = \mathcal{E}_{j|n}(\mathbf{x})\, T(\theta/K). \tag{14.16}$$

Their expression is obtained from Eqs. (14.4) and (14.15):

$$\mathcal{E}_{\phi|x} = -\mathcal{E}_{x|\phi} = -\frac{\partial}{\partial y}\,\mathcal{E}(x,y), \quad \mathcal{E}_{\phi|y} = -\mathcal{E}_{y|\phi} = \frac{\partial}{\partial x}\,\mathcal{E}(x,y),$$

$$\mathcal{E}_{x|x} = -\frac{\partial^2}{\partial y^2}\,\mathcal{E}(x,y), \quad \mathcal{E}_{y|y} = -\frac{\partial^2}{\partial x^2}\,\mathcal{E}(x,y), \quad \mathcal{E}_{x|y} = \mathcal{E}_{y|x} = \frac{\partial^2}{\partial x\,\partial y}\,\mathcal{E}(x,y). \tag{14.17}$$

We denote the values of these functions at the origin by the symbols: $\mathcal{E}_{..|..}(0) = \mathcal{E}^0_{..|..}$. The following values of these quantities follow from our assumptions:

$$\mathcal{E}^0_{\phi|x} = \mathcal{E}^0_{x|\phi} = \mathcal{E}^0_{\phi|y} = \mathcal{E}^0_{y|\phi} = 0, \quad \mathcal{E}^0_{x|y} = 0$$
$$\mathcal{E}^0 = 1, \quad \mathcal{E}^0_{x|x} = \mathcal{E}^0_{y|y} = 1, \quad T(0) = 1. \tag{14.18}$$

Our main objective will be the calculation of the diffusion tensor, for which we need the (dimensional) *Lagrangian correlation tensor of the velocities*:

$$L_{j|n}(t) = \left(\frac{\lambda_\perp}{\tau_c}\right)^2 K^2\, \mathcal{L}_{j|n}(Kt/\tau_c), \tag{14.19}$$

with:

$$\mathcal{L}_{j|n}(\theta) = \int d\mathbf{x}\; \langle v_j(0,0)\, v_n(\mathbf{x},\theta)\, \delta[\mathbf{x} - \mathbf{x}(\theta)] \rangle. \tag{14.20}$$

where $\mathcal{L}_{j|n}(\theta)$ is, by definition, the reduced, dimensionless Lagrangian velocity correlation.

The running difusion tensor $D_{j|n}(t)$ [see Eq. (11.30)] is:

$$D_{j|n}(t) = \int_0^t dt_1\, L_{j|n}(t_1) \equiv \frac{\lambda_\perp^2}{\tau_c}\, \mathcal{D}_{j|n}(Kt/\tau_c), \tag{14.21}$$

where the *dimensionless running diffusion tensor* is:

$$\mathcal{D}_{j|n}(\theta) = K \int_0^{\theta} d\theta_1 \, \mathcal{L}_{j|n}(\theta_1). \tag{14.22}$$

We thus arrived at the really difficult point of the problem. In most previous works the Lagrangian velocity autocorrelation was evaluated by using the standard *Corrsin independence approximation*. This is, however, justified only for weak nonlinearity, more precisely $K \ll 1$. We briefly review the application of this method in Sec. 14.3.

14.3 Collisionless Drift Motion: The Corrsin Approximation

We closely follow here the method developed by WANG et al. 1995. Eqs. (14.11) can be integrated over time:

$$\delta \mathbf{x}(\theta) = \int_0^{\theta} d\theta_1 \, \mathbf{v}[\mathbf{x}(\theta_1), \theta_1], \tag{14.23}$$

where:

$$\delta \mathbf{x}(\theta) = \mathbf{x}(\theta) - \langle \mathbf{x}(\theta) \rangle = \mathbf{x}(\theta). \tag{14.24}$$

We then deduce, as usual, the (dimensionless) mean square displacement (MSD) $\Xi(\theta)$ in the x-direction:

$$\Xi(\theta) \equiv \langle \delta x^2(\theta) \rangle = 2 \int_0^{\theta} d\theta_1 \, (\theta - \theta_1) \, \mathcal{L}_{x|x}(\theta_1), \tag{14.25}$$

where $\mathcal{L}_{x|x}(\theta)$ is the $x - x$ component of the dimensionless *Lagrangian velocity correlation tensor*, Eq. (14.20). We now introduce the *Corrsin approximation* for the evaluation of the Lagrangian correlations. This approximation was explained in detail in Sec. 6.4. Corrsin's independence hypothesis (6.72), amounts to replacing in Eq. (14.20) the exact propagator by its ensemble average; the integrand is then factorized as follows:

$$\mathcal{L}_{j|n}(\theta) = \int d\mathbf{x} \, \langle v_j(0,0) \, v_n[\mathbf{x}, \theta] \rangle \, \langle \delta[\mathbf{x} - \mathbf{x}(\theta)] \rangle. \tag{14.26}$$

We thus obtained the Corrsin relationship between reduced Lagrangian and Eulerian correlations:

$$\mathcal{L}_{j|n}(\theta) = \int d\mathbf{x} \, \mathcal{E}_{j|n}(\mathbf{x}) \mathcal{T}(\theta/K) \, \gamma(\mathbf{x}, \theta). \tag{14.27}$$

$\mathcal{E}_{j|n}(\mathbf{x})\mathcal{T}(\theta/K)$ is the Eulerian velocity autocorrelation defined in Eq. (14.16). The function $\gamma(\mathbf{x}, \theta)$ is interpreted in this context as the probability of finding the particle at the position $\mathbf{x}$ at time θ, starting from $\mathbf{x} = 0$ at time $\theta = 0$. It can be represented as follows, using the well-known formula for the evaluation of the average of the exponential of x in the second cumulant approximation [see Eq. (12.24)]:

$$
\begin{aligned}
\gamma(x, \theta) &= \langle \delta[x - x(\theta)] \rangle \\
&= \int dk \, e^{ik \cdot x} \langle \exp[-ik \cdot x(\theta)] \rangle \\
&= \int dk \, e^{ik \cdot x} \exp\left[-\tfrac{1}{2} k^2 \langle x^2(\theta) \rangle\right] \\
&= \frac{1}{2\pi \, \Xi(\theta)} \exp\left[-\frac{x^2 + y^2}{2 \, \Xi(\theta)}\right].
\end{aligned}
\tag{14.28}
$$

As a result of the homogeneity and isotropy of the Eulerian velocity correlation, it is easily seen that the function γ only depends on the length of the vector x; it then follows (WANG et al. 1995) that the Lagrangian correlation tensor is proportional to the unit tensor:

$$
\mathcal{L}_{j|n}(\theta) = \mathcal{L}(\theta) \, \delta_{jn}.
\tag{14.29}
$$

In order to obtain analytic results, we now assume the following forms for the Eulerian potential correlations (*Ornstein - Uhlenbeck process*):

$$
\mathcal{E}(x) = \exp\left(-\frac{x^2 + y^2}{2}\right),
$$

$$
T(\theta/K) = \exp\left(-|\theta|/K\right) \ \left[= \exp\left(-\frac{|t|}{\tau_c}\right)\right].
\tag{14.30}
$$

The Lagrangian correlation is now easily obtained from Eq. (14.27), combined with (14.2), (14.28) and (14.30); the x-integration is elementary and results (for $\theta > 0$) in:

$$
\mathcal{L}(\theta) = \exp\left(-\theta/K\right) \frac{1}{\left[1 + 2 \int_0^\theta d\theta_1 \, (\theta - \theta_1) \, \mathcal{L}(\theta_1)\right]^2}.
\tag{14.31}
$$

This is the *basic integral equation obeyed by the Lagrangian velocity autocorrelation*. Thus, even in the Corrsin approximation, the determination of this function requires the solution of a highly nonlinear integral equation.

We note that in the limit $K \to 0$, Eq. (14.31) reduces to an explicit expression for $\mathcal{L}(\theta)$ [4]

---

[4] This is easily seen by making the substitution $\theta \to \bar{\theta} = K\theta$; the integral in Eq. (14.31) then acquires a factor K^2.

$$\mathcal{L}(\theta) = \exp\left(-\theta/K\right) = T(\theta/K), \quad K \ll 1. \tag{14.32}$$

For finite values of K, Eq. (14.31) can be solved numerically by successive iterations. The dimensionless autocorrelation function depends on the single dimensionless parameter K, the *Kubo number*, defined in Eq. (14.9).

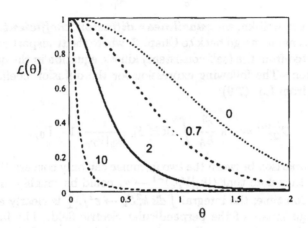

Figure 14.1. Reduced Lagrangian velocity autocorrelation $\mathcal{L}(\theta)$, obtained by numerical integration of Eq. (14.31) for four values of $K = 0$, 0.7, 2, 10.

The numerical solutions of Eq. (14.31) are shown in Fig. 14.1. The bulk of this function becomes wider as K increases. It should be stressed that an increase of K means an increase of the nonlinear effects.

We note that Eq. (14.31) can be converted into an equation for the dimensionless MSD by noting that $\frac{1}{2}d^2\Xi/d\theta^2 = \mathcal{L}$ [see Eq. (14.25)]:

$$\frac{d^2\Xi(\theta)}{d\theta^2} = 2\,\frac{e^{-\theta/K}}{[1 + \Xi(\theta)]^2}, \quad \Xi(0) = 0, \quad \left.\frac{d\Xi(\theta)}{d\theta}\right|_{\tau=0} = 0. \tag{14.33}$$

The solution of this equation then allows us to calculate the (dimensionless) running diffusion coefficient $\mathcal{D}(\theta) = \frac{1}{2}K\,d\Xi/d\theta$, whose asymptotic limit (if it exists!) is the dimensionless diffusion coefficient $\mathcal{D}$.

The basic differential equation (14.33) is nonlinear and cannot, in general, be solved analytically. In the limit $K \ll 1$, the MSD can be neglected in the denominator of Eq. (14.33), and the equation can be trivially integrated, with the following simple result for the dimensionless running diffusion coefficient, defined by Eq. (14.22):

$$\mathcal{D}_{QL}(\theta) = K^2 \left(1 - e^{-\theta/K}\right), \quad \mathcal{D}_{QL} = K^2; \quad K \ll 1 \tag{14.34}$$

which corresponds to the dimensional diffusion coefficient (14.21):

$$D_{QL} = \frac{\lambda_\perp^2}{\tau_c} K^2 = c^2 \frac{\varepsilon^2}{B_0^2 \lambda_\perp^2} \tau_c. \tag{14.35}$$

This is the well-known *quasilinear diffusion coefficient*. In order to justify this name, we go back to Chap. 7, where the transport coefficients were calculated from the (self-consistent) kinetic equation in the quasilinear approximation. The following expression for the diffusion coefficient can be deduced from Eq. (7.9):

$$D_{QL}^{(kin)} = \sqrt{\pi}\, \frac{c^2}{B_0^2} \int d\mathbf{k}\, k_y^2 S_\mathbf{k} \frac{1}{|k_\parallel| V_{Te}} (1 - \Gamma_0). \tag{14.36}$$

The comparison between the two formulae can only concern the general scaling. We thus note that $(|k_\parallel| V_{Te})^{-1} \to \tau_c$ could be roughly understood as a correlation time; the integral $\int d\mathbf{k}\, k_y^2 S_\mathbf{k} \to \varepsilon^2/\lambda_\perp^2$ is clearly a measure of the average square of the perpendicular electric field. The factor $(1 - \Gamma_0)$ is quite interesting: it tells us that the diffusion coefficient vanishes when the finite Larmor radius (FLR) effects are neglected. In the present Langevin model, the motion is driven entirely by the electrostatic drift, which produces a finite Larmor radius: hence, the finite diffusion coefficient. If the latter were neglected, the only remaining mechanism would be the parallel motion combined with parallel collision effects. This leads us to the problem discussed in Chap. 11, where it was found that the motion is indeed subdiffusive, with $D = 0$. In conclusion, the kinetic expression (14.36) and the stochastic expression (14.35) have the same scaling, in particular the characteristic dependence on $\left(E_\perp^2/B_0^2\right)$.

For larger values of K, Eq. (14.33) must be solved numerically. The most interesting result is the dependence of the asymptotic (dimensionless) diffusion coefficient on the Kubo number K, shown in Fig. 14.2 (in log - log representation). For small K the slope of this curve equals 2, as expected from the quasilinear result (14.34). As K increases, the curve bends down and ends up, for large K, with a slope equal to 1, which corresponds to the *Bohm diffusion coefficient*:

$$\mathcal{D}_B \propto K, \quad D_B \propto \frac{\lambda_\perp^2}{\tau_c} K = \frac{c}{B_0} \varepsilon. \tag{14.37}$$

This result was obtained in Chap. 9, from Dupree's theory, based on the DIA-renormalized kinetic equation. Indeed, if in Eq. (9.49) we identify

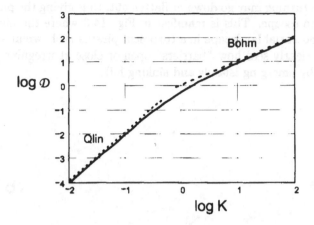

Figure 14.2. Dimensionless diffusion coefficient $\mathcal{D}$ vs. K (log-log representation) (solid line). The dotted line has slope 2, the dash-dotted line has slope 1.

$E_\perp/\overline{k} \to \varepsilon$ and neglect the FLR effect expressed by the Bessel function J_0, we obtain Eq. (14.37).

The Bohm limit is, however, incorrect. Large values of K are outside the domain of validity of Eq. (14.33) itself, not only of its solutions. Indeed, this equation was derived under the assumption of Corrsin's factorization hypothesis. The latter is only ascertained for small values of K (as shown explicitly in the work of WEINSTOCK 1976). It is not a simple matter to go beyond the Corrsin approximation. In the latter, some quite important phenomena are neglected. These will be discussed in the forthcoming sections of the present chapter.

14.4   Qualitative Discussion of Trapping

A particle moving according to the apparently simple Eq. (14.8) describes (in each realization) a very complicated trajectory, because of the intricate electrostatic potential. The situation may be vizualized by plotting at a given instant the potential as a function of x and y. This produces a very rugged landscape, with a lot of irregularly distributed hills and wells. A particle launched in this landscape tries to find its way by following an isopotential line. If it happens to meet a hill, it will start turning around it and will not be able to advance along its initial direction. The same behaviour happens if it crosses a well. The only possibility for it to escape is to find a saddle point near $\phi = 0$. On the other hand, the potential landscape is continuously changing in time; thus a hill around which a particle was turning may go down or flatten out, thus giving the particle the possibility to escape. This is manifest in Fig. 14.3 where the numerically calculated potential landscape in a turbulent plasma is shown at successive times. One clearly sees how "barriers" open or close at irregular positions and times, by emerging islands and sinking hills.

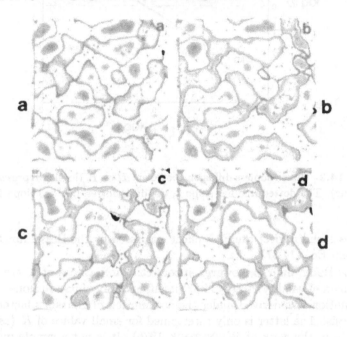

Figure 14.3. Electrostatic potential landscape in a typical turbulent plasma, at successive equally spaced times $a \to d$. (Courtesy of J.H. Misguich and J.-D. Reuss.)

A typical particle orbit has a shape that depends on the strength of the nonlinearity. For very small K it has a rather regular shape, with many kinks, much as the orbit of a molecule in a dilute gas. As K increases, the particle encounters obstacles which trap it; it therefore spends a certain time turning around the hill (or well) before being able to escape and move on an "open" trajectory until it meets another obstacle. Such trajectories resemble superficially the orbit of a particle undergoing "inelastic collisions" with randomly appearing obstacles. This feature is clearly visible in Fig. 14.4, which shows a typical orbit moving in a stochastic potential landscape, such as the one of Fig. 14.3, at high Kubo number. The trapping events can last for a long time, after which the particle may continue along a more or less long unimpeded trajectory, until it meets a new obstacle. It can also be said that the motion in strongly turbulent plasmas has a marked

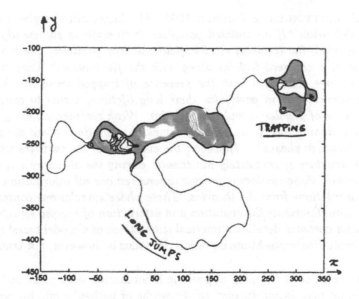

Figure 14.4. Typical trajectory of a particle in a potential landscape of the type of Fig. 14.3, at high amplitude (large K). (Courtesy J. Misguich and M. Vlad)

intermittent character. This will lead to a *non-Gaussian* character of the fluctuating trajectories. A most interesting analogy can be made with the trajectory of a particle performing a *Continuous Time Random Walk (CTRW)* (see Chap. 12). The trajectory of such a particle (in a given realization) is a succession of "free" segments followed by a sojourn in a "trap" where it stays for a finite waiting time before escaping and moving towards another trap, etc. As will be shown below, this analogy can be exploited for obtaining quantitative information about the present type of anomalous transport.

The first mention about the influence on turbulence theories of trapping (in frozen potential landscapes) appears to be in the work of KRAICH-NAN 1970 b, in which he performed numerical simulations in order to test the DIA theory. In his own words, his results showed that "the fidelity of the direct interaction approximation depends strongly on absence of trapping". He did not, however, go beyond this negative statement. The existence of the trapping effect was explicitly manifest in the numerical simulations of PETTINI et al. 1988. Its effect on the diffusion process was, however, not taken into account in that early work, which was erroneously interpreted as being compatible with the Bohm scaling of the diffusion coefficient. The first lucid discussion of the trapping effect appears in the work of CROTTINGER & DUPREE 1992. The latter authors give the following definition "*If an isolated potential fluctuation is sufficiently large and long lived, the resulting electric field will trap particles, causing them to move in a coherent fashion along with the fluctuation.*" They further say: "*We are concerned with the presence of trapped structures because their existence, and in particular their long lifetime, seems to contradict assumptions of popular turbulence theories. Weak turbulence theory, resonance broadening theory, and the DIA make assumptions about the modes having "random phases"... However, the evolution of the particles within a trapped structure is completely nonchaotic during the structure's lifetime. Furthermore, these randomness assumptions remove all information about phase correlations from the theories. These phase correlations contain the information describing the evolution and interaction of trapped structures.*" This paper contains detailed numerical simulations of a model based on the macroscopic Hasegawa-Mima equation. Its focus is, however, not transport theory.

The concept of trapping appears in combination with ideas and techniques from percolation theory[5] in the works of Isichenko and his coworkers (GRUZINOV et al. 1990, ISICHENKO 1991, 1992). A situation of almost frozen turbulence is considered here. The particles that contribute most to the diffusion are those which have long trajectories percolating along the separatrices of the landscape. On this, partly semi-intuitive basis, the diffusion coefficient of a strongly turbulent plasma is shown to depend on the Kubo number more slowly than predicted by the Bohm scaling: $D \propto K^\gamma$, with $\gamma = 0.7$. The strongly trapped particles are not taken into account in this picture.

A first set of numerical simulations by OTTAVIANI 1992 showed that the scaling exponent γ is indeed smaller than 1. More precise simulations (REUSS & MISGUICH 1996, REUSS et al. 1998) provided a value close to Isichenko's $\gamma \approx 0.7$.

The following qualitative discussion may serve as an introduction to the forthcoming developments (MISGUICH et al. 1998 a, OTTAVIANI 1998). We assume that the process can be fitted by a (classical) random walk picture. When K is small, the particles do not "see" the structure of the potential landscape: they move only a small distance before the potential changes (small K is related to small τ_c). The decorrelation is of temporal type: the time step of the random walk equals the correlation time: $t = \tau_c$. The spatial step is the distance covered during τ_c: $l = V\tau_c$. The usual random walk (or "mixing length") estimate [see Eq. (12.25)] then yields: $D = (V\tau_c)^2/\tau_c = (\lambda_\perp^2/\tau_c) K^2$. This is the *quasilinear scaling*. For large K the nonlinearity makes the estimation more difficult. The particle can now

<hr />

[5] A clear presentation of percolation theory is STAUFFER & AHARONY 1992; see also BALESCU 1997 for a shorter presentation.

explore the whole correlated region, hence their space step is $l = \lambda_\perp$: the decorrelation is of spatial type. If the particles are supposed to move freely, the time necessary for covering the correlated region is $t = \lambda_\perp/V$, and this yields a diffusion coefficient estimated as $D = \lambda_\perp^2/(\lambda_\perp/V) = (\lambda_\perp^2/\tau_c)\,K$: This is the *Bohm scaling*. Clearly, this argument does not take trapping into account. Because of the trapping, the particles actually spend a longer time in the correlated zone. Calling $\tau_t(K)$ the characteristic trapping time, which depends on K, the effective time step in the random walk is $t = \lambda_\perp/V + \tau_t(K)$, and the diffusion coeficient is estimated as:

$$D(K) = \frac{\lambda_\perp^2}{\tau_c} \frac{K}{1 + [\tau_t(K)/\tau_c]\,K} \rightarrow \begin{cases} \dfrac{\lambda_\perp^2}{\tau_c} K \quad for \ \dfrac{\tau_t}{\tau_c} K \ll 1 \\[2mm] \dfrac{\lambda_\perp^2}{\tau_t(K)} \quad for \ K \gg 1 \end{cases} \tag{14.38}$$

The trapping time can vary between zero (no trapping) and τ_c (total trapping), we thus obtain:

$$\frac{\lambda_\perp^2}{\tau_c} \frac{K}{1+K} = D_t \leq D(K) \leq D_B = \frac{\lambda_\perp^2}{\tau_c} K. \tag{14.39}$$

This simple argument shows that the Bohm diffusion coefficient is indeed an upper limit, which is only attained when particle trapping can be neglected. The actual calculation of the trapping time is, unfortunately, a very difficult nonlinear problem. Isichenko's theory is one possible approach to this problem. In the remainder of this chapter we present a recently developed approximation method, based on a quite different approach to the treatment of trapping. It has been exposed in a series of papers treating increasingly complicated situations (VLAD et al., 1998, 2000, 2001, 2002, 2004). We start with the simplest problem, closely following VLAD et al. 1998 and VLAD et al., 2004.

14.5 Eulerian Statistics in Subensembles

We consider again the two-dimensional motion of a test particle in presence of a very strong constant magnetic field B under the action of a fluctuating electrostatic potential, in absence of collisions, *i.e.*, the problem defined in Sec. 14.2. We use the same dimensionless coordinates x and θ, defined in Eq. (14.10). The equations of motion in a given realization are (14.11) and the fluctuating field is defined statistically by the Eulerian correlations (14.15) and (14.16).

The first step in the new treatment consists of defining a subdivision of the ensemble of realizations of the fluctuating potential $\phi(\mathbf{x}, \theta)$ into subensembles. We define a *subensemble* S of the global ensemble as the set of all realizations of the potential $\phi(\mathbf{x}, \theta)$ which have a common, fixed value ϕ^0 at the origin, $\theta = 0$, $\mathbf{x} = 0$, together with fixed values of their x- and y-derivatives. The latter define the velocity of the guiding centres through Eqs. (14.12); hence, the subensemble S contains the set of all trajectories of the global ensemble which have, at time $\theta = 0$, hence at $\mathbf{x} = 0$, a fixed value of the potential and of the velocity. Thus, the subensemble S is characterized by the following conditions:

$$S: \quad \phi(\mathbf{0}; 0) = \phi^0, \quad \mathbf{v}(0,0) = \mathbf{v}^0. \tag{14.40}$$

Let $P_0(\phi^0, \mathbf{v}^0)$ be the probability distribution of these quantities in the global ensemble. The following *exact* equation can then be written for the Lagrangian velocity autocorrelation function (14.20),[6] which is a weighted superposition of the averages in all subensembles:

$$L(\theta) = \frac{\lambda_\perp^2}{\tau_c^2} K^2 \int d\phi^0 d\mathbf{v}^0 \, P_0(\phi^0, \mathbf{v}^0) \, \langle v_x[0, 0] \, v_x[\mathbf{x}(\theta), \theta] \rangle^S$$

$$= \frac{\lambda_\perp^2}{\tau_c^2} K^2 \int d\phi^0 d\mathbf{v}^0 \, P_0(\phi^0, \mathbf{v}^0) \, v_x^0 \, \langle v_x[\mathbf{x}(\theta), \theta] \rangle^S$$

$$\equiv \frac{\lambda_\perp^2}{\tau_c^2} K^2 \, \mathcal{L}(\theta). \tag{14.41}$$

It is important to note that $v_x[0, 0]$ is fixed in the subensemble S, hence it can be taken out of the average: the velocity autocorrelation in the subensemble reduces to the simpler average velocity in the same subensemble.

Next, it is assumed that the potential $\phi(\mathbf{x}, \theta)$ is a Gaussian process. The probability distribution of the initial values of the potential and of the velocity is then calculated as follows, using the Fourier transform of the δ-function and the Gaussian property of the potential:

---

[6] We will only be interested here in the radial diffusion process, *i.e.*, in the $x - x$ component of the Lagrangian correlation tensor and of the diffusion tensor. We thus simplify the notation by deleting the subscripts.

$$P_0(\phi^0, \mathbf{v}^0) = \langle\, \delta[\phi^0 - \phi(0,0)]\, \delta[\mathbf{v}^0 - \mathbf{v}(0,0)] \,\rangle$$

$$= (2\pi)^{-3} \int ds\, d\mathbf{q}\, \exp\left(is\phi^0 + iq_x v_x^0 + iq_y v_y^0\right)$$

$$\times \langle \exp\left[-is\phi(0,0) - iq_x v_x(0,0) - iq_y v_y(0,0)\right]\rangle$$

$$= (2\pi)^{-3} \int ds\, d\mathbf{q}\, \exp\left(is\phi^0 + iq_x v_x^0 + iq_y v_y^0\right)$$

$$\times \exp\left\{-\tfrac{1}{2}\left\langle[s\phi(0,0) + q_x v_x(0,0) + q_y v_y(0,0)]^2\right\rangle\right\}$$

$$= (2\pi)^{-3} \int ds\, d\mathbf{q}\, \exp\left(is\phi^0 + iq_x v_x^0 + iq_y v_y^0\right)$$

$$\times \exp\left\{-\tfrac{1}{2}\left[s^2\mathcal{E}^0 + q_x^2\mathcal{E}_{x|x}^0 + q_y^2\mathcal{E}_{y|y}^0\right.\right.$$

$$\left.\left. +2sq_x\mathcal{E}_{\phi|x}^0 + 2sq_y\mathcal{E}_{\phi|y}^0 + 2q_x q_y\mathcal{E}_{x|y}^0\right]\right\}$$

$$= (2\pi)^{-3} \int ds\, d\mathbf{q}\, \exp\left(is\phi^0 + iq_x v_x^0 + iq_y v_y^0\right)\exp\left[-\tfrac{1}{2}\left(s^2 + q_x^2 + q_y^2\right)\right],$$

where use has been made of Eq. (14.18). The remaining integral is trivial:

$$P_0(\phi^0, \mathbf{v}^0) = \frac{1}{(2\pi)^{3/2}}\, \exp\left(-\frac{\phi^{0\,2}}{2} - \frac{v_x^{0\,2} + v_y^{0\,2}}{2}\right). \qquad (14.42)$$

We now calculate the Eulerian autocorrelation functions of the (dimensionless) velocities in the subensemble S:

$$\langle v_j(0,0)\, v_n(\mathbf{x}, \theta)\rangle^S = v_j^0\, \langle v_n(\mathbf{x}, \theta)\rangle^S, \quad j, n = x, y. \qquad (14.43)$$

This shows a significant simplification of the calculations in a subensemble. Given that $v_j(0,0)$ is fixed in the subensemble, it is taken out of the average, and *the calculation of the Eulerian velocity autocorrelation function reduces to the calculation of the average velocity in the subensemble.* The latter, which is a conditional average, is calculated by a simple extension of the procedure used above:

$$\langle v_n(\mathbf{x}, \theta)\rangle^S = \frac{\langle v_n(\mathbf{x}, \theta)\, \delta[\phi^0 - \phi(0,0)]\, \delta[\mathbf{v}^0 - \mathbf{v}(0,0)]\rangle}{\langle \delta[\phi^0 - \phi(0,0)]\, \delta[\mathbf{v}^0 - \mathbf{v}(0,0)]\rangle}$$

$$= \frac{1}{P_0(\phi^0, \mathbf{v}^0)} \int ds\, d\mathbf{q}\, \exp(is\phi^0 + i\mathbf{q}\cdot\mathbf{v}^0)$$

$$\times \langle v_n(\mathbf{x}, \theta)\, \exp[-is\phi(0,0) - i\mathbf{q}\cdot\mathbf{v}(0,0)]\rangle. \qquad (14.44)$$

We evaluate the average under the last integral as follows:

$$\langle v_n(\mathbf{x}, \theta) \, \exp[-is\phi(0,0) - i\mathbf{q} \cdot \mathbf{v}(0,0)] \rangle$$

$$= \frac{\partial}{-i\,\partial a} \, \langle \exp[-is\phi(0,0) - i\mathbf{q} \cdot \mathbf{v}(0,0) - iav_n(\mathbf{x},\theta)] \rangle|_{a=0}$$

$$= \frac{\partial}{-i\,\partial a} \, \exp\left\{-\tfrac{1}{2} \langle [s\phi(0,0) + \mathbf{q} \cdot \mathbf{v}(0,0) + av_n(\mathbf{x},\theta)]^2 \rangle \right\}\Big|_{a=0}$$

$$= \left\{-is\mathcal{E}_{\phi|n}(\mathbf{x}) - iq_x\mathcal{E}_{x|n}(\mathbf{x}) - iq_y\mathcal{E}_{y|n}(\mathbf{x})\right\} T(\theta/K)$$

$$\times \exp\left[-\tfrac{1}{2}\left(s^2 + q_x^2 + q_y^2\right)\right]. \tag{14.45}$$

This result is substituted into Eq. (14.44) and, using (14.42) we find the final result:

$$\langle v_n(\mathbf{x}, \theta) \rangle^S = \left[\phi^0 \, \mathcal{E}_{\phi|n}(\mathbf{x}) + v_x^o \, \mathcal{E}_{x|n}(\mathbf{x}) + v_y^0 \, \mathcal{E}_{y|n}(\mathbf{x})\right] T(\theta/K)$$

$$\equiv v_n^S(\mathbf{x}) \, T(\theta/K). \tag{14.46}$$

We stress an important point: whereas the global average of $\mathbf{v}$ is zero, its average in the subensemble is non-vanishing. The following important property is easily obtained. The average velocity in the subensemble derives from an average potential $\phi^S(\mathbf{x}, \theta)$:

$$\langle \phi(\mathbf{x}, \theta) \rangle^S = \frac{\langle \phi(\mathbf{x}, \theta) \, \delta[\phi^0 - \phi(0,0)] \, \delta[\mathbf{v}^0 - \mathbf{v}(0,0)] \rangle}{\langle \delta[\phi^0 - \phi(0,0)] \, \delta[\mathbf{v}^0 - \mathbf{v}(0,0)] \rangle}$$

$$= \phi^S(\mathbf{x}) T(\theta/K). \tag{14.47}$$

This function is calculated by the same method as above, with the result:

$$\phi^S(\mathbf{x}) = \left[\phi^0 \, \mathcal{E}(\mathbf{x}) + v_x^o \, \mathcal{E}_{x|\phi}(\mathbf{x}) + v_y^0 \, \mathcal{E}_{y|\phi}(\mathbf{x})\right]. \tag{14.48}$$

By calculating the spatial derivatives of the average potential in the subensemble and using Eqs. (14.2) and (14.46) we obtain:

$$-\frac{\partial}{\partial y} \, \phi^S(\mathbf{x}) = v_x^S(\mathbf{x}),$$

$$\frac{\partial}{\partial x} \, \phi^S(\mathbf{x}) = v_y^S(\mathbf{x}). \tag{14.49}$$

These relations have the same form as the microscopic relations (14.11).

14.6 Lagrangian Statistics in Subensembles. The DCT Method

We now *define, in a given subensemble S a deterministic trajectory* $\overline{\mathbf{x}}^S(\theta)$ *as the solution of the Hamiltonian equations generated by the average potential* $\langle \phi(\mathbf{x}, \theta) \rangle^S$. Using Eqs. (14.46) - (14.49), we find:

$$\frac{d\overline{x}^S(\theta)}{d\theta} = -\frac{\partial}{\partial \overline{y}^S} \langle \phi(\overline{\mathbf{x}}^S, \theta) \rangle^S \Big|_{\overline{\mathbf{x}}^S = \overline{\mathbf{x}}^S(\theta)} \equiv v_x^S[\overline{x}^S(\theta), \overline{y}^S(\theta)] \, T(\theta/K),$$

$$\frac{d\overline{y}^S(\theta)}{d\theta} = \frac{\partial}{\partial \overline{x}^S} \langle \phi(\overline{\mathbf{x}}^S, \theta) \rangle^S \Big|_{\overline{\mathbf{x}}^S = \overline{\mathbf{x}}^S(\theta)} \equiv v_y^S[\overline{x}^S(\theta), \overline{y}^S(\theta)] \, T(\theta/K),$$

$$\overline{x}^S(0) = \overline{y}^S(0) = 0. \tag{14.50}$$

This deterministic trajectory will be called the **decorrelation trajectory (DCT)**.

We now consider the Lagrangian velocity autocorrelation (14.41) and introduce our major assumption. We postulate that *the average Lagrangian velocity in the subensemble can be approximated by the average Eulerian velocity in S, calculated along the decorrelation trajectory:*[7]

$$\langle v_n[\mathbf{x}(\theta), \theta] \rangle^S \cong \langle v_n[\overline{\mathbf{x}}^S(\theta), \theta] \rangle^S = v_n^S[\overline{\mathbf{x}}^S(\theta)] \, T(\theta/K). \tag{14.51}$$

The Lagrangian velocity autocorrelation then reduces, in this **DCT approximation** to:

$$\mathcal{L}(\theta) = \int d\phi^0 dv^0 \, P_0(\phi^0, \mathbf{v}^0) \, v_x^0 \, \langle v_x[\mathbf{x}(\theta), \theta] \rangle^S \quad (DCT)$$

$$\cong \int d\phi^0 dv^0 \, P_0(\phi^0, \mathbf{v}^0) \, v_x^0 \, \langle v_x[\overline{\mathbf{x}}^S(\theta), \theta] \rangle^S ,$$

$$= \int d\phi^0 dv^0 \, P_0(\phi^0, \mathbf{v}^0) \, v_x^0 \, v_x^S \, [\overline{x}^S(\theta), \overline{y}^S(\theta)] \, T(\theta/K). \tag{14.52}$$

The main advantage of this approximation is that the Lagrangian average is replaced by an *Eulerian average*, evaluated along the deterministic decorrelation trajectory: a much simpler problem. Instead of solving the equations of motion in a single realization, then averaging over the realizations of this trajectory, we now solve for a single *deterministic* trajectory, and then average over the initial values of ϕ^0 and of $\mathbf{v}^0$.

---

[7] More explicitly, this denotes the expression (14.46) of the *Eulerian* average of **v** in the subensemble, $\langle \mathbf{v}(\mathbf{x}, \theta) \rangle^S$, in which **x** is replaced by the position at time θ of a fictitious point moving along the deterministic decorrelation trajectory, $\overline{\mathbf{x}}^S(\theta)$.

The DCT approximation can be compared to the Corrsin approximation as follows. The exact relation (14.41) can be written as follows in terms of the propagator $\mathcal{G}$ of the Langevin equations of motion (14.11) [see Eqs. (6.15), (6.30)]:

$$\mathcal{L}(\theta) = \int d\phi^0 dv^0 \, P_0(\phi^0, \mathbf{v}^0) \int dx' v_x^0 \, \langle \mathcal{G}(0,0|\mathbf{x}',\theta) \, v_x(\mathbf{x}',\theta) \, \rangle^S. \quad (14.53)$$

If we applied the Corrsin approximation *in the subensemble* S, this would be written as follows [see Eq. (6.72):

$$\mathcal{L}(\theta) = \int d\phi^0 dv^0 \, P_0(\phi^0, \mathbf{v}^0) \int dx' v_x^0 \, \langle \mathcal{G}(0,0|\mathbf{x}',\theta) \, \rangle^S \langle v_x(\mathbf{x}',\theta) \rangle^S,$$
$$(Cor-S). \quad\quad\quad\quad\quad\quad\quad\quad\quad\quad\quad\quad\quad\quad (14.54)$$

Clearly, this expression is *not* equivalent to the global Corrsin approximation. If we now replace the average propagator (in the subensemble) by the deterministic propagator of the equations for the DCT (14.50) we find precisely the DCT expression (14.52):

$$\mathcal{L}(\theta) = \int d\phi^0 dv^0 \, P_0(\phi^0, \mathbf{v}^0) \int dx'^S v_x^0 \, \mathcal{G}^S(0,0|\mathbf{x}'^S,\theta) \, \langle v_x(\mathbf{x}'^S,\theta) \rangle_\phi^S,$$
$$(DCT). \quad\quad\quad\quad\quad\quad\quad\quad\quad\quad\quad\quad\quad\quad (14.55)$$

14.7 The Decorrelation Trajectory

We now consider the determination of the decorrelation trajectory, by solving Eqs. (14.50). We first introduce, for convenience, a new set of dimensionless parameters specifying the subensemble:

$$v_x^0 = u \cos\alpha, \quad v_y^0 = u \sin\alpha, \quad \phi^0 = u\,p. \quad (14.56)$$

The Hamiltonian $\phi^S(\overline{x}^S, \overline{y}^S)$ is then written in the following explicit form, using Eqs. (14.48) and (14.2):

$$\phi^S(\overline{x}^S, \overline{y}^S) = u \, [p + \cos\alpha \, \nabla_y - \sin\alpha \, \nabla_x] \, \mathcal{E}(\overline{x}^S, \overline{y}^S). \quad (14.57)$$

As the initial value of the velocity $\mathbf{v}^0$ is fixed in the subensemble S, we may choose (in this subensemble) a reference frame whose x-axis is oriented along $\mathbf{v}^0$ ($\alpha = 0$). The equations of motion (14.50) are then written as:

$$\frac{d\overline{x}^S(\theta)}{d\theta} = -\,u\,\frac{\partial}{\partial \overline{y}^S}\left[p + \frac{\partial}{\partial \overline{y}^S}\right]\mathcal{E}(\overline{x}^S,\overline{y}^S)\bigg|_{\overline{x}^S=\overline{x}^S(\theta)} \qquad T(\theta/K),$$

$$\frac{d\overline{y}^S(\theta)}{d\theta} = u\,\frac{\partial}{\partial \overline{x}^S}\left[p + \frac{\partial}{\partial \overline{y}^S}\right]\mathcal{E}(\overline{x}^S,\overline{y}^S)\bigg|_{\overline{x}^S=\overline{x}^S(\theta)} \qquad T(\theta/K). \quad (14.58)$$

The factorized form of the potential autocorrelation (14.15) allows us to introduce the following additional change of variables $\theta \to \Theta$:

$$\Theta(\theta) = K \int_0^{\theta/K} d\theta_1\,T(\theta_1), \qquad \frac{d\Theta(\theta)}{d\theta} = T(\theta/K). \qquad (14.59)$$

With the choice (14.30) for $T(\theta/K)$, we obtain:

$$\Theta(\theta) = K\left(1 - e^{-\theta/K}\right), \qquad K < \infty. \qquad (14.60)$$

We note:

$$\Theta(0) = 0, \qquad \Theta(\infty) = K, \qquad K < \infty. \qquad (14.61)$$

These definitions are meaningful only for finite values of K. On the other hand, for any finite θ, the following limiting value in K is obtained:

$$\lim_{K\to\infty}\Theta(\theta) = \theta. \qquad (14.62)$$

We then introduce the new position vector $\mathbf{x}^S(\Theta)$ for the DCT, related to the previous one as follows:

$$\overline{\mathbf{x}}^S(\theta) = \mathbf{x}^S[\Theta(\theta)], \qquad K < \infty \qquad (14.63)$$

Its components obey the following equations:

$$\frac{dx^S(\Theta)}{d\Theta} = -\,u\,\frac{\partial}{\partial y^S}\left[p + \frac{\partial}{\partial y^S}\right]\mathcal{E}(x^S,y^S)\bigg|_{x^S=x^S(\Theta)}$$

$$\frac{dy^S(\Theta)}{d\Theta} = u\,\frac{\partial}{\partial x^S}\left[p + \frac{\partial}{\partial y^S}\right]\mathcal{E}(x^S,y^S)\bigg|_{x^S=x^S(\Theta)},$$

$$x^S(0) = 0, \qquad y^S(0) = 0. \qquad (14.64)$$

This is now a set of *autonomous* equations, deriving from the Θ-independent Hamiltonian:

$$H^S(x^S,y^S) = u\left[p + \frac{\partial}{\partial y^S}\right]\mathcal{E}(x^S,y^S). \qquad (14.65)$$

This quantity, expressed as a function of Θ, is a constant of the motion:

$$H^S[x^S(\Theta), y^S(\Theta)] = H^S(0,0) = \phi^0. \tag{14.66}$$

Eqs. (14.64) are identical to those describing *"frozen turbulence"*, *i.e.* a system in which the fluctuations are purely spatial; the Eulerian potential autocorrelation is independent of time. The latter equations are obtained from (14.58) in the limit $K \to \infty$, $\exp(-\theta/K) = 1$. In this case:

$$\overline{\mathbf{x}}^S(\theta) = \mathbf{x}^S(\theta) \qquad K = \infty. \tag{14.67}$$

The case of *frozen turbulence* thus appears as a convenient reference problem: if the autonomous equations (14.64) are solved, their solution immediately yields the solution of the non-autonomous equations (14.58) for finite K, through a simple change of variables.

In order to obtain explicit numerical results we will use Eq. (14.30) for the Eulerian potential autocorrelation: the spatial part is assumed to have a Gaussian shape, whereas the temporal factor is supposed to be simply exponential.

The "frozen" decorrelation trajectories obtained by numerical integration of Eqs. (14.64) for four values of the parameter p (and taking $u = 1$) are shown in Fig. 14.5. The coordinates as functions of time Θ are shown in Fig. 14.6.

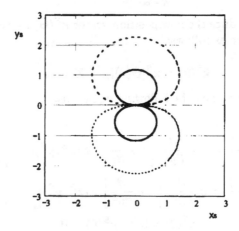

Figure 14.5. Decorrelation trajectories for frozen drift turbulence. Trajectories in the lower half-plane correspond to $p > 0$ [solid: $p = 1$, dashed: $p = 0.25$], those in the upper half-plane to $p < 0$ [solid: $p = -1$, dashed: $p = -0.25$].

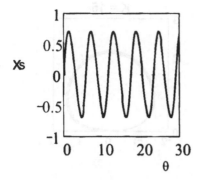
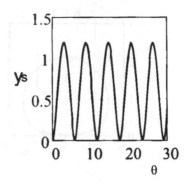

Figure 14.6. The motion along the decorrelation trajectory for frozen turbulence, $x^S(\theta)$, $y^S(\theta)$, for $p = -1$, $u = 1$.

The decorrelation trajectories are thus closed curves (except for $p = 0$, whose trajectory is the open x-axis). The motion along these orbits is periodic in time. This is to be expected in a situation of frozen turbulence: the potential landscape being time-independent, any trapped particle remains trapped forever. Thus, the motion of the fictitious particles along the DCT, for all $|p| > 0$, is made of vortices (eddies); the small ones correspond to large values of the parameter $|p|$, *i.e.* to the top of potential hills (or to the bottom of potential wells). One therefore expects that there is no diffusion possible in this frozen turbulence situation; in other words, the macroscopic evolution should be *subdiffusive*.

We now consider the DCT orbits $\overline{y}^S(\theta)$ in the time-dependent case. Some DCT trajectories obtained by solving Eqs. (14.58) [or by using the solution of (14.64) together with (14.63)] are shown in Fig. 14.7. It clearly appears that the orbit in the $x - y$ plane is the same as in the frozen case. But. for small values of K, it is not entirely covered: the (fictitious) particle comes to a stop before completing a full cycle. This could be predicted from Eqs. (14.58): after a certain time, depending on K, the exponential factor $T(\theta/K)$ goes to zero, and the velocity vanishes. For larger values of K, the (fictitious) particle can complete several cycles before stopping.

This behaviour is clearly shown by plotting $\overline{y}^S(\theta)$ [or $\overline{x}^S(\theta)$] as a function of time (Fig. 14.8). For very small values of K the fictitious particle motion is very quickly damped, before the end of a cycle: one could say that it is "unaware" of the potential landscape. On the contrary, for large values of K the particle goes many times around the cycle before stopping; it thus fully feels the effect of trapping. In the limit of $K \to \infty$ the DCT approaches the "frozen" one, as expected from Eq. (14.67).

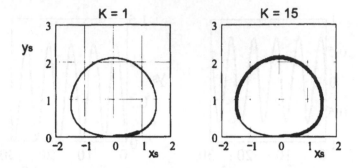

Figure 14.7. DCT orbit for the time-dependent case (thick line) compared to the frozen case (thin line), for two values of K. ($p = -0.25$).

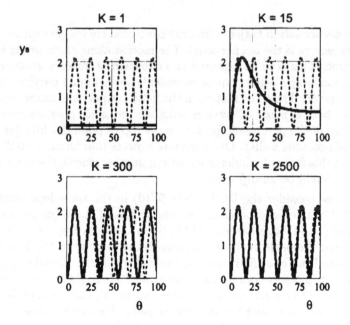

Figure 14.8. DCT orbits $\bar{y}^S(\theta)$ compared to the "frozen" orbits, for four values of K. ($p = -0.25$, $u = 1$)

Another interesting feature can be seen from Fig. 14.8: the period of the oscillations in the frozen case appears to be independent of K: it only depends on p (and on u). This is clearly understood from the fact that

Eqs. (14.64) are independent of K.

At this point the reader might be somewhat puzzled. The DCT are extremely simple objects, along which the (fictitious) particles move in a very simple way (periodic motion, or periodic + stop). This is to be contrasted with the extremely complex motion of the physical particles: the latter do not turn in circles, nor do they stop after a finite time. The DCT should not be understood as real (or even average) trajectories of physical particles. The key for understanding its meaning lies in Eq. (14.43): the velocity along the DCT is proportional to the Lagrangian velocity autocorrelation in the subensemble S. Thus, *the DCT is an indicator of the degree of correlation of the physical particles*. A periodic trajectory means that the particles having the characteristics of the subensemble S remain correlated forever. A stop in the DCT, meaning a vanishing of the DCT velocity, implies that the particles have been decorrelated [as results from Eq. (14.43)]. Note that the decorrelation mechanism in the present, collisionless case is purely *temporal*: it results from the time evolution of the potential, as explained in Sec. 14.4.

We now introduce these trajectories into Eq. (14.52) for the Lagrangian correlation function. Using also Eqs. (14.59) - (14.63), we find the following form of this function:[8]

$$\mathcal{L}(\theta) = G[\Theta(\theta)] \, T(\theta/K), \tag{14.68}$$

where the function G is:

$$G(\Theta) = \frac{1}{\sqrt{2\pi}} \int_{-\infty}^{\infty} dp \int_{0}^{\infty} du \, u^3 \exp\left[-\frac{(1+p^2)\,u^2}{2}\right] v_x^S\left[\mathbf{x}^S(\Theta; u, p)\right]. \tag{14.69}$$

We first note that the solution $\mathbf{x}^S(\Theta)$ of the "frozen turbulence" problem automatically yields the expression of the Lagrangian velocity autocorrelation in the general, time-dependent case. This is due to our assumption about the factorized form of the Eulerian autocorrelation, Eq. (14.15). This factorized form is inherited by the Lagrangian velocity autocorrelation [Eq. (14.68)], but there is an essential difference: both factors, G and T depend on time and on K. Thus the shape of the Lagrangian correlation is the result of the competition between these two factors. Two limiting cases may be considered.

For $K \ll 1$, the function $v_x^S[\mathbf{x}^S(\Theta)] \approx v_x^S[\mathbf{x}^S(\theta)]$ [see Eq. (14.62)]. On the other hand, the exponential factor $\exp(-\theta/K)$ cuts down the Lagrangian correlation in a time much shorter than the characteristic rate of

---

[8] Note that $\mathcal{L}(\theta)$ depends on K (for finite K) through the function $\Theta(\theta)$ and through $T(\theta/K)$.

change of $G[\Theta(\theta)]$; hence the latter can be approximated by its value at the origin, which is found from Eq. (14.64) $[v_x^S(0) \approx u]$ to be simply 1, $G(\Theta) \approx 1$. In the limit $K \ll 1$, the integral in (14.69) is done analytically, and we find:

$$\mathcal{L}(\theta) \approx T(\theta/K) = \exp(-\theta/K), \quad K \ll 1. \tag{14.70}$$

The factor $T(\theta/K)$ is rather trivial: it represents the overall decay of the fluctuations in time, with the characteristic time scale τ_c. We thus recover the result (14.32): the small-K limit is the *quasilinear regime*, as expected.

In the opposite limit, $K \gg 1$, the exponential factor remains very close to 1 for a very long time θ, and the Lagrangian autocorrelation is well approximated by the function $G(\Theta)$. The latter, through its functional dependence on the DCT, contains the information about the structure of the correlated zone. *It determines, in particular, the effect of the trapping,* which is absent from previous treatments of the problem. For very large values of K, the DCT of the time-dependent problem is well approximated by the "frozen" DCT, as can be clearly seen from Eq. (14.67) and Fig. 14.8, thus $G[\Theta(\theta)] \approx G(\theta)$ and $\exp(-\theta/K) \approx 1$. In this limit, the function $G(\theta)$ no longer depends explicitly on K. This function must, however, be calculated numerically. The two limiting behaviours are illustrated in Fig. 14.9. The most remarkable property of the function $G(\theta)$ is the existence of a long negative "tail". The consequences of this behaviour will be discussed below. This feature clearly results from the trapping. A comparison with Fig. 14.1 is striking: the Lagrangian correlation calculated in the Corrsin approximation (which does not account for trapping) possesses no negative tail.

The time integral of the Lagrangian correlation (expressed in terms of the dimensional time t) yields the dimensionless running diffusion coefficient (14.22):

$$\mathcal{D}(\theta) = K \int_0^\theta d\theta_1 \, G[\Theta(\theta_1)] \exp\left(-\frac{\theta_1}{K}\right). \tag{14.71}$$

Changing the integration variable $\theta_1 \to \Theta$, and noting from Eq. (14.60) that $\exp(-\theta/K) \, d\theta = d\Theta$, we find:

$$\mathcal{D}(\theta) = K \, F[\Theta(\theta)], \tag{14.72}$$

with:

$$F(\Theta) = \int_0^\Theta d\Theta_1 \, G(\Theta_1). \tag{14.73}$$

Using Eqs. (14.50), (14.64) and (14.69), $F(\Theta)$ can also be expressed in terms of the DCT trajectory:

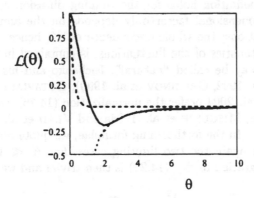

Figure 14.9. Lagrangian velocity autocorrelation [reduced by the factor $(c\varepsilon/B\lambda_\perp)^{-2}$] as a function of θ. Dashed: $\mathcal{L}(\theta) \approx \exp(-\theta/K)$, for $K = 0.2$. Solid: large-K limit: $\mathcal{L}(\theta) \approx G(\theta)$. Dotted: asymptotic form of $G(\theta)$ [Eq. (14.103) below]. (Courtesy M. Vlad and I. Petrisor)

$$F(\Theta) = \frac{1}{\sqrt{2\pi}} \int_0^\Theta d\Theta_1 \int dp\, du\, u^3 \exp\left[-\frac{(1+p^2)\, u^2}{2} \right] x^S(\Theta; u, p).$$

$$(14.74)$$

Finally, the dimensionless diffusion coefficient in the DCT approximation is (for finite K):

$$\mathcal{D}(K) = \mathcal{D}(\theta = \infty) = K\, F(K). \quad K < \infty \qquad (14.75)$$

Before continuing our discussion, we wish to attract attention on a point which may produce some confusion. The dimensional diffusion coefficient in Eq. (14.21) is expressed in units of $(\lambda_\perp^2/\tau_c)$, *i.e.*, a combination of the characteristic length and time of the Eulerian velocity correlation: $D(K) = (\lambda_\perp^2/\tau_c)\, \mathcal{D}(K)$. It is, however, possible to construct a different combination of the available dimensional parameters in order to define a unit of diffusion: from Eq. (14.9) we note that $(\lambda_\perp^2/\tau_c) = (c\varepsilon/B)\, K^{-1}$. Thus, an equivalent definition of the diffusion coefficient is:

$$D(K) = \frac{c\varepsilon}{B_0}\, \mathcal{D}'(K), \qquad (14.76)$$

with:

$$\mathcal{D}'(K) = K^{-1}\mathcal{D}(K) = F(K). \qquad (14.77)$$

(A similar definition holds for the running diffusion coefficient.) In this case, the dimensional factor only depends on the amplitude of the potential fluctuations; the whole dependence on K, hence on the spatio-temporal characteristics of the fluctuations, is contained in $\mathcal{D}'(K)$. Both normalizations may be called "natural". Isichenko and his collaborators (ISICHENKO 1991, 1992, GRUZINOV et al. 1990), OTTAVIANI 1992, 1998 as well as VLAD et al. 2004 prefer the normalization (14.76), whereas REUSS et al. 1996, 1998, MISGUICH et al. 1998a and VLAD et al. 1998 use the definition (14.21). In the forthcoming formulae, we quote both forms.

We consider again the two limiting cases. For $K \ll 1$, we use Eq. (14.70): the integration in Eq. (14.22) is then trivial and we find:

$$D_{QL}(K) = \frac{\lambda_\perp^2}{\tau_c} K^2 = \frac{c\varepsilon}{B_0} K, \quad K \ll 1. \tag{14.78}$$

We thus recover the *quasilinear diffusion coefficient*, Eq. (14.35), scaling as K^2 [according to the normalization (14.21)]. In this case there is no trapping effect: the particles ignore the rugged structure of the potential. This could be expected from the shape of the DCT, as seen in Figs. 14.7 and 14.8: for small K, the fictitious particle comes to a stop long before circling the potential well or hill. The decorrelation is purely temporal.

In the opposite limit, we find:

$$D(K) = \frac{\lambda_\perp^2}{\tau_c} K \int_0^\infty d\theta_1 \, G(\theta_1) = \frac{c\varepsilon}{B_0} \int d\theta_1 \, G(\theta_1) = 0, \quad K \to \infty. \tag{14.79}$$

The vanishing of the integral of $G(\theta)$ is an important property, which can be verified numerically or semi-analytically. Referring to Fig. 14.9, this means that the area enclosed by the positive part of the Lagrangian velocity autocorrelation exactly cancels the area enclosed by its negative tail. We met with a similar situation in Sec. 11.6, Fig. 11.6. We shall come back to this question below. The vanishing of the diffusion coefficient means that in the limit of very large K the system has a *strange, subdiffusive behaviour*. This is to be expected: in this *"frozen turbulence"* situation all particles are trapped in a rigid, static potential and have no possibility of diffusing. This is vividly shown by the shape of the DCT, which tends to a strictly periodic, bounded motion (Figs. 14.5, 14.6).

In between these two limits, the long tail of $G(\theta)$ is cut by the decaying exponential in the integral (14.71). This destroys the cancellation of the two parts of the function $G(\theta)$ and leads to a finite, positive diffusion coefficient. Clearly, the diffusion coefficient depends on the assumed form of the Eulerian potential autocorrelation function $\mathcal{E}(x)\mathcal{T}(\theta/K)$, Eq. (14.15), which is an input for the theory. We first consider the Gaussian form (14.30) which was used throughout the explicit calculations of the present section.

The anomalous diffusion coefficient is calculated from Eqs. (14.74), (14.75) by using a numerical code constructed by M. Vlad. The result is shown in Fig. 14.10.

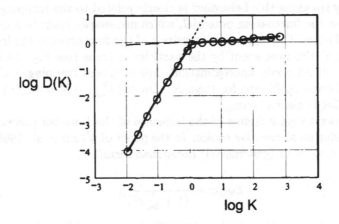

Figure 14.10. Asymptotic diffusion coefficient (in units of λ_1^2/τ_c) as a function of the Kubo number K. The potential autocorrelation is Gaussian, Eq. (14.30). The straight lines correspond to the quasilinear value ($\gamma = 2$) and to the asymptotic DCT value ($\gamma = 0.02$). (Courtesy M. Vlad and I. Petrisor)

For small K the scaling of the diffusion coefficient [according to the definition (14.21)] is: $\mathcal{D}(K) \propto K^2$, in agreement with (14.78). For larger values of K the curve bends toward lower values of the slope, till it reaches the following asymptotic form for $K > 1$, showing a quasi-plateau:

$$D(K) \propto \frac{\lambda^2}{\tau_c} K^\gamma = \frac{c\varepsilon}{B_0} K^{\gamma'}, \quad K \gg 1 \qquad (14.80)$$

with:

$$\gamma' = \gamma - 1. \qquad (14.81)$$

The exponent obtained from Eq. (14.75) is:

$$\gamma = 0.02, \quad \gamma' = -0.98 \quad (Gauss). \qquad (14.82)$$

This behaviour can be understood qualitatively as follows. Mathematically speaking, we may refer to Eq. (14.76) and to Fig. 14.9. As explained above, for small K, the exponential in the integrand dominates and leads to the quasilinear result. As K increases, the exponential $\exp(-\theta/K)$ becomes wider: it cuts down the integral at a point where $G(\theta)$ becomes negative, hence the competition between the two factors results in a smaller diffusion coefficient. The larger K, the wider the region with negative G: this explains the decreasing diffusion coefficient: $D(K) \propto (c\varepsilon/B_0) K^{-\gamma'}$. In the limit $K \to \infty$, $D(K) \to 0$, and we reach the subdiffusive limit (14.79). Physically speaking this behaviour is clearly related to the trapping of the particles in the fluctuating potential, which necessarily leads to a decrease in the diffusion. The larger the intensity ε of the fluctuations, the longer is (on average) the time spent by the particles in traps (see Fig. 14.8). We recall that the Corrsin approximation, which ignores trapping, yields the Bohm diffusion coefficient for large K: $D_B \propto (\lambda_\perp^2/\tau_c) K = (c\varepsilon/B_0)$, *i.e.*, $\gamma' = 0$. This is clearly wrong.

In order to get a feeling of the influence of the Eulerian potential, we now consider an alternative choice. In the paper of VLAD et al. 1998, $\mathcal{E}(\mathbf{x})$ was chosen to be of (generalized) Lorentzian form:

$$\mathcal{E}_V(\mathbf{x}) = \frac{1}{\left(1 + \frac{1}{2n} r^2\right)^n}, \tag{14.83}$$

where n is some positive number, typically, $0.5 < n < 2$ ($n = 1$ corresponds to the ordinary Lorentzian). This choice was motivated by the purpose of comparison with the numerical simulations of MISGUICH et al. 1998 and of VLAD et al. 1998 a. The anomalous diffusion coefficient is calculated as above by using the same numerical code constructed for this purpose by M. Vlad. A typical graph of the reduced diffusion coefficient as a function of K (in log-log representation) obtained in this case, for $n = 0.85$, is shown in Fig. 14.11.

For small K the scaling of the diffusion coefficient [according to the definition (14.21)] is again the quasilinear scaling $D(K) \propto K^2$. For larger values of K the curve bends again toward lower values of the slope, till it reaches the asymptotic form (14.80) for $K \gg 1$. The exponent obtained in this case is, however, significantly larger, but still smaller than 1:

$$\gamma = 0.64, \quad \gamma' = -0.36 \quad (Lorentz, \ n = 0.85). \tag{14.84}$$

These values are in pretty good agreement with the prediction of the semi-quantitative theory of GRUZINOV et al. 1990, and ISICHENKO 1991, 1992, based on percolation theory (see Sec. 14.4). They are also in agreement with the numerical particle simulations of REUSS & MISGUICH 1996, REUSS et al. 1998, MISGUICH et al. 1998 and VLAD et al. 1998 a. In a recent paper, VLAD et al. 2004 studied systematically the dependence of the diffusion coefficient curve $D(K)$ on the Eulerian potential correlation. They considered a function $\mathcal{E}(\mathbf{x})$ that is a linear combination of a Gaussian (14.30) and a generalized Lorentzian (14.83). This study allows us to understand the great difference of the Lorentzian result, compared to the Gaussian case. Indeed, as n increases, the algebraic "tail" of the

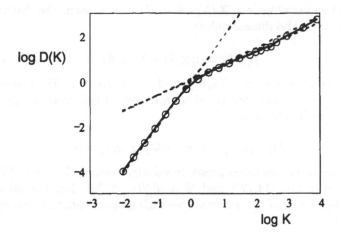

Figure 14.11. Asymptotic diffusion coefficient (in units of $\lambda_\perp^2/\tau_c$) as a function of the Kubo number K. The potential autocorrelation is a generalized Lorentzian, Eq. (14.83), with $n = 0.85$. The left straight line corresponds to the quasilinear value ($\gamma = 2$), the upper asymptotic dashed line to the DCT value ($\gamma = 0.64$), and lower one to the percolation value ($\gamma = 0.7$). (Figure redrawn with permission from data of VLAD et al., *Phys. Rev.* E **58** 7359 (1998). Copyright 1998 by the American Physical Society.)

Eulerian correlation becomes shorter [*i.e.*, $\mathcal{E}(\mathbf{x})$ decreases more steeply at large $|\mathbf{x}|$], its shape approaches more and more the Gaussian behaviour, whose exponential decay is an extreme case. The corresponding decrease of the exponent γ shows that the trapping is stronger, the shorter the tail of $\mathcal{E}(\mathbf{x})$. The "limiting" Gaussian case leads to a quasi-saturation of the diffusion coefficient as a function of K. This question will be discussed in more detail in Sec. 14.9.

14.8 Hybrid Kinetic Equation for Collisionless 2D Drift Motion

In the case of 2D collisionless drift motion, the DCT method can be further developed, thus providing an in-depth analysis of the state of the turbulent plasma. It was shown in Sec. 11.8 that with any V-Langevin equation, a Hybrid Kinetic Equation (HKE) can be associated for the determination of the average distribution function, or *density profile* of the plasma (BALESCU 2000). We thus introduce a distribution function of the guiding centres, which is decomposed into an average part, *i.e.*, the density profile, and a fluctuation. The equations take again the simplest form when the dimensionless variables $\mathbf{x} = \mathbf{X}/\lambda_\perp$, $\theta = Kt/\tau_c$ are used; the distribution function is then also dimensionless:

$$f(\mathbf{x}, \theta) = n(\mathbf{x}, \theta) + \delta f(\mathbf{x}, \theta), \qquad (14.85)$$

where, as usual, $n(\mathbf{x}, \theta) = \langle f(\mathbf{x}, \theta) \rangle$, and $\langle \delta f(\mathbf{x}, \theta) \rangle = 0$. The distribution function obeys a partial differential equation whose characteristic equations are the Langevin equations (14.11):

$$\partial_\theta f(x, y, \theta) + \mathbf{v}(\mathbf{x}, \theta) \cdot \nabla f(x, y, \theta) = 0. \qquad (14.86)$$

The dimensionless incompressible velocity vector field $\mathbf{v}(\mathbf{x}; \theta)$ [$\nabla \cdot \mathbf{v} = 0$] is defined in Eq. (14.12)], and $\nabla = [\partial/\partial x, \partial/\partial y]$. Eq. (14.86) is split as usual, by means of the ensemble averaging operation, into two coupled equations:

$$\partial_\theta n(\mathbf{x}, \theta) = - \langle \mathbf{v}(\mathbf{x}, \theta) \cdot \nabla \delta f(\mathbf{x}, \theta) \rangle \qquad (14.87)$$

$$\partial_\theta \delta f(\mathbf{x}, \theta) + \mathbf{v}(\mathbf{x}, \theta) \cdot \nabla \delta f(\mathbf{x}, \theta) = S(\mathbf{x}, \theta), \qquad (14.88)$$

where the source term is:

$$S(\mathbf{x}, \theta) = -\mathbf{v}(\mathbf{x}, \theta) \cdot \nabla n(\mathbf{x}, \theta) + \nabla \cdot \langle \mathbf{v}(\mathbf{x}, \theta) \delta f(\mathbf{x}, \theta) \rangle. \qquad (14.89)$$

According to the usual strategy, we first solve the initial value problem for Eq. (14.88) by the method of characteristics. The characteristic equations of the first-order partial differential equation (14.88) are written as follows and are solved backwards in time:

$$\frac{d\mathbf{x}(\theta')}{d\theta'} = \mathbf{v}[\mathbf{x}(\theta'), \theta'], \qquad (14.90)$$

with the "initial" condition at time $\theta' = \theta$:

$$\mathbf{x}(\theta) = \mathbf{x}. \qquad (14.91)$$

The solution for the vector $\mathbf{x}(\theta')$ is:

$$\mathbf{x}(\theta') = \mathbf{x} + \int_\theta^{\theta'} d\theta_1 \, \mathbf{v}[\mathbf{x}(\theta_1), \theta_1]. \qquad (14.92)$$

The solution of the partial differential equation (14.88) is then:

$$\delta f(\mathbf{x}, \theta) = \int_0^\theta d\theta_1 \, S[\mathbf{x}(\theta_1), \theta_1] + \delta f[\mathbf{x}, 0]. \qquad (14.93)$$

This result is substituted into Eq. (14.87) for the density profile:

$$\partial_\theta n(\mathbf{x}, \theta) = \nabla \cdot \int_0^\theta d\theta_1 \, \langle \mathbf{v}(\mathbf{x}, \theta) \, \mathbf{v}[\mathbf{x}(\theta_1), \theta_1] \cdot \nabla n[\mathbf{x}(\theta_1), \theta_1] \rangle$$
$$- \nabla \cdot \langle \mathbf{v}(\mathbf{x}, \theta) \, \delta f[\mathbf{x}, 0] \rangle. \tag{14.94}$$

(As usual, the last term in the right hand side of Eq. (14.89) contributes to Eq. (14.94) a term of the form $\langle \mathbf{v} \rangle \langle \mathbf{v} \, \delta f \rangle = 0$.)

This is an exact consequence of Eqs. (14.87), (14.88); it has the following structure, which was already found in Sec. 6.2:

- It is a non-Markovian equation, as it relates the rate of change of the density profile at time θ to its value at all previous times $\theta \geq \theta_1 \geq 0$;
- It is a non-local equation in space, as it relates the rate of change of the density profile at point $\mathbf{x}$ to its value at all neighbouring points $\mathbf{x}(\theta_1)$.

The latter property introduces the difficulty already met in Secs. 6.2 and 11.10. Indeed, the solution $\mathbf{x}(\theta_1)$ is a stochastic quantity, depending on the fluctuating velocity $\mathbf{v}(\mathbf{x}, \theta)$. As a result, the factor $\nabla n[\mathbf{x}(\theta_1), \theta_1]$ cannot be taken out of the average, hence Eq. (14.94) is a differential equation of infinite order for the density profile. This difficulty is avoided by the argument of Secs. 6.2, 11.10. Whenever the length L_n is much greater than the correlation length $\lambda_\perp$ of the fluctuations, the spatial delocalization in Eq. (14.94) can be neglected:

$$n[\mathbf{x}(\theta_1), \theta_1] \approx n[\mathbf{x}(\theta), \theta_1] \equiv n(\mathbf{x}, \theta_1), \quad L_n \gg \lambda_\perp. \tag{14.95}$$

Note that if this condition of weak inhomogeneity is assumed at time $\theta = 0$, it will be even better satisfied at later times, because the diffusive process will flatten the profile even more. With this assumption, Eq. (14.94) reduces to the *closed* non-Markovian equation:

$$\partial_\theta n(\mathbf{x}, \theta) = \nabla \cdot \int_0^\theta d\theta_1 \, \langle \mathbf{v}(\mathbf{x}, \theta) \, \mathbf{v}[\mathbf{x}(\theta_1), \theta_1] \rangle \cdot \nabla n(\mathbf{x}, \theta_1)$$
$$- \nabla \cdot \langle \mathbf{v}(\mathbf{x}, \theta) \, \delta f[\mathbf{x}, 0] \rangle. \tag{14.96}$$

The last term in the right hand side of Eq. (14.96) is the "destruction term", depending on the initial value of the fluctuations [see Eq. (6.35)]. For simplicity, we assume that the initial state is deterministic, $\delta f(\mathbf{x}, 0) = 0$, hence the destruction term is zero.

The memory kernel appearing in the integral of Eq. (14.96) is immediately recognized as the dimensionless Lagrangian velocity autocorrelation tensor [see Eq. (14.20)]:

$$\langle v_i[\mathbf{x}(\theta), \theta]\, v_j[\mathbf{x}(\theta_1), \theta_1]\rangle = \mathcal{L}_{i|j}(\theta, \theta_1) = \mathcal{L}(\theta - \theta_1)\,\delta_{ij}. \qquad (14.97)$$

Finally, Eq. (14.96) reduces to:

$$\partial_\theta n(\mathbf{x}, \theta) = \int_0^\theta d\theta_1\, \mathcal{L}(\theta_1)\, \nabla^2 n(\mathbf{x}, \theta - \theta_1). \qquad (14.98)$$

We thus obtained a typical *non-Markovian diffusion equation* of the form of Eq. (12.96) which appeared in the theory of continuous time random walks (CTRW) discussed in Chap. 12. It is therefore natural to apply the rich package of methods developed in that field to the present problem.

14.9   Hybrid Kinetic Equation and Continuous Time Random Walks

The Lagrangian velocity autocorrelation $\mathcal{L}(\theta)$ was obtained by the DCT method in Sec. 14.7. We first briefly consider the limit of *very small Kubo number*, $K \ll 1$. In this case, $\mathcal{L}(\theta)$ reduces to a simple decaying exponential, Eq. (14.70). This function cuts down exponentially the integration range in Eq. (14.98) at a value $\theta \sim K$, i.e., $t \sim \tau_c$. As the density profile is assumed to be slowly varying over a time scale of order τ_c, the usual process of Markovianization of Eq. (14.98) is justified (see Chap. 6):

- The retardation in the argument of the density profile is neglected: $n(\mathbf{x}, \theta - \theta_1) \approx n(\mathbf{x}, \theta)$;
- The upper limit of the integral over θ_1 is pushed to infinity.

The equation thus reduces to the Markovian equation:

$$\partial_\theta n(\mathbf{x}, \theta) = \mathcal{D}_{QL}\, \nabla^2 n(\mathbf{x}, \theta), \quad K \ll 1. \qquad (14.99)$$

This is an *ordinary diffusion equation,* with a (dimensionless) diffusion coefficient given by the usual expression:

$$\mathcal{D}_{QL} = K \int_0^\infty d\theta\, \mathcal{L}(\theta_1) = K^2, \quad K \ll 1. \qquad (14.100)$$

We thus recover the result (14.78). In the limit of weak turbulence (small Kubo number) the plasma behaves *diffusively,* and the *quasilinear theory* is valid. The density profile evolving from an initial condition concentrated at the origin (the fundamental solution of the diffusion equation) is the well-known *Gaussian packet,* Eq. (12.27) (with $\mathbf{V} = 0$, $d = 2$ and $t = K^{-1}\tau_c\theta$):

$$n_M(r,t) = \frac{1}{4\pi\, D_{QL}\, t}\exp\left(-\frac{r^2}{4 D_{QL} t}\right), \quad K \ll 1 \qquad (14.101)$$

[where $r^2 = x^2 + y^2$, and D_{QL} is defined by Eq. (14.78)]. The present process is thus an ordinary diffusive process [governed by the diffusion equation (14.99)], but with an *anomalous diffusion coefficient*, entirely determined by the characteristics of the turbulence, ε, $\lambda_\perp$, τ_c.

We now consider the opposite limit of *strong turbulence*, $K \gg 1$. In this case the exponential factor in Eqs. (14.58) and (14.68) is $\exp(-\theta/K) \approx 1$. The DCT can then be approximated by its form in the frozen turbulence limit (14.67). Thus the Lagrangian velocity autocorrelation in the strong turbulence (*i.e.*, strong trapping) limit is approximated as:

$$\mathcal{L}(\theta) \approx G(\theta), \quad K \gg 1. \qquad (14.102)$$

We know already from Eq. (14.79) that the plasma behaves *subdiffusively* in this limit. The HKE will provide us with a deeper insight into this process. We therefore need a more detailed analysis of the function $G(\theta)$. The most important property of this function, shown in Fig. 14.9 is the presence of a long negative tail for large times. We will have a better view of its shape in a log-log representation, shown in Fig. 14.12, for the case of a Gaussian potential autocorrelation $\mathcal{E}(\mathbf{x})$.[9]

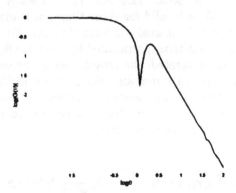

Figure 14.12. The function $G(\theta)$ in log-log representation. The final slope is (-1.92). (Courtesy I. Petrisor)

[9] Note that the numerical calculation of $\mathcal{L}(\theta)$ for large K requires much time and memory. Fig. 14.12 is actually obtained for $K = 20$. The asymptotic regime is reached already for $\log\theta \approx 0.5$; very large values ($\theta > 10^4$) are not plotted, as they are affected by the (slowly) decaying exponential $\exp(-\theta/K)$.

This figure clearly shows that, for sufficiently long times ($\theta \gtrsim 3$), the Lagrangian autocorrelation has a negative tail that is described by an *algebraic* decay. In ordinary representation, the function $G(\theta)$ behaves as (see Fig. 14.9):

$$G_{as}(\theta) \sim -H_0\,\theta^{-\delta}, \quad H_0 = 0.86, \quad \delta = 1.98, \qquad \theta \gtrsim 3. \qquad (14.103)$$

We shall be mainly interested in the long-time behaviour of $G(\theta)$. We therefore define the following analytic approximation for this function:

$$G(\theta) \approx \begin{cases} 1, & \theta < \theta_{\min} \\ G_{as}(\theta), & \theta > \theta_{\min} \end{cases}. \qquad (14.104)$$

The threshold time $\theta_{\min}$ will be determined below.

The situation is quite analogous to the one encountered in our study of the Standard Long Tail Continuous Time Random Walk (SLT-CTRW) in Chap. 12 (see Fig. 12.6). It is therefore natural to apply the methods developed there to our present problem. The non-Markovian diffusion equation is:

$$\partial_\theta n(\mathbf{x},\theta) = \int_0^\theta d\theta_1\, G(\theta_1)\, \nabla^2 n(\mathbf{x}, \theta - \theta_1). \qquad (14.105)$$

We look for an asymptotic solution of this equation, valid for $\theta \gg \theta_{\min}$. We proceed as in Secs. 12.5 and 12.7. Clearly, the asymptotic approximation (14.103) is invalid for short times, as it diverges for $\theta \to 0$: this is why we introduced the analytic form (14.104). Eq. (14.105) is then written as the sum of two terms, separated by the time $\theta_{\min}$. For the long times in which we are interested, the retardation in the density profile can be neglected in the short-time domain, which then contributes a "normal" diffusion term, as shown in the previous section. On the other hand, in the long-time domain, the memory function can be approximated by its asymptotic form (14.103). We thus obtain:

$$\partial_\theta n(\mathbf{x},\theta) = D_0\,\nabla^2 n(\mathbf{x},\theta) - H_0 \int_{\theta_{\min}}^\theta d\theta_1\, \frac{1}{\theta_1^{2-\alpha}}\,\nabla^2 n(\mathbf{x}, \theta - \theta_1), \quad (14.106)$$

and from Eq. (14.103):

$$D_0 = \int_0^{\theta_{\min}} d\theta_1\, G(\theta_1) = \theta_{\min}, \ H_0 = 0.86, \ \mu = 2 - 1.98 = 0.02. \quad (14.107)$$

We introduced here the exponent μ defined as:

$$\mu = 2 - \delta. \tag{14.108}$$

With this notation, Eq. (14.106) has exactly the canonical form of Eq. (12.100) of the SLT-CTRW theory. The exponent μ has a clear physical meaning, as shown below (see also Chap. 12).

In order to determine the value of $\theta_{\min}$, we require that Eq. (14.79), expressing the subdiffusive character of the process, be satisfied:

$$\int_0^\infty d\theta_1 \, G(\theta_1) = D_0 - H_0 \int_{\theta_{\min}}^\infty d\theta_1 \, \frac{1}{\theta_1^{2-\mu}} = 0. \tag{14.109}$$

The integral is easily evaluated, and we find the value:

$$\theta_{\min} = \left[\frac{H_0}{(1-\mu) D_0} \right]^{1/(1-\mu)}. \tag{14.110}$$

With the values (14.107), we find:

$$\theta_{\min} = \left(\frac{H_0}{1-\mu} \right)^{1/(1-\mu)} = 0.87. \tag{14.111}$$

Note that the value of the threshold time could also be determined by the same method as in Sec. 12.6: the second moment $\langle r^2(t) \rangle$ is calculated from Eq. (12.103); one then obtains Eq. (12.104) with $\tau_D = 1$, which is identical with Eq. (14.110). We stress again the important fact that the choice of this cut-off ensures the correct scaling of *all* the moments of the density profile, not only of its second moment: this was shown in detail in Sec. 12.6.

The *analytic form of the asymptotic density profile* was obtained for a general SLT-CTRW in Sec. 12.6: the results (12.82) and (12.85) are directly applicable to the present problem:

$$n(r,\theta) = \theta^{-\mu} \exp\left(-\tfrac{1}{4} q^{2/(2-\mu)} \right), \tag{14.112}$$

where the scaling variable q is:

$$q = \frac{r}{\theta^{\mu/2}}. \tag{14.113}$$

This density profile is quite different from the Gaussian packet $n_M(r,\theta)$ (14.101) obtained from the quasilinear, ordinary Markovian diffusion equation (Fig. 14.13). The homogenization of an initial delta-function $\delta(r)$ is much slower, and the profile develops a long-living tail at large distance. It is important to note that the density profile is independent of the value of $\theta_{\min}$, provided the latter obeys Eq. (14.110).

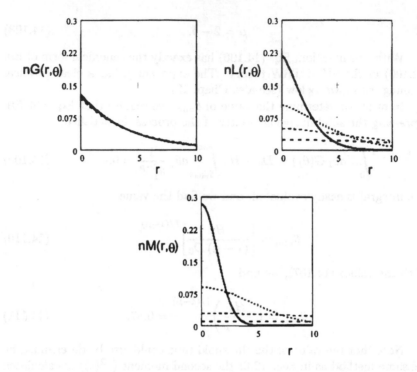

Figure 14.13. Density profiles at various times θ. Solid: $\theta = 1$; Dotted: $\theta = 10$; Dashed: $\theta = 100$; Dash-dotted: $\theta = 1000$. nG: Non-Markovian profile (14.112) for Gaussian $\mathcal{E}$ ($\alpha = 0.02$). nL: Non-Markovian profile (14.112) for Lorentzian $\mathcal{E}$ ($\alpha = 0.64$). nM: Markovian profile (14.102).

We also show in Fig. 14.13 the density profiles obtained for a generalized Lorentzian $\mathcal{E}(\mathbf{x})$, Eq. (14.83) with $n = 0.85$. In that case it turns out that:

$$H_0 = 0.56, \quad \mu = 0.64. \tag{14.114}$$

These profiles are closer to the Markovian profiles, but possess a longer tail. Comparing the profiles obtained from the Gaussian $\mathcal{E}(\mathbf{x})$ and from the Lorentzian we note that the former has a much longer tail, which decreases very slowly compared to the latter. This result may seem counter-intuitive, because the Gaussian correlation $\mathcal{E}(\mathbf{x})$ has a much shorter tail than the Lorentzian. It is, however, understood upon recalling that the Gaussian $\mathcal{E}(\mathbf{x})$ produces a much stronger trapping. As a result, the whole density profile varies much slower in this case.

This feature is more conspicuous in Fig. 14.14: the three profiles are plotted together for large r, at a constant time, viz. $\theta = 5$. The tail produced by the Gaussian $\mathcal{E}(\mathbf{x})$ is everywhere higher than the Lorentzian, which in turn is higher than the Markovian. All these results are in agreement with the discussion of Sec. 14.7 and with Ref. VLAD et al. 2004.

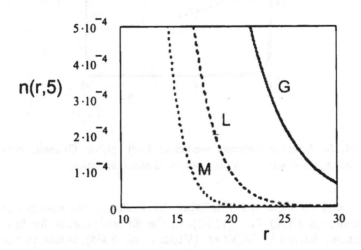

Figure 14.14. Tails of the density profiles at time $\theta = 5$. Solid: Gaussian; Dashed: Lorentzian; Dotted: Markovian.

We now calculate the mean square displacement, by multiplying both sides of Eq. (14.106) by r^2, integrating by parts and using Eq. (14.111):

$$\langle r^2(\theta) \rangle = 4 \frac{H_0}{\mu(1-\alpha)} \theta^\mu. \qquad (14.115)$$

Thus, $\mu = 0.02 < 1$ is the **diffusion exponent**, Eq. (12.76), which is smaller than 1. We thus confirm our previous conclusion: in the limit of very large K, we deal here with *strange diffusion*, more specifically *sub-diffusion*. This explains the result (14.79). The corresponding running diffusion coefficient is:

$$\mathcal{D}(\theta) = \frac{1}{2} \partial_\theta \langle r^2(\theta) \rangle = 2 \frac{H_0}{(1-\mu)} \theta^{\mu-1}, \qquad (14.116)$$

It clearly appears from Fig. 14.15 that $\mathcal{D}(\theta)$ is at all times smaller for the Gaussian $\mathcal{E}(\mathbf{x})$ ($\mu = 0.02$) than for the Lorentzian ($\mu = 0.64$): this is another consequence of the stronger trapping in the former case.

At this point it is very interesting to note that there exists a simple relation between the diffusion exponent α in the subdiffusive regime

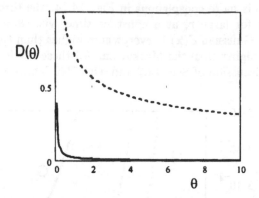

Figure 14.15. Running diffusion coefficient $\mathcal{D}(\theta)$. Solid: Gaussian potential autocorrelation; Dashed: Lorentzian potential autocorrelation.

$(K = \infty)$ and the exponent γ defining the scaling of the asymptotic diffusion coefficient $D(K)$, Eq. (14.80), in the diffusive regime for finite K. This relation, derived by M.VLAD (VLAD et al. 2004), is due to the fact that both $\mathcal{D}(K)$ for finite K and $\mathcal{L}(\theta)$ in the frozen limit $(K = \infty)$ are determined by the same function $G(\theta)$ [or its integral $F(\theta)$].

It is noted that the function $G(\theta)$ *has an asymptotic algebraic tail* (14.103), (14.108)

$$G_{as}(\theta) = -H_0\,\theta^{-\delta} = -H_0\,\theta^{-2+\mu}.$$

In the limit $K \to \infty$ (frozen turbulence), $\mathcal{L}(\theta) = G(\theta)$ and, as just shown, this implies that μ is precisely the diffusion exponent in this limit.

On the other hand, *for finite K*, the diffusion coefficient $\mathcal{D}(K) \propto K^\gamma$ (14.80) is determined by $F(K)$ through Eq. (14.75). With the assumed form for $G_{as}(\theta)$, the function $F(K) \propto K^{-2+\mu+1}$. Then Eq. (14.75) yields $\mathcal{D}(K) \propto K \cdot K^{-1+\mu} = K^\mu$, which then leads to:

$$\gamma = \mu. \tag{14.117}$$

This remarkable relation is verified numerically (in the case of the Gaussian potential correlation) by comparing Eqs. (14.82) and (14.107). It constitutes a very useful test for the results of numerical integrations.

Chapter 15

Diffusion in a Stochastic Magnetic Field

15.1 Statement of the problem

The problem studied in Chap. 14 concerned the motion of a charged test particle in two dimensions, in presence of a stochastic electrostatic field and in absence of collisions. The application of the Decorrelation Trajectory (DCT) method enabled us to study this problem in the regime of strong turbulence, taking into account the phenomena of trapping in the fluctuating potential landscape. In that case the decorrelation mechanism of the particles is purely temporal: it is produced by the potential fluctuations which randomly trap or release the particles. In a frozen (time-independent) potential the particles remain trapped forever: there is no possible diffusion (the process is subdiffusive).

We now address the problem of the *motion of a charged test particle in three dimensions, in presence of a stochastic magnetic field and in presence of collisions*. This problem was already defined in Chap. 11 by Eqs. (11.67) and (11.68). It has been the object of numerous studies by many authors: JOKIPII & PARKER 1969, RECHESTER & ROSENBLUTH 1978, KADOMTSEV & POGUTSE 1979, DIAMOND et al. 1980, KROMMES et al. 1983, ISICHENKO 1991, WHITE & WU 1993, MYRA et al. 1993, RAX & WHITE 1992, SPINEANU et al. 1994, WANG et al. 1995, MISGUICH et al. 1995, 1998 b, REUSS & MISGUICH 1996, VLAD et al. 1996, 2003.

In Chap. 11, however, only a simplified version of this problem was treated, because it could lead to an exact analytic solution. We now treat the complete problem, including the dependence of the magnetic fluctuations on all three spatial coordinates, and retaining the effect of both parallel collisions and perpendicular collisions. For simplicity, however, we consider that the magnetic fluctuations are time-independent.[1] The pres-

---

[1] The inclusion of a time dependence of the magnetic field Eulerian correlations, of the

ence of perpendicular collisions is crucial: it provides a *spatial decorrelation mechanism* that allows a trapped particle to be released from its "prison" even in a frozen magnetic turbulence. This mechanism is thus complementary to the one described in Chap. 14.

A complete treatment of the anomalous transport would involve the combination of both electrostatic and magnetic effects. In specific applications, *e.g.* to controlled fusion, an inhomogeneous and curved (say, toroidal) reference magnetic field should be considered, thus producing additional drift motions. These extensions are left for future works: we believe that the main elements for a treatment of these problems are contained in the previous and the present chapters.

Besides the three-dimensionality and the inclusion of perpendicular collisions, there is an additional difference between the electrostatic and the magnetic problems. In the latter case the parallel collisions introduce a *multiplicative noise* through the product of the fluctuating magnetic field and the parallel collisions. Clearly, an extension of the previous DCT method is needed: we follow here closely the work of VLAD et al., 2003.

15.2   V-Langevin Equations for Magnetic Fluctuations

For convenience, we recall here the main notations introduced in Chap. 11. We consider a charged test particle moving in presence of a magnetic field having a strong constant component (defining the z-direction of a reference frame) and a smaller spatially fluctuating component, perpendicular to the former. Dimensionless coordinates $[\mathbf{x} = (x, y, z)]$, defined by means of the correlation lengths, Eq. (11.47), are used:

$$\mathbf{B}(x, y, z) = B_0[\mathbf{e}_z + \beta\, b_x(x, y, z)\ \mathbf{e}_x + \beta\, b_y(x, y, z)\ \mathbf{e}_y]. \qquad (15.1)$$

The statistical definition of the fluctuating magnetic field has been described in detail in Chap. 11. In particular, the corresponding dimensionless vector potential $\mathbf{A}(\mathbf{x}, z) = \psi(\mathbf{x}, z)\, \mathbf{e}_z$ is assumed to be a Gaussian process, defined by the Eulerian autocorrelation $\mathcal{A}(\mathbf{x})\mathcal{B}_{\parallel}(z)$, (11.54); additional important Eulerian correlations are the mixed field-potential correlations $C_{n|\psi}(\mathbf{x}, z)$, (11.59) and the field-field correlations $\mathcal{B}_{j|n}(\mathbf{x})\mathcal{B}_{\parallel}(z)$, Eq. (11.61). These correlations introduce two characteristic lengths: the *parallel correlation length* $\lambda_{\parallel}$ and the *perpendicular correlation length* $\lambda_{\perp}$, which were used in the scaling of the coordinates.

The collisions are modelled as in Chap. 11 by their influence on the parallel ($v_{\parallel}$), resp. perpendicular ($\mathbf{v}_{\perp}$) velocity [see Eq. (11.77)]:

type of Eq. (14.15), is trivially simple, but makes the final formulae more cumbersome. It also introduces new regimes of diffusion (because of the presence of an additional parameter, the correlation time τ_c). This problem has been treated in VLAD et al. 2003.

$$\langle v_{\parallel}(0)\, v_{\parallel}(t)\rangle = \chi_{\parallel}\nu\,\exp(-\nu|t|),$$

$$\langle v_{j\perp}(0)\, v_{n\perp}(t)\rangle = \delta_{jn}\,\chi_{\perp}\nu\,\exp(-\nu|t|), \quad j,n = x,y \qquad (15.2)$$

where $\chi_{\parallel}$ ($\chi_{\perp}$) are the dimensional parallel (perpendicular) collisional diffusion coefficients defined by Eqs. (11.35) [(11.33)] and ν is the collision frequency. We also introduce a dimensionless time, scaled with the collision frequency:[2]

$$\tau = \nu t. \qquad (15.3)$$

Next, we introduce the following dimensionless diffusion coefficients:[3]

$$\overline{\chi}_{\parallel} = \frac{\chi_{\parallel}}{\nu\,\lambda_{\parallel}^2}, \quad \overline{\chi}_{\perp} = \frac{\chi_{\perp}}{\nu\,\lambda_{\perp}^2}. \qquad (15.4)$$

Note that $\overline{\chi}_{\parallel}$ is the quantity which appeared already in Chap. 11, Eq. (11.92). The dimensionless "collisional" velocities $\eta_{\parallel}$, $\eta_{\perp}$ are defined as follows:

$$\eta_{\parallel}(\tau) = \frac{v_{\parallel}(t)}{\sqrt{\chi_{\parallel}\nu}}, \quad \eta_{\perp}(\tau) = \frac{\mathbf{v}_{\perp}(t)}{\sqrt{\chi_{\perp}\nu}}. \qquad (15.5)$$

Thus,

$$\left\langle \eta_{\parallel}(0)\,\eta_{\parallel}(\tau)\right\rangle_{\parallel} = \exp(-|\tau|), \quad \left\langle \eta_j(0)\,\eta_n(\tau)\right\rangle_{\perp} = \delta_{jn}\exp(-|\tau|). \qquad (15.6)$$

Finally, we introduce the dimensionless number M, which plays the role of the Kubo number in the present problem:

$$M = \beta\,\frac{\lambda_{\parallel}}{\lambda_{\perp}}\,\sqrt{\overline{\chi}_{\parallel}}. \qquad (15.7)$$

Note that M depends on both the characteristics of the random magnetic field (β, $\lambda_{\parallel}$, $\lambda_{\perp}$) and the characteristics of the collisions ($\chi_{\parallel}$, ν).

We now derive from Eqs. (11.67), (11.68) the following equations for the scaled coordinates of the guiding centre of the test particle:

$$\frac{d\mathbf{x}(\tau)}{d\tau} = M\,\mathbf{b}[\mathbf{x}(\tau), z(\tau)]\,\eta_{\parallel}(\tau) + \sqrt{\overline{\chi}_{\perp}}\,\eta_{\perp}(\tau), \qquad (15.8)$$

[2] We choose the notation τ in order to stress the fact that this scaled time is different from the one used in Chap. 14. Here we have no finite correlation time τ_c (the magnetic fluctuations are frozen); instead, the collisions introduce the natural time scale ν^{-1}.

[3] It may be noted that, whereas in a strong magnetic field $\chi_{\perp}/\chi_{\parallel} \ll 1$, the ratio of the dimensionless collisional coefficients $\overline{\chi}_{\perp}/\overline{\chi}_{\parallel}$ may be small or large, depending on the ratio of the correlation lengths.

$$\frac{dz(\tau)}{d\tau} = \sqrt{\overline{\chi}_\parallel}\, \eta_\parallel(\tau). \tag{15.9}$$

These V-Langevin equations are the starting point of our treatment. Their solution depends on three independent dimensionless parameters, viz.: M, $\overline{\chi}_\parallel$, $\overline{\chi}_\perp$.

15.3   Average over Perpendicular Collisions

As usual in anomalous transport theory, our goal is the evaluation of the Lagrangian perpendicular velocity autocorrelation $\mathcal{L}(\tau)$, from which we deduce the anomalous radial (or cross-field) diffusion coefficient. In the present problem, this calculation involves three averaging processes, viz., over the fluctuating magnetic potential $\psi(\mathbf{x}, z)$, over the parallel "collisional" velocity $\eta_\parallel(\tau)$ and over the perpendicular "collisional" velocity $\eta_\perp(\tau)$. Given the assumed statistical independence of the three processes, this operation may be decomposed into several steps. In a first step the third average mentioned above is performed. We consider the effect of the perpendicular collisions alone, by setting $M = 0$. We are then left with:

$$\frac{d\boldsymbol{\xi}(\tau)}{d\tau} = \sqrt{\overline{\chi}_\perp}\, \eta_\perp(\tau). \tag{15.10}$$

Its solution, for $\boldsymbol{\xi}(0) = 0$, is:

$$\boldsymbol{\xi}(\tau) = \sqrt{\overline{\chi}_\perp} \int_0^\tau d\tau_1\, \eta_\perp(\tau_1). \tag{15.11}$$

We introduce the change of variables:

$$\mathbf{x}'(\tau) \equiv \mathbf{x}(\tau) - \boldsymbol{\xi}(\tau), \tag{15.12}$$

and derive from (15.8) the equation:

$$\frac{d\mathbf{x}'(\tau)}{d\tau} = M\, \mathbf{b}[\mathbf{x}'(\tau) + \boldsymbol{\xi}(\tau),\, z(\tau)]\, \eta_\parallel(\tau)$$
$$\equiv \tilde{\mathbf{v}}[\mathbf{x}'(\tau) + \boldsymbol{\xi}(\tau),\, z(\tau), \tau]. \tag{15.13}$$

The correlations of $\boldsymbol{\xi}(\tau)$ are easily obtained from (15.6) (for $\tau > \tau'$):

$$\langle \xi_j(\tau)\xi_n(\tau') \rangle_\perp = \chi_\perp \int_0^\tau d\tau_1 \int_0^{\tau'} d\tau_2\, \exp(-|\tau_1 - \tau_2|)$$
$$= \chi_\perp [e^{-(\tau-\tau')} + e^{-\tau} + e^{-\tau'} + 2\tau - 3] \equiv 2\overline{\chi}_\perp \hat{\mu}(\tau, \tau'), \tag{15.14}$$

and:

$$\left\langle [\xi(\tau)]^2 \right\rangle_\perp = 2\overline{\chi}_\perp (e^{-\tau} + \tau - 1) \equiv 2\overline{\chi}_\perp \, \mu(\tau). \tag{15.15}$$

The function $\mu(\tau)$ is the same as the one introduced in Eq. (11.88). The (Gaussian) probability distribution (PDF) of the perpendicular collisional displacements is thus:

$$P_{1\perp}(\xi, \tau) = \frac{1}{4\pi \, \overline{\chi}_\perp \, \mu(\tau)} \exp\left[-\frac{|\xi|^2}{4\overline{\chi}_\perp \, \mu(\tau)} \right]. \tag{15.16}$$

The autocorrelation of the magnetic potential, averaged over the potential and over the perpendicular velocity[4] is now written as:

$$A(\mathbf{x}, z, \tau) = \left\langle \left\langle \psi(0,0) \, \psi[\mathbf{x} + \xi(\tau), z] \right\rangle_\psi \right\rangle_\perp$$

$$= \int d\xi \, A(\mathbf{x} + \xi, z) \, P_{1\perp}(\xi, \tau). \tag{15.17}$$

We are now ready to start a treatment analogous to the one of Chap. 14.

15.4 The DCT Method for Magnetic Diffusion

The next step is the definition of a set of subensembles S. They are defined as the set of realizations of the trajectories of the guiding centres that have a common fixed value of the magnetic potential, of the magnetic field *and* of the parallel velocity at time $\tau = 0$:

$$S: \quad \psi(0) = \psi^0, \quad \mathbf{b}(0) = \mathbf{b}^0, \quad \eta_\|(0) = \eta^0. \tag{15.18}$$

The dimensionless Lagrangian perpendicular velocity correlation is then obtained by a slight generalization of Eq. (14.41):

$$\mathcal{L}(\tau) = M^2 \int d\psi^0 d\mathbf{b}^0 d\eta^0 \; P(\psi^0, \mathbf{b}^0) P(\eta^0) \, b_x^0 \eta^0$$

$$\times \int d\xi \, \left\langle \left\langle b_x[\mathbf{x} + \xi, z(\tau)] \, \eta_\|(\tau) \right\rangle_\psi \right\rangle_\|^S \, P_{1\perp}(\xi, \tau), \tag{15.19}$$

which is written more briefly as:

$$\mathcal{L}(\tau) = M \int d\psi^0 d\mathbf{b}^0 d\eta^0 \; P(\psi^0, \mathbf{b}^0) P(\eta^0) \, b_x^0 \eta^0 \; \left\langle \tilde{v}_x[\mathbf{x}(\tau), \tau] \right\rangle^S, \tag{15.20}$$

---

[4] This is a kind of mixed Eulerian-Lagrangian average, where $\mathbf{x}$ and z are fixed, but $\xi(\tau)$ is evaluated along its trajectory.

where:

$$P(\psi^0, \mathbf{b}^0) = (2\pi)^{-3/2} \exp[-\tfrac{1}{2}(\psi^{0\,2} + b_x^{0\,2} + b_y^{0\,2})],$$
$$P(\eta^0) = (2\pi)^{-1/2} \exp(-\tfrac{1}{2}\,\eta^{0\,2}). \tag{15.21}$$

By analogy with Eq. (14.44) the Eulerian subensemble average of the velocity is expressed as a conditional average which, because of the statistical independence at $\theta = 0$, factorizes as follows:

$$\langle \tilde{v}_x(\mathbf{x}, \tau) \rangle^S = \int dz\, J_x^S(\mathbf{x}, z, \tau)\, K_\parallel^S(z, \tau). \tag{15.22}$$

The "perpendicular factor" is:

$$J_x^S(\mathbf{x}, z, \tau) = [P(\psi^0, \mathbf{b}^0)]^{-1} M \int d\boldsymbol{\xi}\, P_{1\perp}(\boldsymbol{\xi}, \tau)$$
$$\times \left\langle b_x(\mathbf{x} + \boldsymbol{\xi}, z)\, \delta[\psi^0 - \psi(0,0)]\, \delta[\mathbf{b}^0 - \mathbf{b}(0,0)] \right\rangle_\psi, \tag{15.23}$$

and the "parallel factor":

$$K_\parallel^S(z, \tau) = [P(\eta^0)]^{-1} \left\langle \eta_\parallel(\tau)\, \delta[z - z(\tau)]\, \delta[\eta^0 - \eta_\parallel(0)] \right\rangle_\parallel. \tag{15.24}$$

The calculations for the perpendicular factor are very similar, although somewhat longer than those of Eqs. (14.44) - (14.46): they will not be repeated in detail here. Use is made of the definitions (11.59) and (11.61):

$$J_x^S(\mathbf{x}, z, \tau) = M\, B_\parallel(z) \int d\boldsymbol{\xi}\, P_{1\perp}(\boldsymbol{\xi}, \tau)$$
$$\times \left[\psi^0 C_{\psi|x}(\mathbf{x} + \boldsymbol{\xi}) + b_x^0 B_{x|x}(\mathbf{x} + \boldsymbol{\xi}) + b_y^0 B_{y|x}(\mathbf{x} + \boldsymbol{\xi}) \right]. \tag{15.25}$$

Using also Eq. (15.16), as well as the choice of Gaussian magnetic correlations, (11.63), (11.64), the remaining integral over $\boldsymbol{\xi}$ can also be done analytically, with the result:

$$J_x^S(\mathbf{x}, z, \tau) = M\, B_\parallel(z)\, n_\perp^2(\tau)\, \exp[-\tfrac{1}{2}\, n_\perp(\tau)\, (x^2 + y^2)]$$
$$\times \left\{ -\psi^0 y + b_x^0 [1 - n_\perp(\tau)\, y^2] + b_y^0\, n_\perp(\tau)\, xy \right\}. \tag{15.26}$$

The corresponding y-component of the vector $\mathbf{J}$ is:

$$J_y^S(\mathbf{x}, z, \tau) = M\, B_\parallel(z)\, n_\perp^2(\tau)\, \exp[-\tfrac{1}{2}\, n_\perp(\tau)\, (x^2 + y^2)]$$
$$\times \left\{ \psi^0 x + b_x^0\, n_\perp(\tau)\, xy + b_y^0 [1 - n_\perp(\tau)\, x^2] \right\}. \tag{15.27}$$

We introduced here the factor [see Eq. (15.15)]:

$$n_\perp(\tau) = \frac{1}{1 + \langle \xi^2(\tau) \rangle_\perp} = \frac{1}{1 + 2\overline{\chi}_\perp \, \mu(\tau)}. \tag{15.28}$$

The calculation of the parallel factor $K_\parallel(z,\tau)$ is somewhat more cumbersome. It makes use of the following expressions (for $\tau > 0$):

$$\left\langle \eta_\parallel(0)\,\eta_\parallel(\tau) \right\rangle = \exp(-\tau), \quad \left\langle \eta_\parallel(0)\,z(\tau) \right\rangle = \sqrt{\overline{\chi}_\parallel}\,\varphi(\tau),$$

$$\left\langle \eta_\parallel(\theta)\,z(\tau) \right\rangle = \sqrt{\overline{\chi}_\parallel}\,\varphi(\tau), \quad \langle z^2(\tau) \rangle = 2\overline{\chi}_\parallel\,\mu(\tau). \tag{15.29}$$

[The function $\varphi(\tau)$ has been defined in Eq. (11.88).] Proceeding along the same line as in Sec. 14.5, we find:

$$K_\parallel^S(z,\tau) = -\frac{1}{P(\eta^0)} \left[\sqrt{\overline{\chi}_\parallel}\,\varphi(\tau)\,\frac{\partial}{\partial z} + e^{-\tau}\,\frac{\partial}{\partial \eta^0} \right] P^S(z,\tau) \tag{15.30}$$

where:

$$P^S(z,\tau) = \frac{1}{(2\pi)^2} \int dk \, \exp[ikz - k^2 \overline{\chi}_\parallel \mu(\tau)]$$

$$\times \int dq \, \exp[-\tfrac{1}{2}q^2 + iq\eta^0 - qk\sqrt{\overline{\chi}_\parallel}\varphi(\tau)]$$

$$= \frac{1}{\left[2\pi \langle \delta z^2(\tau) \rangle^S \right]^{1/2}} \exp\left\{ -\frac{\left[z - \langle z(\tau) \rangle^S \right]^2}{2 \langle \delta z^2(\tau) \rangle^S} \right\}. \tag{15.31}$$

The function $P^S(z,\tau)$ thus has the simple physical interpretation of the PDF of parallel displacements in the subensemble S. It is characterized by a non-vanishing average displacement:

$$\langle z(\tau) \rangle_\parallel^S = \eta^0 \sqrt{\overline{\chi}_\parallel}\varphi(\tau), \tag{15.32}$$

and the parallel mean square displacement (MSD):

$$\langle \delta z^2(\tau) \rangle_\parallel^S = \overline{\chi}_\parallel \left[2\mu(\tau) - \varphi^2(\tau) \right]. \tag{15.33}$$

Note that the average displacement in the subensemble depends on S (through η^0), whereas the MSD is the same in all subensembles. The result (15.31) is now substituted into (15.30); the resulting $K_\parallel^S$ is combined with J_n from Eq. (15.26) or (15.27), and put into the integral over z (15.22). The calculations are somewhat lengthy, but only involve elementary Gaussian integrals. The result is:

$$\langle \tilde{v}_x(\mathbf{x}, \tau) \rangle^S \equiv v_x^S(\mathbf{x}, \tau) = M\, n_\perp^2(\tau)\, \exp\left[-\tfrac{1}{2}\, n_\perp(\tau)\, (x^2 + y^2)\right]\, f_\|(\tau, \eta^0)$$
$$\times \left\{ -\psi^0\, y + b_x^0\, [1 - n_\perp(\tau)\, y^2] + b_y^0\, n_\perp(\tau)\, xy \right\}. \tag{15.34}$$

The y-component of the average velocity in the subensemble S is obtained in a similar way:

$$\langle \tilde{v}_y(\mathbf{x}, \tau) \rangle^S \equiv v_y^S(\mathbf{x}, \tau) = M\, n_\perp^2(\tau)\, \exp\left[-\tfrac{1}{2}\, n_\perp(\tau)\, (x^2 + y^2)\right]\, f_\|(\tau, \eta^0)$$
$$\times \left\{ \psi^0\, x + b_x^0\, n_\perp(\tau)\, xy + b_y^0\, [1 - n_\perp(\tau) x^2] \right\}. \tag{15.35}$$

We introduced here the following notation:

$$f_\|(\tau, \eta^0) = n_\|^{1/2}(\tau)\, \eta^0\, \exp\left\{ -\tfrac{1}{2}\, n_\|(\tau) \overline{\chi}_\| \varphi^2(\tau)\, \eta^{0\,2} \right\}$$
$$\times \left[e^{-\tau} - \overline{\chi}_\|\, n_\|(\tau)\, \varphi^3(\tau) \right], \tag{15.36}$$

where:

$$n_\|(\tau) = \frac{1}{1 + \overline{\chi}_\| \left[2\mu(\tau) - \varphi^2(\tau) \right]}. \tag{15.37}$$

It is easily checked that the average velocity field in the subensemble is divergenceless:

$$\nabla \cdot \mathbf{v}^S(\mathbf{x}, \tau) = 0. \tag{15.38}$$

It therefore derives from a Hamiltonian, which is simply the average potential in the subensemble, $\psi^S(\mathbf{x}, \tau) = \langle \psi(\mathbf{x}, \tau) \rangle^S$ calculated by the same method as above:

$$\psi^S(\mathbf{x}, \tau) \equiv \langle \psi(\mathbf{x}, \tau) \rangle^S = M\, n_\perp^2(\tau)\, \exp\left[-\tfrac{1}{2}\, n_\perp(\tau)\, (x^2 + y^2)\right]\, f_\|(\tau, \eta^0)$$
$$\times \left\{ \psi^0 + b_x^0\, n_\perp(\tau)\, y - b_y^0\, n_\perp(\tau)\, x \right\}. \tag{15.39}$$

It is related to the average velocity in the subensemble as follows:

$$v_x^S(\mathbf{x}, \tau) = \frac{\partial \psi^S(x, y, \tau)}{\partial y}, \quad v_y^S(\mathbf{x}, \tau) = -\frac{\partial \psi^S(x, y, \tau)}{\partial x}. \tag{15.40}$$

We now come back to our original goal, the calculation of the Lagrangian velocity correlation, Eq. (15.20). We introduce the *decorrelation trajectory (DCT) approximation* in the same way as in Chap.

14. We approximate the Lagrangian average velocity in the subensemble by the *Eulerian* average velocity calculated along the *deterministic* decorrelation trajectory $\mathbf{x}^S(\tau)$, whose components are the solutions of the following *deterministic* equations of motion:

$$\frac{dx^S(\tau)}{d\tau} = v_x^S[x^S(\tau), y^S(\tau), \tau],$$

$$\frac{dy^S(\tau)}{d\tau} = v_y^S[x^S(\tau), y^S(\tau), \tau]. \tag{15.41}$$

As follows from Eq. (15.40) they form in each subensemble a Hamiltonian system of $1\frac{1}{2}$ degrees of freedom, just as Eqs. (14.50). This fact may seem surprising in the present case, because the original physical system is *dissipative*, as it includes the action of the collisions. The latter has been absorbed into the definition of $\mathbf{v}^S$ through the averaging process.

Thus, the problem takes the same form as in Chap. 14: the calculation of the Lagrangian velocity autocorrelation has been reduced to the calculation of a superposition of Eulerian averages in the subensembles:

$$\mathcal{L}(\tau) = M \int d\psi^0 d\mathbf{b}^0 d\eta^0 \ P(\psi^0, \mathbf{b}^0) P(\eta^0) \ b_x^0 \eta^0 \ v_x^S[x^S(\tau), y^S(\tau), \tau]. \tag{15.42}$$

The (dimensionless) anomalous running diffusion coefficient is then:

$$\mathcal{D}(\tau) = M \int_0^\tau d\tau' \int d\psi^0 d\mathbf{b}^0 d\eta^0 \ P(\psi^0, \mathbf{b}^0) P(\eta^0) \ b_x^0 \eta^0 \ v_x^S[x^S(\tau'), y^S(\tau'), \tau']. \tag{15.43}$$

The only random element remaining in these expressions is the distribution of the *initial values* of the potential, of the magnetic field and of the parallel velocity.

15.5 Limiting cases

a. The limit of infinite perpendicular correlation length

The solution of the DCT equations of motion, Eqs. (15.41) allow us to calculate the Lagrangian velocity autocorrelation (15.42), hence the anomalous diffusion coefficient. These equations are highly nonlinear and can, in general, only be solved numerically. We discuss these numerical results in the next Sections. There are, however, some limiting cases which can be treated analytically. These provide us with useful comparisons with well-known results and with validity tests of the DCT approximation.

The first situation to be discussed here is the case of infinite perpendicular correlation length, $\lambda_\perp = \infty$. This problem was discussed at length in Chap. 11, where an exact solution was derived. We thus consider Eqs. (15.41), combined with (15.34) and (15.35) for the DCT. In the limit $\lambda_\perp = \infty$ the perpendicular correlation length can no longer be used as a scaling factor for the position coordinates [see Eq. (14.10)], we therefore go back to dimensional variables X^S, Y^S. In the right hand sides of the DCT equations, all terms depending on $x^S = \lambda_\perp^{-1} X^S$ and on $y^S = \lambda_\perp^{-1} Y^S$ vanish in this limit. Similarly, Eq. (15.4) shows that $\overline{\chi}_\perp = 0$, hence from (15.28): $n_\perp(\tau) = 1$. The change $dx^S(\tau)/d\tau \rightarrow dX^S(t)/dt$ introduces a factor $\lambda_\perp \nu$ in the right hand side, whose prefactor becomes:

$$\lambda_\perp \nu M = \beta \sqrt{\chi_\| \nu} = \sqrt{\epsilon}, \tag{15.44}$$

where ϵ is the parameter introduced in Sec. 11.6, Eq. (11.91): it has the dimension of a velocity squared. The limiting equations for the DCT thus reduce to:

$$\frac{dX^S(t)}{dt} = \sqrt{\epsilon}\, u\, f_\|(\nu t, \eta^0),$$

$$\frac{dY^S(t)}{dt} = 0. \tag{15.45}$$

These are trivially simple linear equations that can immediately be solved by quadrature (although the integral of $f_\|(\nu t, \eta^0)$ cannot be done analytically).

The DCT is now one-dimensional and has the very simple shape shown in Fig. 15.1. $X^S(\theta)$ grows from zero to a maximum, after which it decays slowly to zero with a long tail. The maximum corresponds to the position of the zero of the velocity $f_\|(\tau)$ (in the present case, $\tau_r = 2.734$).

We now calculate the Lagrangian velocity autocorrelation using Eq. (15.42). We note that in the present limit the DCT velocity does not depend on $X^S(\tau), Y^S(\tau)$, hence the solution of the equation (15.45) is not required, and the integrations (involving only Gaussian functions) can be done explicitly, using Eq. (15.36):[5]

---

[5] In the paper by VLAD et al. 2003, $\mathcal{L}^\infty$ (D^∞) are denoted by $\mathcal{L}_0$ (D_0), referring to the fact that $M = 0$. We prefer to use here the superscript "∞", referring more precisely to the limit $\lambda_\perp = \infty$. Indeed, M may vanish for several reasons, as can be seen from Eq. (15.7), e.g. $\beta = 0$, $\lambda_\| = 0$, $\chi_\| = 0$ or $\lambda_\perp = \infty$. Here we choose specifically $\lambda_\perp = \infty$, keeping the other quantities finite.

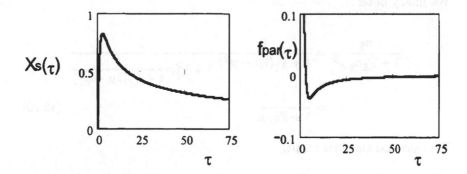

Figure 15.1. The time dependence of the position $X^S(\tau)$ and of the velocity $f_\parallel(\tau)$ of a fictitious particle on the decorrelation trajectory in the limit $\lambda_\perp = \infty$. ($\epsilon = \eta^0 = u = 1$; $\overline{X}_\parallel = 0.1$).

$$\mathcal{L}^\infty(\tau) = \epsilon \int_{-\infty}^{\infty} dp\, e^{-p^2/2} \frac{1}{2} \int_0^\infty du\, u\, e^{-u^2/2}$$

$$\times \frac{1}{\sqrt{2\pi}} \int_{-\infty}^{\infty} d\eta^0 e^{-\eta^{0\,2}/2}\, u^2 \eta^0\, f_\parallel(\tau, \eta^0). \qquad (15.46)$$

The result is:

$$\mathcal{L}^\infty(\tau) = \epsilon \frac{n_\parallel^{1/2}(\tau)}{\left[1 + \overline{X}_\parallel n_\parallel(\tau)\, \varphi^2(\tau)\right]^{3/2}} \left\{ e^{-\tau} - \overline{X}_\parallel n_\parallel(\tau)\, \varphi^3(\tau) \right\}. \qquad (15.47)$$

We now note [using the definitions (11.88)]:

$$\{...\} = e^{-\tau} - \overline{X}_\parallel n_\parallel\, \varphi^2\, \varphi$$

$$= e^{-\tau} \left[1 + \overline{X}_\parallel n_\parallel\, \varphi^2\right] - \overline{X}_\parallel n_\parallel\, \varphi^2. \qquad (15.48)$$

Thus:

$$\mathcal{L}^\infty(\tau) = \epsilon \frac{n_\parallel^{1/2}}{\left[1 + \overline{X}_\parallel n_\parallel\, \varphi^2\right]^{1/2}} \left\{ e^{-\tau} - \overline{X}_\parallel \frac{n_\parallel}{1 + \overline{X}_\parallel n_\parallel\, \varphi^2}\, \varphi^2 \right\}. \qquad (15.49)$$

We finally note:

$$\frac{n_\parallel}{1 + \overline{\chi}_\parallel n_\parallel \, \varphi^2} = \frac{1}{1 + \overline{\chi}_\parallel(2\mu - \varphi^2)} \frac{1}{1 + \overline{\chi}_\parallel^2 \varphi \dfrac{1}{1 + \overline{\chi}_\parallel(2\mu - \varphi^2)}}$$

$$= \frac{1}{1 + 2\overline{\chi}_\parallel \, \mu}. \tag{15.50}$$

We thus find the final result:

$$\mathcal{L}^\infty(\tau) = \epsilon \frac{1}{\left[1 + 2\overline{\chi}_\parallel \, \mu(\tau)\right]^{1/2}} \left\{ e^{-\tau} - \overline{\chi}_\parallel \frac{\varphi^2(\tau)}{1 + 2\overline{\chi}_\parallel \, \mu(\tau)} \right\}. \tag{15.51}$$

This is precisely equal to the expression (11.90) obtained by the exact solution of the Langevin equations (11.69) - (11.71). Thus, *in the limit* $\lambda_\perp \to \infty$, *the DCT method yields the exact result for the Lagrangian velocity autocorrelation, hence for the subdiffusive coefficient.*

b. The Collisionless limit

We now consider the limit in which the collision frequency ν tends to zero. In this case the (dimensional) collisional diffusivities behave as follows [see Eqs. (11.35), (11.33)]:

$$\chi_\parallel \nu = \frac{V_T^2}{2\nu} \nu = \tfrac{1}{2} V_T^2 : \text{ finite}$$

$$\chi_\perp = \frac{\nu^2}{\Omega^2} \frac{V_T^2}{2\nu} = 0. \tag{15.52}$$

As a result, there is no spatial decorrelation due to perpendicular collisions. The mean free path $\lambda_{mfp} = V_T/\nu$ being very long, the perpendicular coordinates $X(t), Y(t)$ (starting from the origin) vary very slowly and thus remain close to zero. We thus recover, for a different reason, the same situation as in the previous subsection: the DCT equations of motion are Eqs. (15.45). The Lagrangian velocity autocorrelation is again given by Eq. (15.51), which must be evaluated for $\nu = 0$. Upon expansion in powers of ν, we find:

$$n_\parallel(\nu t) = \left[1 + \frac{2}{3} \frac{V_T^2}{\lambda_\parallel^2} \nu t^3 + \dots\right]^{-1} = 1, \quad (\nu = 0) \tag{15.53}$$

$$\frac{n_\parallel(\nu t)}{1 + \overline{\chi}_\parallel \, n_\parallel(\nu t)\, \phi^2(\nu t)} = \frac{1}{1 + \frac{V_T^2}{2\lambda_\parallel^2} t^2}. \quad (\nu = 0). \qquad (15.54)$$

Hence, from Eq. (15.49):

$$\mathcal{L}_{JP} = \frac{\beta^2 V_T^2}{2} \left(1 + \frac{V_T^2}{2\lambda_\parallel^2} t^2 \right)^{-3/2}. \qquad (15.55)$$

Remarkably, in this collisionless limit, the Lagrangian velocity autocorrelation is a definite positive function. It has a long algebraic tail, decaying as t^{-3} for long times. It can be seen in Fig. 11.6 that, as the collisionality decreases, the positive part of the Lagrangian velocity autocorrelation becomes wider, whereas its negative tail becomes thinner. Nevertheless, the contribution of the latter still compensates the positive one, and the regime is subdiffusive as long as $\nu \neq 0$. But when $\nu = 0$, the negative tail disappears and the regime suddenly becomes *diffusive*. The diffusion coefficient can be calculated by using the known formula (DWIGHT 1961, eq. 857.02):

$$\int_0^\infty dx \, \frac{1}{(ax^2 + bx + c)^{3/2}} = \frac{1}{b\sqrt{c} + c\sqrt{a}}. \qquad (15.56)$$

We thus find the result:

$$D_{JP} = \int_0^\infty dt \, \mathcal{L}_{JP}(t) = \frac{1}{\sqrt{2}} \beta^2 V_T \cdot \lambda_\parallel. \qquad (15.57)$$

This result has long been known to astrophysicists: it was derived semi-intuitively by JOKIPII & PARKER 1969 (see also WANG et al. 1995).

c. The Kadomtsev-Pogutse limit

In 1979 Kadomtsev and Pogutse (KP) derived semi-qualitatively an approximation for the anomalous cross-field diffusion coefficient. This approximation is essentially a weak-nonlinearity regime, in which the magnetic field fluctuations are non-chaotic. It will be shown that the KP diffusion coefficient is obtained from the DCT formalism if the following relations are assumed, for a collisional plasma:

$$\overline{\chi}_\parallel \gg \overline{\chi}_\perp > 0, \qquad \frac{\lambda_\parallel}{\lambda_\perp} \ll 1. \qquad (15.58)$$

A precise prescription is needed for the manner in which these multiple limits are taken. Mathematically speaking, we consider the following limiting procedure, expressed in terms of dimensionless quantities:

$$\frac{\lambda_\parallel}{\lambda_\perp} \to 0, \quad \frac{\nu}{\Omega} \to 0, \quad \chi_\parallel : finite, \quad \gamma \equiv \frac{\chi_\perp}{\chi_\parallel} = \left(\frac{\nu}{\Omega}\frac{\lambda_\perp}{\lambda_\parallel}\right)^2 : finite.$$

$$(15.59)$$

Physically, these conditions are realized in a weakly collisional plasma in presence of a strong reference magnetic field, with magnetic fluctuations of long perpendicular correlation length. As a result, the quantities $x = X/\lambda_\perp$, $y = Y/\lambda_\perp$ can be neglected in the right hand side of the general DCT equation of motion. We then find for the DCT velocity an expression equal to Eq. (15.45), corrected by a factor $n_\perp^2(\nu t)$, which is finite as a result of the prescription (15.59). This leads to the following form of the reduced Lagrangian velocity autocorrelation $\overline{\mathcal{L}}(\tau) = \epsilon^{-1}\mathcal{L}(\nu t)$:

$$\overline{\mathcal{L}}_{KP}(\tau) = n_\perp^2(\tau)\,\overline{\mathcal{L}}^\infty(\tau), \qquad (15.60)$$

where $\mathcal{L}^\infty(\tau)$ is the subdiffusive Lagrangian velocity autocorrelation defined in Eq. (15.46).

It is well known that (for $\nu \neq 0$) the subdiffusive Lagrangian velocity autocorrelation leads to a zero diffusion coefficient (subdiffusion); thus:

$$\overline{D}^\infty = \int_0^\infty d\tau\,\overline{\mathcal{L}}^\infty(\tau) = 0, \qquad (15.61)$$

(where the reduced diffusion coefficients are defined as $\overline{D} = D/\beta^2\overline{\chi}_\parallel$). Because of the factor $n_\perp^2(\tau)$, the integral of $\overline{\mathcal{L}}_{KP}(\tau)$ no longer vanishes, and yields a finite diffusion coefficient. It can be estimated analytically by using the following approximation, based on the shape of the function $n_\perp(\tau)$. The latter starts from the value 1 at $\tau = 0$, and tends monotonically to zero for large τ. A rough approximation consists of replacing it by a step function:

$$n_\perp(\tau) \approx \begin{cases} 1, & \tau < \tau_d \\ 0, & \tau > \tau_d \end{cases}. \qquad (15.62)$$

A "natural" value of the cut-off time τ_d is:

$$\tau_d = \frac{1}{2\gamma\,\chi_\parallel} = \frac{1}{2\chi_\perp}. \qquad (15.63)$$

It then follows that the diffusion coefficient is approximated as:

$$\overline{D}_{KP} = \int_0^\infty d\tau\,\overline{\mathcal{L}}_{KP}(\tau) \approx \int_0^{\tau_d} d\tau\,\overline{\mathcal{L}}^\infty(\tau). \qquad (15.64)$$

Because of Eq. (15.61), this expression can be replaced by:

$$\overline{D}_{KP} = -\int_{\tau_d}^{\infty} d\tau\, \overline{\mathcal{L}}^{\infty}(\tau). \tag{15.65}$$

The advantage of this form is that $\overline{\mathcal{L}}^{\infty}(\tau)$ has a very simple asymptotic form [valid for τ larger than the coordinate of the minimum of $\overline{\mathcal{L}}^{\infty}(\tau)$], which can be integrated analytically:

$$\overline{D}_{KP} = -\int_{\theta_d}^{\infty} d\tau\, \frac{-\chi_{\parallel}}{\left[1 + 2\chi_{\parallel}\tau\right]^{3/2}} = \frac{1}{\left(1 + 2\chi_{\parallel}\tau_d\right)^{1/2}}. \tag{15.66}$$

Combining this result with Eq. (15.63), we find:

$$\overline{D}_{KP} = \left(\frac{\gamma}{1+\gamma}\right)^{1/2}. \tag{15.67}$$

In the usual situation when γ is small, we find the simpler formula:

$$\overline{D}_{KP} = \sqrt{\gamma}, \quad \gamma \ll 1. \tag{15.68}$$

By going back to dimensional quantities, one finds the result:

$$D_{KP} = \beta^2 \frac{\lambda_{\parallel}}{\lambda_{\perp}} \sqrt{\overline{\chi}_{\parallel}\overline{\chi}_{\perp}}, \tag{15.69}$$

which is (up to a factor 2) the well-known *Kadomtsev-Pogutse formula*, first obtained by KADOMTSEV & POGUTSE 1979, and rederived by a different method (based on the Corrsin approximation) by WANG et al. 1995.

It will be noted that the derivation of the KP diffusion coefficient given here (as well as the previous derivations) is based on a very drastic approximation, Eqs. (15.62), (15.63). Given the uncertainty of the validity of this approximation, we compared numerically the $\overline{D}_{KP}$ with the "exact" diffusion coefficient in the limit (15.59), obtained by using Eq. (15.60):

$$\overline{D}_{ex} = \int_{0}^{\infty} d\tau\, n_{\perp}^{2}(\tau)\, \overline{\mathcal{L}}_{\infty}(\tau). \tag{15.70}$$

The first results showed that there may exist a rather large discrepancy between $\overline{D}_{KP}$ and $\overline{D}_{ex}$. Consider, for instance the typical choice of parameters: $\overline{\chi}_{\parallel} = 1$, $\gamma = 0.01$. We find (up to the second decimal):

$$\overline{D}_{KP} = 0.10, \quad \overline{D}_{ex} = 0.19. \tag{15.71}$$

The origin of the discrepancy is in the invalidity of Eq. (15.62) for the exact autocorrelation in this case. For the KP approximation we may check numerically that:

$$\int_0^{\tau_d} d\tau \, \overline{\mathcal{L}}_\infty(\tau) = -\int_{\tau_d}^\infty d\tau \, \overline{\mathcal{L}}_\infty(\tau) = 0.10 = \overline{D}_{KP}. \qquad (15.72)$$

The corresponding values for the exact autocorrelation are:

$$\int_0^{\tau_d} d\tau \, \overline{\mathcal{L}}_{ex}(\tau) = 0.195, \quad -\int_{\tau_d}^\infty d\tau \, \overline{\mathcal{L}}_{ex}(\tau) = 0.007. \qquad (15.73)$$

Thus, the main contribution to the diffusion coefficient (with the present choice of parameters) comes from the range $\tau < \tau_d$, which is excluded from the KP approximation.

This situation may, however, be different for other values of the parameters. It is, indeed, striking that the approximate KP diffusion coefficient depends on the single parameter γ, whereas the exact one depends on *both parameters* $\overline{\chi}_\parallel$ and γ. We compared the exact $\overline{D}_{ex}$, Eq. (15.70) and the approximate $\overline{D}_{KP}$, Eq. (15.68), (which is constant) for various values of $\overline{\chi}_\parallel$, for a series of fixed values of γ (Fig. 15.2). The exact curve is a decreasing function of $\overline{\chi}_\parallel$, which crosses the approximate $\overline{D}_{KP}$.

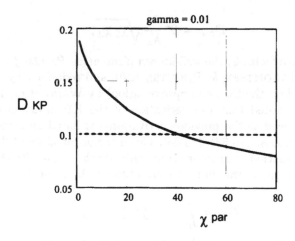

Figure 15.2. The "exact" diffusion coefficient in the Kadomtsev-Pogutse limit, Eq. (15.70) compared with the approximate KP formula Eq. (15.68). $[\chi^{par} \equiv \overline{\chi}_\parallel]$.

Thus, for each value of γ there is one value of $\overline{\chi}_\parallel$ for which the approximate and the exact $\overline{D}_{KP}$ are equal. On each side of this value there is a "range of confidence" for $\overline{\chi}_\parallel$ where the approximate KP formula is

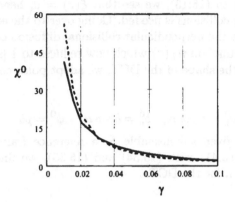

Figure 15.3. Values of $\overline{\chi}_\|$ for which $\overline{D}_{KP} = \overline{D}_{ez}$ (solid line). The dotted line is the fit, Eq. (15.74). $[\chi^0 \equiv \overline{\chi}_\|^0]$.

acceptable. The value of $\overline{\chi}_\|$ which gives an exact match as a function of γ is shown in Fig. 15.3. It is interesting to note that this function is quite well approximated by a power law:

$$\overline{\chi}_\|^0 = 0.056\, \gamma^{-3/2}. \tag{15.74}$$

These remarks show the limited validity of the Kadomtsev-Pogutse formula (not of its derivation from DCT!).

15.6   The Subdiffusive Regime

Before considering the completely general case, we continue our investigation with another limiting, but less trivial situation: we assume that the dimensionless perpendicular collisional diffusion is zero: $\overline{\chi}_\perp = 0$. The V-Langevin equations (15.8), (15.9) then reduce to:

$$\frac{d\mathbf{x}(\tau)}{d\tau} = M\, \mathbf{b}[\mathbf{x}(\tau), z(\tau)]\, \eta_\|(\tau),$$

$$\frac{dz(\tau)}{d\tau} = \sqrt{\overline{\chi}_\|}\, \eta_\|(\tau). \tag{15.75}$$

These equations are similar to (11.69) - (11.71), but the dependence of the fluctuating magnetic field on all three coordinates is retained. In other words, we now assume that the perpendicular correlation length is finite: $\lambda_\perp \neq \infty$. The important consequence of this fact is that Eqs. (15.75) are *nonlinear*. The situation is, however, simpler than the general case:

comparing (15.75) to (15.13), we see that $\xi(\tau) = 0$, hence no averaging over perpendicular collisions is needed. Going over to the equations for the DCT, we note that the perpendicular collisional diffusion coefficient enters only through the function $n_\perp(\tau)$ which now reduces to 1 [see Eq. (15.28). In order to study the shape of the DCT, we adopt polar coordinates, as in Eq. (14.56):

$$b_x^0 = b \cos \alpha, \quad b_y^0 = b \sin \alpha, \quad \psi^0 = pb, \tag{15.76}$$

and choose, in a given subensemble S, a reference frame with $\alpha = 0$. Combining Eq. (15.41) with (15.34) and (15.35), we find the following equations of motion for the DCT:

$$\frac{dx^S(\tau)}{d\tau} = M\, b\, f_\parallel(\tau, \eta^0)\, \exp\left[-\tfrac{1}{2}\left(x^{S\,2}(\tau) + y^{S\,2}(\tau)\right)\right]$$
$$\times \left\{ 1 - y^S(\tau)\, \left[p + y^S(\tau)\right] \right\}, \tag{15.77}$$

$$\frac{dy^S(\tau)}{d\tau} = M\, b\, f_\parallel(\tau, \eta^0)\, \exp\left[-\tfrac{1}{2}\left(x^{S\,2}(\tau) + y^{S\,2}(\tau)\right)\right]$$
$$\times x^S(\tau)\, \left\{ p + y^S(\tau) \right\}. \tag{15.78}$$

These nonlinear equations must be integrated numerically. The solutions depend on two "external" dimensionless parameters: M, $\overline{\chi}_\parallel$, and on four parameters specifying the subensemble: p, b, α, η^0, a rather large parameter space! We consider here a restricted set of trajectories by selecting values of the parameters allowing a good visualization of the phenomena. The values of four of these parameters will be fixed in the forthcoming figures: $M = 10$, $b = \eta^0 = 1$, $\alpha = 0$. The parameters influencing the shape of the trajectories are the values of the initial potential, p (as in Chap. 14) and the value of $\overline{\chi}_\parallel$. In the forthcoming figures we choose $\overline{\chi}_\parallel = 0.1$.

A first property must be stressed, as it controls the motion in the most general case. The function $f_\parallel(\tau)$ starts from a finite positive value at $\tau = 0$,[6] then vanishes at a time $\tau = \tau_r$ and changes sign; it passes through a minimum after which it tends monotonously to zero with a long (algebraic) tail (Fig. 15.1). For $\overline{\chi}_\parallel = 0.1$, the value of $\tau_r = 2.734$. This property implies that the velocity of the fictitious particle moving along the DCT vanishes at τ_r and then changes sign [see Eqs. (15.77), (15.78)]; τ_r will therefore be called the *return time*.

We show in Fig. 15.4 the decorrelation trajectories for four typical values of the initial potential p. For the larger values of p, the DCT's

---

[6] This statement holds for $\eta^0 > 0$ [see Eq. (15.36)], it is easily adapted to the opposite case. For simplicity we no longer write explicitly the argument η^0 in the notation: $f_\parallel(\theta) \equiv f_\parallel(\theta, \eta^0)$.

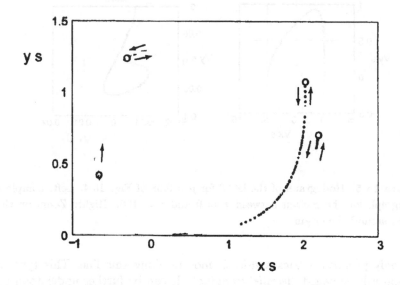

Figure 15.4. Decorrelation trajectories for $M = 10$, $\chi_{\parallel} = 0.1$ and four values of p: solid : $p = 0.05$, dotted : $p = 0.08$, dashed : $p = 1$, dash-dotted : $p = 1.2$. The small circles represent the position of the fictitious particle at the return time $\tau = \tau_r$.

arc closed curves, resembling those of Fig. 14.5 of the electrostatic case. The motion along these trajectories is, however, quite different. Starting from the origin, the fictitious particle performs a number of cycles counter-clockwise, up to the return time. At $\tau = \tau_r$ it stops, and then retraces its steps until it reaches the origin $x^S = y^S = 0$ as τ tends to infinity. These trajectories correspond to permanent trapping.

For smaller values of p (red and blue curves in Fig. 15.4), the fictitious particle "tries" to follow a closed trajectory similar (but larger) than those of Sec. 14.7, but the time for completing a cycle is longer than the return time. They therefore stop at $\tau = \tau_r$ and return to the origin. Note that the first (counter-clockwise) stage of the motion is performed in the relatively short time $\tau_r = 2.734$, whereas it takes an infinite time for the particle to return to the origin. The fictitious particles thus never stop definitively at any finite time. According to our discussion of Chap. 14, this means that the physical particles remain correlated for all times: there is no detrapping mechanism in this case. Intuitively, one clearly understands this kind of motion. Particles confined to a single dimension (the magnetic field line) and undergoing the effect of a rugged magnetic field and of parallel collisions

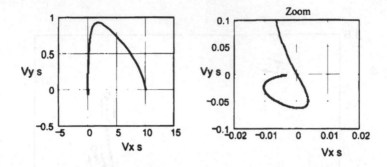

Figure 15.5. Hodograph of the DCT for $p = 0.08$ of Fig. 15.4. Left: complete hodograph for the motion between $\tau = 0$ and $\tau = 100$. Right: Zoom on the region around the origin.

can only perform a back-and-forth motion along this line. This type of motion will be called *"parallel trapping"*. It can be further understood by analyzing more closely the DCT.

We choose for illustration the trajectory of Fig. 15.4 corresponding

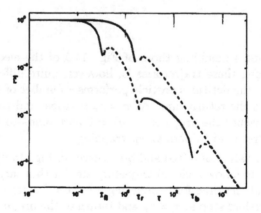

Figure 15.6. Logarithm of the absolute value of the Lagrangian velocity autocorrelation $\log|\mathcal{L}(\tau)|$ vs. $\log \tau$ [solid line]. The upper curve represents the analytic solution for the same function in the limit $\lambda_\perp = \infty$, Eq. (11.90). The dashed lines represent regions where $\mathcal{L}(\tau) < 0$. (Reprinted with permission from VLAD et al. *Phys. Rev.* E **67** 026406 (2003). Copyright 2003 by the American Physical Society.)

to $p = 0.08$ and represent in Fig. 15.5 its hodograph. As appears from Eqs. (15.77), (15.78), the hodograph starts at the point $v_x^S = M = 10$, $v_y^S = 0$ and quickly performs the big loop shown on the figure at left until it reaches the value $v_x^S = v_y^S = 0$ at $\tau = \tau_r$. Thereafter it goes "a little bit" beyond the origin in the negative half-plane and stays there for the remaining time. It is quite interesting to look more closely at this small appendix (figure at right): it shows a nice journey which slowly approaches the origin (*i.e.*, the final stop) as $\tau \to \infty$.

We now use the solutions for the DCT in order to calculate the Lagrangian velocity autocorrelation $\mathcal{L}(\theta)$ by Eq. (15.42) and the anomalous running diffusion coefficient $D(\theta)$ by Eq. (15.43). This operation requires a rather complicated numerical operation, which has been performed by using a programme developed by M. Vlad.

In Fig. 15.6 the function $\log |\mathcal{L}(\tau)|$ is plotted vs. $\log \tau$, and compared to the analytic solution (11.90) for $\lambda_\perp = \infty$. A quite interesting difference appears between the two cases. Both functions vanish at the return time τ_r, as expected from a brief look at Eq. (15.42), as $v_x^S[x^S(\tau_r), y^S(\tau_r), \tau_r] = 0$.[7] But the function $\mathcal{L}(\tau)$ has two *additional* zeroes (accompanied by a change of sign) for finite $\lambda_\perp$, one at $\tau_{fl} = M^{-1} < \tau_r$ and the other at $\tau_b > \tau_r$. This is a typical effect of the nonlinearity. Note that no special feature appears at τ_{fl} or at τ_b in the shape of the DCT's; on the contrary, τ_r is a characteristic feature of the trajectories, as seen above.

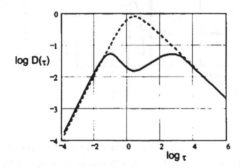

Figure 15.7. Reduced anomalous running diffusion coefficient (solid line) compared with its limiting value for $\lambda_\perp = \infty$, Eq. (11.96) (dashed line), in log- log representation. (Reprinted with permission from VLAD et al. *Phys. Rev.* E **67** 026406 (2003). Copyright 2003 by the American Physical Society.)

---

[7] Note that, although $f_\parallel(\tau, \eta^0)$, Eq. (15.36) depends on the subensemble S through η^0, its *zero* is determined by the equation: $\exp(-\tau) - \overline{\chi}_\parallel n_\parallel(\tau)\varphi^3(\tau) = 0$, which is independent of the subensemble.

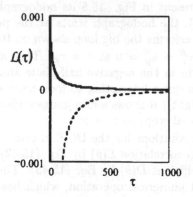

Figure 15.8. Lagrangian velocity autocorrelations $\mathcal{L}(\tau)$ (solid) and $\mathcal{L}^{\infty}(\tau)$ (dashed).

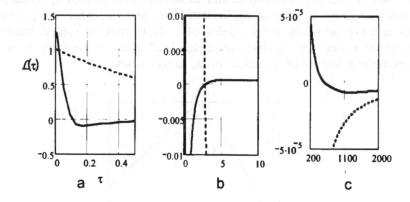

Figure 15.9. Zooms on the Lagrangian velocity autocorrelation functions $\mathcal{L}(\tau)$ and $\mathcal{L}^{\infty}(\tau)$. a): Zero of $\mathcal{L}(\tau)$ at τ_{fl}; b): Common zero of $\mathcal{L}(\tau)$ and $\mathcal{L}^{\infty}(\tau)$ at τ_r; c): Zero of $\mathcal{L}(\tau)$ at τ_b.

We also consider the running diffusion coefficient, Eq. (15.43), represented in Fig. 15.7 and compare it with its exact value for $\lambda_{\perp} = \infty$, Eq. (11.96). Clearly the zeroes of $\mathcal{L}(\tau)$ are now reflected in extrema of the running diffusion coefficient. We thus note that the nonlinear $\mathcal{D}(\tau)$ very quickly departs from the ascending $\mathcal{D}^{\infty}(\tau)$ and starts decreasing at a time that could be called the "flight time", $\tau_{fl} = \nu\lambda_{\perp}/V = M^{-1}$ (where $V = \beta\sqrt{\chi_{\parallel}\nu}$ is a typical parallel velocity). The decrease is due to the contribution of DCT's with large p which perform many turns in a short time (see the discussion above) and lead to a progressive mixing in the integral (15.43). At the return time τ_r the velocities change sign simultaneously in all subensembles, and the processes are reversed in both cases: $\mathcal{D}(\tau)$ goes through a minimum, whereas $\mathcal{D}^\infty(\tau)$ goes through a maximum. At a later time τ_b, $\mathcal{D}(\tau)$ joins again $\mathcal{D}^\infty(\tau)$: from there on the two functions become identical and tend finally to zero as $\tau^{-1/2}$ [see Eq. (11.99)]. We thus reached an important conclusion: *The nonlinearity does not affect the asymptotic subdiffusive character of the transport in absence of perpendicular collisions.*

It should be noted that the log-log representation may lead to a strongly distorted visualization of the functions. Thus, if $\mathcal{L}(\tau)$ is plotted over a sufficiently long time, the features described above are barely visible (Fig. 15.8). In order to see the zeroes and the sign changes we need to plot strong zooms in the relevant regions (Fig. 15.9).

15.7   The Diffusive Regime

We now consider the general case of the DCT method, for which $\eta_\perp \neq 0$. In this case there appears a new feature, *i.e.*, the true decorrelation of the physical particles from the magnetic field lines. Even in the case of a frozen magnetic field landscape, the perpendicular collisions allow the particles to jump from one field line to another, thus producing a cross-field motion leading to a non-zero anomalous perpendicular diffusion coefficient. In contrast with the problem of electrostatic drift turbulence treated in Chap. 14, the decorrelation is no longer produced by the random evolution of the potential landscape (*temporal decorrelation*) but rather by the true unsticking of the particles followed by their cross-field jump (*spatial decorrelation*). The nonlinearity thus produces a succession of *trapping events*, both *parallel* (due to parallel collisions, as in the previous section) and *perpendicular* (due to the ruggedness of the magnetic field), followed by *decorrelation events* produced by the perpendicular collisions. The result is, in each realization, a very complex physical trajectory, of the type pictured in Fig. 14.4. The DCT method allows us to extract from this complicated situation the relevant information necessary for the calculation of the anomalous diffusion coefficient.

We start as before with an analysis of the decorrelation trajectories, which now obey, instead of Eqs. (15.77), (15.78), the following deterministic equations of motion (for $\alpha = 0$):

$$\frac{dx^S(\tau)}{d\tau} = M\, b\, n_\perp^2(\tau)\, \exp\left[-\tfrac{1}{2}\, n_\perp(\tau)\,(x^{S\,2}(\tau) + y^{S\,2}(\tau))\right]\, f_\parallel(\tau, \eta^0)$$
$$\times \left\{ 1 - y^S(\tau)\left[p + n_\perp(\tau)\, y^S(\tau)\right] \right\}, \tag{15.79}$$

$$\frac{dy^S(\tau)}{d\tau} = M\, b\, n_\perp^2(\tau) \exp\left[-\tfrac{1}{2} n_\perp(\tau)\left(x^{S\,2}(\tau) + y^{S\,2}(\tau)\right)\right] f_\parallel(\tau, \eta^0)$$
$$\times x^S(\tau)\left[p + n_\perp(\tau)\, y^S(\tau)\right]. \tag{15.80}$$

The solutions now depend on three "external" dimensionless parameters: M, $\overline{\chi}_\parallel$, $\overline{\chi}_\perp$, and on four parameters specifying the subensemble: p, b, α, η^0. We again consider a restricted set of trajectories by selecting the same values of the parameters as in the previous section, adding a fixed value for $\overline{\chi}_\perp$. We thus take: $M = 10$, $b = \eta^0 = 1$, $\alpha = 0$. The main parameters influencing the general shape of the trajectories are the values of the initial potential, p (as in Chap. 14) and the relative values of $\overline{\chi}_\parallel$ and $\overline{\chi}_\perp$. Note that in a strong magnetic field the physical (dimensional) collisional diffusivities are very different: $\chi_\parallel \gg \chi_\perp$; but the ratio of the dimensionless diffusivities $\overline{\chi}_\perp/\overline{\chi}_\parallel$ may become large if the correlation lengths are such that $\lambda_\parallel^2/\lambda_\perp^2 \gg 1$. In the forthcoming figures we choose: $\overline{\chi}_\parallel = 0.1$, $\overline{\chi}_\perp = 1$.

In Fig. 15.10 four typical trajectories are shown [they are obtained by numerical integration of Eqs. (15.79), (15.80)]. For large values of p (e.g., $p = 1.8$), i.e. near the top of a potential hill, they resemble those of Fig. 15.4: the fictitious particle follows an almost closed trajectory. But, in contrast to the latter case, it shows a tendency of leaving this initial circle: the trajectory is actually spiral-like. The particle finally almost comes to a stop after a finite time, of order τ_r. For $p = 1.8$, the particle stops after $1\frac{1}{2}$ turns. For $p = 0.06$ the particle "tries" to turn around the hill, but after a finite time its trajectory changes its curvature and finally stops. For smaller values, such as $p = 0.005$, the trajectory is definitely an open curve (of finite length). These features clearly demonstrate the process of *spatial decorrelation* produced by the perpendicular collisions of the test particle with the background particles. This phenomenon is translated in the DCT by the appearance of spiral trajectories (for large p) or of open trajectories (for small p). The effect is more marked, the larger the reduced perpendicular diffusivity $\overline{\chi}_\perp$ relative to $\overline{\chi}_\parallel$. Another significant difference is that, contrary to the purely parallel motion, the stopping point of the fictitious particle is no longer the origin, but lies anywhere in the $x^S - y^S$ plane, at a point determined by the values of the parameters.

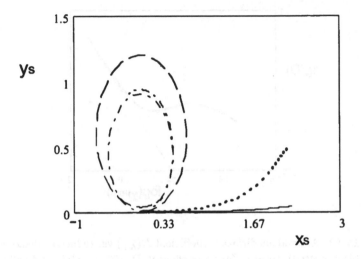

Figure 15.10. Decorrelation trajectories for $M = 10$, $\overline{\chi}_\parallel = 0.1$, $\overline{\chi}_\perp = 1$. Solid: $p = 0.005$, Dotted: $p = 0.06$, Dashed: $p = 1.15$, Dash-dotted: $p = 1.8$.

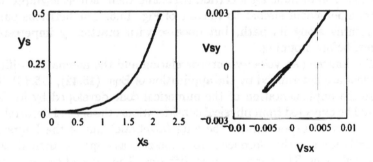

Figure 15.11. DCT trajectory corresponding to the blue curve ($p = 0.06$) of Fig. 15.10 (left). Enlargement of the corresponding hodograph near the origin (right).

What about the parallel trapping that appeared clearly in the previous section? A closer analysis reveals its presence here too, but with a much less important role when $\overline{\chi}_\parallel < \overline{\chi}_\perp$. Indeed, a look at the hodograph (*e.g.* of the blue trajectory of Fig. 15.10) shows that $v_x^S \approx v_y^S \approx 0$ for $\tau > \tau_r$. But a strong zoom on the region near the origin (Fig. 15.11) shows that there is actually a non-zero velocity with negative sign beyond τ_r: the velocity

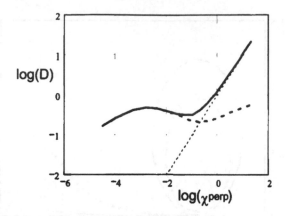

Figure 15.12. Anomalous diffusion coefficient $D(\overline{\chi}_\perp)$ vs. reduced collisional diffusivity (dash-dotted), total diffusion coefficient $D_{tot}(\overline{\chi}_\perp)$ (solid) and collisional diffusivity $\overline{\chi}_\perp$ (dotted). (log-log representation). [$\chi^{perp} \equiv \overline{\chi}_\perp$]. (Reprinted with permission from VLAD et al. *Phys. Rev.* E **67** 026406 (2003). Copyright 2003 by the American Physical Society.)

grows in absolute value for a certain time, and then returns asymptotically to zero, a behaviour similar to the one of Fig. 15.5. The fictitious particle thus returns along its path, but does so with practically imperceptible slowness before stopping.

The Lagrangian velocity autocorrelation and the anomalous diffusion coefficient are determined by the application of Eqs. (15.42), (15.43). They require the implementation of the numerical code developed by M. Vlad. As could be expected from physical considerations, the spatial decorrelation process cuts off exponentially the long (algebraic) tail of the Lagrangian autocorelation and therefore leads to a non-zero asymptotic anomalous diffusion coefficient. The regime is now *diffusive*. The value of the asymptotic diffusion coefficient depends, of course, on the parameters M, $\overline{\chi}_\parallel$, $\overline{\chi}_\perp$.

It is not our purpose here to browse through the whole parameter space. We limit ourselves to a single analysis, quite interesting in itself, and which illustrates the type of results obtainable by the DCT method. We fix the values of the parameters: $M = 10$, $\overline{\chi}_\parallel = 1$ and study the influence of the perpendicular diffusivity $\overline{\chi}_\perp$ on the anomalous diffusion coefficient $D(\overline{\chi}_\perp)$ [in units of $\lambda_\perp^2 \nu$, like $\overline{\chi}_\perp$, Eq. (15.4)]. This function is plotted in Fig. 15.12, together with the total cross-field diffusion coefficient, *i.e.*,

$$D_{tot}(\overline{\chi}_\perp) = D(\overline{\chi}_\perp) + \overline{\chi}_\perp \qquad (15.81)$$

The function $D(\overline{\chi}_\perp)$ has a very peculiar behaviour. For very small

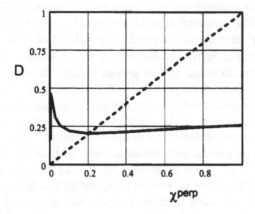

Figure 15.13. Anomalous diffusion coefficient $D(\overline{\chi}_\perp)$ (solid) and collisional diffusivity (dashed) vs. $\overline{\chi}_\perp \cdot [\chi^{perp} \equiv \overline{\chi}_\perp]$.

$\overline{\chi}_\perp$, it increases very quickly, becoming larger than $\overline{\chi}_\perp$ by several orders of magnitude. But for $\overline{\chi}_\perp \approx 10^{-3}$, $D(\overline{\chi}_\perp)$ reaches a maximum after which it decreases. This unexpected feature continues up to a minimum. Beyond the latter the anomalous diffusion coefficient increases again, but does so more slowly than the collisional one. As a result, for $\overline{\chi}_\perp \gtrsim 3$, it becomes just a small correction to $\overline{\chi}_\perp$.

In order to put things into perspective, we also plot in Fig. 15.13 the function $D(\overline{\chi}_\perp)$ vs. $\overline{\chi}_\perp$ (not in log-log representation!). Its behaviour could be understood as the combined action of perpendicular trapping and parallel trapping. A typical DCT in the range of parameters corresponding to the decreasing part of $D(\overline{\chi}_\perp)$, together with its hodograph (enlarged near the origin), is shown in Fig. 15.14.

This is a typical open trajectory (similar to the blue DCT of Fig. 15.10), but in a situation where $\overline{\chi}_\parallel$ is large compared to $\overline{\chi}_\perp$. This produces a significant backward turn of the trajectory for $\tau > \tau_r = 2.753$. *The trapping action of the parallel collisions is thus important. It is, however combined with the decorrelating action of the perpendicular collisions:* the fictitious particle does not retrace its previous path, but gradually departs from the first part of the DCT. The velocity reaches high values for $\tau < \tau_r$, then vanishes and changes sign. The hodograph exhibits the familiar loop with small velocities:[8] this loop lasts for an infinite time $\tau > \tau_r$. The fictitious particle, after being launched along its backward track finally reaches very slowly a stopping point (different from the origin).

---

[8] Note the difference in the scales of the hodographs in Figs. 15.11 and 15.14: the velocities are much larger in the latter case.

In view of this analysis, we may interpret the shape of Figs. 15.12 and 15.13 as follows. The perpendicular collisions act as a very efficient "seed": Even a very tiny amount of perpendicular collisional diffusivity suffices for changing the subdiffusive process into a diffusive one.[9] For these very small $\overline{\chi}_\perp$, the amplification of the perpendicular diffusion is enormous, reaching several orders of magnitude.[10] But after an initial rising part of the curve, the nonlinearity starts playing an important role, in particular by introducing the action of the parallel collisions. If these were alone, they would produce subdiffusion by parallel trapping. In combination with the decorrelating perpendicular collisions, they tend to diminish the effect of the latter on the diffusion coefficient, hence the decreasing part of the function $D(\overline{\chi}_\perp)$. This effect is, however, transient. For increasing values of $\overline{\chi}_\perp$, the anomalous D starts increasing again, but does so slower than $\overline{\chi}_\perp$ itself. As a result, its *relative* importance becomes smaller. When $\overline{\chi}_\perp \gtrsim \overline{\chi}_\parallel$, the anomalous contribution to the diffusion coefficient becomes a small correction to the total diffusivity.

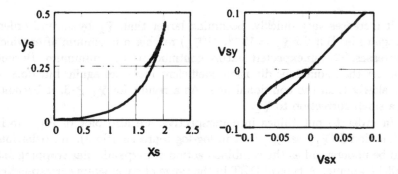

Figure 15.14. Decorrelation trajectory (left) and the corresponding hodograph, enlarged near the origin (right) for: $M = 10$, $\overline{\chi}_\parallel = 1$, $\overline{\chi}_\perp = 0.1$, $p = 0.06$, $\eta^0 = 1$.

---

[9] For this reason, the subdiffusive regime actually appears as an interesting mathematical limit which is, however, never reached in reality. There always exists a small perpendicular collisional diffusivity, even in very strong (but finite) magnetic fields.

[10] Note that in a strong magnetic field one always has $\chi_\perp \ll \chi_\parallel$; provided $\lambda_\parallel/\lambda_\perp$ is not very large, we will also have $\overline{\chi}_\perp \ll \overline{\chi}_\parallel$.

Chapter 16

Drift Waves and Zonal Flows

16.1 Qualitative Overlook (I)

An important breakthrough in fusion physics was the discovery of an unexpected operation regime of the tokamak, which led to a substantial improvement in the confinement time. This regime has been called the *H-mode* (or High-confinement mode) by opposition with the "ordinary" *L-mode* (or Low-confinement mode) (WAGNER et al. 1982). The explanation of this effect is the formation of a *transport barrier* at a certain distance from the centre of the magnetic axis. At this barrier the turbulence level propagating from the core is substantially reduced. As a result the transport processes are also reduced, leading to the observed increase in the confinement time. Clearly, this experimental discovery triggered an intense effort in experimental, computational and theoretical plasma physics in order to understand the possible mechanisms of the *L-H transition*. An enormous amount of literature has appeared in the past twenty years: it is impossible to cover the whole field within the present monograph. A very clear review, together with an extensive list of references (prior to 2000) will be found in the paper of TERRY 2000. A collection of more recent results will be found in DIAMOND et al. 2001. The aspects related to the L-H bifurcation are covered in the review paper of TERRY 2000, in KIM & DIAMOND 2003 and more extensively in the works of S.-I and K. Itoh and their coworkers (ITOH et al. 1999, ITOH & KAWAI 2002, YOSHIZAWA et al. 2003). We outline here the general ideas of the production of transport barriers.

Consider a plasma in a state of drift wave turbulence. The electrostatic potential (as well as related quantities) have a rugged structure of the type shown in Fig. 14.3, which is, moreover, continuously changing in time (see Chaps. 13, 14). Suppose that in a certain region there exists a systematic *shear flow*, which we first assume stationary. This flow will carry along *nonuniformly* the drift wave structures, thus deforming them. This is shown schematically in Fig. 16.1. After some time the structure is

383

torn up and fragmented into smaller substructures. This means that the various regions of the initial structure loose their correlation because of the drag by the shear flow. As a result, *an overall increase of the wave vector of the structure is produced in the direction perpendicular to the shear flow.*

Figure 16.1. Deformation and fragmentation of a drift wave potential structure by a stationary shear flow.

In a real situation, as encountered in a tokamak (but also in the terrestrial atmosphere), the situation is more complex. It will be shown that under certain circumstances, the drift wave turbulence is able to generate spontaneously a shear flow in the poloidal direction. The latter is also random, but characterized by a correlation length that is much longer in the y-direction (perpendicular to both the magnetic field and the density gradient, *i.e.*, the poloidal direction in a tokamak) than the one of the original drift wave turbulence. This large-scale turbulent flow is called a *zonal flow*. In order to avoid confusion, let us stress that we use here a slightly more general definition of the term "zonal flow" compared to other authors, who limit this name to flows with strictly vanishing poloidal wave vector, $k_y = 0$. We understand under this term a structure whose poloidal wave vector k_y is much smaller than the radial one $k_y \ll k_x$, *but not necessarily zero*. Its effect is again, understandably, a tearing apart of the drift wave structures in the radial direction and their fragmentation. This effect is very strikingly seen in the massively parallel numerical simulations by LIN et al. 1998, showing the structure of the potential in a section of a tokamak (Fig. 16.2).

The mechanism of the generation of a zonal flow will be studied below. Before going into the detailed treatment, we must define our starting point.

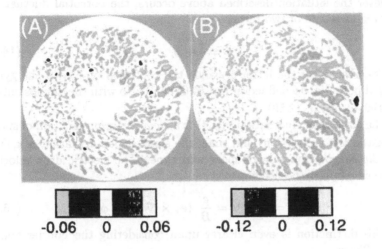

Figure 16.2. Poloidal contour plots of fluctuation potential $e\phi/T_i$ in the steady state of a nonlinear global simulation. *Left:* in presence of zonal flow; *Right:* with zonal flows suppressed. (Reprinted with permission from LIN et al. *Science* **281** 1835 (1998). Copyright 1998 AAAS.)

16.2   The Modified Hasegawa-Mima Equation

The zonal flow problem is naturally discussed at the macroscopic level. The basic equations are therefore the plasmadynamical balance equations (2.2) - (2.4). In all existing theoretical treatments, the problem is simplified by the assumptions of Sec. 2.5 and 2.6, which we briefly recall here: negligible electron mass, constant electron temperature, zero ion temperature, neglect of collisional viscosity. With these assumptions the plasma is described by the reduced Hasegawa-Wakatani equations (with $\mu = 0$) (2.86), (2.90) or the Hasegawa-Mima equation (2.98). These equations require, however, a modification in the present problem, for the following reason. It appears from the previous qualitative discussion that in this case it is important to distinguish two types of drift wave structures according to their extension in the poloidal ($y-$) direction. In other words, the spectrum of potential fluctuations has a peculiar shape. It contains two peaks, one corresponding to *large-scale (or long-wavelength) fluctuations* and another one to *small-scale (or short-wavelength) fluctuations;* these two peaks are widely separated.

This distinction requires a more detailed notation than in previous matters. We introduce the following convention: for the (dimensional) fluctuating electrostatic potential, we drop the letter "δ"; thus: $\delta\phi \rightarrow \phi$.

Whenever the situation described above occurs, the potential fluctuation is written as a sum of two terms:

$$\phi = \overline{\phi} + \tilde{\phi}. \tag{16.1}$$

The absence of the letter δ avoids the confusion of $\overline{\phi}$ with the gyro-averaged potential $\overline{\delta\phi}$ defined in Eq. (4.66) and of $\tilde{\phi}$ with the dimensionless potential $\delta\tilde{\phi}$ of Eq. (2.59).

The large-scale fluctuation $\overline{\phi}(\mathbf{x}, t)$ is a *slowly varying* function of space (at least in the y-direction) and time; the small-scale fluctuation $\tilde{\phi}(\mathbf{x}, t)$ is a *rapidly varying* function of space and time. The electric drift velocity (2.53) can similarly be decomposed as:

$$\delta\mathbf{u}^E = \overline{\mathbf{V}} + \tilde{\mathbf{V}} = \frac{c}{B} \ (\mathbf{e}_z \times \nabla)\, (\overline{\phi} + \tilde{\phi}). \tag{16.2}$$

This distinction is even clearer upon considering the Fourier transforms.[1] Thus:

$$\overline{\phi}(\mathbf{x}, t) = \int d\mathbf{q}\ e^{i\mathbf{q}\cdot\mathbf{x}}\ \overline{\phi}(\mathbf{q}, t), \qquad \rho_s q \ll 1 \,. \tag{16.3}$$

This expression means that only wave-vectors much smaller than ρ_s^{-1} contribute significantly to $\overline{\phi}(\mathbf{q}, t)$. On the other hand,

$$\tilde{\phi}(\mathbf{x}, t) = \int d\mathbf{k}\ e^{i\mathbf{k}\cdot\mathbf{x}}\ \tilde{\phi}(\mathbf{k}, t), \qquad \rho_s k \gg 1 \,. \tag{16.4}$$

It may also be said that the Fourier transform of the potential is a function with two quite distinct and widely separated maxima, one at $q \ll \rho_s^{-1}$ and the other at $k \gg \rho_s^{-1}$. Whenever such a significant separation exists between small and large scales, a *partial ensemble average* $\langle ... \rangle_\sim$ can be introduced, defined as an average over small scale quantities. Thus, in particular:

$$\overline{\phi} = \langle \phi \rangle_\sim \,. \tag{16.5}$$

This partial average operation allows us to study separately the short- and the long-scale fluctuations. It possesses the usual properties:

$$\langle \overline{\phi} \rangle_\sim = \overline{\phi}, \qquad \langle \tilde{\phi} \rangle_\sim = 0. \tag{16.6}$$

We now derive separate equations of evolution for the two components. In Chap. 2, the Hasegawa-Mima equation was derived from the Hasegawa-Wakatani equation by assuming that the electron density fluctuations are

---

[1] It should be kept in mind that small wave vectors in Fourier space correspond to long scale quantities in physical space. In order not to overburden the notations, we use the same symbols for $\phi(\mathbf{x}, t)$ and its Fourier transform $\phi(\mathbf{q}, t)$, which are recognizable from their arguments (or from the context).

adiabatic. It must be noted, however, that the adiabatic assumption (2.96) is only applicable to the small-scale density fluctuations (SMOLYAKOV et al. 2000). Thus, the adiabatic assumption can only be applied to the small-scale part of Eq. (2.86). Using the new notations, we thus obtain for the small-scale density fluctuations:

$$\left(\partial_t + \overline{\mathbf{V}} \cdot \nabla_\perp\right) \frac{\tilde{n}}{n_0} + \tilde{\mathbf{V}} \cdot \nabla_\perp \ln n_0 = -\overline{\sigma} \, \nabla_z^2 \left(\frac{e\tilde{\phi}}{T_e} - \frac{\tilde{n}}{n_0}\right). \qquad (16.7)$$

(In obtaining this result, we used the obvious identity: $\tilde{\mathbf{V}} \cdot \nabla_\perp \tilde{n}_{ad} \equiv 0$, and the fact that $\overline{\mathbf{V}} \cdot \nabla_\perp \ln n_0$ is a large-scale quantity). After implementation of the adiabatic assumption, this equation reduces to the following one, replacing Eq. (2.86):

$$\left[\partial_t + \overline{\mathbf{V}} \cdot \nabla_\perp\right] \frac{e\tilde{\phi}}{T_e} + \tilde{\mathbf{V}} \cdot \nabla_\perp \ln n_0 = 0. \qquad (16.8)$$

Eq. (2.90) (with $\mu = 0$) remains unchanged. Subtracting it from (16.8) we obtain:

$$\left(\partial_t + \overline{\mathbf{V}} \cdot \nabla_\perp\right) \frac{e\tilde{\phi}}{T_e} + V_{ds} \nabla_y \frac{e\tilde{\phi}}{T_e}$$
$$-\rho_s^2 \left(\partial_t + \overline{\mathbf{V}} \cdot \nabla_\perp + \tilde{\mathbf{V}} \cdot \nabla_\perp\right) \nabla_\perp^2 \left(\frac{e\tilde{\phi}}{T_e} + \frac{e\overline{\phi}}{T_e}\right) = 0, \qquad (16.9)$$

where V_{ds} is the diamagnetic drift velocity, defined in Eq. (2.27). This will be called the *Modified Hasegawa-Mima equation*. It was derived by SMOLYAKOV & DIAMOND 1999 and SMOLYAKOV et al. 2000 (using a less explicit result of MATTOR & DIAMOND 1994).[2] It will constitute the basis for our further developments.

Note that in both the Hasegawa-Mima and in the modified Hasegawa-Mima equations, the dynamics has been reduced to two dimensions. Hence, we simplify the notations by suppressing the subscript "$\perp$" and defining both position coordinates and wave vectors as two-dimensional vectors:

$$\mathbf{x} = (x, y), \quad \nabla = \nabla_\perp = (\nabla_x, \nabla_y), \quad \mathbf{k} = (k_x, k_y), \quad \mathbf{q} = (q_x, q_y). \quad (16.10)$$

Eq. (16.9) contains both small-scale and large-scale quantities. Its large-scale part is extracted by the operation of partial averaging, using Eqs. (16.6):

---

[2] Note that the equation derived in SMOLYAKOV & DIAMOND 1999 is incomplete: it misses the term $\tilde{\mathbf{V}} \cdot \nabla_\perp$ in the first parenthesis of (16.9) and the term $e\overline{\phi}/T_e$ in the second parenthesis. The correct equation was derived in SMOLYAKOV et al. 2000.

$$\partial_t \, \nabla^2 \overline{\phi} = - \left\langle \left(\tilde{\mathbf{V}} \cdot \nabla \right) \nabla^2 \tilde{\phi} \right\rangle_{\sim} . \tag{16.11}$$

This equation describes the generation of the large-scale potential by nonlinear coupling of small-scale quantities. The term $\left(\overline{\mathbf{V}} \cdot \nabla \right) \nabla^2 \overline{\phi}$ is of higher order in the large scale gradients. By introducing the Poisson bracket (2.58) we obtain the compact form:

$$\partial_t \, \nabla^2 \overline{\phi} = - \frac{c}{B} \left\langle \left[\tilde{\phi}, \, \nabla^2 \tilde{\phi} \right]_P \right\rangle_{\sim} . \tag{16.12}$$

The small-scale part of Eq. (16.9) is best treated by using the Fourier representation (16.3), (16.4). We use the same convention as above: k represents a *large* wave vector, q is a *small* one, $q \ll k$. The Fourier transformation is easily obtained. In the result the large-scale contributions are eliminated by dropping terms containing $\tilde{\phi}(\mathbf{q})$ and $\overline{\phi}(\mathbf{k})$. We then easily obtain:

$$\partial_t \left(1 + \rho_s^2 k^2 \right) \, \tilde{\phi}(\mathbf{k}) + i \, k_y V_{ds} \, \tilde{\phi}(\mathbf{k})$$

$$+ \frac{c}{B} \int d\mathbf{q} \, \mathbf{e}_z \cdot (\mathbf{k} \times \mathbf{q}) \, \left[1 + \rho_s^2 |\mathbf{k} - \mathbf{q}|^2 \right] \, \overline{\phi}(\mathbf{q}) \, \tilde{\phi}(\mathbf{k} - \mathbf{q}) = 0. \tag{16.13}$$

Note that the complete term in square brackets is: $\left[1 + \rho_s^2 |\mathbf{k} - \mathbf{q}|^2 - \rho_s^2 q^2 \right]$; the term $\rho_s^2 q^2$ can, however, be neglected because of Eq. (16.3).

16.3   The Wave Liouville Equation

The small-scale potential fluctuation $\tilde{\phi}(\mathbf{k})$ never appears isolated in the expressions of the macroscopic, observable quantities. Indeed, such quantities are partial averages, and $\left\langle \tilde{\phi}(\mathbf{k}) \right\rangle_{\sim} = 0$. The interesting quantities are quadratic functionals of the potential, of the type appearing in Eq. (16.11). It is therefore convenient to derive an equation for such a functional, to which any other can be related. Such an equation would have a nice form if it described a conserved quantity. The electrostatic potential energy density, proportional to $|\tilde{\phi}(\mathbf{k})|^2$ would be a candidate. Actually, if $\overline{\phi}(\mathbf{q}) = 0$, Eq. (16.13) would reduce to a simple harmonic oscillator equation, for which the energy $\omega_d |\tilde{\phi}(\mathbf{k})|^2$ is a conserved quantity. Here $\omega_d(\mathbf{k})$ is the *drift wave frequency* (3.13):

$$\omega_d(\mathbf{k}) = \frac{k_y V_{ds}}{1 + \rho_s^2 k^2}, \tag{16.14}$$

where V_{ds} is the diamagnetic drift velocity, Eq. (3.7).

It is, however, physically clear that the energy of the small-scale fluctuations cannot be separately conserved, because of its interaction with the large-scale ones. The following question then arises: is it possible to find a quadratic functional of the small-scale potential fluctuations alone, that is conserved by Eq. (16.13)? This would be an *adiabatic invariant*. It was shown in SMOLYAKOV & DIAMOND 1999 and in SMOLYAKOV et al. 2000 that the answer is affirmative [see also MUHM et al. 1992, LEBEDEV et al. 1995].

We note that Eq. (16.13) actually involves the following small-scale quantity:

$$\widehat{\phi}(\mathbf{k}) = (1 + \rho_s^2 k^2)\, \widetilde{\phi}(\mathbf{k}). \tag{16.15}$$

Its equation of evolution is:

$$\partial_t \widehat{\phi}(\mathbf{k}) + i\omega_d(\mathbf{k})\widehat{\phi}(\mathbf{k})$$
$$+ \frac{c}{B} \int d\mathbf{q}\; \overline{\phi}(\mathbf{q})\, (\mathbf{e}_z \times i\mathbf{q})\cdot i\mathbf{k}\; \widehat{\phi}(\mathbf{k} - \mathbf{q}) = 0. \tag{16.16}$$

We now define a quantity that is similar to the **k**-spectrum $S_{\mathbf{k}}$ introduced in Eq. (7.78). There is, however, an important difference: the new quantity is defined by a *partial average*. It therefore retains a slow, large-scale dependence on **x**. This large-scale modulation is accounted for by introducing a quantity analogous to the *Wigner function* of quantum statistical mechanics (see, *e.g.*, BALESCU 1975):

$$N(\mathbf{x}, \mathbf{k}, t) = \int d\mathbf{p}\; e^{i\mathbf{p}\cdot\mathbf{x}}\, \left\langle \widehat{\phi}(\mathbf{k})\, \widehat{\phi}(-\mathbf{k}+\mathbf{p}) \right\rangle_{\sim}, \qquad \rho_s p \ll 1. \tag{16.17}$$

Its equation of evolution is easily obtained from (16.16):

$$\partial_t N(\mathbf{x}, \mathbf{k}, t) = - \int d\mathbf{p}\; e^{i\mathbf{p}\cdot\mathbf{x}}\, \left\{ i\,[\omega_d(\mathbf{k}) + \omega_d(-\mathbf{k}+\mathbf{p})]\left\langle \widehat{\phi}(\mathbf{k})\, \widehat{\phi}(-\mathbf{k}+\mathbf{p}) \right\rangle_{\sim} \right.$$
$$+ \frac{c}{B} \int d\mathbf{q}\; \overline{\phi}(\mathbf{q})\, (\mathbf{e}_z \times i\mathbf{q})\cdot \left[i\mathbf{k} \left\langle \widehat{\phi}(\mathbf{k}-\mathbf{q})\widehat{\phi}(-\mathbf{k}+\mathbf{p}) \right\rangle_{\sim} \right.$$
$$\left.\left. - i(\mathbf{k}-\mathbf{p}) \left\langle \widehat{\phi}(\mathbf{k})\, \widehat{\phi}(-\mathbf{k}+\mathbf{p}-\mathbf{q}) \right\rangle_{\sim} \right] \right\}. \tag{16.18}$$

The first term in the right hand side is transformed by noting that $\omega_d(-\mathbf{k}) = -\omega_d(\mathbf{k})$ and that $p \ll k$, hence the parenthesis can be expanded as $[...] \approx \mathbf{p} \cdot [\partial\omega_d(\mathbf{k})/\partial\mathbf{k}]$. This term thus yields a contribution:

$$-\frac{\partial\omega_d(\mathbf{k})}{\partial\mathbf{k}} \cdot \frac{\partial N(\mathbf{x}, \mathbf{k}, t)}{\partial\mathbf{x}}.$$

In the second term we first invert the partial averages:

$$\left\langle \hat{\phi}(\mathbf{k} - \mathbf{q})\, \hat{\phi}(-\mathbf{k} + \mathbf{p}) \right\rangle_{\sim} = (2\pi)^{-2} \int d\mathbf{x}'\, e^{-i(\mathbf{p}-\mathbf{q})\cdot\mathbf{x}'} N(\mathbf{x}', \mathbf{k} - \mathbf{q}),$$

$$\left\langle \hat{\phi}(\mathbf{k})\, \hat{\phi}(-\mathbf{k} + \mathbf{p} - \mathbf{q}) \right\rangle_{\sim} = (2\pi)^{-2} \int d\mathbf{x}'\, e^{-i(\mathbf{p}-\mathbf{q})\cdot\mathbf{x}'} N(\mathbf{x}', \mathbf{k}).$$

These expresions are introduced in the square brackets above, and the integrations are easily performed. One uses again the smallness of q: $N(\mathbf{k} - \mathbf{q}) - N(\mathbf{k}) \approx -\mathbf{q} \cdot [\partial N(\mathbf{k})/\partial \mathbf{k}]$ and finally obtains for the contribution of the second term:

$$\frac{\partial[\mathbf{k}\cdot\overline{\mathbf{V}}(\mathbf{x})]}{\partial \mathbf{x}} \cdot \frac{\partial N(\mathbf{x}, \mathbf{k}, t)}{\partial \mathbf{k}} - \frac{\partial[\mathbf{k}\cdot\overline{\mathbf{V}}(\mathbf{x})]}{\partial \mathbf{k}} \cdot \frac{\partial N(\mathbf{x}, \mathbf{k}, t)}{\partial \mathbf{x}}.$$

Collecting the partial results, we finally obtain:

$$\partial_t N(\mathbf{x}, \mathbf{k}, t) = \left\{ \frac{\partial}{\partial \mathbf{x}} \left[\omega_d(\mathbf{k}) + \mathbf{k}\cdot\overline{\mathbf{V}}(\mathbf{x}, t) \right] \right\} \cdot \frac{\partial N(\mathbf{x}, \mathbf{k}, t)}{\partial \mathbf{k}}$$
$$- \left\{ \frac{\partial}{\partial \mathbf{k}} \left[\omega_d(\mathbf{k}) + \mathbf{k}\cdot\overline{\mathbf{V}}(\mathbf{x}, t) \right] \right\} \cdot \frac{\partial N(\mathbf{x}, \mathbf{k}, t)}{\partial \mathbf{x}}. \qquad (16.19)$$

Eq. (16.19) is (rather improperly) called the "*Wave kinetic equation*" in the literature. It has a remarkable Hamiltonian structure.[3] Indeed, k plays the role of a momentum canonically conjugate to x. The Hamiltonian is:

$$H(\mathbf{x}, \mathbf{k}, t) = \omega_d(\mathbf{k}) + \mathbf{k}\cdot\overline{\mathbf{V}}(\mathbf{x}, t). \qquad (16.20)$$

Eq. (16.19) is thus simply the Liouville equation associated with this Hamiltonian [see Eq. (4.1)]:

$$\partial_t N = [H, N]_P \equiv \mathcal{L}_W\, N. \qquad (16.21)$$

$[...]_P$ is the usual Poisson bracket, and $\mathcal{L}_W$ will be called the *Wave Liouvillian*. For these reasons we prefer to call Eq. (16.19) the *Wave Liouville Equation*. It determines the slow dynamics of the action function. Eq. (16.19) tells us that the action density $N(\mathbf{x}, \mathbf{k}, t)$ is constant along any trajectory derived from the Hamiltonian: this is the conservation property we were looking for. A true kinetic equation (in the sense of statistical mechanics) will be derived in Sec. 16.6.

[3] For a clear discussion of the Hamiltonian aspects of this problem in a simpler case see Chap. 4 of STIX 1992.

16.4 Zonal Flows and Reynolds Stress

Before starting the detailed analysis of the Wave Liouville Equation, we note that the generation of zonal flows is actually a very direct consequence of the momentum balance in a fluid or a plasma. This mechanism was first considered for a plasma by BIGLARI et al. 1990, and more explicitly by DIAMOND & KIM 1991 (see also TERRY 2000, KORSHOLM et al. 2001). From the centre-of-mass momentum balance equation, neglecting the collisional dissipative pressure tensor $\overleftrightarrow{\pi}$, we obtain the equation of motion for an incompressible fluid ($\nabla \cdot \mathbf{u} = 0$):

$$\partial_t u_m + \nabla_n(u_n u_m) + \nabla_m P = \frac{1}{c\rho}(\mathbf{j} \times \mathbf{B})_m. \qquad (16.22)$$

We now consider the magnetic-surface average (*i.e.*, the average over y in slab geometry) of the component u_y (for simplicity, we do not use a different notation for this average):

$$\partial_t u_y + \nabla_x(u_x u_y) = 0. \qquad (16.23)$$

Next, we assume as in Sec. 16.2 that the plasma is turbulent and that the fluctuations are of two kinds with scale lengths widely separated. The partial average over the small scale fluctuations of Eq. (16.23) is then:

$$\partial_t \bar{u}_y + \nabla_x \langle \tilde{u}_x \tilde{u}_y \rangle_\sim = 0. \qquad (16.24)$$

We note here the appearance of the **Reynolds stress** tensor:

$$R_{nm} = \langle \tilde{u}_n \tilde{u}_m \rangle_\sim . \qquad (16.25)$$

This quantity is well-known in neutral (Navier-Stokes) fluid turbulence. It is related to the *anomalous momentum flux*. It is expressed as a correlation of two (small-scale) fluctuating quantities. Its analogy with the particle flux (5.13) is striking: the latter contains the correlation of the electric field and of the density fluctuations.

Before commenting on this equation, we note that it is in perfect agreement with Eq. (16.12) for the large-scale potential fluctuations. Using $\mathbf{u} = (c/B)(\mathbf{e}_z \times \nabla \phi)$, Eq. (16.23) yields:

$$\partial_t \nabla_y \phi = \frac{c}{B} \left[(\nabla_y \phi)(\nabla_x \nabla_y \phi) - (\nabla_x \phi)(\nabla_y \nabla_y \phi) \right],$$

and a similar equation for $\partial_t \nabla_x \phi$. By a second spatial differentiation and a combination of the two resulting equations we obtain:

$$\partial_t \nabla^2 \phi = \frac{c}{B} \left[(\nabla_y \phi)(\nabla_x \nabla^2 \phi) - (\nabla_x \phi)(\nabla_y \nabla^2 \phi) \right],$$

which is precisely Eq. (16.12).

Eq. (16.24) tells us that - if the signs turn out as they should - the x-(radial) gradient of the R_{xy} component of the Reynolds stress may generate a component of the large-scale velocity in the perpendicular (y-, poloidal) direction, *i.e.*, a **zonal flow**.

We now note that the Reynolds stress (16.25) is related to the wave action density (16.17). In order to take account of the slow large-scale variation of this quantity, we must introduce a "Wigner function":

$$R_{xy} = -\frac{c^2}{B^2} \left\langle \nabla_x \tilde{\phi} \, \nabla_y \tilde{\phi} \right\rangle_\sim$$

$$= -\frac{c^2}{B^2} \int dk \int dp \, e^{ip\cdot x} \left\langle ik_x \, i(-k+p)_y \, \tilde{\phi}_k \tilde{\phi}_{-k+p} \right\rangle_\sim$$

$$\approx -\frac{c^2}{B^2} \int dk \, \frac{k_x k_y}{(1+\rho_s^2 k^2)^2} \, N(x, k, t). \tag{16.26}$$

(We limited ourselves to the dominant order for $p \ll k$.) Thus, the equation for the large-scale average zonal flow is coupled to the drift wave action density:

$$\partial_t \, \nabla_x \overline{\phi}(x, t) = \frac{c}{B} \, \nabla_x \int dk \, \frac{k_x k_y}{(1+\rho_s^2 k^2)^2} \, N(x, k, t). \tag{16.27}$$

We thus obtained a closed set of coupled equations, (16.19) and (16.27), for the physical quantities $N(x, k, t)$ and $\overline{\phi}(x, t)$ [or $\overline{V}(x, t)$].

16.5   Qualitative Overlook (II)

Having defined the relevant physical quantities of the problem, we now return to the qualitative discussion of Sec. 16.1 and consider in some more detail the mechanisms generating zonal flows and transport barriers. There appear various interpretations in the literature. Thus, TERRY 2000 in his qualitative presentation reasons mainly on the basis of Navier-Stokes turbulence. He argues that the effect of the zonal flow is mainly a shortening of the "eddy lifetime" (associated with the correlation time). At any given scale, an eddy looses coherence because of advection by other eddies. As the eddy lifetime decreases, by the time an eddy rotates, its energy is transferred to other eddies: the level of turbulence thus decreases. This argument does not, however, tell us in any detail the fate of the drift waves, or of the zonal flows.

The arguments of Diamond and his collaborators (DIAMOND et al. 1998, 2001) are more elaborate. The fragmentation produced by the zonal flows results in an *increase* of the mean square of the radial wave-vector $\langle k_r^2 \rangle$, which is a measure of the inverse square of the correlation length λ_r^{-2}

in the radial direction. This is clearly seen in Fig. 16.2. As a result, the drift wave frequency $\omega_d(\mathbf{k})$, Eq. (3.13), will *decrease*. It follows from the conservation of the action density [4] $N(\mathbf{k}) = \varepsilon(\mathbf{k})/\omega_d(\mathbf{k})$ that the *drift wave energy* $\varepsilon(\mathbf{k})$ *must also decrease*. The total energy being conserved, it follows that *the zonal flow energy increases*. We are thus in presence of an *instability*. Of course, this instability will be saturated after some time by some mechanism (to be studied separately). The net result is the appearance of a new state in which the level of turbulence, hence the anomalous radial transport is significantly reduced (or "suppressed"): a *transport barrier* has been created.

DIAMOND et al. 1998, 2001 then consider a drift wave packet propagating in a zonal flow field and calculate the rate of change of the energy by using the conservation of the action $N(\mathbf{k}, t) = N(\mathbf{k}, 0)$, thus writing $d\varepsilon(\mathbf{k})/dt = N(\mathbf{k}, 0)\, d\omega_d(\mathbf{k})/dt$ and obtain an expression in terms of the work on the mean flow. In passing they use the "eikonal equation" for $dk_r/dt = -\partial(k_\theta \overline{V})/\partial x \equiv -k_\theta \overline{V}'$. In order to complete the calculation they use the correct wave kinetic equation (16.19) (which refers to the *correct* action!) and "*the methodology of quasilinear theory*" in order to come out with a diffusion coefficient in $\mathbf{k}$-space:

$$D_K^D = k_\theta^2 \sum_{\mathbf{q}} q^2 \left| \tilde{V}'_{E\mathbf{q}} \right|^2 R^D(\mathbf{k}, \mathbf{q}), \qquad (16.28)$$

where the Diamond propagator is:

$$R^D(\mathbf{k}, \mathbf{q}) = \frac{\gamma(\mathbf{k})}{[qV^g(\mathbf{k})]^2 + \gamma^2(\mathbf{k})}. \qquad (16.29)$$

D_K^D has indeed the form of a renormalized quasilinear diffusion coefficient, Eq. (9.19), with a broadened resonance. The action thus presumably obeys a diffusion equation (which is not written down!), thus explaining the growth of $\langle k_r^2 \rangle$ and leading to a growth rate of the zonal flow instability $\gamma(\mathbf{k})$, hence to the decrease of the drift wave transport. In our opinion, although the result is (approximately) correct, the derivation is not quite consistent and does not cover all aspects of the process.

The authors of DIAMOND et al. 2001 are perfectly aware that this quasilinear calculation is of limited validity. There is one important aspect of turbulence that is absent from the reasoning described above: the effect of *trapping*. It was stressed in Chap. 14 that this effect plays a primary role in understanding the turbulent transport. In particular, it invalidates the "quasilinear methodology" except in the limit of weak turbulence. The potential landscape appearing in the simulations of LIN et al. 1998 (Fig. 16.2) is indeed very rugged, and therefore produces important trapping.

---

[4] This "traditional" definition of the wave action should actually be refined in the present case, see Eq. (16.17).

The diffusion process must therefore be treated by a theory taking this effect into account. The trapping effect may produce more or less long-lived coherent structures in the turbulent plasma. DIAMOND et al. 2001 extend their initial argument by supposing such coherent structures to be actors in non-local transport schemes based on the avalanche concept from self-organized criticality (SOC) theory (see Chap. 17); this idea is further developed in DIAMOND & MALKOV 2002.

Following a different line of thinking, several recent works study the detailed form of the nonlinear coherent structures. Thus, SMOLYAKOV et al. 2000 treat the wave Liouville equation by a multiple-time perturbation technique which in first approximation yields the quasilinear result. In the next approximation the resulting equation admits localized solutions of the kink-soliton type. Coherent nonlinear structures that are exact solutions of a model system of one-dimensional equations for the drift wave - zonal flow system are very elegantly determined by KAW et al. 2002. These correspond to wave trapping which finally produces solitons, shocks or other similar localized structures. These may be responsible for some of the non-Gaussian features of transport mentioned above.

In the present chapter we take a different approach to the interpretation of the trapping effect on the global transport. We base our treatment on the *decorrelation trajectory (DCT) method*: this will be done in detail in the remainder of this chapter. The fate of the amplified zonal flows and their final stabilization is much less understood: several interpretations have been advanced but no final conclusion seems to appear yet. This matter will not be discussed here. We rather concentrate on the detailed study of the diffusive process that may generate poloidally extended wave packets, identified with zonal flows.

16.6   The Wave Kinetic Equation

It was shown in Sec. 16.3 that the wave action density obeys an equation of evolution (16.21) which has the structure of a Liouville equation, deriving from a Hamiltonian (16.20). It is therefore natural to treat this equation by the methods of nonequilibrium statistical mechanics (BALESCU 1997) in order to derive a true Wave Kinetic equation. The procedure appears as a straightforward generalization of the operations of Sec. 14.8. We follow here closely our paper BALESCU 2003.

The Hamiltonian structure suggests quite naturally the following interpretation. The complex set of waves in the turbulent plasma can be described as a set of *quasiparticles* (LANDAU & LIFSHITZ 1951; a very clear discussion of this question is given in STIX 1992). A quasiparticle represents a *wave packet*, characterized by the position of its interference maximum, $\mathbf{x}$, and an associated wave vector $\mathbf{k}$, playing the role of the canonically conjugate momentum.[5] We may thus define a phase space $(\mathbf{x}, \mathbf{k})$ which, in the present case, is four-dimensional: it is the framework for the description of the motion of a given quasiparticle, $\mathbf{x}(t)$, $\mathbf{k}(t)$. Next, we introduce the statistical mechanical description (BALESCU 1997) by considering an ensemble of realizations of this one-quasiparticle system. The system is then represented by a (one-quasiparticle) distribution function $N(\mathbf{x}, \mathbf{k}, t)$ whose evolution is dictated by the Liouville equation. This distribution function is none other than the action density.

We now note that the action density is the result of a *partial averaging* over the small-scale fluctuations, Eq. (16.17). It remains, however, a fluctuating quantity in the large-scale domain. This feature is clearly seen in Eqs. (16.19) and (16.20). The Hamiltonian contains the fluctuating term $\mathbf{k} \cdot \overline{\mathbf{V}}(\mathbf{x}, t)$. Hence the Wave Liouville Equation is a *stochastic differential equation*, i.e., it is of the same nature as a *hybrid kinetic equation* introduced in Secs. 11.8 and 14.8. The associated characteristic equations are therefore typical *V-Langevin equations* of the kind studied in Chap. 14.

The first step in the treatment of the Wave Liouville equation is to separate, as usual, an average part and a fluctuating part in the distribution function:

$$N(\mathbf{x}, \mathbf{k}, t) = n(\mathbf{x}, \mathbf{k}, t) + \delta N(\mathbf{x}, \mathbf{k}, t),$$
$$n(\mathbf{x}, \mathbf{k}, t) = \langle N(\mathbf{x}, \mathbf{k}, t) \rangle, \quad \langle \delta N(\mathbf{x}, \mathbf{k}, t) \rangle = 0. \tag{16.30}$$

Note that the average here is over the *large-scale* potential fluctuations. We also separate the Wave Liouvillian operator:

$$\mathcal{L}_W = \mathcal{L}_W^0 + \delta \mathcal{L}_W, \tag{16.31}$$

where:

$$\mathcal{L}_W^0 = -\mathbf{V}^g(\mathbf{k}) \cdot \frac{\partial}{\partial \mathbf{x}}, \tag{16.32}$$

$$\delta \mathcal{L}_W = -\overline{\mathbf{V}}(\mathbf{x}, t) \cdot \frac{\partial}{\partial \mathbf{x}} - \overline{\mathbf{W}}(\mathbf{x}, \mathbf{k}, t) \cdot \frac{\partial}{\partial \mathbf{k}}. \tag{16.33}$$

Here $\mathbf{V}^g(\mathbf{k})$ is the (unperturbed) group velocity of the drift waves:[6]

---

[5] Some authors justify this point of view by invoking quantum mechanics. In that case the particle representation of waves is a fundamental one. Thus, the photons can be described either by their position $\mathbf{x}$, or by their momentum $\hbar\mathbf{k}$. But this is somewhat artificial, as we are dealing here with a purely classical picture ($\hbar = 0$). The Hamiltonian structure does not require quantum mechanics.

[6] We assume here that the drift wave frequency is independent of $\mathbf{x}$, as the temperature is constant, and the density gradient is also taken to be approximately constant (locally).

$$V^g(\mathbf{k}) = \frac{\partial \omega_d(\mathbf{k})}{\partial \mathbf{k}}$$

$$= \frac{V_{ds}}{(1 + \rho_s^2 k^2)^2} \left\{ -2\rho_s^2 k_x k_y \ \mathbf{e}_x + \left[1 + \rho_s^2 \left(k_x^2 - k_y^2 \right) \right] \ \mathbf{e}_y \right\}. \qquad (16.34)$$

It represents the velocity of propagation of the wave packet, or quasiparticle, in absence of turbulence. In (16.33) we introduced a shorthand notation for the fluctuating "velocity" of the wave vector:

$$\overline{W}(\mathbf{x}, \mathbf{k}, t) = -\frac{\partial}{\partial \mathbf{x}} \ [\mathbf{k} \cdot \overline{\mathbf{V}}(\mathbf{x}, t)]. \qquad (16.35)$$

Averaging the Wave Liouville equation (16.21) we obtain:[7]

$$\partial_t n - \mathcal{L}_W^0 \ n = \langle \delta \mathcal{L}_W \ \delta N \rangle, \qquad (16.36)$$

$$\partial_t \ \delta N - \mathcal{L}_W^0 \ \delta N - \delta \mathcal{L}_W \ \delta N = \delta \mathcal{L}_W \ n. \qquad (16.37)$$

These equations have a familiar form [see Eqs. (14.86), (14.87)]. In the left hand side of (16.37) we recognize a linear term describing free propagation of the drift waves in the inhomogeneous medium, and a nonlinear term describing wave-wave interactions. The right hand side is a source term, providing the coupling to the average equation.

The solution strategy is the same as in Chap. 14: we first solve Eq. (16.37) by the method of characteristics, then substitute the result into (16.36) in order to obtain a closed equation for the average distribution function. The characteristic equations of the Wave Liouville equation (16.37) must be solved backwards in time (see Sec. 14.8):[8]

$$\frac{d\mathbf{x}(t|t')}{dt'} = \mathbf{V}^g[\mathbf{k}(t|t')] + \overline{\mathbf{V}}[\mathbf{x}(t|t'), t'], \qquad (16.38)$$

$$\frac{d\mathbf{k}(t|t')}{dt'} = \overline{\mathbf{W}}[\mathbf{x}(t|t'), \mathbf{k}(t|t'), t']. \qquad (16.39)$$

The "initial" condition is:

$$\mathbf{x}(t|t) = \mathbf{x}, \quad \mathbf{k}(t|t) = \mathbf{k}. \qquad (16.40)$$

The solution of these Langevin equations in a given realization is:

---

[7] In the right hand side of the equation for the fluctuation we omit, as usual, writing the term $+ \langle \delta \mathcal{L}_W \ \delta N \rangle$ which does not contribute to the average equation.

[8] These equations will be discussed in more detail in Sec. 16.7.

$$\mathbf{x}(t|t') = \mathbf{x} - \int_{t'}^{t} dt_1 \left\{ \mathbf{V}^g[\mathbf{k}(t|t_1)] + \overline{\mathbf{V}}[\mathbf{x}(t|t_1), t_1] \right\},$$

$$\mathbf{k}(t|t') = \mathbf{k} - \int_{t'}^{t} dt_1 \ \overline{\mathbf{W}}[\mathbf{x}(t|t_1), \mathbf{k}(t|t_1), t_1]. \tag{16.41}$$

The solution of the partial differential equation (16.37) is then:

$$\delta N(\mathbf{x}, \mathbf{k}, t) = \int_0^t dt_1 \ \delta \mathcal{L}_W[\mathbf{x}(t|t_1), \mathbf{k}(t|t_1), t_1] \ n[\mathbf{x}(t|t_1), \mathbf{k}(t|t_1), t_1]$$

$$+ \ \delta N[\mathbf{x}(t|0), \mathbf{k}(t|0), 0]. \tag{16.42}$$

For simplicity we assume that at time zero the fluctuation is zero: $\delta N(\mathbf{x}, \mathbf{k}, 0) = 0$. Substitution into Eq. (16.36) yields:

$$\partial_t n(\mathbf{x}, \mathbf{k}, t) - \mathcal{L}_W^0 \ n(\mathbf{x}, \mathbf{k}, t)$$

$$= \int_0^t dt_1 \ \langle \delta \mathcal{L}_W[\mathbf{x}, \mathbf{k}, t] \delta \mathcal{L}_W[\mathbf{x}(t|t_1), \mathbf{k}(t|t_1), t_1] \ n[\mathbf{x}(t|t_1), \mathbf{k}(t|t_1), t_1] \rangle.$$

$$\tag{16.43}$$

We obtained a *non-Markovian, non-local equation* of the same type as Eq. (14.94). Note that, although $\delta \mathcal{L}_W$ (16.33) is a linear combination of $\partial/\partial \mathbf{x}$ and $\partial/\partial \mathbf{k}$, Eq. (16.6) *is not a diffusion equation*. Indeed, the density profile in the right hand side is evaluated at $\mathbf{x}(t|t_1), \mathbf{k}(t|t_1)$, which are stochastic quantities [Eq. (16.41)]. The same difficulty which appeared in Secs. 6.2, 11.10 and 14.8 is present here too. The non-Gaussian features mentioned in Sec. 16.5 are presumably hidden in this equation, which is an exact consequence of the Wave Liouville equation.

In order to proceed further, we assume as in Sec. 14.8 that the correlation lengths λ_x, λ_y are much smaller than the macroscopic gradient lengths, L_n, L_{nk}, both in $\mathbf{x}$ and in $\mathbf{k}$. As we are dealing here with large-scale fluctuations, defined by the condition $\langle q \rangle \rho_s \ll 1$, this implies the assumption: $\rho_s \ll \lambda_x, \lambda_y \ll L_n, L_{nk}$, which is not unreasonable. This argument allows us to eliminate the delocalization, and use the following approximation in the integrand of Eq. (16.6):

$$n[\mathbf{x}(t|t_1), \mathbf{k}(t|t_1), t_1] \approx n[\mathbf{x}, \mathbf{k}, t_1]. \tag{16.44}$$

Eq. (16.6) thus becomes:

$$\partial_t n(\mathbf{x}, \mathbf{k}, t) - \mathcal{L}_W^0 \ n(\mathbf{x}, \mathbf{k}, t)$$

$$= \int_0^t dt_1 \ \langle \delta \mathcal{L}_W[\mathbf{x}, \mathbf{k}, t] \ \delta \mathcal{L}_W[\mathbf{x}(t|t_1), \mathbf{k}(t|t_1), t_1] \rangle \ n[\mathbf{x}, \mathbf{k}, t_1]. \tag{16.45}$$

This is a *closed, non-Markovian equation of evolution for the average distribution function.* Using the definition (16.33), this equation is written explicitly as follows:[9]

$$
\partial_t n(\mathbf{x}, \mathbf{k}, t) + \mathbf{V}^g(\mathbf{k}) \cdot \frac{\partial}{\partial \mathbf{x}} \, n(\mathbf{x}, \mathbf{k}, t)
$$

$$
= \int_0^t dt_1 \left\{ \frac{\partial}{\partial x_r} \, \mathsf{L}_{r|s}(t - t_1) \, \frac{\partial}{\partial x_s} + \frac{\partial}{\partial x_r} \, \mathsf{L}_{r|k_s}(t - t_1) \, \frac{\partial}{\partial k_s} \right.
$$

$$
\left. + \frac{\partial}{\partial k_r} \, \mathsf{L}_{k_r|s}(t - t_1) \, \frac{\partial}{\partial x_s} + \frac{\partial}{\partial k_r} \, \mathsf{L}_{k_r|k_s}(t - t_1) \, \frac{\partial}{\partial k_s} \right\} n(\mathbf{x}, \mathbf{k}, t_1).
$$

$$
(16.46)
$$

This will be called the (true) **Wave Kinetic Equation** (BALESCU 2003). It is a *non-Markovian "advection-double-diffusion" equation,* describing two coupled diffusion processes, in x-space and in k-space, combined with propagation (advection) in x-space. Eq. (16.6) contains the precise formulation of what Diamond and his group call the "random walk in k-space". These authors, however, consider neither the coupling of this process with the diffusion in x-space, nor the non-Markovian character of these processes.

Eq. (16.6) introduces four 2×2 Lagrangian correlation tensors, defined as follows:

$$
\mathsf{L}_{r|s}(t - t_1) = \left\langle \overline{V}_r[\mathbf{x}, t] \, \overline{V}_s[\mathbf{x}(t|t_1), t_1] \right\rangle, \tag{16.47}
$$

$$
\mathsf{L}_{k_r|s}(t - t_1) = \left\langle \overline{V}_r[\mathbf{x}, t] \, \overline{W}_s[\mathbf{x}(t|t_1), \mathbf{k}(t|t_1), t_1] \right\rangle, \tag{16.48}
$$

$$
\mathsf{L}_{r|k_s}(t - t_1) = \left\langle \overline{W}_r[\mathbf{x}, \mathbf{k}, t] \, \overline{V}_s[\mathbf{x}(t|t_1), t_1] \right\rangle, \tag{16.49}
$$

$$
\mathsf{L}_{k_r|k_s}(t - t_1) = \left\langle \overline{W}_r[\mathbf{x}, \mathbf{k}, t] \, \overline{W}_s[\mathbf{x}(t|t_1), \mathbf{k}(t|t_1), t_1] \right\rangle. \tag{16.50}
$$

Assuming stationary turbulence, these functions only depend on the difference of the two times. If Eq. (16.6) could be Markovianized, it would introduce four asymptotic diffusion tensors, defined as usual:

$$
\mathsf{D}_{k_r|k_s} = \int_0^\infty dt \, \mathsf{L}_{k_r|k_s}(t), \tag{16.51}
$$

and similar definitions for the three other tensors. The Markovianization process is, however, not necessarily justified in general (see Sec. 14.9).

---

[9] In obtaining this equation we used the following commutation properties: $\overline{\mathbf{V}} \cdot (\partial/\partial \mathbf{x}) F = (\partial/\partial \mathbf{x}) \cdot (\overline{\mathbf{V}} F)$ and $\overline{\mathbf{W}} \cdot (\partial/\partial \mathbf{k}) F = (\partial/\partial \mathbf{k}) \cdot (\overline{\mathbf{W}} F)$, which both result from $(\partial/\partial \mathbf{x}) \cdot \overline{\mathbf{V}} = 0$.

16.7   The DCT Method for Zonal Flows

As follows from the discussion of Sec. 14.4, an essential role in the nature of turbulent phenomena in a regime such as shown in Fig. 16.2 is played by the trapping processes in the rugged potential landscape. It is therefore natural to study these phenomena by using the idea of the *decorrelation trajectory*, developed in the previous chapters. This method requires an extension, which will be the subject of the present chapter.

In order to take into account the different status of the radial ($x-$) and the poloidal ($y-$) directions as described above, we consider an *anisotropic* primary Eulerian potential correlation. Instead of the function $\mathcal{E}(X, Y, t)$ considered in Chap. 14, which introduced a single correlation length $\lambda_\perp$, we now introduce a function depending on two different correlation lengths; thus, denoting by X, Y the dimensional coordinates:[10]

$$\mathcal{E}(X, Y) \approx 0, \quad for \ |X| \gg \lambda_x,$$
$$\mathcal{E}(X, Y) \approx 0, \quad for \ |Y| \gg \lambda_y. \tag{16.52}$$

In the "zonal flow situation" it will be assumed that $\lambda_y \gg \lambda_x$. This assumption introduces an additional dimensionless parameter into the problem. As a result, the natural dimensionless variables are:[11]

$$x = \frac{X}{\lambda_x}, \quad y = \frac{Y}{\lambda_y}, \quad \theta = \frac{t}{\tau_c}, \quad \phi \to \varepsilon\phi. \tag{16.53}$$

In the explicit numerical calculations of the forthcoming sections we will use the following specific form (anisotropic Ornstein-Uhlenbeck process), replacing Eq. (14.30):

$$\mathcal{E}(x, y) = \exp\left(-\frac{x^2 + y^2}{2}\right) = \exp\left(-\frac{X^2}{2\lambda_x^2} - \frac{Y^2}{2\lambda_y^2}\right),$$
$$T(\theta) = \exp\left(-\theta\right) = \exp\left(-\frac{t}{\tau_c}\right). \tag{16.54}$$

The V-Langevin equations (16.38), (16.39) (in the forward time direction) are:

$$\frac{dx(\theta)}{d\theta} = K_d v_x^g[\mathbf{k}(\theta)] - K \left.\frac{\partial\phi(x, y, \theta)}{\partial y}\right|_{\mathbf{x}=\mathbf{x}(\theta)}$$
$$\equiv K_d v_x^g[\mathbf{k}(\theta)] + K v_x[\mathbf{x}(\theta), \theta], \tag{16.55}$$

[10] In the paper BALESCU 2003 only the isotropic situation $\lambda_x = \lambda_y = \lambda_\perp$ was considered.
[11] Note that the reduced time θ is different from the one used in Chap. 14: it lacks a factor K.

$$\frac{dy(\theta)}{d\theta} = K_d v_y^g[\mathbf{k}(\theta)] + K \left. \frac{\partial \phi(x,y,\theta)}{\partial x} \right|_{\mathbf{x}=\mathbf{x}(\theta)}$$

$$\equiv K_d v_y^g[\mathbf{k}(\theta)] + K v_y[\mathbf{x}(\theta), \theta], \tag{16.56}$$

$$\frac{dk_x(\theta)}{d\theta} = K k_x(\theta) \left. \frac{\partial^2 \phi(x,y,\theta)}{\partial x \partial y} \right|_{\mathbf{x}=\mathbf{x}(\theta)} - K \Lambda^{-1} k_y(\theta) \left. \frac{\partial^2 \phi(x,y,\theta)}{\partial x^2} \right|_{\mathbf{x}=\mathbf{x}(\theta)}$$

$$\equiv K w_x[\mathbf{x}(\theta), \mathbf{k}(\theta), \theta], \tag{16.57}$$

$$\frac{dk_y(\theta)}{d\theta} = K \Lambda k_x(\theta) \left. \frac{\partial^2 \phi(x,y,\theta)}{\partial y^2} \right|_{\mathbf{x}=\mathbf{x}(\theta)} - K k_y(\theta) \left. \frac{\partial^2 \phi(x,y,\theta)}{\partial x \partial y} \right|_{\mathbf{x}=\mathbf{x}(\theta)}$$

$$\equiv K w_y[\mathbf{x}(\theta), \mathbf{k}(\theta), \theta]. \tag{16.58}$$

These equations will be solved with the following initial condition:

$$\mathbf{x}(0) = \mathbf{0}, \quad \mathbf{k}(0) = \mathbf{k}^0. \tag{16.59}$$

It is important to take an initial wave vector of non-zero length: an initial $|\mathbf{k}^0| = 0$ would remain zero for all times. Moreover, physically, the dimensionless $k \geq 1$. In Chap. 14 there appeared a single characteristic dimensionless parameter, the Kubo number, related to the characteristics of the turbulence. In the present problem, there appear two additional dimensionless parameters, one related to the group velocity (containing the gradient length) and one related to the anisotropy; the Kubo number is also conveniently modified. We thus use in the present chapter the following definitions.

- *Kubo number*:

$$K = \frac{c\varepsilon}{B_0} \frac{\tau_c}{\lambda_x \lambda_y}, \tag{16.60}$$

- *Diamagnetic Kubo number*:

$$K_d = \frac{\tau_c}{\lambda_x} V_{ds}, \tag{16.61}$$

- *Anisotropy parameter*:

$$\Lambda = \frac{\lambda_x}{\lambda_y}. \tag{16.62}$$

The three scaled velocities $\mathbf{v}^g$, $\mathbf{v}$ and $\mathbf{w}$ are defined by Eqs. (16.55) - (16.58); they are related to the corresponding dimensional quantities as follows:

$$V_x^g = \frac{\lambda_x}{\tau_c} K_d \, v_x^g, \qquad V_y^g = \frac{\lambda_y}{\tau_c} K_d \, v_y^g,$$

$$\overline{V}_x = \frac{\lambda_x}{\tau_c} K \, v_x, \qquad \overline{V}_y = \frac{\lambda_y}{\tau_c} K \, v_y,$$

$$\overline{W}_x = \frac{1}{\rho_s \tau_c} K \, w_x, \qquad \overline{W}_y = \frac{1}{\rho_s \tau_c} K \, w_y. \qquad (16.63)$$

(For compactness, we drop the overbar on the dimensionless large-scale fluctuating velocities). The components of the dimensionless unperturbed group velocity are:

$$v_x^g(\mathbf{k}) = -\frac{2k_x k_y}{(1 + k_x^2 + k_y^2)^2}, \qquad v_y^g(\mathbf{k}) = \Lambda \, \frac{1 + k_x^2 - k_y^2}{(1 + k_x^2 + k_y^2)^2}. \qquad (16.64)$$

The transport problem requires the evaluation of the four Lagrangian correlations (16.47) - (16.50). We shall concentrate here on the last one, which introduces all the new features; the three others are treated in a completely similar fashion. Its corresponding dimensionless form is defined in analogy with Eq. (14.19):

$$\mathsf{L}_{k_r|k_s}(t) = \frac{1}{(\rho_s \tau_c)^2} \, \mathcal{L}_{k_r|k_s}(t/\tau_c). \qquad (16.65)$$

Using Eq. (16.50) we obtain:[12]

$$\mathcal{L}_{k_r|k_s}(\theta) = K^2 \int d\mathbf{x} \, d\mathbf{k} \, \left\langle w_r(0, \mathbf{k}^0, 0) \, w_s(\mathbf{x}, \mathbf{k}, \theta) \, \delta[\mathbf{x} - \mathbf{x}(\theta)] \delta[\mathbf{k} - \mathbf{k}(\theta)] \right\rangle.$$

$$(16.66)$$

In order to prepare the way for the DCT method, we first need to define the Eulerian statistics of the problem. Note that all fluctuating quantities (including $\mathbf{v}$ and $\mathbf{w}$) are derived from the potential ϕ, hence the primary Eulerian correlation has the same form as in Eq. (14.15), but with an anisotropic function $\mathcal{E}(\mathbf{x})$, as explained above [see Eq. (16.54)]:

$$\langle \phi(0,0) \, \phi(\mathbf{x}, \theta) \rangle = \mathcal{E}(\mathbf{x}) \, T(\theta). \qquad (16.67)$$

---

[12] It should be kept in mind that $\mathcal{L}_{r|s}^{KK}(\theta)$ also depends on $\mathbf{k}^0$, which is not written down explicitly.

In addition to the quantities calculated in Eqs. (14.16), (14.2), we also need now the second derivatives of ϕ, involved in **w**. We introduce the matrix u_{rs}:

$$u_{rs}(\mathbf{x}, \theta) = \frac{\partial}{\partial x_r} \, v_s(\mathbf{x}, \theta), \qquad (16.68)$$

such that:

$$w_x(\mathbf{x}, \mathbf{k}, \theta) = - k_x \, u_{xx}(\mathbf{x}, \theta) - \Lambda^{-1} k_y \, u_{xy}(\mathbf{x}, \theta),$$
$$w_y(\mathbf{x}, \mathbf{k}, \theta) = - \Lambda \, k_x \, u_{yx}(\mathbf{x}, \theta) - k_y \, u_{yy}(\mathbf{x}, \theta). \qquad (16.69)$$

We recall here, for convenience, the definitions (14.16) and add the additional required correlations:

$$\langle \phi(0,0) \, v_j(\mathbf{x}, \theta) \rangle = \mathcal{E}_{\phi|j}(\mathbf{x}) \, T(\theta),$$
$$\langle v_r(0,0) \, v_j(\mathbf{x}, \theta) \rangle = \mathcal{E}_{r|j}(\mathbf{x}) \, T(\theta),$$
$$\langle u_{rs}(0,0) \, v_j(\mathbf{x}, \theta) \rangle = \mathcal{E}_{rs|j}(\mathbf{x}) \, T(\theta),$$
$$\langle \phi(0,0) \, u_{jm}(\mathbf{x}, \theta) \rangle = \mathcal{E}_{\phi|jm}(\mathbf{x}) \, T(\theta),$$
$$\langle v_r(0,0) \, u_{jm}(\mathbf{x}, \theta) \rangle = \mathcal{E}_{r|jm}(\mathbf{x}) \, T(\theta),$$
$$\langle u_{rs}(0,0) \, u_{jm}(\mathbf{x}, \theta) \rangle = \mathcal{E}_{rs|jm}(\mathbf{x}) \, T(\theta). \qquad (16.70)$$

We have the following relations [including Eq. (14.2)]:

$$\mathcal{E}_{\phi|x} = -\frac{\partial}{\partial y} \, \mathcal{E}, \qquad \mathcal{E}_{\phi|y} = \frac{\partial}{\partial x} \, \mathcal{E} ,$$

$$\mathcal{E}_{x|x} = -\frac{\partial^2}{\partial y^2} \, \mathcal{E}, \quad \mathcal{E}_{y|y} = -\frac{\partial^2}{\partial x^2} \, \mathcal{E}, \quad \mathcal{E}_{x|y} = \mathcal{E}_{y|x} = \frac{\partial^2}{\partial x \partial y} \, \mathcal{E},$$

$$\mathcal{E}_{rs|j} = -\frac{\partial}{\partial x_r} \, \mathcal{E}_{s|j}, \quad \mathcal{E}_{\phi|jm} = \frac{\partial}{\partial x_j} \, \mathcal{E}_{\phi|m},$$

$$\mathcal{E}_{r|jm} = \frac{\partial}{\partial x_j} \, \mathcal{E}_{r|m}, \quad \mathcal{E}_{rs|jm} = \frac{\partial}{\partial x_j} \, \mathcal{E}_{rs|m}. \qquad (16.71)$$

We note the following symmetries:

$$\mathcal{E}_{x|y} = \mathcal{E}_{y|x}, \quad \mathcal{E}_{\phi|xx} = -\mathcal{E}_{\phi|yy},$$
$$\mathcal{E}_{xx|xx} = \mathcal{E}_{yy|yy} = -\mathcal{E}_{xx|yy} = -\mathcal{E}_{yy|xx} = -\mathcal{E}_{xy|yx} = -\mathcal{E}_{yx|xy},$$
$$\mathcal{E}_{yy|xy} = \mathcal{E}_{xy|yy} = -\mathcal{E}_{xx|xy} = -\mathcal{E}_{xy|xx},$$
$$\mathcal{E}_{xx|yx} = \mathcal{E}_{yx|xx} = -\mathcal{E}_{yy|yx} = -\mathcal{E}_{yx|yy}. \qquad (16.72)$$

We define some combinations of these Eulerian correlations:

$$H_x = k_x^0 \mathcal{E}_{xx|\phi} + \Lambda^{-1} k_y^0 \mathcal{E}_{xy|\phi} \ ,$$
$$H_y = \Lambda k_x^0 \mathcal{E}_{yx|\phi} + k_y^0 \mathcal{E}_{yy|\phi} \ . \tag{16.73}$$

$$A_x = k_x \mathcal{E}_{\phi|xx} + \Lambda^{-1} k_y \mathcal{E}_{\phi|xy} \ ,$$
$$A_y = \Lambda k_x \mathcal{E}_{\phi|yx} + k_y \mathcal{E}_{\phi|yy} \ . \tag{16.74}$$

$$B_{r|x} = k_x \mathcal{E}_{r|xx} + \Lambda^{-1} k_y \mathcal{E}_{r|xy},$$
$$B_{r|y} = \Lambda k_x \mathcal{E}_{r|yx} + k_y \mathcal{E}_{r|yy} \ . \tag{16.75}$$

$$C_{x|s} = k_x^0 \mathcal{E}_{xx|s} + \Lambda^{-1} k_y^0 \mathcal{E}_{xy|s} \ ,$$
$$C_{y|s} = \Lambda k_x^0 \mathcal{E}_{yx|s} + k_y^0 \mathcal{E}_{yy|s} \tag{16.76}$$

$$A_{x|x} = k_x^0 k_x \mathcal{E}_{xx|xx} + \Lambda^{-1} k_x^0 k_y \mathcal{E}_{xx|xy} + \Lambda^{-1} k_y^0 k_x \mathcal{E}_{xy|xx} + \Lambda^{-2} k_y^0 k_y \mathcal{E}_{xy|xy},$$
$$A_{x|y} = \Lambda k_x^0 k_x \mathcal{E}_{yx|xx} + k_x^0 k_y \mathcal{E}_{yx|xy} + k_y^0 k_x \mathcal{E}_{yy|xx} + \Lambda^{-1} k_y^0 k_y \mathcal{E}_{yy|xy} = A_{y|x},$$
$$A_{y|y} = \ = \Lambda^2 k_x^0 k_x \mathcal{E}_{yx|yx} + \Lambda k_x^0 k_y \mathcal{E}_{yx|yy} + \Lambda k_y^0 k_x \mathcal{E}_{yy|yx} + k_y^0 k_y \mathcal{E}_{yy|yy}. \tag{16.77}$$

We denote by a superscript ^{0} the value of a quantity at the origin, thus:

$$\mathcal{E}^0_{..|..} = \mathcal{E}_{..|..}(\mathbf{x} = 0), \quad A_r^0 = A_r(\mathbf{x} = 0, \mathbf{k} = \mathbf{k}^0), \quad etc. \tag{16.78}$$

Next, we introduce:

$$\Delta^0 = A_{x|x}^0 A_{y|y}^0 - A_{x|y}^{0\ 2} \ , \tag{16.79}$$

$$D^0 = \Delta^0 - A_{x|x}^0 A_y^{0\ 2} - A_{y|y}^0 A_x^{0\ 2} + A_{x|y}^0 A_x^0 A_y^0 \ . \tag{16.80}$$

The following combinations will appear in the final results:

$$F_x^0 = A_{y|y}^0 A_x^0 - A_{x|y}^0 A_y^0 \,,$$

$$F_y^0 = A_{x|x}^0 A_y^0 - A_{x|y}^0 A_x^0 \,,$$

$$J_x^0 = \left(A_{y|y}^0 - A_y^{0\,2} \right) \Delta^0 - A_{x|y}^0 A_{y|y}^0 A_x^0 A_y^0 \,,$$

$$J_y^0 = \left(A_{x|x}^0 - A_x^{0\,2} \right) \Delta^0 - A_{x|x}^0 A_{x|y}^0 A_x^0 A_y^0 \,,$$

$$L^0 = - A_{x|y}^0 \Delta^0 + A_{x|x}^0 A_{y|y}^0 A_x^0 A_y^0 \,. \tag{16.81}$$

We now develop the extension of the *decorrelation trajectory (DCT)* method, following the line of Secs. 14.5 and 14.6. The basic step is the decomposition of the ensemble of realizations of the turbulent ensemble into subensembles. This decomposition is now different from the previous one, because not only the potential ϕ and its first derivatives (through **v**), but also its second derivatives (through **w**) enter the theory. We therefore define the subensemble S as the set of realizations in which the potential ϕ, the two components of the velocity v_r, and the two components of the "**k**-velocity" w_r have a given value at time 0:

$$S: \quad \phi(0,0) = \phi^0, \quad v_r(0,0) = v_r^0, \quad -k_s^0 u_{rs}(0,0) = w_r^0 \,. \tag{16.82}$$

Let $P_0(\phi^0, \mathbf{v}^0, \mathbf{w}^0)$ be the probability distribution of these initial values; then the dimensionless KK-Lagrangian correlation (16.66) is [see Eqs. (14.41) and (14.43)]:

$$\mathcal{L}_{k_x|k_x}(\theta) = K^2 \int d\phi^0 d\mathbf{v}^0 d\mathbf{w}^0 \; P_0(\phi^0, \mathbf{v}^0, \mathbf{w}^0)$$
$$\times w_x^0 \; \langle w_x[\mathbf{x}(\theta), \mathbf{k}(\theta), \theta] \rangle^S \,, \tag{16.83}$$

where $\langle ... \rangle^S$ denotes the average in the subensemble. The first goal is the calculation of the distribution function P_0. In analogy with Sec. 14.5, we obtain:

$$P_0(\phi^0, \mathbf{v}^0, \mathbf{w}^0)$$
$$= \big\langle \delta[\phi^0 - \phi(0,0)] \, \delta^2[v_m^0 - v_m(0,0)]$$
$$\delta[w_x^0 + k_x^0 u_{xx}(0,0) + \Lambda^{-1} k_y^0 u_{xy}(0,0)] \, \delta[w_y^0 + \Lambda k_x^0 u_{yx}(0,0) + k_y^0 u_{yy}(0,0)] \big\rangle$$
$$= (2\pi)^{-3} \exp\left[-\tfrac{1}{2} \left(v_x^{0\,2} + v_y^{0\,2} \right) \right] \int ds \, dp \exp\left(is\phi^0 + ip_n w_n^0 \right)$$
$$\times \exp\Big[-\tfrac{1}{2} \big\langle \big\{ s\phi(0,0) - p_x \left[k_x^0 u_{xx}(0,0) + \Lambda^{-1} k_y^0 u_{xy}(0,0) \right] $$
$$- p_y \left[\Lambda k_x^0 u_{yx}(0,0) + k_y^0 u_{yy}(0,0) \right] \big\}^2 \big\rangle \Big] \tag{16.84}$$

Note that $\mathcal{E}^0 = 1$, $\mathcal{E}^0_{r|s} = \delta_{rs}$, $\mathcal{E}^0_{\phi|r} = \mathcal{E}^0_{r|ns} = 0$. The remaining integrations are performed as in Chap. 14. The result is rather more complicated than Eq. (14.42):

$$
P_0(\phi^0, \mathbf{v}^0, \mathbf{w}^0) = \frac{1}{\sqrt{(2\pi)^5 D^0}} \exp\left\{ -\frac{1}{2} \left[\frac{\Delta^0}{D^0} \phi^{0\,2} + v_x^{0\,2} + v_y^{0\,2} \right.\right.
$$
$$
+ \frac{J_x^0}{D^0\Delta^0} w_x^{0\,2} + \frac{J_y^0}{D^0\Delta^0} w_y^{0\,2} + \frac{2L^0}{D^0\Delta^0} w_x^0 w_y^0
$$
$$
\left.\left. + \frac{2F_x^0}{D^0} \phi^0 w_x^0 + \frac{2F_y^0}{D^0} \phi^0 w_y^0 \right] \right\} \tag{16.85}
$$

Note that this expression, as well as the following expressions for the velocities, are formally identical with those of the case $\Lambda = 1$ (BALESCU 2003). The whole dependence on Λ is in the coefficients J^0, L^0, F^0, etc.

We now calculate the average velocities in the subensemble, by an extension of the method described in Sec. 14.5:[13]

$$
\langle v_n(\mathbf{x}, \theta) \rangle^S = \left[P_0(\phi^0, \mathbf{v}^0, \mathbf{w}^0) \right]^{-1} \langle v_n(\mathbf{x}, \theta)\, \delta[\phi^0 - \phi(0,0)]\, \delta^2[v_m^0 - v_m(0,0)]
$$
$$
\times\, \delta[w_x^0 + k_x^0 u_{xx}(0,0) + \Lambda^{-1} k_y^0 u_{xy}(0,0)]\, \delta[w_y^0 + \Lambda k_x^0 u_{yx}(0,0) + k_y^0 u_{yy}(0,0)] \rangle
$$
$$
\equiv v_n^S(\mathbf{x})\, T(\theta). \tag{16.86}
$$

A lengthy, but elementary calculation yields:

$$
v_n^S(\mathbf{x}) = \frac{1}{D^0} \left(\Delta^0 \mathcal{E}_{\phi|n} - F_x^0 C_{x|n} - F_y^0 C_{y|n} \right) \phi^0
$$
$$
+ \mathcal{E}_{x|n}\, v_x^0 + \mathcal{E}_{y|n}\, v_y^0
$$
$$
+ \frac{1}{D^0\Delta^0} \left(\Delta^0 F_x^0 \mathcal{E}_{\phi|n} - J_x^0\, C_{x|n} - L^0\, C_{y|n} \right) w_x^0
$$
$$
+ \frac{1}{D^0\Delta^0} \left(\Delta^0 F_y^0 \mathcal{E}_{\phi|n} - L^0\, C_{x|n} - J_y^0\, C_{y|n} \right) w_y^0. \tag{16.87}
$$

The average **k**-velocity in the subensemble is obtained in the same way:

$$
\langle w_n(\mathbf{x}, \mathbf{k}, \theta) \rangle^S = w_n^S(\mathbf{x}, \mathbf{k})\, T(\theta), \tag{16.88}
$$

with

---

[13] It should be noted that the average velocity v^S is *not* the same as the one defined in Eq. (14.44): the subensemble used here is narrower than the one of Sec. 14.5, as it requires fixed values of $\mathbf{w}^0$.

$$w_n^S(\mathbf{x}, \mathbf{k}) = \frac{1}{D^0} \left(-\Delta^0 A_n + F_x^0 A_{x|n} + F_y^0 A_{y|n} \right) \phi^0$$

$$- B_{x|n} \, v_x^0 - B_{y|n} \, v_y^0$$

$$+ \frac{1}{D^0 \Delta^0} \left(-\Delta^0 F_x^0 \, A_n + J_x^0 A_{x|n} + L^0 A_{y|n} \right) w_x^0$$

$$+ \frac{1}{D^0 \Delta^0} \left(-\Delta^0 F_y^0 \, A_n + L^0 A_{x|n} + J_y^0 A_{y|n} \right) w_y^0. \tag{16.89}$$

A strong test of these expressions is: $v_n^S(0, \mathbf{k}^0) = v_n^0$, $w_n^S(0, \mathbf{k}^0) = w_n^0$.

This deterministic DCT is now introduced in the expressions of the four Lagrangian correlation functions (16.47) - (16.50) [see Eqs. (16.83) and (14.52)]. These quantities, evaluated in the *decorrelation trajectory approximation* are:

$$\mathcal{L}_{j|n}(\theta) = K^2 \int d\phi^0 dv^0 dw^0 \, P_0(\phi^0, \mathbf{v}^0, \mathbf{w}^0) \, v_j^0 \, v_n^S[\mathbf{x}^S(\theta), \theta)] \, T(\theta),$$

$$\mathcal{L}_{j|k_n}(\theta) = K^2 \int d\phi^0 dv^0 dw^0 \, P_0(\phi^0, \mathbf{v}^0, \mathbf{w}^0) \, v_j^0 \, w_n^S[\mathbf{x}^S(\theta), \mathbf{k}^S(\theta), \theta)] \, T(\theta),$$

$$\mathcal{L}_{k_j|n}(\theta) = K^2 \int d\phi^0 dv^0 dw^0 \, P_0(\phi^0, \mathbf{v}^0, \mathbf{w}^0) \, w_j^0 \, v_n^S[\mathbf{x}^S(\theta), \theta)] \, T(\theta),$$

$$\mathcal{L}_{k_j|k_n}(\theta) = K^2 \int d\phi^0 dv^0 dw^0 \, P_0(\phi^0, \mathbf{v}^0, \mathbf{w}^0) \, w_j^0 \, w_n^S[\mathbf{x}^S(\theta), \mathbf{k}^S(\theta), \theta)] \, T(\theta). \tag{16.90}$$

The corresponding running diffusion coefficients are:

$$\mathcal{D}_{..|..}(\theta) = \int_0^\theta d\theta_1 \, \mathcal{L}_{..|..}(\theta_1). \tag{16.91}$$

We stress again the great advantage of the DCT approximation: the calculation of the Lagrangian correlations is replaced by the simpler problem of the calculation of Eulerian averages, evaluated along the deterministic decorrelation trajectory. The present problem is, however, significantly more complicated in the present case than in the case of the simple drift wave turbulence treated in Sec. 14.6. The complication comes from the intimate coupling of the equations for $\mathbf{x}^S(\theta)$ and $\mathbf{k}^S(\theta)$. Any attempt to treat separately one or the other variable would be an oversimplification. The Lagrangian correlations are now 5-fold integrals, which makes their numerical evaluation more difficult. The final result depends on more parameters, viz. the Kubo number K, the diamagnetic Kubo number K_d, the anisotropy parameter Λ and also the initial wave vector $\mathbf{k}^0$.

16.8   The Decorrelation Trajectories

We now analyze some typical DCT trajectories, obtained by numerical integration of the deterministic DCT equations of motion, which we rewrite here in a compact form:[14]

$$\frac{d\mathbf{x}}{d\theta} = K_d \, \mathbf{v}^g(\mathbf{k}) + K \, \mathbf{v}(\mathbf{x}) \, T(\theta), \qquad (16.92)$$

$$\frac{d\mathbf{k}}{d\theta} = K \, \mathbf{w}(\mathbf{x},\mathbf{k}) \, T(\theta), \qquad (16.93)$$

where the symbols were defined in Eqs. (16.54), (16.64), (16.87), (16.89).

A specified trajectory depends on the Kubo numbers K and K_d, on the anisotropy ratio Λ, on the five parameters defining the subensemble S (ϕ^0, $\mathbf{v}^0$, $\mathbf{w}^0$) and on the initial value of the wave vector $\mathbf{k}^0$. For the purpose of illustration, we choose the case of a subensemble with a moderate value of $\phi^0 = 0.2$. The weight of the corresponding trajectories is significant in the expression of the correlation functions [because of the factor P_0 in Eqs. (16.90)], and the qualitative features of the DCT are clearly exhibited in this case. The values of the initial velocities $\mathbf{v}^0$, $\mathbf{w}^0$ are chosen arbitrarily. We thus define the subensemble S chosen in the forthcoming figures by the following values of the parameters: $\phi^0 = 0.2$, $v_x^0 = v_y^0 = -0.2$, $w_x^0 = w_y^0 = 1$. The initial wave vector is chosen as: $k_x^0 = \sqrt{0.2}$, $k_y^0 = \sqrt{0.8}$ ($k^0\,{}^2 = 1$). It thus represents a wave packet that is elongated in the radial ($x-$) direction.[15]

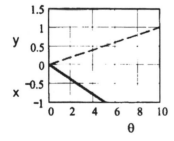
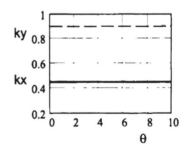

Figure 16.3. Unperturbed motion of the fictitious DCT quasiparticle, for $K = 0$. Left: Position components $x(\theta)$ [solid], $y(\theta)$ [dotted]. Right: Wave vector components $k_x(\theta)$ [solid], $k_y(\theta)$ [dotted].

[14] For convenience we suppress here and in forthcoming formulae the superscript S on the coordinates $\mathbf{x}^S$ and the wave vectors $\mathbf{k}^S$.
[15] In the paper BALESCU 2003, one can find figures of DCT trajectories corresponding to other choices of parameters.

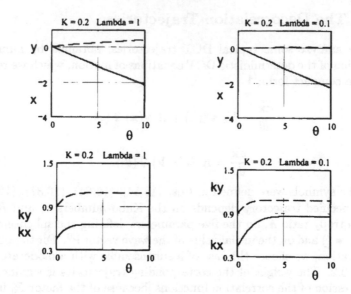

Figure 16.4. Upper graph: Components of the decorrelation position $x(\theta)$ [solid] and $y(\theta)$ [dotted]; Lower graph: Components of the wave vector $k_x(\theta)$ [solid] and $k_y(\theta)$ [dotted]. $K = 0.2$, $\Lambda = 1$, 0.1. Subensemble parameters are given in the main text.

We first show in Fig. 16.3 the unperturbed trajectory of a (fictitious) quasiparticle and of its wave vector moving along a DCT in absence of turbulence. As expected, it moves in a straight line in an oblique direction, with the group velocity $\mathbf{v}^g(\mathbf{k}^0)$ that is constant because the wave vector $\mathbf{k}$ remains constant ($K\mathbf{w}^S = 0$).

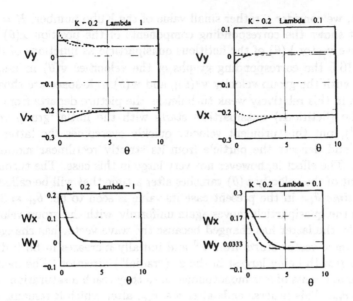

Figure 16.5. Upper graphs: Components of the total x-velocity: $V_n(\theta) = K_d v_n^g(\theta) + K v_n(\theta)$; $V_x(\theta)$: solid, $V_y(\theta)$: dashed. Also shown are the components of $\mathbf{V}^g(\theta) = K v^g(\theta)$. Lower graphs: Components of the k-velocity $W_n(\theta) = K w_n(\theta)$. $K = 0.2$, $\Lambda = 1$, 0.1.

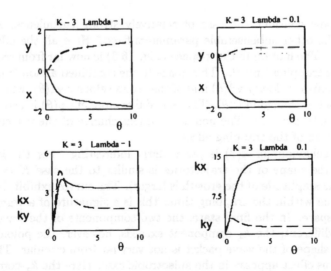

Figure 16.6. Same quantities as in Fig. 16.4, for $K = 3$. All other parameters are unchanged.

Next, we consider a rather small value of the Kubo number, $K = 0.2$. Fig. 16.4 shows the corresponding components of the position $\mathbf{x}(\theta)$ and of the wave vector $\mathbf{k}(\theta)$ of the fictitious quasiparticle as functions of time. In Fig. 16.5 the corresponding graphs of the velocities $\mathbf{v}(\theta)$ in x-space [together with the group velocity $\mathbf{v}^g(\theta)$], and $\mathbf{w}(\theta)$ in k-space are shown.

Even in this relatively weak turbulence, the picture departs from Fig. 16.3. The (fictitious) quasiparticle starts with the initial group velocity $\mathbf{v}^g(\mathbf{k}^0)$, but the turbulent velocity quickly overcomes the latter [see Fig.16.5] and deflects the particle from its strictly rectilinear motion of Fig. 16.3. The effect is, however not very large in this case. The turbulent component of the velocity $\mathbf{v}(\theta)$ vanishes after a time that will be called the *trapping time* θ_{tr}; in the present case its value is seen to be: $\theta_{tr} \approx 3$, after which the quasiparticle moves again uniformly, with the group velocity. Meanwhile, the latter has changed because the wave vector has changed.

The wave vector starts from $\mathbf{k}^0$ and initially increases in both directions, the growth being largest in the x- ("radial") direction. The increase of both components of $\mathbf{k}$ is monotonous, until they reach a saturation value after $\theta \approx \theta_{tr}$. This process ends after $\theta \approx \theta_{tr}$, after which $\mathbf{k}$ remains constant: $\mathbf{k}(\theta) \rightarrow \mathbf{k}^\infty \neq \mathbf{k}^0$ [and $\mathbf{w}(\theta) \rightarrow 0$]. This change of the wave vector corresponds to a *refraction process*. Note that even at this small value of K, the influence of the anisotropy is significant. For $\Lambda = 0.1$, the growth of the x-component of the wave vector is stronger than in the isotropic case. As a result, the ratio $|k_y/k_x|$ is smaller in the asymptotic state, meaning that the wave packet is less elongated in the radial direction than it was at the start.

We now consider the case of relatively strong turbulence, $K = 3$, keeping the other subensemble parameters fixed. Here all the effects are magnified. The motion in the x-space (Fig. 16.6) is now far from rectilinear during the trapping time θ_r. This is due to the combined action (Fig. 16.7) of the turbulent velocity $\mathbf{v}^S(\theta)$ and of the group velocity $\mathbf{v}^g(\theta)$, which varies because of the change in $\mathbf{k}(\theta)$. The poloidal component $y(\theta)$ is less affected in the anisotropic case. The non-monotonous change of the velocity $\mathbf{V}(\theta)$ is a signature of the trapping effect.

The behaviour of $\mathbf{k}(\theta)$ is particularly interesting. For the isotropic situation the shape of the trajectories is similar to the case $K = 0.2$ (although the amplitude of the growth is larger). The curves exhibit, however, a maximum within the trapping time: this is a signature of a *trapping effect* in k-space. In the final state, the two components of the wave vector are very different; the k_x-component exceeds, however, the poloidal one. The final shape of the wave packet is not very far from circular. The most conspicuous effect appears in the anisotropic case. Here the k_x-component increases rapidly, and crosses the k_y-component. In the final state ($\theta > \theta_{tr}$) the ratio $|k_y/k_x| \ll 1$. Thus, the shape of the wave packet has changed radically, from a moderate elongation in the x-direction to a large elonga-

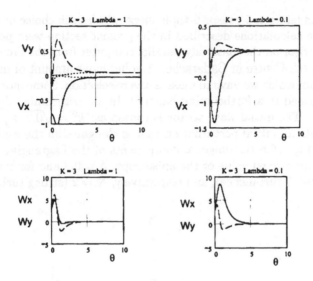

Figure 16.7. Same quantities as in Fig. 16.5, for $K = 3$. All other parameters are unchanged.

tion in the y-direction. This can be interpreted as the *fragmentation of the wave packet in the radial direction and the generation of a zonal flow*. This picture corresponds precisely to the qualitative discussion of Sec. 16.1. We see that the anisotropy has a very important influence in this process.

The results obtained in the four cases discussed here refer to the behaviour of the quasiparticles in a *single subensemble*. They should be considered as an indication that zonal flow generation is possible under certain conditions, especially in the case of an anisotropic strongly turbulent state. It must be stressed, however, that the behaviour of the quasiparticles can be very different in different subensembles $(\phi^0, \mathbf{v}^0, \mathbf{w}^0)$ and for different values of the external parameters (K, K_d, Λ).[16] A final statement about the zonal flow generation process can only be made when all the subensembles are superposed in the calculation of the diffusion coefficients. This is the object of the next Section.

16.9 Diffusion Coefficients

The calculation of the Lagrangian velocity correlations, Eq. (16.90), and of the corresponding running diffusion coefficients (16.91) require very long

[16] See, *e.g.*, the figures of BALESCU 2003, corresponding to very different values of the parameters.

computations (sixteen 5-, resp. 6-tuple integrals for each choice of parameters). All the calculations described in the present section were performed by Iulian Petrişor on the massively parallel computer from the French CEA Fusion Research Centre in Cadarache. A voluminous amount of data were obtained, from which we have to make a severe selection. Some more results will be published in a forthcoming paper.[17] In all forthcoming figures we take $K_d = 1$. The initial wave vector is chosen as: $k_x^0 = \sqrt{0.2}$, $k_y^0 = \sqrt{0.8}$: this represents an initial perturbation that is elongated in the x-direction. We show in Fig. 16.8 the diagonal components of the Lagrangian velocity correlations for a fixed value of the anisotropy, $\Lambda = 0.2$ and for two values of $K = 0.2$ (weak turbulence), and respectively, $K = 2$ (strong turbulence).

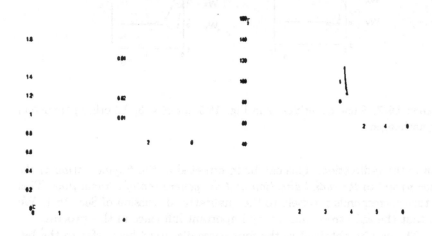

Figure 16.8. Diagonal Lagrange velocity correlations *vs.* time, for $\Lambda = 0.2$. Left figure: $K = 0.2$, Right figure: $K = 2$. The "large" curve is $\mathcal{L}_{k_x|k_x}(\theta)$; the curves in the insert represent $\mathcal{L}_{k_y|k_y}(\theta)$ [dashed], $\mathcal{L}_{x|x}(\theta)$ [solid] and $\mathcal{L}_{y|y}(\theta)$ [dash-dotted].

The first obvious feature is the much larger value of $\mathcal{L}_{k_x|k_x}$ compared to the three other quantities (we therefore included a zoom in order to visualize the latter). The shapes of all these functions are very similar to what could be expected by comparison with Fig. 14.9. For small K, the correlations decay almost exponentially, whereas for large K the trapping effect produces a minimum. As a result, the long time tail of this function is negative; its contribution will be subtracted from the short-time contribution in the integral (16.51) defining the running diffusion coefficient. The decrease of the latter is visible in Fig. 16.9.

[17] Persons interested in more detailed data may address directly Dr. Petrisor at ipetri@central.ucv.ro

The behaviour of the running diffusion coefficients at fixed Λ is thus quite similar to the one described in Sec. 14.7 for the simple electrostatic problem. They start linearly in time for short time (ballistic regime); for small K they tend monotonously toward an asymptotic constant, the asymptotic diffusion coefficients $\mathcal{D}_{..|..}$ (Fig. 16.9). For large values of K, the coefficients are larger; they go through a maximum, and then decrease, as a result of trapping, before reaching their asymptotic value. For very large values of K the system tends toward a subdiffusive regime.

In order to get a global picture, we show in Fig. 16.10 all sixteen running diffusion coefficients. Note that all diagonal coefficients are definite positive for all times; the non-diagonal ones have various signs. This is in agreement with non-equilibrium thermodynamics, which requires only the diagonal transport coefficients to be definite positive, in order not to violate the second law. We also note that, for the chosen values of K and Λ, the coefficient $\mathcal{D}_{k_x|k_x}(\theta)$ is by one or two orders of magnitude larger than all the others at all times.

We now consider the influence of the anisotropy factor on the two most relevant quantities for our purpose: the diagonal coefficients $\mathcal{D}_{k_x|k_x}(\theta)$ and $\mathcal{D}_{k_y|k_y}(\theta)$. In Figs. 16.11 these functions are represented for a fixed value of $K = 0.2$ (weak turbulence) and four values of Λ varying between $\Lambda = 1$ (isotropy) and $\Lambda = 0.1$ (strong anisotropy: $\lambda_x < \lambda_y$). The effect of the anisotropy is most striking in these two cases. For an isotropic system, the two coefficients are not very different (0.06, resp. 0.04); but for decreasing values of Λ (increasing anisotropy) the two coefficients vary in

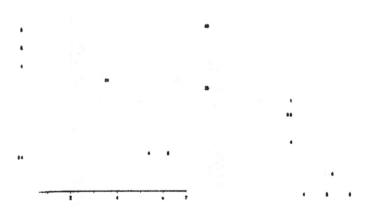

Figure 16.9. Diagonal running diffusion coefficients vs. time. $\Lambda = 0.2$. Left picture: $K = 0.2$, right picture: $K = 2$. The "large" curve represents $\mathcal{D}_{k_x|k_x}(\theta)$; the curves in the insert represent $\mathcal{D}_{k_y|k_y}(\theta)$ [dashed], $\mathcal{D}_{x|x}(\theta)$ [solid] and $\mathcal{D}_{y|y}(\theta)$ [dash-dotted].

opposite directions. $\mathcal{D}_{k_x|k_x}(\theta)$ increases strongly, reaching asymptotically a value more than ten times larger than in the isotropic case (6.96); on the contrary, $\mathcal{D}_{k_y|k_y}(\theta)$ *decreases*, reaching the value 0.03 for $\Lambda = 0.1$.

This effect is equally visible in the case of strong turbulence, $K = 2$, represented in Figs. 16.12. But here an interesting new feature appears, most clearly visible in the right Fig. 16.12. In the isotropic case ($\Lambda = 1$), the curve has the usual shape exhibiting a maximum, after which it is drawn down by the effect of *trapping*, finally reaching its finite asymptotic value. For small values of Λ, a competition appears between the effect of trapping which tends to draw the curve downwards, and the effect of the anisotropy which, in this case, draws the curve upwards. In the asymptotic state, we still note that $\mathcal{D}_{k_x|k_x} > \mathcal{D}_{k_y|k_y}$.

In Figs. 16.13 we show the asymptotic diffusion coefficients $\mathcal{D}_{k_x|k_x}$, resp. $\mathcal{D}_{k_y|k_y}$, as functions of K, for four values of Λ, in log-log representation. Each of the curves has the same general behaviour as in the electrostatic case, Fig. 14.10. They all start with the initial slope $= 2$, which is the signature of the quasilinear regime. For $K > 1$ the curves become almost flat (slope ≈ 0.1):[18] this is quite comparable to the electrostatic behaviour with a Gaussian Eulerian potential autocorrelation. The new feature (as

[18] We cannot ascertain that this is really the final value of the slope for large K. In order to be quite sure, the computations should be extended for even larger values; but these calculations are consuming very long times, which are out of reach of our computer facilities.

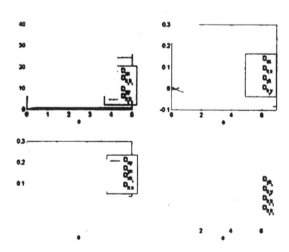

Figure 16.10. The sixteen components of the running diffusion tensor $\mathcal{D}_{..|..}(\theta)$, for $K = 2$, $\Lambda = 0.2$. Note the different scales on the four graphs.

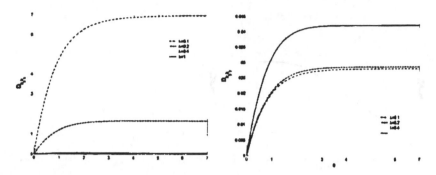

Figure 16.11. Running diffusion coefficients $\mathcal{D}_{k_x|k_x}(\theta)$ (left picture) and $\mathcal{D}_{k_y|k_y}(\theta)$ for $K = 0.2$. Four values of Λ are represented: $\Lambda = 1$ (solid), $\Lambda = 0.4$ (dotted), $\Lambda = 0.2$ (dashed), $\Lambda = 0.1$ (dash-dotted).

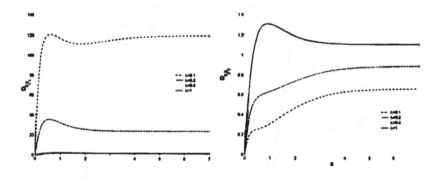

Figure 16.12. Running diffusion coefficient $\mathcal{D}_{k_x|k_x}(\theta)$ (left picture) and $\mathcal{D}_{k_y|k_y}(\theta)$ (right picture) for $K = 2$ and four values of Λ. Same symbols as in Fig. 16.11.

expected from the preceding analysis) is the following: As Λ varies from 1 to 0.1, the curves in the left Fig. 16.13 ($\log \mathcal{D}_{k_x|k_x}$) move *upwards*, while the curves in the right Fig. 16.13 ($\log \mathcal{D}_{k_y|k_y}$) move *downwards*. As a result, the ratio $\mathcal{D}_{k_x|k_x}/\mathcal{D}_{k_y|k_y}$ increases dramatically for strong anisotropy, for all values of K (Fig. 16.14).

This figure encompasses the main result of the present chapter. *In an anisotropic state of turbulence, the diffusion coefficient $\mathcal{D}_{k_x|k_x}$ can become as much as 250 times larger than $\mathcal{D}_{k_y|k_y}$. This effect produces precisely the fragmentation of structures in the x- (radial) direction and their elongation in the y- (poloidal) direction.* This is a formal confirmation of the situation

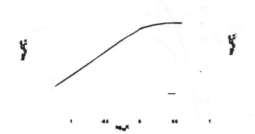

Figure 16.13. Asymptotic diffusion coefficients $\mathcal{D}_{k_x|k_x}$ (left picture) and $\mathcal{D}_{k_y|k_y}$ as functions of K for four values of Λ [$\Lambda = 1$ (solid), $\Lambda = 0.4$ (dotted), $\Lambda = 0.2$ (dash-dotted), $\Lambda = 0.1$ (dashed)]; $\log - \log$ representation.

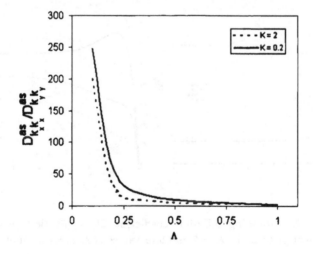

Figure 16.14. Ratio $\mathcal{D}_{k_x|k_x}/\mathcal{D}_{k_y|k_y}$ as a function of Λ, for two values $K = 0.2$, $K = 2$.

represented schematically in Fig. 16.1. We thus exhibit here *an internal mechanism of generation of a zonal flow (in our definition of this term)*.

We limit here our discussion of this problem. The mechanism described here is probably not the only possible one. On the other hand, the diffusion cannot go on indefinitely (lest $\langle k_x^2(\theta) \rangle$ would grow indefinitely). A nonlinear saturation mechanism must stop this growth. This complex problem is still controversial; we shall not discuss it here.

Chapter 17

Nonlocal Transport

17.1 Confinement Degradation

The refinement of experimental diagnostics in fusion devices, as well as the steady increase of computing power of late generation computers during the last decades led to a more and more detailed description of transport processes in high-temperature magnetically confined plasmas. The emerging picture is, not surprisingly, becoming more and more complex. A number of unexpected features appeared in these studies. In the present chapter we will present some of the most striking experimental results, and then briefly analyze some of the theoretical attempts at explaining at least some aspects of the latter.

A first result, which was discovered in the "intermediate" generation of tokamaks, and confirmed ever since, is the *confinement degradation* phenomenon. It is customary to characterize the practical performance of a fusion device by a global scalar quantity τ_E, defined by writing the following energy balance equation (ITOH et al. 1999, STROTH 1998):

$$\frac{dW}{dt} = S_E - \frac{W}{\tau_E}, \qquad (17.1)$$

where W is the total plasma energy (the integral of the internal energy density over the relevant volume of plasma), and the source term S_E represents the external power supply to the plasma. The second term in the right hand side represents (very schematically) the energy losses due to transport across the confining magnetic field. In a steady regime, the *energy confinement time* τ_E is thus defined as:

$$\tau_E = \frac{W}{S_E}. \qquad (17.2)$$

This quantity contains (in a very concealed way) information about the transport mechanisms present in the plasma. An analogous definition

417

holds for the *particle confinement time* τ_P, in terms of the total particle number N and of the external particle supply S_P:

$$\tau_P = \frac{N}{S_P}. \tag{17.3}$$

As progressively stronger auxiliary external heating sources were applied to tokamaks, be it as neutral beam injection or as radiofrequency heating procedures of various kinds, the *energy confinement time* τ_E appeared to become shorter and shorter, as compared to the prediction from the Ohmic heating regime of the early tokamaks. One would have hoped that the total energy would increase proportionally to, or faster than the heating source intensity; but the contrary appeared to be true. Moreover, the scaling of τ with the size of the system, predicted as $\tau_{P,E} \propto a^2$ ($a =$ minor radius of the torus) by the neoclassical transport theory was found to be wrong: experiments indicate rather a slower increase with the size of the system: $\tau \propto a^\beta$, $\beta < 2$. All this was bad news for fusion. In spite of partial explanations, the theoretical analysis of this confinement degradation process - which is strongly linked to anomalous transport theory - is still a largely open problem. Fortunately, subsequent experimental investigations led to the discovery of new regimes of "enhanced confinement", the most celebrated one being the *H-mode*. The latter is connected with the formation of a *transport barrier*, a problem discussed in Chap. 16.

17.2   Perturbative Transport Experiments

The most important results for anomalous transport in fusion devices came from the systematic development and application of *perturbative techniques*, whose principle was described in the Introduction (Sec. 1.4). The first studies of this kind provided a more complete and precise picture of the transport processes in fusion plasmas. It can be said that by this technique the transport equations in fusion plasmas were inscribed in the framework of nonequilibrium thermodynamics (DE GROOT & MAZUR 1984). These techniques and the first results are very clearly described and analyzed in LOPES CARDOZO 1995.

The big surprise appeared in a paper of the team of the University of Texas at Austin, directed by K. W. Gentle and working on the TEXT tokamak (GENTLE et al. 1995). The paper is provocatively entitled: *An experimental counterexample to the local transport paradigm*. The result was confirmed by the TFTR team of Princeton (KISSICK et al. 1996) and subsequently by many other works. A very clear discussion is given by CALLEN & KISSICK 1997.

The basic experiment can be described as follows. A radiating impurity, in particular carbon atoms, is rapidly injected into the plasma edge of a tokamak by the laser ablation technique. The latter allows precise timing, sharp onset and controllable size of the perturbation. The result is an initial sharp drop of the electron temperature at the edge. The fate of this perturbation is then followed in time at various radial distances from the magnetic axis. The result of the experiment is shown in Fig. 17.1.

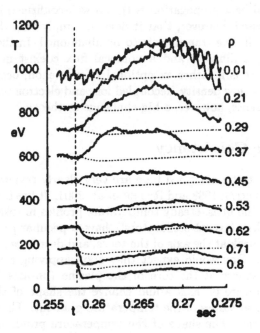

Figure 17.1. Data for the electron temperature from a number of radii across the plasma, following the cold pulse. Predictions from a conventional transport model are shown as dashed lines. (Reprinted with permission from GENTLE *et al.* *Phys. Plasmas* 2 2292 (1995). Copyright 1995, American Institute of Physics.)

From the edge (lowest curve) up to $\rho = 0.5$, everything seems "normal". The sudden cooling at the edge is progressively diffusing inward, and the temperature slowly reverts towards its initial value. But the surprise appears at small radii, $\rho < 0.5$: the plasma reacts *very rapidly* (almost instantaneously!) to the cold pulse by a *strong rise of the temperature*. It is found that edge *heating* causes a complementary response: the central temperature decreases in this case. No diffusive process can produce such a rapid propagation of the signal to the large distance separating the core from the edge. We are thus in presence of a *nonlocal transport effect*. The latter is defined (CALLEN & KISSICK 1997) as "any 'prompt' perturbation response at a large distance in comparison with the distance from the perturbation source over which diffusive local transport changes can propagate in the elapsed time Δt after the perturbation, $\sim \sqrt{\chi\, \Delta t}$".

The change of polarity of the perturbation is the other mysterious feature coming out of these experiments. It does not appear, however, to be a universal feature. Experiments at JET and RTP tokamaks have produced core cooling in response to edge cooling.

The nonlocal heat propagation is thus a well-confirmed phenomenon. It should be stressed, however, that it does not imply that the usual paradigm of local, diffusive transport is to be abandoned. In the experiment described above, a diffusive and a nonlocal flux coexist in the plasma. Nonlocal transport is a rather exceptional phenomenon, occurring under certain conditions (low density, decoupled ion- and electron temperatures). Its existence presents, however, a big challenge to theory.

17.3   Profile Consistency

The experiments to be described here are very clearly reviewed in two recent papers by STROTH 1998 and RYTER et al. 2001. It has been known for many years that the quasi-steady temperature profiles in tokamaks and in other fusion devices (except stellarators!) have a peculiar property. Their shape presents a peak at the axis of the torus, and decreases monotonically toward the edge. One would expect that upon applying a well localized auxiliary power source at the core of the plasma (*on-axis heating*)[1] the increased temperature on axis would change the shape of the profile, in particular, the local values of the temperature gradient. This is not, however, what happens. The shape of the temperature profile is remarkably independent of the amount of injected power in a given machine: the curves are simply shifted according to the edge temperature. This behaviour is found in most tokamaks, of various sizes (from 0.7 m to 3.0 m), with various particle densities and various magnetic fields. A typical illustration is given in Fig. 17.2, chosen among six similar examples given in RYTER et al. 2001. This phenomenon has been called *"profile consistency"*, or *"profile resilience"*, or *"profile stiffness"*.

A qualitative interpretation can be described as follows. The robust temperature profile is the result of some nonlinear process of self-organization. The latter introduces the necessity of a *critical temperature gradient* $(\nabla T)_c$. When the deviation of the temperature gradient (produced by heating) exceeds the critical value, the heat transport strongly increases and brings back the profile to its initial value. The existence of

---

[1] This localized heating can be realized, for instance, by electron cyclotron heating (ECH). By this method, energy originating from an external source is injected with a well defined frequency. The latter is chosen to equal the electron Larmor eigenfrequency of the plasma at a well defined position [determined by the value of the magnetic field at that position: $|\Omega_e(r)| = eB(r)/m_e c$]. The energy is thus deposited only on the electrons, at a precise location (within a narrow range).

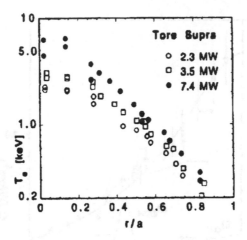

Figure 17.2. T_e profile for central heating for the Tore Supra tokamak (Cadarache), for three values of the injected power. (From RYTER et al 2001)

critical values of the temperature gradient should not be surprising. We know from Sec. 3.6 that the ITG mode (and the similar ETG mode) becomes unstable above a critical value of η_i (hence of ∇T_i), thus enhancing the transport. The detailed mechanism of the self-organization process remains, however, to be explained.

It was, of course, natural to perform also experiments in which the heating is produced *off-axis*: this is quite possible with modern technology, such as electron cyclotron heating, as explained above. It was found that the profile consistency applies just as well in this case: the temperature profiles found here are identical with those obtained with on-axis heating! The first experiments of this type were done by LUCE et al. 1992, PETTY & LUCE 1994 on the DIII-D tokamak (San Diego). This result is not easily interpreted. If the transport were of diffusive type, a simulation of the diffusion equation shows that the temperature profile would be almost flat in the region located between the magnetic axis and the point where energy is injected; outside of this region there would be a monotonous decrease (see Sec. 17.7). Instead of this, one finds the peaked "consistent" profile.

This means that energy is very quickly transported to the core, *i.e. in the direction of increasing temperatures*. This *uphill transport* thus excludes any diffusive mechanism. It would require a convective process, which should be very rapid. We thus find a very strong connection with the problem of nonlocal transport described in the previous section. The fact that this type of transport can go both downhill and uphill is an additional enigma!

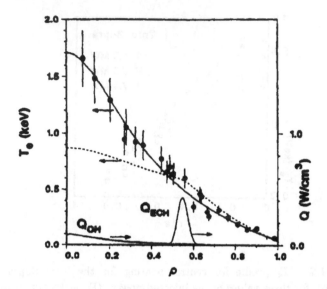

Figure 17.3. Electron temperature radial profile obtained by off-axis power deposition. The power deposition profile is shown as Q_{ECH}. The dashed curve is a simulation based on a diffusive transport model. (Reprinted with permission from LUCE et al. *Phys. Rev. Lett.* 68 52 (1992). Copyright 1992 by the American Institute of Physics.)

17.4   Attempts at a Theoretical Explanation

Since the experimental discovery of the unusual phenomena described above there have been a large number of attempts at a theoretical interpretation: we cannot quote them all here.

An idea which has attracted many authors and has been extensively exploited in many papers is the *self-organized critical state (SOC)* (BAK et al. 1987, 1988). This concept describes nonlinear dynamical systems in which the individual degrees of freedom keep each other in a more or less stable balance. This concerted action leads the system to a SOC, which is supposed to be independent of the details of the dynamics. The paradigm introduced by the original authors is the *sandpile*.[2] Initially such a system would have an arbitrary shape. As additional sand grains are added at random positions, the pile would gradually evolve to a marginally stable state, the self-organized critical state (SOC). This has the character of a stable attractor, towards which the system tends to evolve, whatever its initial condition. Additional added grains modify the local slope, making

---

[2] Bak, Tang and Wiesenfeld call the sandpile "the Ising model" of SOC!

it stiffer: this triggers an instability. When the local gradient exceeds the critical value, determined by the SOC state, the additional grains will fall under the action of gravity, perturbing the slope at neighbouring points. Additional instabilities are thus produced at these points, resulting in more grains falling. This effect is gradually transmitted to the whole system, producing an *avalanche*. The avalanches may have any size and duration; there is no intrinsic length scale in the system. This leads typically to probability distribution functions (PDF) in the form of inverse power laws. The dynamics of the sandpiles is then studied, in two or three dimensions, by means of a *cellular automaton* code, with oversimplified rules of evolution.

It has been suggested that the SOC concept could provide a model of the nonlocal transport processes in a tokamak. Among the numerous papers published on the subject, we quote two recent ones: CARRERAS et al. 1999, SANCHEZ et al. 2001. In these papers the dynamics of the plasma is replaced by a running sandpile. The instability gradients are represented by the local sandpile slope. When heat (= additional sand grains) is provided, and the local gradient is critical, an avalanche is started (= nonlocal transport).[3] The main advantage of the sandpile model is its extreme simplicity, which allows long time calculations.

In successive papers, the sandpile model has been refined and extended. An important addition was the decomposition of the heat flux into two coexisting and mutually interacting components (or channels): a *diffusive* channel and an *avalanche* channel (SANCHEZ et al. 2001): this feature is clearly apparent in the experimental results, as illustrated in Fig. 17.1.

Whereas the idea of a SOC is attractive, its illustration by a sandpile appears (to me!) oversimplified and not very convincing. The limitations of this picture were recognized by the authors themselves. There exist more physically based ways of modelling nonlocal transport: these will lead us into the realm of *continuous time random walks (CTRW)* discussed in Chap. 12. Although concepts like Lévy walks are mentioned in passing in the very early works on SOC, the first explicit attempts of introducing CTRW-like ideas in the problem appear in CARRERAS et al. 1999 and ZA-SLAVSKY et al. 2000. But what appears (to me!) to be a real breakthrough is the very recent work published in 2004, which will be discussed in Sec. 17.6.

17.5 Lévy Flights and Nonlocal Transport

In order to introduce our subject, we revisit Chap. 12. It was shown there that the concept of CTRW leads to a detailed description, not only

---

[3] CARRERAS et al. 1999 introduce also the possibility of *trapping*: in some sites the sand grains are "deeply buried" and will only move when more than one additional grain is added.

of "usual" diffusive processes, but also of "strange", *i.e.*, subdiffusive and superdiffusive processes. In that chapter particular attention was given to subdiffusive processes (in particular, the SLT-CTRW process): these are related to the trapping effects studied in Chaps. 14 - 16. The same formalism applies also to superdiffusive processes, briefly mentioned in Sec. 12.3. The latter are obvious candidates for a description of nonlocal transport. They are generated by random walks in which the particles perform, with non-zero probability, very large jumps. Such *Lévy flights* are a convenient representation of "avalanches". The direct application of the CTRW formalism presents, however, some difficulties which will be discussed now.

As shown in Sec. 12.3, a CTRW is defined [in its simplest form, Eq. (12.57)] by the PDF of the spatial jumps, $f(\mathbf{x})$, and by the PDF of waiting times $\psi(t)$ [or by their Fourier- and Laplace-transforms $\hat{f}(\mathbf{k})$, resp.$\tilde{\psi}(s)$]. We consider here only processes that have vanishing average velocity. We wish to stress the following important point.

Consider first a process defined by Eqs. (12.69), (12.70), in which $\hat{f}(\mathbf{k})$ *is analytic, and* $\tilde{\psi}(s)$ *is non-analytic*, and which is represented, for small values of the arguments by:

$$\left\{\begin{array}{ll} \hat{f}(\mathbf{k}) = 1 - (2d)^{-1}k^2 + ... & k \to 0 \\ \tilde{\psi}(s) = 1 - s^\beta + ..., & 0 < \beta < 1, \quad s \to 0 \end{array}\right. \quad \text{(Subdiffusion)}. \quad (17.4)$$

This process, which was called the *SLT-CTRW* process, is always *subdiffusive*. It leads to an equation of evolution for the density profile $n(\mathbf{x}, t)$, (12.96), that is *local in space* [*i.e.*, $\partial_t n(\mathbf{x}, t)$ depends on $n(\mathbf{x}, t')$ evaluated at the same point $\mathbf{x}$], but *nonlocal in time*, *i.e.*, *non-Markovian* [the rate of change of $n(\mathbf{x}, t)$ depends on the past history]. We rewrite it here for convenience (12.96):

$$\partial_t n(\mathbf{x}, t) = (2d)^{-1} \int_0^t dt_1 \; \phi(t_1) \; \nabla^2 n(\mathbf{x}, t - t_1). \quad (17.5)$$

Consider, on the other hand, a process defined by *a non-analytic* $\hat{f}(\mathbf{k})$, *and an analytic* $\tilde{\psi}(s)$, and which is represented by Eqs. (12.64) and (12.59):

$$\left\{\begin{array}{ll} \hat{f}(\mathbf{k}) = 1 - k^\alpha + ..., & 0 < \alpha < 2, \quad k \to 0 \\ \tilde{\psi}(s) = 1 - s + ... & s \to 0 \end{array}\right. \quad \text{(Superdiffusion)}. \quad (17.6)$$

This process leads to a density profile whose F-L transform is given in Eq. (12.65), *i.e.*, a *Lévy distribution*, a paradigm of a *superdiffusive process*. The corresponding asymptotic equation of evolution is obtained from Eq. (12.54), noting that (12.60) $\tilde{\phi}(s) = 1 - ...$:

$$s\,\widehat{\overline{n}}(\mathbf{k}, s) - 1 = -\left[1 - \widehat{f}(\mathbf{k})\right]\,\widehat{\overline{n}}(\mathbf{k}, s),$$

hence:

$$\partial_t n(\mathbf{x}, t) = -n(\mathbf{x}, t) + \int d^d x'\ f(\mathbf{x} - \mathbf{x}')\ n(\mathbf{x}', t). \tag{17.7}$$

Thus, this *pure Lévy process* obeys an equation that is *nonlocal in space, but local (Markovian) in time*. We now study some properties of this process. For simplicity, we limit ourselves to one spatial dimension ($d = 1$), which is actually of interest in the problem of radial diffusion in a toroidal geometry.

The Montroll-Weiss equation (12.57) yields the asymptotic ($s \to 0$, $k \to 0$) solution in Laplace-Fourier form (12.65); it is inverse-transformed to obtain the well-known Lévy distribution studied in Appendix D, Eq. (D.5):

$$n_{\alpha,1}(x, t) = (2\pi)^{-1} \int dk\ \exp\left(-ikx - |k|^\alpha t\right). \tag{17.8}$$

This integral can, in general, not be evaluated analytically. But we may obtain its scaling properties, as was done in Sec. 12.5. Performing the substitution $k \to p = kt^{1/\alpha}$ in the integral, we obtain:

$$n_{\alpha,1}(x, t) = t^{-1/\alpha}\ F_{\alpha,1}\left(\frac{x}{t^{1/\alpha}}\right). \tag{17.9}$$

We thus obtained a *scaling relation*, analogous to Eq. (12.82) of the SLT-case. The function $F_{\alpha,1}$ cannot be evaluated analytically:

$$F_{\alpha,1}(q) = (2\pi)^{-1} \int dp\ \exp\left(-ipq - |p|^\alpha\right). \tag{17.10}$$

The form (17.9) allows us to obtain the scaling of the mean square displacement (MSD):

$$\langle x^2(t)\rangle_{\alpha,1} = \int dx\ x^2\ t^{-1/\alpha}\ F_{\alpha,1}(xt^{-1/\alpha})$$

$$= t^{2/\alpha} \int d\xi\ \xi^2\ F_{\alpha,1}(\xi),$$

where $\xi = xt^{-1/\alpha}$. Thus:

$$\langle x^2(t)\rangle_{\alpha,1} = M_{\alpha,1}\ t^{2/\alpha}. \tag{17.11}$$

Remembering that $0 < \alpha < 2$, we see that the process is *superdiffusive* in this whole range of α, with a diffusion exponent (12.76) $\mu = 2/\alpha > 1$.

Clearly, when $\alpha = 2$, *i.e.* when the structure function $f(k)$ is Gaussian, the normal diffusive behaviour is recovered.

The two cases (17.4) and (17.6) produce opposite effects on the diffusion. We now combine the two cases: we should witness a *competition between subdiffusion and superdiffusion*. We thus assume the asymptotic forms:[4]

$$\begin{cases} \widehat{f}(k) = 1 - |k|^\alpha + ..., & 0 < \alpha < 2, \quad k \to 0 \\ \widetilde{\psi}(s) = 1 - s^\beta + ..., & 0 < \beta < 1, \quad s \to 0 \end{cases} \tag{17.12}$$

The Montroll-Weiss equation then yields, to dominant order:

$$\widehat{\widetilde{n}}_{\alpha,\beta}(k, s) = \frac{s^{\beta-1}}{s^\beta + |k|^\alpha} = s^{\beta-1} \int_0^\infty d\xi \, \exp\left[-\xi \left(s^\beta + |k|^\alpha\right)\right]. \tag{17.13}$$

The inverse Laplace-Fourier transform cannot, in general, be done analytically. But we can again determine the scaling behaviour of the density profile. In the expression of the inverse Fourier transform we use successively the substitutions: $k \to k_1 = \xi^{1/\alpha} k$ and $\xi \to \xi_1 = s^\beta \xi$:

$$\begin{aligned} \widetilde{n}_{\alpha,\beta}(x, s) &= (2\pi)^{-1} \int dk \int d\xi \, s^{\beta-1} \, \exp\left(-ikx - |k|^\alpha \xi - s^\beta \xi\right) \\ &= \int d\xi \, s^{\beta-1} \, \exp(-s^\beta \xi) \, \xi^{-1/\alpha} \, G(\xi^{-1/\alpha} x) \end{aligned}$$

and, finally:

$$\widetilde{n}_{\alpha,\beta}(x, s) = s^{(\beta-\alpha)/\alpha} \, H(x s^{\beta/\alpha}). \tag{17.14}$$

The functions appearing here are:

$$G(y) = (2\pi)^{-1} \int dk_1 \exp\left(-ik_1 \xi^{-1/\alpha} x - |k_1|^\alpha\right),$$

$$H(z) = \int d\xi_1 \, \xi_1^{-1/\alpha} \, e^{-\xi_1} \, F(\xi_1^{-1/\alpha} z). \tag{17.15}$$

From here on, the scaling of the density profile can be obtained by inverse Laplace transformation:

[4] We use here the notation γ for the exponent of s, instead of α, which was used for the SLT, Eq. (12.67), in order to avoid confusion with the diffusion index α. In the latter case, we have seen that $\alpha = \gamma$; this is no longer so in the present case. We also set $\tau_D = 1$, and $B = 1$ in Eq. (12.67), for simplicity.

hence:

$$n_{\alpha,\beta}(x,t) = \int ds \; e^{st} \; s^{(\beta-\alpha)/\alpha} \; H(xs^{\beta/\alpha}).$$

Taking $s_1 = xs^{\beta/\alpha}$ as an integration variable we find:

$$n_{\alpha,\beta}(x,t) = t^{-\beta/\alpha} \; F_{\alpha,\beta}(p), \tag{17.16}$$

with the scaling variable:

$$p = \frac{x}{t^{\beta/\alpha}}, \tag{17.17}$$

and:

$$F_{\alpha,\beta}(p) = \frac{\alpha}{\beta} \, p^{-1} \int ds_1 \; \exp\left[\left(\frac{s_1}{p}\right)^{\alpha/\beta}\right] H(s_1). \tag{17.18}$$

These equations generalize Eqs. (12.82), (12.80) and (12.81) obtained for the SLT-CTRW. The latter are recovered for $\alpha = 2$. We finally obtain the scaling of the MSD:

$$\left\langle x^2(t) \right\rangle_{\alpha,\beta} = \int dx \; x^2 \; t^{-\beta/\alpha} \; F_{\alpha,\beta}(p).$$

Substituting $x \to p$, we find

$$\left\langle x^2(t) \right\rangle_{\alpha,\beta} = M_{\alpha,\beta} \; t^\mu, \qquad \mu = \frac{2\beta}{\alpha}. \tag{17.19}$$

We now clearly see the competition between the exponents β (favouring subdiffusion) and α (favouring superdiffusion). The diffusion exponent μ is determined by the ratio of β to α. The situation is summarized as follows:

$\alpha > 2\beta$	SUBDIFFUSIVE	
$\alpha = 2\beta$	DIFFUSIVE	(17.20)
$\alpha < 2\beta$	SUPERDIFFUSIVE	

We recall the ranges:

$$0 < \alpha < 2, \qquad 0 < \beta < 1. \tag{17.21}$$

The regular Gaussian diffusive process is recovered for $\alpha = 2$, $\beta = 1$, $\mu = 1$; the SLT subdiffusive regime is obtained for $\alpha = 2$: in this case $\mu = \beta < 1$. It must thus be underscored that *a Lévy jump PDF does not, by itself, produce a superdiffusive regime: the latter requires also an appropriate form of the waiting time PDF*. A striking example is provided by the case $\alpha = 2\beta$ in which both the step size PDF and the waiting time PDF are non-standard, but which produce a diffusive regime.

We now consider the equation of evolution obeyed by the combined Lévy-SLT randomwalk defined by Eq. (17.12): it is easily obtained from Eq. (17.13) in Laplace-Fourier representation:

$$s\,\widehat{\widehat{n}}(\mathbf{k}, s) - 1 = -\,s^{1-\beta}\,|k|^{\alpha}\,\widehat{\widehat{n}}(\mathbf{k}, s). \qquad (17.22)$$

The appearance of the exponents α, β suggests that the inverse F-L transform of this equation could be represented by a *fractional differential equation*. This point of view was developed by a number of authors (CARRERAS et al 1999 for the present problem in its "sandpile version"; an excellent review paper is METZLER & KLAFTER 2000. A very clear presentation appears in the quite recent work of DEL-CASTILLO-NEGRETE et al. 2004). I continue using here, however, the formalism of integral equations, of which the fractional differential formulation is a limiting case.

Using the methodology of Sec. 12.6, the inverse Laplace transformation of Eq. (17.22) yields:

$$\partial_t\widehat{n}(k, t) = -\tfrac{1}{2}|k|^{\alpha}\,\widehat{n}(k, t) + \frac{1-\beta}{\Gamma(\beta)}\,|k|^{\alpha}\int dt_1\ t_1^{-2+\beta}\,\widehat{n}(k, t - t_1). \qquad (17.23)$$

[For $\alpha = 2$, the non-Markovian diffusion equation (12.100) is recovered.] This is a particular case of the Montroll-Shlesinger master equation (12.53). We now ask the following important question:

Can a master equation of the form (17.23) be derived from microscopic particle dynamics?

As explained at the beginning of Chap. 12, all random walk models are based on a deliberate abandonment of the laws of dynamics of complex systems, in favour of a purely stochastic picture, in which the motion of the particles is determined by transition probabilities. It would be very satisfactory if it could be shown that these models can be related to well-defined approximations of exact dynamical kinetic equations or, at least, of semi-dynamical hybrid kinetic equations. It was shown in Secs. 12.8 and 14.9 that the answer is affirmative in the case of the purely subdiffusive SLT-CTRW (*i.e.*, the case $\alpha = 2$). We now study the general mixed case: $0 < \alpha < 2,\ 0 < \beta < 1$.

Consider a V-Langevin equation of type (11.43) (in one dimension):

$$\frac{dx(t)}{dt} = v[x(t), t], \qquad (17.24)$$

where $v[x(t), t]$ is a random function, statistically defined *a priori*. A hybrid kinetic equation (HKE) can always be associated with this equation, as shown in Sec. 11.8, Eq. (11.111):

$$\partial_t f(x, t) + v(x, t) \, \nabla f(x, t) = 0. \tag{17.25}$$

(We use the compact notation: $\nabla \equiv \partial/\partial x$.) The V-Langevin equation is simply the characteristic equation of the HKE. The stochastic function f is decomposed as usual into an average part and a fluctuation: $f(x, t) = n(x, t) + \delta f(x, t)$, the HKE is separated into two equations, and the fluctuation δf is eliminated by the procedure of Sec. 14.8, with the result:

$$\partial_t n(x, t)$$
$$= \nabla \int_0^t dt_1 \, \langle v(x, t) \, v[x(t_1), t_1] \, \nabla n[x(t_1), t_1] \, \rangle \tag{17.26}$$

(the term involving the initial condition of δf is taken to be zero). Here appears again the difficulty encountered in all previous cases of this type (Secs. 6.2, 14.8). The density profile entering the integral depends on the random variable $x(t_1)$, it can therefore not be taken out of the average. The Fourier transform of the density profile is (using the solution of the Langevin equation with $x(t) = x$):

$$n[x(t_1), t_1] = \int dq \, \exp[iqx(t_1)] \, \widehat{n}(k, t_1)$$
$$= \int dq \, \exp\left\{ iq \left[x + \int_t^{t_1} dt_2 \, v[x(t_2), t_2] \right] \right\} \widehat{n}(k, t_1) \tag{17.27}$$

Upon Fourier inversion of Eq. (17.26) and substitution of (17.27), we find

$$\partial_t \widehat{n}(k, t) = \frac{i^2}{2\pi} \int_0^t dt_1 \int dx \int dq \, k \, e^{-ikx} \, q \, e^{iqx}$$
$$\times \left\langle v(x, t) \, v[x(t_1), t_1] \, \exp\left\{ iq \int_t^{t_1} dt_2 \, v[x(t_2), t_2] \right\} \right\rangle \widehat{n}(k, t_1). \tag{17.28}$$

In all previously treated cases we assumed (under certain conditions) that a *localization* was appropriate; this means that we assumed $x(t_1) \approx x$, hence $\exp\left\{ iq \left[x + \int_t^{t_1} dt_2 \, v[x(t_2), t_2] \right] \right\} \approx \exp(iqx)$, the exponential under the integral in Eq. (17.28) is put equal to 1, and we obtain:

$$\partial_t \widehat{n}(k, t) = -\frac{1}{2\pi} \int_0^t dt_1 \int dx \int dq \, kq \, e^{-ik(x-q)}$$
$$\times \langle v(x, t) \, v[x(t_1), t_1] \, \rangle \, \widehat{n}(k, t_1). \tag{17.29}$$

In the case of homogeneous turbulence, the Lagrangian velocity correlation is independent of x, and we obtain after integration over x and over q:

$$\partial_t \hat{n}(k,t) = -k^2 \int_0^t dt_1 \; \langle \, v(x,t) \, v[x(t_1), t_1] \, \rangle \; \hat{n}(k, t_1), \qquad (17.30)$$

which is none other than the Fourier transform of the *non-Markovian diffusion equation* (14.98). Two comments are important about this result.

a) It has been shown in Sec. 14.9 that, under certain conditions, *the subdiffusive SLT-CTRW reduces to a non-Markovian diffusion equation of the type (17.30)*: any continuous time random walk with $\alpha = 2$ and $0 < \beta < 1$ can be reduced asymptotically to this equation. In other words, *the localization of the kinetic equation is compatible with any random walk model of the subdiffusive SLT-CTRW class.*

b) There appears an important difference between Eq. (17.30) and Eq. (17.23) which corresponds to a mixed Lévy random walk, including the case of superdiffusion. Disregarding the first term in the right hand side of the latter, which is irrelevant for long times [see Eq. (12.99)], the coefficient in front of the time integral is $|k|^\alpha$ instead of k^2. This implies that *a superdiffusive CTRW is incompatible with the localization of the hybrid kinetic equation (17.25) or the V-Langevin equation (17.24).*

This introduces a serious difficulty. Any attempt of justification of a Lévy-type model from microscopic dynamics must involve a calculation of the very complicated average in Eq. (17.28). A series expansion of the exponential will not help: it will yield an infinite series involving all (Lagrangian!) moments of the velocity. This series contains only integral powers of the wave vector q. The only possibility of obtaining an equation of the type (17.23) is to show that in some circumstances the average involved in Eq. (17.28) is a *non-analytic function of q*. This is not excluded, but it would be extremely difficult to prove it. In conclusion, the microscopic justification of a superdiffusive Lévy-type random walk is, at present, an open problem. Nonetheless, it appears that such a process is supported by the experimental evidence. We therefore continue its study.

17.6   The van Milligen-Sanchez-Carreras Model

In an attempt to explain the strange experimental findings described in the beginning of the present chapter, the *Lévy flights* are a key element. Such processes permit very long jumps with non-negligible probability. They are, moreover, integrated in a well-studied mathematical structure, much more so than the rather naive avalanches on a sandpile.

It should first be clear that a pure Lévy flight model could not by itself be used for treating the present problem. In order to model nonlocal transport in a fusion plasma some key elements must be added to the model. This was done in the beautiful work of VAN MILLIGEN, SANCHEZ & CARRERAS 2004 a, 2004 b, henceforth abbreviated as "*vMSC*". These papers deal with particle transport; an extension to heat transport is announced for forthcoming work. We give here a brief presentation of the vMSC work; additional details will be found in the original papers.

A first element that should be included is a *boundary condition at finite distance*, more precisely at $X = 0$ and $X = 2a$, a being the minor radius of the torus. For convenience, all lengths appearing in the problem will be scaled with $2a$, in particular, the dimensionless coordinate $x = X/2a$ will be used everywhere. The boundary condition chosen is very simple: *any particle which, along its random walk, reaches the boundary either at $x = 0$ or at $x = 1$, is lost to the system.* Let us note that the role of the boundary is not trivial. In a Gaussian diffusive process there exists an intrinsic characteristic length, $\sigma = \sqrt{MSD}$; whenever $\sigma \ll 1$ the effect of the boundary inside the system is negligible, because of the exponential cutoff of the PDF. In a Lévy-type CTRW, however, there is no characteristic length, all scales contribute to the evolution, up to the total length of the system; whatever happens at the edge can therefore influence the evolution in the core.

As a consequence of the chosen boundary condition, the number of particles in the system is no longer conserved. If a steady non-equilibrium state is to be reached, the loss of particles must be compensated by a *source* of particles which fuels the system. In practice, this is achieved by experimental methods such as gas puffing or pellet injection in a tokamak. This source is described by an additional term in the master equation: $S(x, t)$, which may depend on position and on time.

At the present stage of the description, we take as a starting point the general master equation (12.51) derived by VAN MILLIGEN et al. 2004 a, in one dimension, restricted to the spatial range[5] $0 \leq x \leq 1$, and supplemented with a source term:

$$\partial_t n(x, t) = -\int_0^t dt' \phi(t - t'; x) \, n(x, t')$$

$$+ \int_0^1 dx' \, f(x - x'; x', t) \int_0^t dt' \phi(t - t'; x') \, n(x', t') + S(x, t). \quad (17.31)$$

The steady state, if any, should therefore obey the following equation, obtained by setting $\partial_t n(x, t) = 0$, and taking a time-independent source:

[5] Note that in this representation the magnetic axis is located at $x = 0.5$.

$$S(x) = n(x) \int_0^t dt' \phi(t - t'; x)$$

$$- \int_0^1 dx' \, f(x - x'; x', t) \, n(x') \int_0^t dt' \phi(t - t'; x') . \tag{17.32}$$

In order to satisfy this equation, the right hand side must be independent of t: this sets a constraint on the jump PDF f and on the function ϕ (hence on the waiting time PDF ψ). A simple way of satisfying the constraint is to take the following forms:[6]

$$f(x - x'; x', t) \to f(x - x'; x'),$$
$$\phi(t - t'; x') = \frac{1}{\tau_D} \, \delta(t - t'),$$
$$\psi(t - t'; x') = \frac{1}{\tau_D} \, \exp\left(-\frac{t - t'}{\tau_D}\right). \tag{17.33}$$

In terms of the exponents defined in Sec. 17.5, this implies $\beta = 1$. At this stage, the master equation is reduced to the form:

$$\partial_t n(x, t) = \frac{1}{\tau_D} \int_0^1 dx' \, f(x - x'; x') \, n(x', t)$$
$$- \frac{1}{\tau_D} \, n(x, t) + S(x). \tag{17.34}$$

Note that the effect of the choice (17.33) is a *Markovianization* of the master equation, which remains, however, nonlocal in space.[7] As was shown in Sec. 17.5, the latter property is indispensable for a description of Lévy flights.

Last but not least, an explanation of the experimental evidence discussed in Secs. 17.1 - 17.3 requires the inclusion of the following features. *There should be a coexistence of normal transport and nonlocal transport. The separation of the two regimes and the transition from one to another should be controlled by a critical value of the density gradient.*

It is at this point that the inhomogeneous nature of the step size PDF (*i.e.*, its dependence on x') [at variance with the homogeneous $f(x - x')$ of Eq. (12.53)] plays a crucial role. The freedom provided by this feature

---

[6] It was shown in vMSC that a more general solution is: $\phi(t - t', x') = g(x') \, \delta(t - t')$, $\psi(t - t', x') = g(x') \, \exp[-g(x') \, (t - t')]$. This position-dependent waiting time PDF is, however, not used in the sequel.

[7] This feature is complementary to the nature of the master equation for the subdiffusive SLT-CTRW process, Eq. (12.96), which is non-Markovian in time, but local in space.

includes, in particular, the possibility of a dependence on x' through the density profile itself, or through its gradient. Let us introduce the following notation for the density gradient:

$$\nabla n(x', t) \equiv \left. \frac{\partial n(x, t)}{\partial x} \right|_{x=x'}. \tag{17.35}$$

Note that this choice makes the jump PDF time-dependent:[8]

$$f(x - x'; x', t) = f[x - x'; \ \nabla n(x', t)]. \tag{17.36}$$

More important, *the choice (17.36) makes the master equation (17.34) nonlinear.* This opens up a wealth of new possibilities, but it also makes its analytical solution impossible. Another important new feature is the fact that the probability of a jump of size $x - x'$ depends on the local environment (instead of being homogeneous in space). Eq. (17.34) is now written as follows:

$$\partial_t n(x, t) = \frac{1}{\tau_D} \int_0^1 dx' \ f[x - x'; \ \nabla n(x', t)] \ n(x', t)$$

$$- \frac{1}{\tau_D} n(x, t) + S(x). \tag{17.37}$$

In order to specify the function f, we now introduce the crucial element: a (positive) *critical gradient* $(\nabla n)_c > 0$. We wish to express the idea that, *wherever the local density gradient is smaller than* $(\nabla n)_c$, *the transport is normal, diffusive; the corresponding jump PDF is a Gaussian (i.e.,* $\alpha = 2$). *On the other hand, when the local gradient is steeper than the critical value, the transport is superdiffusive, the jump PDF is a Lévy PDF; for convenience, it is taken to be a Cauchy PDF, i.e.,* $\alpha = 1$. In order to implement this idea, we introduce the following notations. The following Lévy distributions are defined in terms of the notation of Appendix D, Eqs. (D.3), (D.5):

$$P_L(x, \alpha, \sigma) = (2\pi)^{-1} \int dk \ \exp(-ikx - \sigma^\alpha |k|^\alpha)$$

$$= \int dk \ \exp(-ikx) \ \hat{n}_\alpha(\sigma k). \tag{17.38}$$

We know that $P_L(x, 2, \sigma)$ is a Gaussian PDF, and $P_L(x, 1, \sigma)$ is a Cauchy PDF, Eq. (D.8). We also introduce the abbreviation:

---

[8] The time dependence introduced by $\nabla n(x', t)$ does not preclude the existence of a stationary state, because in the latter the density profile is independent of time.

$$\zeta(x',t) = \Theta[\ |\nabla n(x',t)| - (\nabla n)_c],\qquad (17.39)$$

where $\Theta(x)$ is the Heaviside function. The final choice for the step size PDF is:

$$f_{MSC}[x - x'; \nabla n(x't)] = \zeta(x',t)\ P_L(x - x', 1, \sigma_1)$$
$$+ [1 - \zeta(x',t)]\ P_L(x - x', 2, \sigma_2).\qquad (17.40)$$

The "minimal" model is completed, for simplicity, by assuming the source to be constant in space and time:[9] $S(x,t) \to S$. The final master equation of the model is thus:

$$\partial_t n(x,t) = \frac{1}{\tau_D} \int_0^1 dx'\ f_{MSC}[x - x';\ \nabla n(x',t)\ ]\ n(x',t)$$
$$- \frac{1}{\tau_D}\ n(x,t) + S.\qquad (17.41)$$

The most outstanding feature of the vMSC model is the coexistence of two regimes, or *channels*: a *diffusive channel*, for $|\nabla n(x',t)| < (\nabla n)_c$, and a *superdiffusive channel* for $|\nabla n(x',t)| > (\nabla n)_c$, separated by a critical value of the density gradient. Note that all the ingredients of the sandpile model are present in the vMSC model too. But the mathematical framework is now made much more rigorous, by replacing the "avalanches" by Lévy flights.

17.7   Results of the vMSC Model

a. Steady State Density Profile

We now discuss the results of the numerical solution of the master equation (17.41) obtained by VAN MILLIGEN et al. 2004 a, b. The first of these results concerns the particle density profile $n(x)$ in the steady state. The values used for the parameters are: $(\nabla n)_c = 50$, $\sigma_1 = 0.04$, $\sigma_2 = 0.02$, $\tau_D = 1$. The shape of the profile varies for different values of the (constant) particle source S. Its generic form is sketched in Fig. 17.4.

The profile exhibits two separate regions. The central region[10] has a parabolic shape; the outer region (on both sides of the former) has a practically linear shape, with a constant value of the gradient.

In the central region the transport is diffusive: here the superdiffusive channel is inactive, because the gradient is everywhere smaller than the

---

[9] This condition will be dropped below, *i.e.*, in VAN MILLIGEN et al. 2004 b.
[10] Remember: the position of the magnetic axis is at $x = 0.5$.

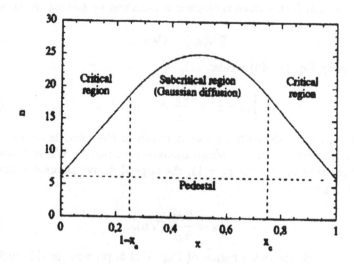

Figure 17.4. Typical steady state density profile obtained by numerical integration of Eq. (17.41). (Reprinted with permission from VAN MILLIGEN et al. *Phys. Plasmas* **11** 2272 (2004). Copyright 2004, American Institute of Physics.)

critical one. This can be shown analytically. We know that in the diffusive case ($\beta = 2$), the master equation reduces to the diffusion equation (12.29), to which a source term is added. The stationary solution of the latter is obtained from:

$$D\,\frac{d^2 n(x)}{dx^2} = -S, \qquad D = \frac{\sigma_2^2}{\tau_D}. \tag{17.42}$$

Integrating this equation, with the boundary condition $\nabla n(0.5) = 0$, we obtain:

$$\frac{dn(x)}{dx} = -\frac{\tau_D}{\sigma_2^2}\,S\,(x - 0.5). \tag{17.43}$$

At the boundaries $x = 0$, $x = 1$ there remains a residual density ("the pedestal"), which cannot be determined by the model: it will henceforth be disregarded. An additional integration of Eq. (17.43) [with $n(0) = n(1) = 0$] yields the parabolic profile visible in Fig. 17.4:

$$n(x) = \frac{\tau_D}{2\sigma_2^2}\,S\,x(1 - x). \tag{17.44}$$

As we know from Eq. (17.40), this solution is only valid as long as the local value of the gradient does not exceed the critical value ∇n_c. The position at which this value is reached is obtained by solving the condition:

$$\nabla n(x_c) = (\nabla n)_c, \qquad (17.45)$$

which, using Eq. (17.43) yields:

$$\left| x_c - \frac{1}{2} \right| = \frac{\sigma_2^2}{\tau_D} \, S^{-1} \, (\nabla n)_c \,. \qquad (17.46)$$

This position depends on the strength of the particle source. For a sufficiently weak source the whole domain is diffusive. When S reaches a value such that $x_c = 0$ (or $x_c = 1$), the superdiffusive channel is activated at the periphery:

$$S_c = \frac{2\sigma_2^2}{\tau_D} \, (\nabla n)_c. \qquad (17.47)$$

For $S > S_c$ the outer region of Fig. 17.4 is present. In this region the gradient of the density profile turns out to be constant: this is a result of the nonlinear self-regulation of the system.

b. Confinement Degradation

We now consider the problem of the scaling of the particle confinement time τ_P, which can be obtained analytically. In order to calculate this quantity by Eq. (17.3), we need to evaluate the total number of particles in the system in the steady state. When $S < S_c$, the transport is diffusive everywhere, and we obtain by straightforward integration of Eq. (17.44):

$$N = \frac{\tau_D}{12 \, \sigma_2^2} \, S, \qquad S < S_c. \qquad (17.48)$$

In the opposite limit, when $S \gg S_c$, the transport is superdiffusive everywhere,[11] and we obtain Eq. (17.41) with $f_{MSC} \to P_L(x - x', \beta, \sigma_1)$[12] and taking $S(x) \equiv S_0 \, s(x)$:

$$n(x) \approx \int_{-\infty}^{\infty} dx' \, P_L(x - x') \, n(x') + \tau_D \, S_0 \, s(x),$$

or, in Fourier representation:

$$\widehat{n}(k) = \tau_D \, \frac{S_0 \, \widehat{s}(k)}{1 - \widehat{P}_L(k)}. \qquad (17.49)$$

---

[11] Except in a small range around $x = 0.5$, in order to ensure that $\nabla n(0.5) = 0$. This yields a density profile consisting of two straight lines, with a small round junction at $x = 0.5$, see Fig. 17.6.
[12] In subsequent formulae we do not write down explicitly the arguments β, σ_1.

Using Eq. (D.5), and expanding the Lévy exponential for $\sigma_1 \ll 1$ we obtain:

$$n(x) = \frac{\tau_D}{2\pi} S_0 \int dk \, e^{ikx} \, \frac{\widetilde{s}(k)}{1 - \exp(-\sigma_1^\alpha \, |k|^\alpha)}$$

$$\approx \frac{\tau_D}{2\pi} S_0 \int dk \, e^{ikx} \, \widetilde{s}(k) \, \sigma_1^{-\alpha} \, |k|^{-\alpha}. \tag{17.50}$$

Integrating once more over x, we get the total number of particles in the steady state:

$$N = \frac{\tau_D \, S_0}{\sigma_1^\alpha} \, F \tag{17.51}$$

where the constant F (which is irrelevant for our purpose) is:

$$F = (2\pi)^{-1} \int_0^1 dx \int dk \, e^{ikx} \, \widetilde{s}(k) \, |k|^{-\alpha}.$$

We now come back to the general case in which there is a possible co-existence of the diffusive and the superdiffusive channels. The total number of particles is obtained by matching the diffusive result (17.48) with the integral taken over the superdiffusive region (if any). One thus obtains:

$$N = \begin{cases} \dfrac{\tau_D}{12 \, \sigma_2^2} \, S, & S < S_c \\[2mm] \dfrac{1}{4} (\nabla n)_c - \dfrac{\sigma_2^4}{3\tau_D^2} \, (\nabla n)_c^3 \, S^{-2}, & S \geqslant S_c \end{cases} \tag{17.52}$$

The confinement time is now readily obtained from Eq. (17.3). It should be noted, however, that the latter cannot exceed the limit provided by Eq. (17.51) for the extreme case of a pure Lévy dynamics (with $\beta = 1$). Thus:

$$\tau_P = \begin{cases} \dfrac{\tau_D}{12 \, \sigma_2^2} \, , & S < S_c \\[3mm] \max\left\{ \dfrac{1}{4} (\nabla n)_c \, S^{-1} - \dfrac{\sigma_2^4}{3\tau_D^2} \, (\nabla n)_c^3 \, S^{-3}, \, \dfrac{\tau_D}{\sigma_1} \, F \right\}, & S \geqslant S_c \end{cases} \tag{17.53}$$

This formula vividly expresses the degradation of the confinement with increasing strength of the particle source (*i.e.*, the fuelling rate). Plotting $\tau_P(S)$ we obtain Fig. 17.5.

For practical purposes (such as designing future tokamaks), an interesting information is the scaling of the confinement time with the size of

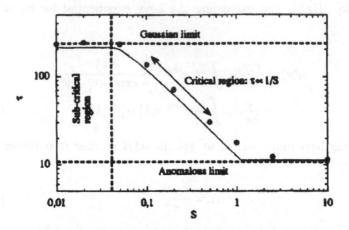

Figure 17.5. Particle confinement time as a function of the fuelling rate. (Reprinted with permission from VAN MILLIGEN et al. *Phys. Plasmas* **11** 2272 (2004). Copyright 2004, American Institute of Physics.)

the system. Remember that in our treatment all lengths were made non-dimensional upon division by $2a$, the minor diameter of the torus. Reverting to dimensional quantities, we note in particular that the parameters σ_i $(i = 1, 2)$ are expressed as $\sigma_i = \Sigma_i/2a$. We thus obtain the following scalings:

$$\tau_P \propto \begin{cases} a^2, & S < S_c \\ a^\beta & (\beta = 1), & S \gg S_c \end{cases} \tag{17.54}$$

The scaling $\tau_P \propto a^2$ is the well-known *neoclassical scaling* (see, *e.g.*, BALESCU 1988 b, Sec. 18.5). The vMSC model predicts a slower increase of the confinement time with the size of the system, $\tau_P \propto a^\alpha$, $0 < \alpha < 2$ (in the present implementation, $\alpha = 1$). It thus correctly describes all the qualitative features of the experiments

c. Perturbative Transport

Whenever the superdiffusive channel is active, the possibility exists for a particle to make large "instantaneous" jumps. This was the main motivation of adopting a Lévy-type CTRW in the construction of the vMSC model. It is, however, necessary to check the presence of nonlocal transport (as defined in Sec. 17.2) as a prediction of the final model.

Not much detail is given in VAN MILLIGEN et al. 2004 a about this problem. The authors give, however, the result of a simulation of a "cold pulse" initiated at the boundary of the stationary plasma at a given time. This perturbation propagates inward at a rate nearly three orders of magnitude faster than expected from the global confinement time. The propagation is definitely not diffusive, but rather nearly ballistic. This result is certainly in qualitative agreement with expectations. There exist, however, no experimental data concerning perturbative particle propagation in a tokamak. In order to simulate an experimental result such as Fig. 17.1, we must expect further developments (as announced by the authors) on coupled particle and heat transport. The change of polarity in the core of the plasma still remains a conundrum!

d. Profile Consistency

in the second paper of their series, VANMILLIGEN et al 2004 b address the problem of profile consistency, as defined in Sec. 17.3, in particular the puzzling problem of "uphill transport". It is very satisfactory to find that the same vMSC model also provides an explanation of this phenomenon. The main difference with the cases treated in Sec. 17.6 is that we now consider *localized particle sources* $S(x)$ instead of the homogeneous sources S_0 treated there. Specifically, the following form of the fueling function is adopted:

$$S(x) = \frac{S_0}{2} \frac{1}{\sqrt{2\pi}w} \left\{ \exp\left[-\frac{(x-a)^2}{2w^2}\right] + \exp\left[-\frac{[x-(1-a)]^2}{2w^2}\right] \right\}. \quad (17.55)$$

In all simulations $w = 0.025$ (except in Fig. 17.6, where $w = 0.05$).[13] The master equation must be solved with the boundary condition requiring the total flux crossing the boundary to equal the integrated source.

In a first type of problems, the authors consider *on-axis fuelling*, *i.e.*, $a = 0.5$. A first run, with $S_0 = 0.2$ yields the result of Fig. 17.6.

The density profile is peaked on axis; it has practically the same shape as obtained in Sec. 17.6: a rounded (parabolic) region near the maximum, and a linear shape outside. As the value of S_0 considered here is rather high (compared to the value leading to Fig. 17.4), nearly all the plasma is supercritical. The significant fact here is that a fuelling $S(x)$ localized on the axis leads to practically the same result as a homogeneous S_0. This insensitivity to the details of $S(x)$ is a first manifestation of the profile consistency phenomenon.

We now consider *off-axis fuelling*, ($a = 0.3$). As a reference case, the authors consider first what happens in a purely diffusive process, *i.e.*, setting $\zeta(x',t) \equiv 0$ in Eq. (17.40). The result is shown in Fig. 17.7.

[13] The necessity of taking a half-sum of two Gaussians results from the fact that in the present "spread-out" representation $0 \le x \le 1$ of the torus, the magnetic axis is at $x = 0.5$, and the positions $x = a$ and $x = 1 - a$ are equivalent: they represent the same magnetic surface.

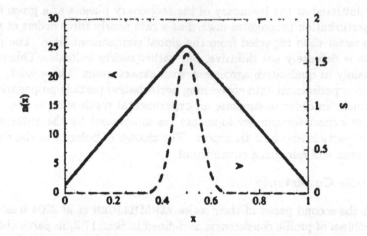

Figure 17.6. Density profile with on-axis fuelling ($a = 0.5$), $S_0 = 0.2$. (Reprinted with permission from VAN MILLIGEN et al., Phys. Plasmas **11**, 3787 (2004). Copyright 2004 American Institute of Physics.)

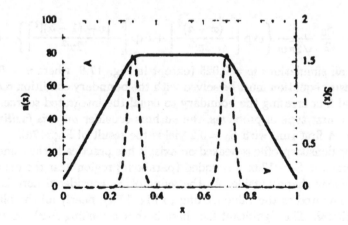

Figure 17.7. Density profile $n(x)$ with off-axis fuelling ($a = 0.3$), $S_0 = 0.2$, for a purely diffusive system ($\zeta = 0$). (Reprinted with permission from VAN MILLIGEN et al. *Phys. Plasmas* **11** 3787 (2004). Copyright 2004, American Institute of Physics.)

The remarkable feature here is the *flat profile in the core of the plasma*. This can be easily understood. In the interior region the source is zero.

Hence the diffusion equation $\partial_t n = D \nabla^2 n$ implies that in the steady state $\nabla^2 n = 0$: the curvature vanishes in the inner region. As a result, the density gradient is constant there. But $\nabla n(0.5) = 0$ by symmetry; hence $\nabla n(x) = 0$, hence also the particle flux $\Gamma = 0$ in the whole inner region, where $S(x) = 0$. The particle flux is non-zero only in the outer region, beyond the location of the local fuel injection: it is directed towards the boundary ("downhill flux"), in agreement with the second law of thermodynamics.

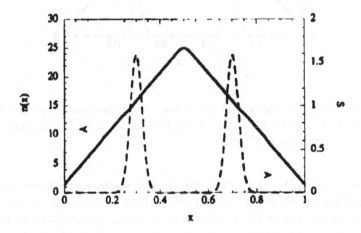

Figure 17.8. Density profile $n(x)$ with strong off-axis fuelling ($a = 0.3$), $S_0 = 0.2$, solution of the complete generalized master equation. (Reprinted with permission from VAN MILLIGEN et al. *Phys. Plasmas* 11 3787 (2004). Copyright 2004, American Institute of Physics.)

We now consider the same problem of *off-axis fuelling*, by solving the full generalized master equation (17.41), with $\zeta \neq 0$. The surprising - but satisfactory! - result is that the density profile obtained in this case, Fig. 17.8, is precisely the same as in Fig. 17.6, *i.e.*, in the case of on-axis fuelling. This result is in perfect qualitative agreement with the experimental findings of Luce and Petty, described in Sec. 17.3. It confirms the principle of profile consistency and, consequently, the process of *uphill transport* necessary in order to create the peaked profile that is absent in the purely diffusive case illustrated in Fig. 17.7. The explanation provided by the vMSC model is the following. The strong fuelling source activates (for a certain time) the superdiffusive channel beyond it. The latter allows a non-negligible fraction of particles to make large excursions in both directions, in particular towards the central axis, thus filling up the inner region. This

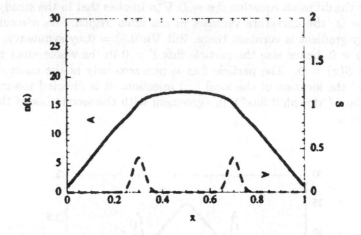

Figure 17.9. Same as Figure 17.8, for weaker off-axis fuelling. (Reprinted with permission from VAN MILLIGEN et al., Phys. Plasmas 11, 3787 (2004). Copyright 2004 American Institute of Physics.)

process is one of self-organization, due to the nonlinear character of the generalized master equation. At this value of the fuelling rate we are in a regime where practically the whole system is in the supercritical region.

This interpretation is strengthened by running a simulation with smaller values of the fuelling intensity S_0. As shown in Fig. 17.9, the steady profile is now almost flat in the centre, thus resembling Fig. 17.7. This is due to a wider diffusive region in the central region, as exemplified in Fig. 17.4. The vMSC model thus describes both the coexistence and the relative dominance of one or the other of the diffusive and the superdiffusive regimes.

Previous authors have been tempted to stick to an advection-diffusion interpretation of the peaked profiles obtained in off-axis experiments, by adding an additional inward-directed convective, hence, non-diffusive term ("inward pinch") to the diffusion equation (17.42). But it turns out that such an empirically "parachuted" convective velocity should have different values for different values of the effective diffusion coefficient, or even be absent in some cases. No universal explanation of the phenomena is possible with such a patch-up (see VAN MILLIGEN et al. 2004 b, where more details about this problem will be found).

In conclusion, we have seen that the vMSC model contains in a single, simple and clear construction the ingredients necessary for the explanation of most of the mysterious phenomena related to non-local transport in magnetically confined plasmas. This model will certainly be further developed in the future.

There remains, though, one puzzling question: *what is the microscopic basis of the vMSC model?* Given the successful explanations which it provides, I feel convinced that such a basis exists; the problem is, however, very difficult. As was stressed in the beginning of Chaps. 11 and 12, the random walk models, in particular, the CTRW, result from replacing the exact laws of dynamics by a set of purely probabilistic rules. By doing this we discovered the "strange" regimes of subdiffusion and of superdiffusion. In the *subdiffusive* case, it is not too difficult to understand how a *slowing-down mechanism* of the diffusion is produced by the *trapping processes*. This is a purely dynamical effect, adequately described by starting from a (very irregular) Hamiltonian, with numerous and time-varying peaks and troughs, as shown in Chap. 14. It is, on the other hand, very difficult to imagine a molecular dynamical explanation of a mechanism that would lead to something like Lévy flights. The profile consistency principle observed in confined plasmas also poses a difficult problem from the molecular point of view. It has a clear connection to Prigogine's concept of *nonequilibrium dissipative structures* (NICOLIS & PRIGOGINE 1977). The latter is, however, also based on a macroscopic or a semi-dynamical description (Langevin) combined with a stochastic description of fluctuations. It offers no clue to Lévy flights, which remain a challenging open conundrum.

Appendix A

The Electrostatic Approximation

We may wonder under what conditions the electrostatic approximation is valid in a real plasma. Actually, the "electrostatic approximation" of plasma physics is not really static, because the electric field perturbation $\delta \mathbf{E}(x,t)$ is time-dependent. But then, Maxwell's equations tell us that there should also exist a magnetic field perturbation! It may be checked that the solution of the plasmadynamical quantities obtained here are not exactly compatible with the Ampère equation (1.16) (which was dismissed). A rough, but simple criterion of validity can be derived if we assume that:

- a) the plasma is weakly inhomogeneous;
- b) the frequency range of interest is very low ($\omega \approx 0$).

Then, the momentum balance equations (2.3), linearized and summed over the species, reduce to the balance of forces:

$$-\boldsymbol{\nabla}\, \delta p + c^{-1} \delta(\mathbf{j} \times \mathbf{B}) = 0. \tag{A.1}$$

The total pressure perturbation δp reduces under the assumption: $\delta T_i = \delta T_e = 0$ to:

$$\delta p = (T_{e0} + T_{i0})\, \delta n_e. \tag{A.2}$$

Thus, (A.1) (with $\boldsymbol{\nabla} \to i\mathbf{k}$) becomes:

$$-i\mathbf{k}\,(T_{e0} + T_{i0})\,\delta n_e + c^{-1}(\delta \mathbf{j} \times \mathbf{B}) = 0,$$

or, using the Ampère equation: (with $\omega \approx 0$):

$$-\mathbf{k}\,(T_{e0} + T_{i0})\delta n_e + (4\pi)^{-1}(\mathbf{k} \times \delta\mathbf{B}) \times \mathbf{B} = 0. \tag{A.3}$$

From here, we find the following order-of-magnitude estimate:

$$\frac{|\delta\mathbf{B}|/B}{\delta n_e/n} \approx \frac{1}{2}\frac{n(T_{e0} + T_{i0})}{B^2/8\pi} \equiv \frac{1}{2}\beta. \tag{A.4}$$

The parameter β is the ratio of the plasma pressure to the magnetic pressure: it is an important parameter in fusion physics. In a tokamak this is usually a small number; for instance, assuming the values of Table 2.1, $\beta \approx 0.4$. Hence, in many cases the magnetic perturbations can be neglected. They may, however, be locally important, hence they should not be totally dismissed.

Appendix B

The Cauchy Integral

It is surprising that a concept of crucial importance in mathematical physics is treated only cursorily in western mathematical textbooks. To the best of my knowledge, the only detailed treatments of this matter are contained in three russian monographs, two of which have been translated into English: MIKHLIN 1950, MUSKHELISHVILI 1953 (the most exhaustive treatment), and GAKHOV 1958. A shorter treatment is also found in the textbooks of SMIRNOV 1975 and of LAVRENTIEV & CHABAT 1972. A rather long appendix is devoted to this problem in my own monograph BALESCU 1963. As most of these references are rather old or of difficult access, I summarize in the present Appendix the most important properties of this object, following closely my 1963 text.

We consider a class of complex functions $\varphi(x)$ defined by the following three properties.

a) $\varphi(x)$ *is defined for all real values of* x: $-\infty \leq x \leq +\infty$.

b) $\varphi(x)$ *satisfies a Lipschitz condition* (also called *Hölder condition*):

$$|\varphi(x_2) - \varphi(x_1)| \leq A |x_2 - x_1|^{\mu}, \quad A > 0, \quad 0 \leq \mu \leq 1 \tag{B.1}$$

(this condition is stronger than mere continuity).

c) $\varphi(x)$ *has the following behaviour at infinity:* $\varphi(x)$ tends to zero in such a way that:

$$|\varphi(x)| < A |x|^{-\nu}, \quad A > 0, \quad \nu > 0. \tag{B.2}$$

We now consider a *complex variable* z, and the expression $\Phi(z)$, defined as:

$$\Phi(z) = \frac{1}{2\pi} \int_{-\infty}^{\infty} dx \, \frac{\varphi(x)}{x - z}. \tag{B.3}$$

The function $\Phi(z)$ is called a *Cauchy integral*. Eq. (B.3) can be regarded as an operation which associates with each function $\varphi(x)$, defined on the real axis, a function $\Phi(z)$ defined in the whole complex plane. In

446

some texts, $\Phi(z)$ is called the *Hilbert transform* of $\varphi(x)$. It is easily shown that:

$\Phi(z)$ *is a regular (holomorphic) function in any domain D which does not include points of the real axis; it tends to zero for* $|z| \to \infty$.

In particular, $\Phi(z)$ is regular in the whole upper half-plane and in the whole lower half-plane (the latter regions will henceforth be called, respectively, S_+ and S_-). The integral (B.3) has, however, no meaning as it stands for real z. The following property is the fundamental basis of the whole concept:

When z tends toward the real value ζ *by taking only values in* S_+, *the function* $\Phi(z)$ *tends toward a definite limit* $\Phi^{(+)}(\zeta)$.

When z tends toward the real value ζ *by taking only values in* S_-, *the function* $\Phi(z)$ *tends toward a definite limit* $\Phi^{(-)}(\zeta)$.

The limits $\Phi^{(+)}(\zeta)$ *and* $\Phi^{(-)}(\zeta)$ *are generally different.*

Consider first the behaviour for $z \to \zeta$ from above. Equivalently, we set $z = \zeta + i\epsilon$, with $\epsilon > 0$, and consider the limit $\epsilon \to 0$. The limit can be calculated by slightly deforming the contour of integration in Eq. (B.3) into the contour Γ_+ of Fig. B.1 a.

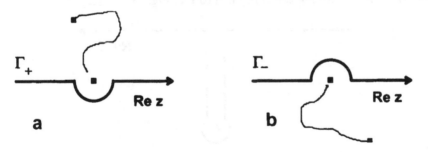

Figure B.1. Contours of integration in Eq. (B.3). a: for $\Phi^{(+)}$; b: for $\Phi^{(-)}$.

The limit is correctly taken if the topological relationship between the point z and the contour of integration is preserved, *i.e.*, if in the process $z \to \zeta$, the point z always remains *above* the contour. We now define the concept of the *Cauchy principal part* as follows:

$$P \int_{-\infty}^{\infty} dx\, \frac{\varphi(x)}{x-\zeta} = \left\{ \int_{-\infty}^{\zeta-\epsilon} dx\, \frac{\varphi(x)}{x-\zeta} + \int_{\zeta+\epsilon}^{\infty} dx\, \frac{\varphi(x)}{x-\zeta} \right\}, \quad \epsilon \to +0. \quad \text{(B.4)}$$

The integration along the contour Γ_+ is now easily performed: there is a first term equal to $\frac{1}{2}(2\pi i)$ times the residue of the integrand at the pole

$x = \zeta$ (this is the contribution of the dent around ζ) plus a principal part integral:

$$\Phi^{(+)}(\zeta) = \frac{1}{2}i\,\varphi(\zeta) + \frac{1}{2\pi}P\int_{-\infty}^{\infty} dx\,\frac{\varphi(x)}{x-\zeta}, \quad \zeta\ real. \qquad (B.5)$$

A similar calculation (see Fig. B.1 b) leads to the following limiting value for $z \to \zeta - i\epsilon,\ \epsilon > 0$:

$$\Phi^{(-)}(\zeta) = -\frac{1}{2}i\,\varphi(\zeta) + \frac{1}{2\pi}P\int_{-\infty}^{\infty} dx\,\frac{\varphi(x)}{x-\zeta}. \qquad (B.6)$$

These equations, called the **Plemelj formulae** are fundamental for the whole theory. Another equivalent and very useful form is:

$$\Phi^{(+)}(\zeta) - \Phi^{(-)}(\zeta) = \varphi(\zeta), \quad \zeta\ real, \qquad (B.7)$$

$$\Phi^{(+)}(\zeta) + \Phi^{(-)}(\zeta) = \frac{1}{\pi}P\int_{-\infty}^{\infty} dx\,\frac{\varphi(x)}{x-\zeta}, \quad \zeta\ real. \qquad (B.8)$$

The function $\Phi^{(+)}(\zeta)$ can be *continued analytically* into the lower half-plane S_-. The resulting function is different from $\Phi^{(-)}(\zeta)$. Indeed, it is obtained by deforming the contour into $\bar{\Gamma}_+$, Fig. B.2.

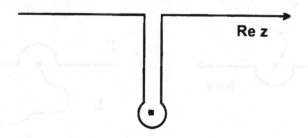

Figure B.2. Contour of integration for the analytical continuation of $\Phi^{(+)}$ into the lower half-plane.

The resulting integral is easily evaluated by the method of residues. We thus obtain a definition of the analytical function $\Phi^{(+)}(z)$ for the whole range of complex values of z:

$$\Phi^{(+)}(z) = \frac{1}{2\pi}\int_{-\infty}^{\infty} dx\,\frac{\varphi(x)}{x-z}, \quad z \in S_+$$

$$\Phi^{(+)}(\zeta) = \frac{1}{2\pi}P\int_{-\infty}^{\infty} dx\,\frac{\varphi(x)}{x-\zeta} + \frac{1}{2}i\,\varphi(\zeta), \quad \zeta\ real,$$

$$\Phi^{(+)}(z) = \frac{1}{2\pi}\int_{-\infty}^{\infty} dx\,\frac{\varphi(x)}{x-z} + i\,\varphi(z) = \Phi^{(-)}(z) + i\,\varphi(z), \quad z \in S_-. \qquad (B.9)$$

It should be stressed that the function $\Phi^{(+)}(z)$ is regular in the upper half-plane, but *its analytical continuation into S_- has singularities in the lower half-plane*. These singularities are the same as those of the continuation of the function $\varphi(z)$ into S_-.

Noting that Eq. (B.5) is equivalent to:

$$\Phi^{(+)}(\zeta) = \frac{1}{2\pi} \int_{-\infty}^{\infty} dx\, \frac{\varphi(x)}{x - \zeta - i\epsilon}, \quad \epsilon \to +0; \qquad \text{(B.10)}$$

a comparison with (B.5) yields the following important relation:

$$\frac{1}{y - i\epsilon} = \mathcal{P}\frac{1}{y} + \pi i\, \delta(y), \quad y\ \text{real}, \ \epsilon \to +0. \qquad \text{(B.11)}$$

Appendix C

The Plasma Dispersion Function

The *Plasma Dispersion Function* is defined as follows:

$$Z(z) = \frac{1}{\sqrt{\pi}} \int_{-\infty}^{\infty} du \frac{1}{u - z} e^{-u^2}. \tag{C.1}$$

The integration over u is over the whole real axis. We recognize here a *Cauchy integral* of the class discussed in Appendix B. This equation only has a meaning when considered as a function of a *complex* variable z. Moreover, Eq. (C.1) defines *two different functions of z*, according to whether z lies in the upper half-plane (S_+) or in the lower half-plane (S_-). In all the plasma physics applications $Z(z)$ is defined by (C.1) *for $z \in S_+$*. As such, it is a holomorphic function in the upper half plane. A particular interest lies in the limit of $Z(z)$ as z approaches the real axis from above. We thus define the following notation:

$$z = \zeta + i\epsilon, \quad \zeta : real, \quad \epsilon > 0. \tag{C.2}$$

The plasma dispersion function for *real* values of ζ is then defined as:

$$Z(\zeta) = \frac{1}{\sqrt{\pi}} \int_{-\infty}^{\infty} du \frac{1}{u - \zeta - i\epsilon} e^{-u^2}, \quad \epsilon \to +0. \tag{C.3}$$

Using the limiting distribution $(u - \zeta - i\epsilon)^{-1}$ given by Eq. (B.11), we find, more explicitly:

$$Z(\zeta) = \frac{1}{\sqrt{\pi}} P \int_{-\infty}^{\infty} du \frac{1}{u - \zeta} e^{-u^2} + i\sqrt{\pi} e^{-\zeta^2}. \tag{C.4}$$

The real and imaginary parts of this function of the real variable ζ are plotted in Fig. C.1. Extensive tables of the plasma dispersion function, both for real and for complex values of z have been published by FRIED & CONTE 1961.

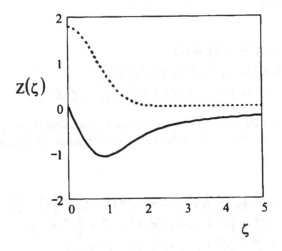

Figure C.1. Real part (solid line) and Imaginary part (dotted line) of the plasma dispersion function $Z(\zeta)$ for real ζ.

Before discussing some of its properties, we introduce also the *generalized plasma dispersion functions*, defined as follows:

$$Z_m(\zeta) = \frac{1}{\sqrt{\pi}} \int_{-\infty}^{\infty} du \frac{1}{u - \zeta - i\epsilon} u^m e^{-u^2}, \quad \epsilon \to +0. \qquad (C.5)$$

Clearly: $Z(\zeta) = Z_0(\zeta)$.

The following recurrence relation is quite useful (SMITH 1984):

$$\zeta Z_m(\zeta) = Z_{m+1}(\zeta), \quad m = 2p + 1, \qquad (C.6)$$

The following relations involve the derivative of $Z(\zeta)$:

$$Z'_m(\zeta) = m Z_{m-1}(\zeta) - 2 Z_{m+1}(\zeta), \quad m > 0,$$
$$Z'_0(\zeta) = -2 Z_1(\zeta). \qquad (C.7)$$

We also note:

$$\zeta Z'_m(\zeta) = m Z_m(\zeta) - 2 Z_{m+2}(\zeta). \qquad (C.8)$$

In particular, we find:

$$Z_2(\zeta) = \zeta + \zeta^2 Z(\zeta),$$
$$Z_4(\zeta) = \tfrac{1}{2}\zeta + \zeta^3 + \zeta^4 Z(\zeta),$$
$$Z_6(\zeta) = \tfrac{3}{4}\zeta + \tfrac{1}{2}\zeta^3 + \zeta^5 + \zeta^6 Z(\zeta). \qquad (C.9)$$

452   *The Plasma Dispersion Function*

We also have:

$$Z_2(\zeta) = -\tfrac{1}{2}\zeta\, Z'(\zeta),$$
$$Z_4(\zeta) = \tfrac{1}{4}\left[-\zeta\, Z'(\zeta) + \zeta^2 Z''(\zeta)\right],$$
$$Z_6(\zeta) = \tfrac{1}{8}\left[-3\zeta\, Z'(\zeta) + 3\zeta^2 Z''(\zeta) - \zeta^3 Z'''(\zeta)\right]. \qquad \text{(C.10)}$$

We now consider the very frequently used approximations of the plasma dispersion function. For *small values of* ζ one obtains the following Taylor expansions:

$$Z_0(\zeta) = -2\zeta + \tfrac{4}{3}\zeta^3 - \tfrac{8}{15}\zeta^5 + ... + i\sqrt{\pi}\left(1 - \zeta^2 + \tfrac{1}{2}\zeta^4 - \tfrac{1}{6}\zeta^6 + ...\right),$$
$$Z_2(\zeta) = \zeta - 2\zeta^3 + \tfrac{4}{3}\zeta^5 + ... + i\sqrt{\pi}\left(\zeta^2 - \zeta^4 + \tfrac{1}{2}\zeta^6 + ...\right),$$
$$Z_4(\zeta) = \tfrac{1}{2}\zeta + \zeta^3 - 2\zeta^5 + ... + i\sqrt{\pi}\left(\zeta^4 - \zeta^6 + ...\right),$$
$$Z_6(\zeta) = \tfrac{3}{4}\zeta + \tfrac{1}{2}\zeta^3 + \zeta^5 + ... + i\sqrt{\pi}\left(\zeta^6 + ...\right), \qquad \zeta \ll 1. \qquad \text{(C.11)}$$

For *large values of* ζ, the imaginary part of these functions is exponentially small [see Eq. (C.6)], and the following asymptotic expansion is obtained:

$$Z_0(\zeta) = -\frac{1}{\zeta} - \frac{1}{2\zeta^3} - \frac{3}{4\zeta^5} - \frac{15}{8\zeta^7} - \frac{105}{16\zeta^9} - ...,$$
$$Z_2(\zeta) = -\frac{1}{2\zeta} - \frac{3}{4\zeta^3} - \frac{15}{8\zeta^5} - \frac{105}{16\zeta^7} - ...,$$
$$Z_4(\zeta) = -\frac{3}{4\zeta} - \frac{15}{8\zeta^3} - \frac{105}{16\zeta^5} - ...,$$
$$Z_6(\zeta) = -\frac{15}{8\zeta} - \frac{105}{16\zeta^3} - ... \qquad \zeta \gg 1. \qquad \text{(C.12)}$$

Additional properties of the plasma dispersion function and its generalizations can be found in the papers by PERATT 1984, SMITH 1984 and SUMMERS & THORNE 1991.

Appendix D

Lévy Probability Distributions

Let $n(\mathbf{x})$ be a probability distribution function (PDF) which, for simplicity, is assumed to be an even function of $\mathbf{x}$, and let $\tilde{n}(\mathbf{k})$ be its Fourier transform, or characteristic function. Clearly, $\tilde{n}(\mathbf{k}) = \tilde{n}(k)$, where $k = |\mathbf{k}|$. A *stable law* is defined as a PDF whose characteristic function (CF) has the following form-invariance property:

$$\tilde{n}(a_1\mathbf{k})\,\tilde{n}(a_2\mathbf{k}) = \tilde{n}(a\mathbf{k}), \qquad (\text{D.1})$$

where a is a positive constant, depending on the arbitrary positive constants a_1, a_2. In order to solve this functional equation, we set $\tilde{n}(\mathbf{k}) = \exp[B(k)]$, hence (D.1) becomes:

$$B(a_1 k) + B(a_2 k) = B(a k).$$

It is easily seen that the solution is a power law:

$$B(k) = -C\,k^{\beta},$$

with arbitrary real numbers C, β. We must then satisfy:

$$-C\left(a_1^{\beta}\,k^{\beta} + a_2^{\beta}\,k^{\beta}\right) = -C\left(a^{\beta}\,k^{\beta}\right).$$

This equation determines the constant a:

$$a = \left(a_1^{\beta} + a_2^{\beta}\right)^{1/\beta}. \qquad (\text{D.2})$$

Thus, the stable laws must be of the form: $\tilde{n}(\mathbf{k}) = \exp(-C\,k^{\beta})$; we must, however, limit the possible values of the constants. For convergence, C must be positive; for regularity in $k = 0$, β must also be positive. Moreover, it can be shown that when $\beta > 2$, the inverse Fourier transform, *i.e.*, the PDF $n(\mathbf{x})$ is not definite positive, hence is not an acceptable PDF. In conclusion, *a stable law is characterized by a characteristic function of the form:*

$$\tilde{n}(\mathbf{k}) = \exp(-C\,k^\beta), \quad C > 0, \quad 0 < \beta \leq 2. \tag{D.3}$$

The corresponding PDF is then:

$$n(\mathbf{x}) = (2\pi)^{-d} \int d^d\mathbf{k}\, e^{-i\mathbf{k}\cdot\mathbf{x}} \exp(-C\,k^\beta). \tag{D.4}$$

Any PDF of this form is called a *Lévy distribution function*. These functions were first considered, for arbitrary β, by Cauchy in 1853; he was not aware, however, that for $\beta > 2$ they become negative for some values of x. Their systematic study was initiated by LEVY 1937.[1] In recent years the importance of these functions is being realized in plasma physics, but also in statistical physics, condensed matter physics and many other fields. A detailed discussion of their properties and a set of approximation formulae and algorithms for their numerical calculation is found in MONTROLL & BENDLER 1984 (the present appendix follows closely this paper).

For simplicity, we only discuss here the one-dimensional distributions, $d = 1$, and set $C = 1$; the really important parameter is the exponent β:

$$n_\beta(x) = (2\pi)^{-1} \int dk\, \exp(-ikx - |k|^\beta), \quad 0 < \beta \leq 2. \tag{D.5}$$

These functions are properly normalized:

$$\int dx\, n_\beta(x) = \int dk\, \delta(k)\, \exp(-|k|^\beta) = 1. \tag{D.6}$$

A closed analytical form for $n_\beta(x)$ exists only for a few special values of β. The most celebrated one corresponds to $\beta = 2$:

$$n_2(x) = (2\pi)^{-1} \int dk\, e^{-ikx}\, e^{-k^2} = (4\pi)^{-1/2} \exp(-\frac{x^2}{4}). \tag{D.7}$$

This is just the *Gaussian* or *normal PDF*. Its special importance stems from the central limit theorem discussed in Sec. 12.1. Another simple case is $\beta = 1$, which yields the *Lorentz*, or *Cauchy* PDF:

$$n_1(x) = \frac{1}{\pi} \frac{1}{1 + x^2}. \tag{D.8}$$

Other special cases are $\beta = 1/2$, related to the Fresnel integrals, and $b = 2/3$, related to the Whittaker function. These are the only cases in which the Lévy PDF can be related to known special functions.

---

[1] It may be added, for completeness that the most general class of stable distribution is a 3-parameter class of functions, in which k^β, Eq. (D.3), is replaced by $k^\beta \exp(i\pi\,\gamma/2)$. We shall, however, never need the case $\gamma \neq 0$ in the problems treated here.

The value at the origin, $n_\beta(0)$ can be calculated analytically:

$$n_\beta(0) = \frac{1}{\pi} \int_0^\infty dk \, \exp(-k^\beta) = \frac{1}{\pi\beta} \int_0^\infty dy \, y^{(1-\beta)/\beta} e^{-y} = \frac{1}{\pi\beta} \Gamma\left(\frac{1}{\beta}\right).$$
(D.9)

This value is very slowly varying for large β, but grows very steeply for $\beta < 0.5$ (see Fig. D.1).

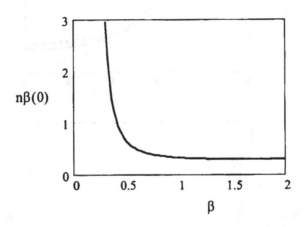

Figure D.1. The value at the origin of the Lévy distributions, $n_\beta(0)$

In Fig. D.2a we show three Lévy functions for $1 \le \beta \le 2$, as well as the function for $\beta = 4$. It is clearly seen that the latter becomes negative for $x > 3.5$, and is therefore not acceptable as a PDF. Fig. D.2b is an enlarged picture of these functions, for $x > 3$: it exhibits the long tails of these functions, compared with the sharp (exponential) decay of the Gaussian ($\beta = 2$).

In Fig. D.3 we show the corresponding graphs for the range $\beta < 1$ (calculated numerically). We see that for $\beta < 0.5$ the shape of the curves changes dramatically. In accordance with the result shown in Fig. D.1, the value at the origin quickly becomes very large.

The following series representation for small x is obtained by expanding $\exp(-ikx)$ in the integrand of Eq. (D.5):

$$n_\beta(x) = \frac{1}{\pi\beta} \left[\Gamma\left(\frac{1}{\beta}\right) - \frac{1}{2!} \Gamma\left(\frac{3}{\beta}\right) x^2 + \frac{1}{4!} \Gamma\left(\frac{5}{\beta}\right) x^4 + ... \right].$$
(D.10)

The series converges for all values of x when $1 < \beta < 2$; it diverges for all x when $0 < \beta < 1$. As a result, the approximation of the domain of

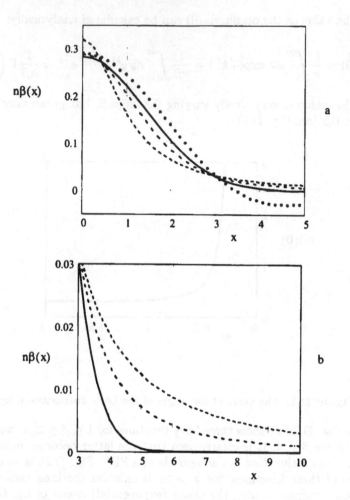

Figure D.2.   (a) Lévy distributions $n_\beta(x)$ for $\beta \geq 1$. Dashed: $\beta = 1$, dash-dotted: $\beta = 1.5$, solid: $\beta = 2$, dotted: $\beta = 4$. (b) is an enlarged view of the tail.

small x of the Lévy distributions with $\beta < 0.5$ is very difficult, because of the rapid variation of the functions.

A widely used expansion for large x is due to WINTNER 1941:

$$n_\beta(x) = \frac{1}{\pi} \sum_{n=1}^{\infty} \frac{(-)^{n+1}}{n!} \sin\left(\tfrac{1}{2}\pi\beta n\right) \Gamma(1 + \beta n) \frac{1}{|x|^{1+\beta n}}. \qquad (D.11)$$

For large x, the dominant term is:

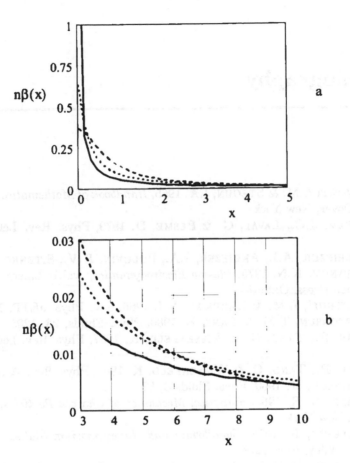

Figure D.3. (a) Lévy distributions $n_\beta(x)$ for $\beta < 1$. Dashed: $\beta = 0.8$, dotted: $\beta = 0.5$, solid: $\beta = 0.3$. (b) is an enlarged view of the tail.

$$n_\beta(x) \sim \frac{\sin\left(\frac{1}{2}\pi\beta\right)\Gamma(1+\beta)}{\pi}\frac{1}{|x|^{1+\beta}}. \tag{D.12}$$

The single term (D.12) is a reasonable approximation for not too small β.

Additional approximation and interpolation formulae can be found in the paper of MONTROLL & BENDLER 1984.

Bibliography

ABRAMOWITZ, M. & STEGUN, I.A. 1965, *Handbook of Mathematical Functions*, Dover, New York

ADAM, J.C., LAVAL, G. & PESME, D. 1979, Phys. Rev. Lett., **43**, 1671

AKHIEZER, A.I., AKHIEZER, I.A., POLOVIN, R.V., SITENKO, A.G. & STEPANOV, K.N. 1975, *Plasma Electrodynamics*, vol.1: Linear theory, Pergamon Press, Oxford

ALTSHUL', L.M. & KARPMAN, V.I. 1966, Sov. Phys. JETP, **22**, 361

ANTONSEN, T.M. & LANE, B. 1980, Phys. Fluids, **23**, 1205

BAK, P., TANG, C. & WIESENFELD, K. 1987, Phys. Rev. Lett., **59**, 381

BAK, P., TANG, C. & WIESENFELD, K. 1988, Phys. Rev. **A 38**, 364

BALESCU, R. 1960, Phys. Fluids **3**, 52

BALESCU, R. 1963, *Statistical Mechanics of Charged Particles*, Interscience, New York

BALESCU, R. 1975, *Equilibrium and Nonequilibrium Statistical Mechanics*, Wiley, New York

BALESCU, R. 1988 a, *Transport Processes in Plasmas*, vol. 1: *Classical Transport*, North Holland, Amsterdam

BALESCU, R. 1988 b, *Transport Processes in Plasmas*, vol. 2: *Neoclassical Transport*, North Holland, Amsterdam

BALESCU, R. 1990, Phys. Fluids B **2**, 2100

BALESCU, R. 1992, Phys. Fluids, B **4**, 91

BALESCU, R. 1995, Phys. Rev. E **51**, 4807

BALESCU, R. 1997, *Statistical Dynamics: Matter out of Equilibrium*, Imperial College Press, London

BALESCU, R. 2000, Plasma Phys. Contr. Fusion, **42**, B1

BALESCU, R. 2003, Phys. Rev. E **68**, 046409

BALESCU, R. & MISGUICH, J.H. 1984, in: *Statistical Physics and Chaos in Fusion Plasmas* (C.W. Horton, Jr. & L.E. Reichl, eds), Wiley, New York, p. 295

BALESCU, R., VANDENEIJNDEN, E. & WEYSSOW, B. 1993, J. Plasma Phys. **50**, 425

BALESCU, R., WANG, H.D. & MISGUICH, J.H. 1994, Phys. Plasmas, 1, 3826

BALL, R.C., HAVLIN, S. & WEISS, G.H. 1987, J. Phys., A 20, 4055

BATCHELOR, G.K. 1956, *The Theory of Homogeneous Turbulence*, Cambridge Univ. Press, Cambridge, UK

BATEMAN, G. 1978, *MHD Instabilities*, M.I.T. Press, Cambridge, Mass.

BERNSTEIN, I.B. & CATTO, P.J. 1985, Phys. Fluids, 28, 1342

BIGLARI, H., DIAMOND, P.H. & TERRY, P. 1990, Phys. Fluids, B 2, 1

BINNEY, J.J., DOWRICK, N.J., FISHER, A.J. & NEWMAN, M.E.J. 1992, *The Theory of Critical Phenomena*, Clarendon Press, Oxford

BOGOLIUBOV, N.N. & SHIRKOV, D.V. 1980, *Introduction to the Theory of Quantized Fields*, 3rd ed., Wiley, New York

BOOKER, H.G. 1984, *Cold Plasma Waves*, M. Nijhoff, Dordrecht

BOUCHAUD, J.P. & GEORGES, A. 1990, Phys. Reports, 195, 127

BOUTROS-GHALI, T. & DUPREE, T.H. 1981, Phys. Fluids 24, 1839

BOWMAN, J.C. 1992, *Realizable Markovian Statistical Closures: General theory and application to Drift Wave Turbulence*, Ph.D. Thesis, Princeton Univ.

BRAGINSKII, S.I. 1965, in: *Reviews of Plasma Physics*, vol. 1, ed. M.A. Leontovich, Consultants Bur., New York, p. 205

BRAMBILLA, M. 1998, *Kinetic Theory of Plasma Waves*, Clarendon Press, Oxford

BURRELL, K.H., GENTLE, K.W., LUHMANN JR, N.C., MARMAR, E.S., MURAKAMI, M., SCHOENBERG, K.F., TANG, W.M. & ZARNSTORFF, M.C. 1990, Phys. Fluids, B 2, 2904

BYKOV, A.M. & TOPTYGUIN, I.N. 1992, Sov. Phys. JETP, 74, 462

CAIRNS, R.A. 1985, *Plasma Physics*, Blackie, Glasgow

CALLEN, J.D. & KISSICK, M.W. 1997, Plasma Phys. Contr. Fusion, 39, B173

CAMARGO, S.J. & TASSO, H. 1992, Phys. Fluids, B 4, 1199

CAP, F.F. 1976, *Handbook of Plasma Instabilities*, vol. 1, Academic Press, New York

CARATI, D., CHRIAA, K. & BALESCU, R. 1994, Phys. Rev. E 50, 1444

CARRERAS, B.A., LYNCH, V.E., NEWMAN, D.E. & ZASLAVSKY, G.M. 1999, Phys. Rev. E 60, 4770

CATTO, P.J., TANG, W.M. & BALDWIN, D.E. 1981, Plasma Phys., 23, 639

CERCIGNANI, C. 1988, *The Boltzmann Equation and its Applications*, Springer, Berlin

CHAPMAN, S. and COWLING, T.G. 1952, *The Mathematical Theory of Non-uniform gases* (2nd. ed.), Cambridge Univ. Press

CHEN, F. 1974, *Introduction to Plasma Physics*, Plenum, New York

CHURCHILL, R.V. 1972, *Operational Mathematics*, McGraw Hill, New York

COFFEY, W.T., KALMYKOV, YU.P. & WALDRON, J.T. 1996, *The Langevin Equation - with applications in Physics, Chemistry and Electrical Engineering*, World Scientific, Singapore

COLLINS, J. 1984, *Renormalization*, Cambridge Univ. Press, Cambridge

CORRSIN, S. 1959, in: *Proceed. Sympos. on Atmospheric Diffusion and Air Pollution, Oxford 1958*, (F.N. Frenkiel & P.A. Sheppard, eds.), Acad. Press, New York, p. 161

CORRSIN, S. 1962, in: *Mécanique de la turbulence*, Colloq. Internat. Marseille, CNRS, Paris, p. 161

CROTTINGER, J.A. & DUPREE, T.H. 1992, Phys. Fluids, B **4**, 2854

CSANADY, G.T. 1973, *Turbulent Diffusion in the Environment*, Reidel

DAVIDSON, R.C. 1972, *Methods in Nonlinear Plasma Theory*, Acad. Press, New York

DE GROOT, S.R. & MAZUR, P. 1984, *Nonequilibrium Thermodynamics*, reprint, Dover, New York

DEL-CASTILLO-NEGRETE, D., CARRERAS, B.E. & LYNCH, V.E. 2004, Phys. Plasmas, **11**, 3854

DENDY, R.O. 1990, *Plasma Dynamics*, Clarendon Press, Oxford

DIAMOND, P.H., CHAMPEAUX, S., MALKOV, M., DAS, A., GRUZINOV, I., ROSENBLUTH, M.N., HOLLAND, C., WECHT, B., SMOLYAKOV, A., HINTON, F.L., LIN, Z. & HAHM, T.S. 2001, Nucl. Fusion, **41**, 1067

DIAMOND, P.H., DUPREE, T.H. & TETREAULT, D.J. 1980, Phys. Rev. Lett., **45**, 562

DIAMOND, P.H. & KIM, Y.B. 1991, Phys. Fluids B **3**, 1626

DIAMOND, P.H. & MALKOV, M. 2002, Physica Scripta, T **98**, 63

DIAMOND, P.H., ROSENBLUTH, M.N., HINTON, F.L., MALKOV, M., FLEISCHER, J. & SMOLYAKOV, A., 1998, Proceed. 17th IAEA Conf. Fusion Energy, Yokohama, (Published by IAEA, Vienna), p. IAEA-CN-69/TH3/1

DOETSCH, G. 1943, *Die Laplace Transformation*, Dover, New York

DOETSCH, G. 1955, *Handbuch der Laplace Transformation* (2 volumes, in German) Birkhäuser, Basel

DOLAN, T.J. 1982, *Fusion Research*, Pergamon Press, Oxford

DRUMMOND,W.E. & PINES, D. 1962, Nucl. Fusion Suppl., **3**, 1049

DRUMMOND,W.E. & PINES, D. 1964, Ann. Phys. (NY), **28**, 478

DUBIN, D.H., KROMMES, J.A., OBERMAN, C. & LEE, W.W. 1983, Phys. Fluids, **26**, 3524

DUBOIS, D.F. 1976, Phys. Fluids, **19**, 1764

DUBOIS, D.F. & ESPEDAL, M. 1978, Plasma Phys., **20**, 1209

DUBOIS, D.F. & PESME, D. 1984, Phys. Fluids, **27**, 218

DuBois, D.F. & Rose, H.A. 1981, Phys. Rev. **A 24**, 1476

Duchs, D.F. 1989, JET report JET-R(89)13

Dupree, T.H. 1966, Phys. Fluids, **9**, 1773

Dupree, T.H. 1967, Phys. Fluids, **10**, 1049

Dupree, T.H. 1970, Phys. Rev. Lett., **25**, 789

Dupree, T.H. 1972, Phys. Fluids, **15**, 334

Dupree, T.H. 1978, Phys. Fluids, **21**, 783

Dupree, T.H. & Tetreault, D.J. 1978, Phys. Fluids, **21**, 425

Dupree, T.H., Wagner, C.E. & Manheimer, W.M. 1975, Phys. Fluids, **18**, 1167

Dwight, H.B. 1961, *Tables of Integrals and other Mathematical Data* (4th ed.), Macmillan, New York

Einstein, A. 1905, Ann. d. Physik, **17**, 549

Einstein, A. 1906, Ann. d. Physik, **19**, 371

Engelmann, F. & Morrone, T. 1972, Comm. Plasma Phys. Contr. Fusion, **1**, 75

Ferziger, J.H. and Kaper, H.G. 1972, *Mathematical Theory of Transport Processes in Gases*, North Holland, Amsterdam

Forster, D., Nelson, D.R. & Stephen, M.J. 1976, Phys. Rev. Lett., **36**, 867

Forster, D., Nelson, D.R. & Stephen, M.J. 1977, Phys. Rev., **A 16**, 732

Fournier, J.D. & Frisch, U. 1983, Phys. Rev. **A 28**, 1000

Fournier, J.D., Sulem, P.L. & Pouquet, M. 1982, J. Phys. **A 15**, 1393

Fried, B.D. & Conte, S.D. 1961, *The Plasma Dispersion Function*, Academic Press, New York

Gakhov, F.D. 1958, *Kraevye Zadachi* (Boundary Problems), in Russian, Fizmatgiz, Moscow

Galeev, A.A. & Karpman, V.N. 1963, Sov. Phys. JETP, **17**, 403

Galeev, A.A., Oraevskii, V.N. & Sagdeev, R.Z. 1963, Sov. Phys. JETP, **17**, 615

Galeev, A.A. & Sagdeev, R.Z. 1968, Sov. Phys. JETP, **26**, 233

Gardiner, C.W. 1985, *Handbook of Stochastic Methods*, 2nd ed., Springer, Berlin

Gary, S.P. & Abraham-Shrauner, B. 1981, J. Plasma Phys., **25**, 145

Gary, S.P. & Sanderson, J.J. 1978, Phys. Fluids, **21**, 1181

Gary, S.P. & Sanderson, J.J. 1979, Phys. Fluids, **22**, 1500

Gaspard, P. 1998, *Chaos, Scattering and Statistical Mechanics*, Cambridge Univ. Press, Cambridge, UK

Gentle, K.W., Bravenec, R.V., Cima, G., Gasquet, H., Hallock, G.A., Phillips, P.E., Ross, D.W., Rowan, W.L., Wootton,

A.J., CROWLEY, T.P., HEARD, J., OUROUA, A., SCHOCH, P.M. & WATTS, C. 1995, Phys. Plasmas, **2**, 2292

GOLDSTEIN, H. 1980, *Classical Mechanics*, revised Ed., Addison-Wesley, New York

GOLDSTON, R.J. & RUTHERFORD, P.H. 1995, *Introduction to Plasma Physics*, Inst. of Physics, Bristol

GRAD, H. 1949, Commun. Pure & Appl. Math., **2**, 311

GRADSHTEIN, I.S. & RYZHIK, M. 1980, *Table of Integrals, Series and Products*, Academic Press, New York, 4th ed. (also available on CD-Rom)

GREEN, M.S. 1951, J. Chem. Phys., **19**, 1036

GRUZINOV, A.V., ISICHENKO, M.B. & KALDA, YA.L. 1990, Sov. Phys. JETP, **70**, 263

GUREVICH, A.V., ZYBIN, K.P. & ISTOMIN, YA.N. 1983, Sov. Phys. JETP, **57**, 51

GUREVICH, A.V., ZYBIN, K.P. & ISTOMIN, YA.N. 1987, Nucl. Fusion, **27**, 453

HAGAN, W.K. & FRIEMAN, E.A. 1985, Phys. Fluids, **28**, 2641

HAMAGUCHI, S. & HORTON, W. 1990 a, Phys. Fluids, **B2**, 1833

HAMAGUCHI, S. & HORTON, W. 1990 b, Report IFSR #462 (Inst. for Fusion Studies, Univ. of Texas at Austin)

HASEGAWA, A. & MIMA, K. 1978, Phys. Fluids, **21**, 87

HASEGAWA, A. & WAKATANI, M. 1983, Phys. Rev. Lett. **50**, 682

HAVLIN, S. & BENAVRAHAM, D. 1987, Adv. Phys. **36**, 695

HAZELTINE, R.D. & MEISS, J.D. 1992, *Plasma Confinement*, Addison-Wesley, New York

HINTON, F.L. and HAZELTINE, R.D. 1976, Rev. Mod. Phys., **48**, 239

HIRSCHFELDER, J.O., CURTISS, C.F. and BIRD, R.B. 1954, *Molecular Theory of Gases and Liquids*, Wiley, New York

HIRSHMAN, S.P. & SIGMAR, D.J. 1981, Nucl. Fusion, **21**, 1079

HORTON, C.W. 1985, in: *Handbook of Plasma Physics*, edited by M.N. Rosenbluth & R.Z. Sagdeev, North Holland, Amsterdam, vol. 2, p. 383

HORTON, W. 1999, Rev. Mod. Phys., **71**, 735

HORTON, JR., W. & CHOI, D.I. 1979, Phys. Rep., **49**, 273

HORTON, W., CHOI, D.-I. & TANG, W.M. 1981, Phys. Fluids, **24**, 1077

HORTON, W., ESTES, R.D. & BISKAMP, D. 1980, Plasma Phys., **22**, 663

HORTON, W., HU, B., DONG, J.Q. & ZHU, P. 2003, New J. Phys., **5**, 14.1

HU, G., KROMMES, J.A. & BOWMAN, J.C. 1997, Phys. Plasmas, **4**, 2116

HUI, B.H. & DUPREE, T.H. 1975, Phys. Fluids, **18**, 236

ICHIMARU, S. 1992, *Statistical Plasma Physics,* vol. 1, Addison-Wesley, New York

ISICHENKO, M.B. 1991, Plasma Phys. Contr. Fusion, **33**, 795, 809

ISICHENKO, M.B. 1992, Rev. Mod. Phys., **64**, 961

ITOH, K., ITOH, S-I. & FUKUYAMA, A. 1999, *Transport and Structural Formation in Plasmas,* Inst. of Phys. Publ., Bristol

ITOH, S.I. & KAWAI, Y. 2002, *Bifurcation Phenomena in Plasma,* Kyushu Univ., Fukuoka

JOKIPII, R.J. & PARKER, E.N. 1969, Astrophys. J., **155**, 777

JONES, F. 1982, *Partial Differential Equations,* 4th ed., Springer, New York

KADOMTSEV, B.B. 1965, *Plasma Turbulence,* Academic Press, New York

KADOMTSEV, B.B. & PETVIASHVILI, V.I. 1963, Sov. Phys. JETP, **16**, 1578

KADOMTSEV, B.B. & POGUTSE, O.P. 1970, Phys. Rev. Lett., **25**, 1155

KADOMTSEV, B.B. & POGUTSE, O.P. 1971, Phys. Fluids, **14**, 2470

KADOMTSEV, B.B. & POGUTSE, O.P. 1979, in: *Plasma Phys. & Contr. Nucl. Fusion Research,* Proceed. 7th Intern. Conf., Innsbruck, 1978, IAEA, Vienna

KAW, P., SINGH, R. & DIAMOND, P.H. 2002, Plasma Phys. Contr. Fusion, **44**, 51

KIM, E-J. & DIAMOND, P.H. 2003, Phys. Plasmas,

KISSICK, M.V., CALLEN, J.D., FREDRICKSON, E.D., JANOS, A.C. & TAYLOR, G. 1996, Nucl. Fusion, **36**, 1691

KLAFTER, J. & ZUMOFEN, G. 1994, J. Phys. Chem., **98**, 7366

KLIMONTOVICH, YU.L. 1967, *Statistical Theory of Nonequilibrium Processes in Plasmas,* Pergamon Press, Oxford

KLIMONTOVICH, YU.L. 1982, *Kinetic Theory of Nonideal Gases and Nonideal Plasmas,* Pergamon Press, Oxford

KONO, M. & ICHIKAWA, Y.H. 1973, Prog. Theor. Phys. **49**, 754

KORN, G.A. & KORN, T.M. 1968, *Mathematical Handbook for Scientists and Engineers,* McGraw Hill, New York

KORSHOLM, S.B., MICHELSEN, P.K., NAULIN, V., RASMUSSEN, J.J., GARCIA, L., CARRERAS, B.A. & LYNCH, V.E., 2001, Plasma Phys. and Contr. Fusion, **43**, 1377

KRAICHNAN, R.H. 1959, J. Fluid Mech., **5**, 497

KRAICHNAN, R.H. 1961, J. Math. Phys. **2**, 124

KRAICHNAN, R.H. 1970 a, J. Fluid Mech., **41**, 189

KRAICHNAN, R.H. 1970 b, Phys. Fluids, **13**, 22

KRAICHNAN, R.H. 1982, Phys. Rev. **A 25**, 3281

KRALL, N.A. & TRIVELPIECE, A.W., 1986, *Principles of Plasma Physics,* (reprint), San Francisco Press, San Francisco

KROMMES, J.A. 1978, Progr. Theor. Phys., Suppl., **64**, 138

KROMMES, J.A. 1984, *Handbook of Plama Physics*, vol. 2 (M.N. Rosenbluth and R.Z. Sagdeev, eds.), North Holl., Amsterdam, p. 183

KROMMES, J.A. 1993, Phys. Fluids, B **5**, 1066

KROMMES, J.A. 1996, Phys. Rev. E **53**, 4865

KROMMES, J.A. 1999, Plasma Phys. Control. Fusion, **41**, A641

KROMMES, J.A. & KIM, C-B. 1988, Phys. Fluids, **31**, 869

KROMMES, J.A. & KLEVA, R.G. 1979, Phys. Fluids, **22**, 2168

KROMMES, J.A., LEE, W.W. & OBERMAN, C. 1986, Phys. Fluids, **29**, 2421

KROMMES, J.A. & OBERMAN, C. 1976, J. Plasma Phys., **16**, 229

KROMMES, J.A, OBERMAN, C. & KLEVA, R.G. 1983, J. Plasma Phys., **30**, 11

KROMMES, J.A. & SIMILON, P. 1980, Phys. Fluids, **23**, 1553

KRUSKAL, M.D. 1962, J. Math. Phys., **3**, 806

KUBO, R. 1957, J. Phys. Soc. Japan, **12**, 570

LANDAU, L. 1946, J. Phys. USSR, **10**, 25

LANDAU, L.D. & LIFSHITZ, E.M. 1951, *Electrodynamics of Continuous Media*, Pergamon, London

LANGEVIN, P. 1908, C.R. Acad. Sciences Paris, **146**, 530

LAVAL, G. 1993, Phys. Fluids, B **5**, 711

LAVAL, G. & PESME, D. 1980, Phys. Lett. **80 A**, 266

LAVAL, G. & PESME, D. 1983, Phys. Fluids, **26**, 52, 66

LAVRENTIEV, D. & SHABAT, B. 1972, *Méthodes de la théorie des fonctions d'une variable complexe*, Ed. de Moscou, Moscow

LEBEDEV, V.B., DIAMOND, P.H., SHAPIRO, V.D. & SOLOVIEV, G.I. 1995, Phys. Plasmas, **2**, 4420

LEE, W.W. 1983, Phys. Fluids, **26**, 556

LENARD, A. 1960, Ann. Phys. (N.Y.), **3**, 390

LEVY, P. 1937, *Théorie de l'addition des variables aléatoires* (in French), Gauthier-Villars, Paris

LIANG, Y.M. & DIAMOND, P.H. 1993, Comm. Plasma Phys. Contr. Fusion, **15**, 139

LIEWER, P. 1985, Nucl. Fusion, **25**, 543

LIN, Z., HAHM, T.S., LEE, W.W., TANG, W.M. & WHITE, R.B. 1998, Science, **281**, 1835

LITTLEJOHN, R.G. 1981, Phys. Fluids, **24**, 1730

LITTLEJOHN, R.G. 1984, Phys. Fluids, **27**, 976

LONGCOPE, D.W. & SUDAN, R.N. 1991, Phys. Fluids, B **3**, 1945

LOPES CARDOZO, N.J. 1995, Plasma Phys. Contr. Fusion, **37**, 799

LUCE, T.C., PETTY, C.C. & DE HAAS, J.C.M. 1992, Phys. Rev. Lett., **68**, 52

LUMLEY, J.L. 1962, in: *Mécanique de la turbulence*, Colloq. Internat. Marseille, CNRS, Paris, p. 17

MA, S.K. 1976, *The Modern Theory of Critical Phenomena*, Benjamin, Reading

MANHEIMER, W.M. & LASHMORE-DAVIES, C.A. 1989, *MHD and Microinstabilities in Confined Plasmas*, Addison-Wesley, New York

MARCINKIEWICZ,J. 1939, Math. Z., **44**, 612

MARTIN, P.C., SIGGIA, E.D. & ROSE, H.A. 1973, Phys. Lett., **A 8**, 423

MATTOR, N. & DIAMOND, P.H. 1994, Phys. Plasmas, **1**, 4002

McCOMB, W.D. 1992, *The Physics of Fluid Turbulence*, Clarendon Press, Oxford

METZLER, R. & KLAFTER, J. 2000, Phys. Reports, **339**, 1

MIKHAILOVSKII, A.B. 1974, *Theory of Plasma Instabilities*, vol. 2, Consultants Bureau, New York

MIKHAILOVSKII, A.B. 1998, *Instabilities in a Confined Plasma*, Inst. of Physics Publ., Bristol

MIKHLIN, S.G. 1950, *Singular Integral Equations*, Amer. Math. Soc. Translations, no. 24

MILLIONCHTCHIKOV, M.D. 1941, C.R. Acad. Sc. URSS, **32**, 615

MISGUICH, J.H. & BALESCU, R. 1975 a, Physica, **79 C**, 373

MISGUICH, J.H. & BALESCU, R. 1975 b, J. Plasma Phys., **13**, 385, 419, 429

MISGUICH, J.H. & BALESCU, R. 1978, Plasma Phys., **20**, 781

MISGUICH, J.H. & BALESCU, R. 1982, Plasma Phys., **24**, 289

MISGUICH, J.H., BALESCU, R., PECSELI, H.L., MIKKELSEN, T., LARSEN, S.E. & QIU, X.-M. 1987, Phys. Plasmas Contr. Fusion, **29**, 825

MISGUICH, J.H., REUSS, J.-D., VLAD, M. & SPINEANU, F. 1998 a, Physicalia Mag., **20**, 103

MISGUICH, J.H., REUSS, J.D., VLAD, M. & SPINEANU, F. 1998 b, Phys. Lett. **A 241**, 94

MISGUICH, J.H., VLAD, M., SPINEANU, F. & BALESCU, R. 1995, Comments Plasma Phys. Contr. Fusion, **17**, 45

MONIN, A.S. & YAGLOM, A.M. 1971,1975, *Statistical Fluid Mechanics: Mechanics of Turbulence*, 2 vols., MIT Press, Cambridge, Mass.

MONTGOMERY, D. 1975 in: *Plasma Physics* (edited by C. De Witt & J. Peyraud), p. 427, Gordon & Breach, New York

MONTROLL, E.W. & BENDLER 1984, J. Stat. Phys., **34**, 129

MONTROLL, E.W. & SHLESINGER, M.F. 1984, in: *Studies of Statistical Mechanics* (J.L. Lebowitz & E.W. Montroll, editors), vol. 11, p. 5, North Holland, Amsterdam

MONTROLL, E.W. & WEISS, G.H. 1965, J. Math. Phys., **6**, 167

MUHM, A., PUKHOV, A.M., SPATSCHEK, K.H. & TSYTOVICH, V. 1992, Phys. Fluids B **4**, 336

MUSKHELISHVILI, N.I. 1953, *Singular Integral Equations*, Noordhoff, Groningen

MYRA, J.R., CATTO, P.J., MYNICK, H.E. & DUVALL, D.E. 1993, Phys. Fluids B **5**, 1160

NICHOLSON, D.R. 1983, *Introduction to Plasma Physics*, Wiley, New York

NICOLIS, G. & PRIGOGINE, I. 1977, *Self-organization in Nonequilibrium Systems*, Wiley, New York

NISHIKAWA, K. & WAKATANI, M. 1990, *Plasma Physics: Basic Theory with Fusion Applications*, Springer, Berlin

NORDMAN, H. & WEILAND, J. 1989, Nucl. Fusion, **29**, 251

ORSZAG, S.A. 1970, J. Fluid Mech., **41**, 363

ORSZAG, S.A. 1973, in: *Fluid Dynamics, Dynamique des Fluides*, (R. Balian & J.-L. Peube, Editors), Cours d'été de l'Ecole de Physique Théorique, Les Houches 1973, Gordon & Breach, New York

ORSZAG, S.A. & KRAICHNAN, R.H. 1967, Phys. Fluids, **10**, 1720

OTTAVIANI, M. 1992, Europhys. Lett., **20**, 111

OTTAVIANI, M. 1998, Physicalia Mag., **20**, 95

OTTAVIANI, M., BOWMAN, J.C. & KROMMES, J.A. 1991, Phys. Fluids B **3**, 2186

PEIERLS, R. 1965, *Quantum Theory of Solids*, Clarendon Press, Oxford

PELLETIER, G. 1980, J. Plasma Phys., **24**, 287, 421

PERATT, A.L. 1984, J. Math. Phys., **25**, 466

PETTINI, M., VULPIANI, A., MISGUICH, J.H., DE LEENER, M., ORBAN, J. & BALESCU, R. 1988, Phys. Rev. A **38**, 344

PETTY, C.C. & LUCE, T.C. 1994, Nucl. Fusion, **34**, 121

PFIRSCH, D. & SCHLUTER, A. 1962, Report MPI/PA/7/62 (Max Planck Institut für Physik und Astrophysik, Garching)

PHYTHIAN, R. 1969, J. Phys. (General Phys.), A **2**, 181

PINES, D. & BOHM, D. 1952, Phys. Rev. **85**, 338

POPE, G.S. 2000, *Turbulent Flows*, Cambridge Univ. Press, Cambridge, UK

POUQUET, A., FOURNIER, J.D. & SULEM, P.L. 1978, J. de Physique Lettres, **39**, 199

PRIEST, E.R. 1982, *Solar Magnetohydrodynamics*, Reidel, Dordrecht

PROUDMAN, I. & REID, W.H. 1954, Phil. Trans. Roy. Soc. (London), A **247**, 163

PRUDNIKOV, A.P., BYCHKOV, YU.A. & MARICHEV, O.I. 1988, *Integrals and Series*, vil. **1**, Gordon and Breach, New York

RAJAGOPAL, A.K. & SUDARSHAN, E.C.G. 1974, Phys. Rev. A **10**, 1852

RAX, J.M. & WHITE, R.B. 1992, Phys. Rev. Lett., **68**, 1523

RECHESTER, A.A. & ROSENBLUTH, M.N. 1978, Phys. Rev. Lett., **40**, 38

RESIBOIS, P. & DE LEENER, M. 1977, *Classical Kinetic Theory of Fluids*, Wiley, New York

REUSS, J.-D. & MISGUICH, J.H. 1996, Phys. Rev. E **54**, 1857

REUSS, J.-D., VLAD, M. & MISGUICH, J.H. 1998, Phys. Lett. A **241**, 94

ROMANELLI, F. 1989, Phys. Fluids, B **1**, 1018

ROSE, H.A. 1982, Phys. Rev. Lett., **48**, 260

ROSENBLUTH, M.N. & HINTON, F.L. 1998, Phys. Rev. Lett., **80**, 724

ROSS, D. 1989, Comm. Plasma Phys. Contr. Fusion, **12**, 155

ROSS, D. 1992, Plasma Phys. Contr. Fusion, **34**, 137

ROSTOKER, N. 1961, Nucl. Fusion, **1**, 101

RUDAKOV, L.I. & TSYTOVICH, V.N. 1971, Plasma Phys., **13**, 213

RUTHERFORD, P.H. & FRIEMAN, E.A. 1968, Phys. Fluids, **11**, 569

RYTER, F., ANGIONI, C., BEURSKENS, M., CIRANT, S., HOANG, G.T., HOGEWEIJ, G.M.D., IMBEAUX, F., JACCHIA, A., MANTICA, P., SUTTORP, W. & TARDINI. G. 2001, Plasma Phys. Contr. Fusion, **43**, A323

SAGDEEV, R.Z. & GALEEV, A.A. 1969, *Nonlinear Plasma Theory*, Benjamin, New York

SALAT, A. 1988, Phys. Fluids, **31**, 1499

SANCHEZ, R., NEWMAN, D.E. & CARRERAS, B. 2001, Nucl. Fusion, **41**, 247

SHLESINGER, M.F. 1996, in *Encyclopaedia of Applied Physics*, vol. **16**, p. 45

SHLESINGER, M.F., ZASLAVSKY, G.M. AND KLAFTER, J. 1993, Nature, **363**, 31

SIMILON, P. 1981, *A Renormalized Theory of Drift Wave Turbulence in Sheared Geometry*, Ph. D. Thesis, Princeton Univ.

SITENKO, A.G. 1982, *Fluctuations and Nonlinear Wave Interactions in Plasmas*, Pergamon Press, Oxford

SITENKO, A.G. & SOSENKO, P.P., in: Proceed. Int. Conf. Plasma Physics (Kiev, 1987), vol. **1**, p. 486, World Scientific, Singapore

SMIRNOV, V. 1975, *Cours de Mathématiques Supérieures*, Tome IV, 1-ère partie, (French translation), Mir, Moscow

SMITH, G.R. 1984, Phys. Fluids, **27**, 1499

SMOLYAKOV, A.I. & DIAMOND, P.H. 1999, Phys. Plasmas, **6**, 4410

SMOLYAKOV, A.I., DIAMOND, P.H. & MALKOV, M. 2000, Phys. Rev. Lett., **84**, 491

SMOLYAKOV, A.I. & HIROSE, A. 1993, Plasma Phys. Contr. Fusion, **35**, 1765

STAUFFER, D. & AHARONY, A. 1992, *Introduction to Percolation Theory* (2nd. ed.), Taylor & Francis, London

STIX, T.H. 1992, *Waves in Plasmas*, Amer. Inst. Phys., New York

STRINGER, T.E. 1991, Plasma Phys. Contr. Fusion, **33**, 1715

468   *Bibliography*

STROTH, U. 1998, Plasma Phys. Contr. Fusion, **40**, 9

STUECKELBERG, E.C.G. & PETERMANN, A. 1953, Helv. Phys. Acta, **26**, 499

SUMMERS, D. & THORNE, R.M. 1991, Phys. Fluids B **3**, 1835

SUZUKI, M. 1984, in: *Statistical Physics & Chaos in Fusion Plasmas,* (C.W. Horton & L. Reichl, eds.), p. 311, Wiley, New York

SWANSON, D.G. 1989, *Plasma Waves,* Acad. Press, New York

TANGE, T., INOUE, S., ITOH, K. & NISHIKAWA, K. 1979, J. Phys. Soc. Japan, **46**, 266

TATSUMI, T. 1957, Proc. Roy. Soc. (London), **A 239**, 16

TAYLOR, G.I. 1922, Proc. London Math. Soc., **A 20**, 196

TAYLOR, J.B. & HASTIE, R.J. 1968, Plasma Phys., **10**, 479

TAYLOR, J.B. & MCNAMARA, B. 1971, Phys. Fluids, **14**, 1492

TERRY, P. 2000, Rev. Mod. Phys., **72**, 109

TERRY, P. & DIAMOND, P.H. 1984, in: *Statistical Physics & Chaos in Fusion Plasmas,* (C.W. Horton & L. Reichl, eds.), p. 335, Wiley, New York

TFR GROUP & TRUC, A. 1984, Plasma Phys. Contr. Fusion, **22**, 817

TSUNODA, S.I., DOVEIL, F. & MALMBERG, J.H. 1987, Phys. Rev. Lett., **58**, 1112

TSYTOVICH, V.N. 1977, *Theory of Turbulent Plasma,* Consultants Bur., New York

VACLAVIK, J. 1975, J. Plasma Phys., **14**, 315

VANDEN EIJNDEN, E. 1992, *Contribution à la théorie du transport anormal dans un plasma magnétisé,* Mémoire de Licence, Univ. Libre de Bruxelles

VANDEN EIJNDEN, E. 1997, *Contribution to the Statistical Theory of Turbulence. Application to Anomalous Transport in Plasmas,* Ph.D. thesis, Université Libre de Bruxelles, Brussels, Belgium

VAN KAMPEN, N.G. 1981, *Stochastic Processes in Physics and Chemistry,* North Holland, Amsterdam

VAN MILLIGEN, B.PH., CARRERAS, B.A. & SANCHEZ, R. 2004 b, Phys. Plasmas, **11**, 3787

VAN MILLIGEN, B.PH., SANCHEZ, R. & CARRERAS, B.A. 2004 a, Phys. Plasmas, **11**, 2272

VEDENOV, A.A., VELIKHOV, E.P. & SAGDEEV, R.Z. 1961, Nucl. Fusion, **1**, 82

VEDENOV, A.A., VELIKHOV, E.P. & SAGDEEV, R.Z. 1962, Nucl. Fusion Suppl., **2**, 465

VLAD, M., REUSS, J.-D., SPINEANU, F. & MISGUICH, J.H. 1998 a, J. Plasma Phys., **59**, 707

VLAD, M., SPINEANU, F., MISGUICH, J.H. & BALESCU, R. 1996, Phys. Rev. E **54**, 791

VLAD, M., SPINEANU, F., MISGUICH, J.H. & BALESCU, R. 1998, Phys. Rev. E **58**, 7359

VLAD, M., SPINEANU, F., MISGUICH, J.H. & BALESCU, R. 2000, Phys. Rev. E **61**, 3023

VLAD, M., SPINEANU, F., MISGUICH, J.H. & BALESCU, R. 2001, Phys. Rev. E **63**, 066304

VLAD, M., SPINEANU, F., MISGUICH, J.H. & BALESCU, R. 2002, Nucl. Fusion, **42**, 157

VLAD, M., SPINEANU, F., MISGUICH, J.H. & BALESCU, R. 2003, Phys. Rev. E **67**, 026406

VLAD, M., SPINEANU, F., MISGUICH, J.H., REUSS, J.D., BALESCU, R., ITOH, K. & ITOH, S.I. 2004, Plasma Phys. Contr. Fusion, **46**, 1051

WAGNER, F. et al. 1982, Phys. Rev. Lett., **49**, 1408

WAGNER, F. & STROTH, U. 1993, Plasma Phys. Contr. Fusion, **35**, 1321

WANG, H.D., VLAD, M., VANDEN EIJNDEN, E., SPINEANU, F., MISGUICH, J.H. & BALESCU, R. 1995, Phys. Rev. E **51**, 4844

WEILAND, J. 1999, *Collective modes in Inhomogeneous Plasmas*, Inst. of Phys. Publ., Bristol

WEINSTOCK, J. 1969, Phys. Fluids, **12**, 1045

WEINSTOCK, J. 1970, Phys. Fluids, **13**, 2308

WEINSTOCK, J. 1972, Phys. Fluids, **15**, 454

WEINSTOCK, J. 1976, Phys. Fluids, **19**, 11

WEISS, G.H. 1994, *Aspects and Applications of Random Walks*, North Holland, Amsterdam

WENTZEL, D.G. & TIDMAN, D.A. 1968, *Plasma Instabilities in Astrophysics*, Gordon & Breach, New York

WEYSSOW, B. 1990, *Méthodes Hamiltoniennes en Coordonnées Non Canoniques*, Ph.D. thesis, Univ. Libre de Bruxelles

WHITE, R.B. & WU, Y. 1993, Plasma Phys. Contr. Fusion, **35**, 595

WIDDER, D.V. 1946, *The Laplace Transform*, Princeton University Press, Princeton, N.J.

WILSON, K.C. 1971, Phys. Rev., B **4**, 3174, 3184

WILSON, K.C. 1972, Phys. Rev. Lett., **28**, 548

WILSON, K.C. & FISHER, M.E. 1972, Phys. Rev. Lett., **28**, 240

WINTNER, A. 1941, Duke Math. J., **8**, 678

WONG, H.V.1982, Phys. Fluids, **25**, 1811

YAGI, M. & HORTON, C.W. 1994, Phys. Plasmas, **1**, 2135

YAKHOT, V. & ORSZAG, S.A. 1986, J. Sci. Comp., **1**, 3; Phys. Rev. Lett., **57**, 1722

YANG, S.C. & CHOI, D.I. 1985, Phys. Lett., **108 A**, 25

YOSHIZAWA, A., ITOH, K. & ITOH, S-I. 2003, *Plasma and Fluid Turbulence: Theory and Modelling*, Inst. of Phys. Publ., Bristol

ZAGORODNY, A. & WEILAND, J. 1999, Phys. Plasmas, **6**, 2359

ZASLAVSKY, G.M., EDELMAN, M., WEITZNER, H., CARRERAS, B., MCKEE, G., BRAVENEC, R. & FONCK, R. 2000, Phys. Plasmas, **7**, 3691

ZHANG, W.Y. 1988, *Contribution to the Theory of Drift Wave Turbulence in Magnetically Confined Plasmas*, Ph.D. thesis, Univ. Libre de Bruxelles

ZHANG, W.Y. & BALESCU, R. 1987, Plasma Phys. Contr. Fus., **29**, 993, 1019

ZIMBARDO, G., VELTRI, P. & MALARA, F. 1984, J. Plasma Phys., **32**, 141

Index